Excel

Get the Results You Want!

YEAR 12 CHEMISTRY

Geoffrey Thickett

PASCAL PRESS

CONTENTS

MODULE 5: EQUILIBRIUM REACTIONS

CHAPTER 1: STATIC AND DYNAMIC EQUILIBRIUM 4

1 Modelling static and dynamic equilibria 4
2 Investigating reversibility in chemical reactions 7
3 Non-equilibrium systems 9
4 Rate and energy considerations in reversible reactions 10
Chapter syllabus checklist 12
HSC Exam-type questions 13
Answers 15

CHAPTER 2: FACTORS THAT AFFECT EQUILIBRIUM 17

1 Factors causing changes in equilibria 17
2 Investigating disturbances in equilibrium systems 22
Chapter syllabus checklist 27
HSC Exam-type questions 28
Answers 30

CHAPTER 3: CALCULATING THE EQUILIBRIUM CONSTANT (K_{eq}) 32

1 Equilibrium constants for homogeneous systems 32
2 Equilibrium constant calculations 33
3 Temperature and pressure effects on equilibrium 35
4 Investigating equilibrium constants 37
5 Equilibrium constants for a variety of equilibria 38
Chapter syllabus checklist 41
HSC Exam-type questions 42
Answers 44

CHAPTER 4: SOLUTION EQUILIBRIA 46

1 Dissolution equilibria 46
2 Solubility of polar molecular compounds 48
3 Solubility rules 49
4 Solubility product constant 51
Chapter syllabus checklist 53
HSC Exam-type questions 54
Answers 56

MODULE 6: ACID–BASE REACTIONS

CHAPTER 5: PROPERTIES OF ACIDS AND BASES 57

1 Inorganic acids and bases 57
2 Indicators 59
3 Reactions of acids with bases, carbonates and metals 62
4 Applications of neutralisation reactions 64
5 Enthalpy of neutralisation 66
6 Overview of acid–base theories 66
Chapter syllabus checklist 70
HSC Exam-type questions 71
Answers 72

CHAPTER 6: USING BRØNSTED–LOWRY THEORY 73

1 Investigating the pH of solutions 73
2 pH and pOH calculations 74
3 Strength of acids and bases 77
Chapter syllabus checklist 86
HSC Exam-type questions 87
Answers 88

CHAPTER 7: QUANTITATIVE ANALYSIS 90

1 Titrations 90
2 Titration graphs 93
3 Acid dissociation constants 97
4 Applications of acid–base analysis 100
5 Buffers 104
Chapter syllabus checklist 107
HSC Exam-type questions 108
Answers 109

MODULE 7: ORGANIC CHEMISTRY

CHAPTER 8: NOMENCLATURE 111

1 Hydrocarbons 111
2 Alcohols and carboxylic acids 114
3 Aldehydes and ketones 115

4 Amines and amides ... 116
5 Halogenated organic compounds ... 117
6 Isomers ... 118
Chapter syllabus checklist ... 120
HSC Exam-type questions ... 121
Answers ... 122

CHAPTER 9: HYDROCARBONS ... 124

1 Homologous series of hydrocarbons ... 124
2 Molecular shapes ... 125
3 Intermolecular and intramolecular bonding ... 126
4 Safety in handling and disposal of organic substances ... 128
5 Obtaining and using hydrocarbons ... 129
Chapter syllabus checklist ... 132
HSC Exam-type questions ... 133
Answers ... 134

CHAPTER 10: PRODUCTS OF REACTIONS INVOLVING HYDROCARBONS ... 136

1 Hydrocarbon reactions ... 136
2 Modelling reactions ... 138
Chapter syllabus checklist ... 140
HSC Exam-type questions ... 141
Answers ... 142

CHAPTER 11: ALCOHOLS ... 143

1 Structure and properties of alcohols ... 143
2 Combustion of alcohols ... 145
3 Reactions of alcohols ... 147
4 Production of alcohols ... 150
5 Fuel production ... 152
Chapter syllabus checklist ... 154
HSC Exam-type questions ... 155
Answers ... 156

CHAPTER 12: REACTIONS OF ORGANIC ACIDS AND BASES ... 158

1 Functional group properties ... 158
2 Esters ... 161
3 Soaps and detergents ... 165
4 Chemical synthesis ... 168
Chapter syllabus checklist ... 169
HSC Exam-type questions ... 170
Answers ... 171

CHAPTER 13: POLYMERS ... 173

1 Addition polymers ... 173
2 Condensation polymers ... 178
Chapter syllabus checklist ... 181
HSC Exam-type questions ... 182
Answers ... 184

MODULE 8: APPLYING CHEMICAL IDEAS

CHAPTER 14: ANALYSIS OF INORGANIC SUBSTANCES ... 186

1 Monitoring the environment ... 186
2 Qualitative analysis of ions ... 189
3 Quantitative analysis of ions ... 195
4 Instrumental analytical techniques ... 198
Chapter syllabus checklist ... 204
HSC Exam-type questions ... 205
Answers ... 207

CHAPTER 15: ANALYSIS OF ORGANIC SUBSTANCES ... 209

1 Qualitative investigations of functional groups ... 209
2 Instrumental analysis of organic compounds ... 211
Chapter syllabus checklist ... 219
HSC Exam-type questions ... 220
Answers ... 222

CHAPTER 16: CHEMICAL SYNTHESIS AND DESIGN ... 224

1 Chemical synthesis ... 224
2 Chemical industries ... 226
Chapter syllabus checklist ... 237
HSC Exam-type questions ... 238
Answers ... 240

SAMPLE HSC EXAMINATION PAPERS ... 242

Sample HSC Examination 1 ... 242
Sample HSC Examination 2 ... 246
Answers ... 252

Index ... 259
Data sheet ... 264

CHAPTER 1 STATIC AND DYNAMIC EQUILIBRIUM

MODULE 5 EQUILIBRIUM REACTIONS

INQUIRY QUESTION:

What happens when chemical reactions do not go through to completion?

Many chemical reactions proceed to completion and no reactants remain if the stoichiometric ratio is achieved. The combustion of ethanol is an example of a reaction that proceeds to completion. However, some reactions do not proceed to completion. The final state consists of a mixture of reactants and products. The reaction of hydrogen and nitrogen to form ammonia is an example of such a reaction. Some of these reactions are examined in this chapter.

1 Modelling static and dynamic equilibrium

» Students model static and dynamic equilibrium and analyse the differences between open and closed systems.

Static balance

➔ In physical systems you will be familiar with the concept of *static balance*. A beaker resting on the laboratory bench does not move because the net force acting on it is zero. The gravitational force acting downwards on the beaker is exactly balanced by the upwards reaction force of the bench on the beaker, as illustrated in Figure 1.1.

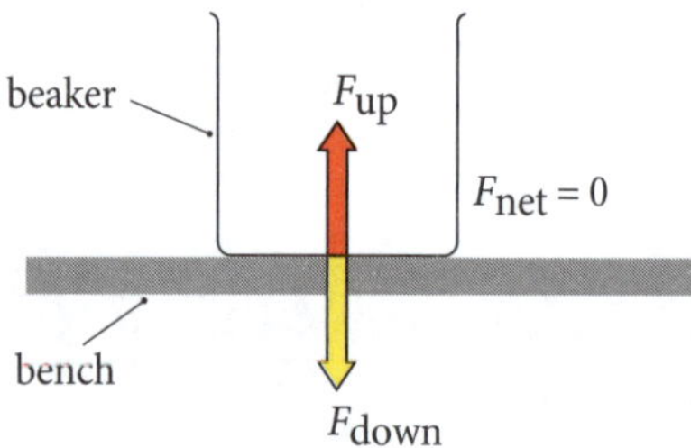

Figure 1.1 Static balance

Dynamic balance

➔ A body can be in a state of *dynamic balance* when it moves at constant velocity. An example of this is the terminal velocity reached by a steel ball-bearing falling in a cylinder of highly viscous oil (see Figure 1.2). The ball initially accelerates down due to gravity but the resistive forces of the fluid acting upwards slow the falling ball until the net force acting on the ball is zero. From that time onwards the ball is in a state of dynamic balance as the opposing forces acting on it are balanced.

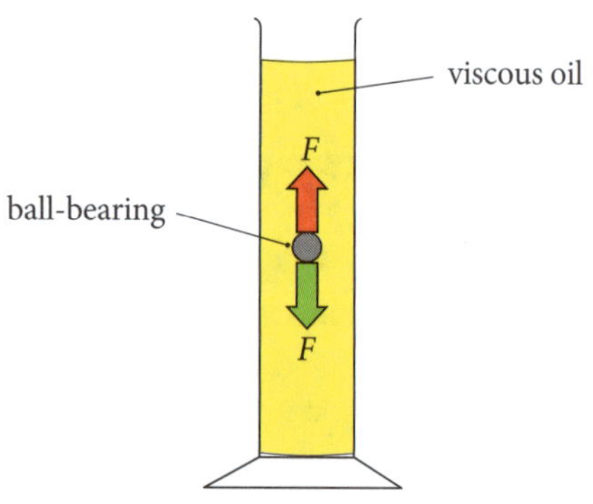

Figure 1.2 Dynamic balance

➔ In chemical systems reactions may be reversible or irreversible. Irreversible reactions are examples of static balance or static equilibrium. In chemical systems the particles are always in constant motion at a fixed temperature. These particles may move from place to place. This is called translational motion. They may also vibrate or they may rotate around an axis. Particle motion exists at all temperatures (except absolute zero). Atoms or molecules in solids undergo vibrational motion.

➔ Many chemical systems exhibit dynamic balance or dynamic equilibrium. In such dynamic systems the reaction occurs in both the forward and reverse directions. In a dynamic equilibrium, reactants and products are present in the system at equilibrium and changes continue to take place at the molecular level. However, in a static equilibrium system the reaction proceeds in only one direction. This can happen when one reactant is totally consumed and so the forward reaction can no longer occur and a dynamic balance cannot be achieved. In a static equilibrium, both forward and backward reactions have come to a halt as no changes occur at the molecular level. Figure 1.3 illustrates these ideas.

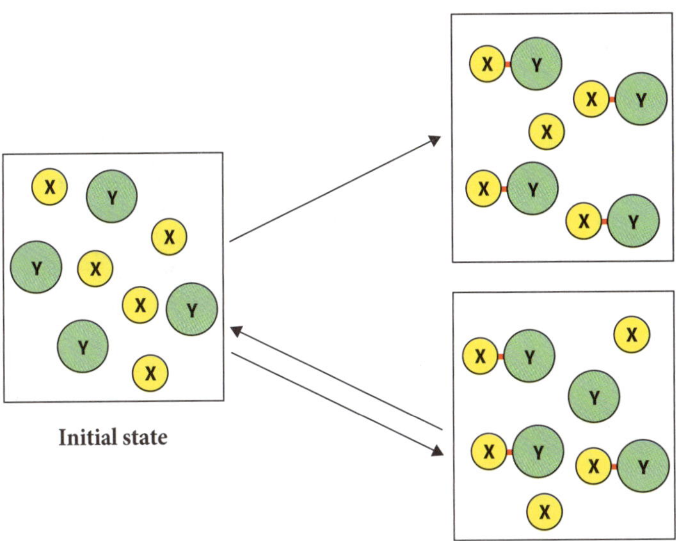

Figure 1.3 Modelling static and dynamic equilibrium

- Figure 1.4 shows two equal-volume glass vessels connected by a tap. One vessel contains helium gas at 100 kPa and 298 K. The second vessel contains neon gas at 100 kPa and 298 K. Each vessel has a pressure gauge to monitor gas pressure. The tap is opened and the interconnected vessels are observed over time. The pressure gauges (P) show no change in pressure over time but later a chemical analysis of the gases in each vessel shows equal amounts of the two gases in each. At this time the system is in a state of **dynamic equilibrium**, with molecules of each gas diffusing back and forth at equal rates between the two vessels.

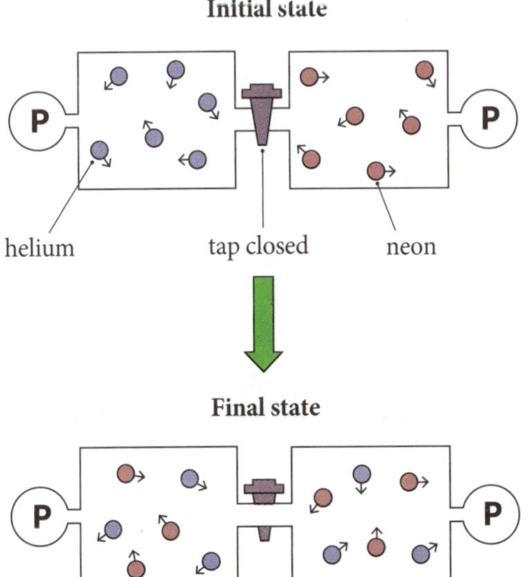

Figure 1.4 Attainment of dynamic equilibrium

dynamic equilibrium: a state of dynamic balance in which the forward and reverse processes occur at the same rate

Closed and open systems

- The attainment of equilibrium depends on whether the system is open or closed.
- In an unopened bottle of soda water (see Figure 1.5), the carbon dioxide gas or water vapour cannot escape. These gases are in a state of dynamic equilibrium between the dissolved or liquid state and the gaseous state at a constant temperature. If the cap is removed then a rapid effervescence is observed. There is no longer a dynamic equilibrium as carbon dioxide and water vapour escape into the surroundings. This system is now an **open system**.

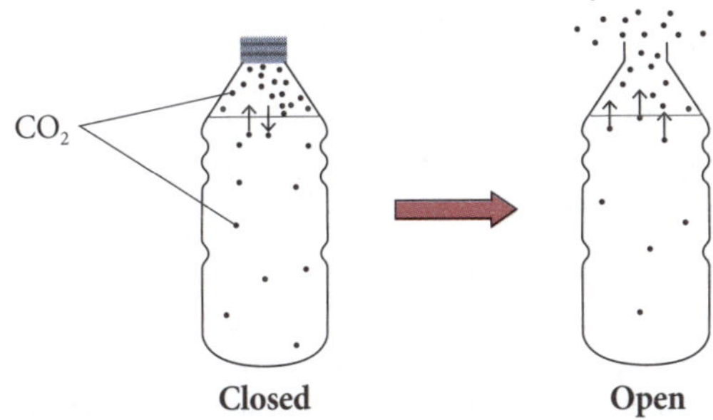

Figure 1.5 Closed and open bottle of soda water

open system: a system in which one or more components can escape

- When a small chip of calcium carbonate is added to a beaker of dilute hydrochloric acid, an effervescence (fizzing) is observed as the calcium carbonate dissolves (see Figure 1.6). The gas that is released into the air is carbon dioxide.

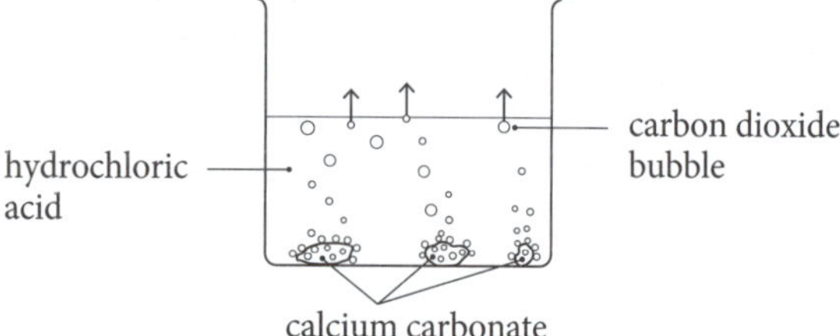

Figure 1.6 Calcium carbonate dissolves in hydrochloric acid

- If an excess of acid is used, all the calcium carbonate dissolves. We can write a balanced equation for this reaction:

$$CaCO_3(s) + 2HCl(aq) \rightarrow CaCl_2(aq) + H_2O(l) + CO_2(g)$$

- This reaction is an example of one that proceeds to completion if the correct proportion of reactants are present or an excess of acid is present. The calcium carbonate is totally consumed. The gaseous carbon dioxide escapes as well as some water vapour from the acid solution.
- In a heterogeneous system involving solids and gases, dynamic equilibrium cannot be achieved if the gas can escape from the system. Consider the thermal decomposition of calcium carbonate to form calcium oxide and carbon dioxide gas. The reaction is endothermic ($\Delta H > 0$).
  - If the vessel is open the reaction proceeds to completion as the carbon dioxide escapes. All the calcium carbonate will decompose on heating in the open vessel:

    $$CaCO_3(s) \rightarrow CaO(s) + CO_2(g); \Delta H = +178 \text{ kJ/mol}$$

    The calcium oxide (lime) is used in the manufacture of cement and in waste water treatment. Historically lime kilns (Figure 1.7) used burning layers of coking coal to decompose limestone ($CaCO_3$). The flue gases escaped in this open system.

    An increase in temperature and the loss of carbon dioxide to the surroundings is used to manufacture lime. In modern industry, the process is performed in high-temperature kilns (1000 °C).
  - In a closed vessel the reaction comes to a dynamic equilibrium in which the amounts of each component no longer change. Reversible arrows ($\leftrightarrows$) are used in equations to show dynamic equilibrium reactions:

    $$CaCO_3(s) \leftrightarrows CaO(s) + CO_2(g)$$
- Chemical reactions are therefore reversible in closed systems. The conversion of reactants into products is called the *forward reaction*. The conversion of products back into reactants is called the *reverse or back reaction*.

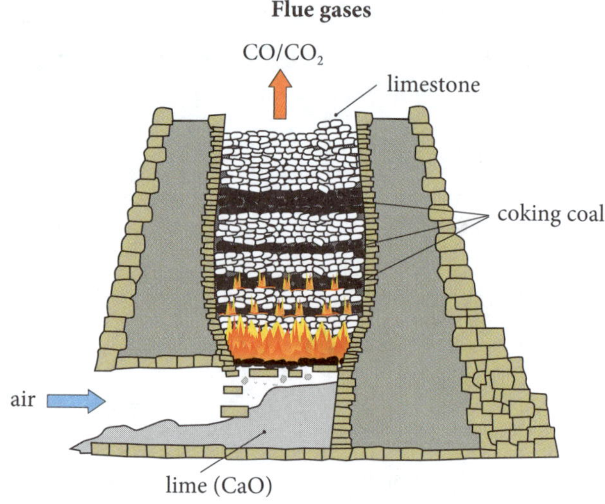

Figure 1.7 Structure of historical lime kiln

closed system: a system in which components cannot escape

EXAMPLE 1

A chemist placed iodine crystals in a Petri dish containing a mixture of water and ethanol. The lid was placed on the dish and a time-lapse video was made to follow the dissolution of the iodine in the water/ethanol solvent. As the iodine dissolved the solution turned an orange-brown, as shown in Figure 1.8. Eventually the colour intensity no longer changed as the solution became saturated in iodine molecules. Explain how the chemist could use a radioactive iodine isotope to demonstrate that, when the colour of the solution no longer changed, the system was in a state of dynamic balance rather than static balance.

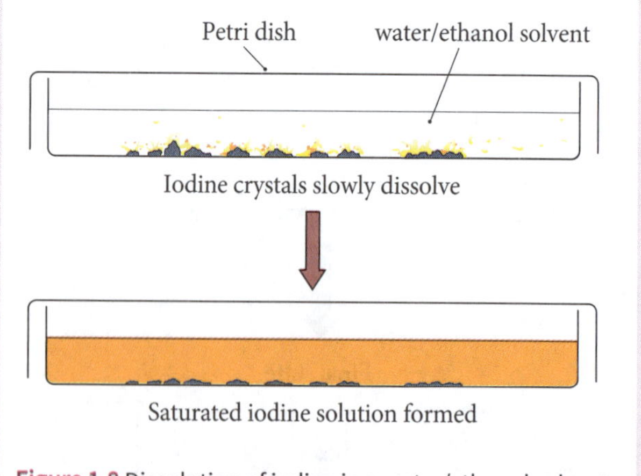

Figure 1.8 Dissolution of iodine in a water/ethanol solvent

Radioactive iodine has the same physical and chemical properties as the non-radioactive isotope

Answer:

A saturated water/ethanol solution of radioactive iodine (e.g. I-123) is prepared. This isotope is a beta and gamma emitter. The radiation can be detected using a Geiger counter.

A few millilitres of this saturated radioactive iodine solution are now added to the saturated orange-brown solution in the Petri dish. There is no change in colour over time.

Some hours later the chemist removed some iodine crystals from the dish. They were washed and dried. The Geiger counter showed that the crystals had become radioactive.

The chemist concluded that some radioactive I-123 molecules had left the saturated solution and crystallised on the surfaces of other crystals. As the colour intensity of the solution had not changed then the system must be in a state of dynamic balance. The non-radioactive iodine crystals undergo dissolution at the same rate as radioactive molecules crystallise:

$$I_2(s) \leftrightharpoons I_2(aq)$$

2 Investigating reversibility in chemical reactions

» Students conduct practical investigations to analyse the reversibility of chemical reactions, for example: cobalt (II) chloride hydrated and dehydrated; iron (III) nitrate and potassium thiocyanate; burning magnesium; and burning steel wool.

→ The following practical investigations allow students to investigate reversible and non-reversible reactions.

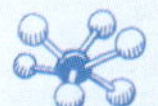

FIRSTHAND INVESTIGATION 1

Hydrated cobalt (II) chloride reactions

Aim

to use hydrated cobalt (II) chloride to demonstrate reaction reversibility

Method

Part A

1 Dissolve 1 g of cobalt (II) chloride hexahydrate ($CoCl_2{\cdot}6H_2O$) crystals in 10 mL of water to produce a pink solution.
2 Add dropwise concentrated hydrochloric acid (ensure safety glasses are worn) to the pink solution until a colour change occurs.
3 Add water dropwise to the blue solution until a colour change occurs.
4 Repeat steps 2 and 3.

Part B

5 Place purple cobalt (II) chloride hexahydrate crystals in a test tube and use a test tube holder to heat the tube gently over a blue Bunsen flame till the crystals are dehydrated.
6 Allow the tube to cool back to room temperature in a test tube rack.
7 Add drops of water to the cooled solid and observe the colour change.

Sample observations

Part A: The pink solution turns blue on addition of hydrochloric acid and returns to pink on addition of water. The changes are reversible.
Part B: The purple solid turns blue on heating and the blue solid turns purple once more when drops of water are added.

Explanation

Part A

The aqueous equilibrium reaction is:

$$\underset{\text{pink}}{Co(H_2O)_6^{2+}(aq)} + 4Cl^-(aq) \leftrightharpoons \underset{\text{blue}}{CoCl_4^{2-}(aq)} + 6H_2O(l)$$

The addition of hydrochloric acid increases the chloride ion concentration and the forward reaction is promoted. The pink hexaaquacobalt (II) ion is converted into the blue tetrachlorocobalt (II) ion. The addition of water promotes the reverse reaction in which the blue ion is converted back to the pink ion. Figure 1.9 summarises these observations.

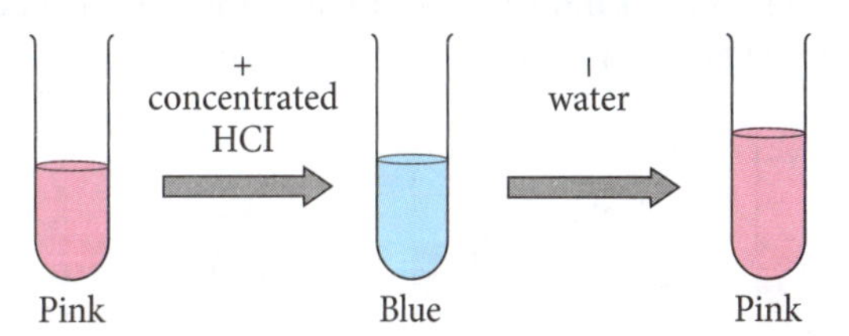

Figure 1.9 Cobalt (II) chloride equilibrium

Part B

The heating of the purple crystals causes dehydration and the solid turns blue as water is lost. The addition of water to the blue solid turns it back to purple as hydration occurs. Figure 1.10 encapsulates these observations.

$$\underset{\text{purple}}{CoCl_2{\cdot}6H_2O(s)} \leftrightharpoons \underset{\text{blue}}{CoCl_2(s)} + 6H_2O(g)$$

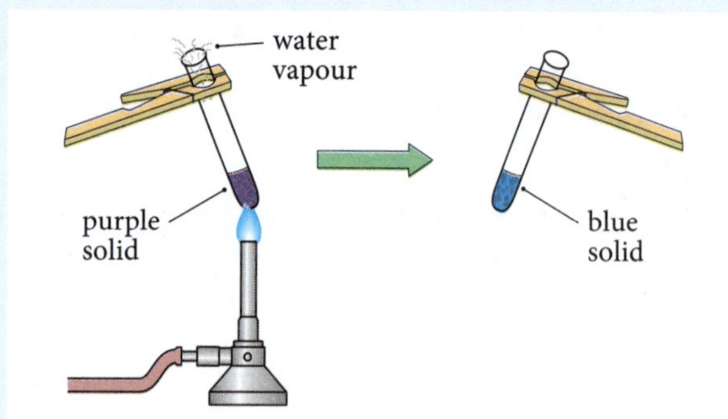

Figure 1.10 Dehydration of $CoCl_2{\cdot}6H_2O$

FIRSTHAND INVESTIGATION 2

Iron (III) thiocyanate equilibrium

Aim

to investigate reaction reversibility in iron (III) thiocyanate solutions

Method

1. Iron (III) nitrate solution and potassium thiocyanate solutions are mixed in a beaker. Dilute the mixture until the colour is orange-red and divide this solution into three clean test tubes.
2. Use tube 1 as a control.
3. To tube 2, add a few crystals of iron (III) nitrate ($Fe(NO_3)_3$) and agitate to dissolve the crystals. To tube 3, add drops of 1 mol/L sodium hydroxide solution and mix thoroughly.

Sample observations

The control tube 1 stays orange-red and tube 2 turns more red. Tube 3 turns more yellow as a fine brown precipitate forms. Figure 1.11 shows these colour changes.

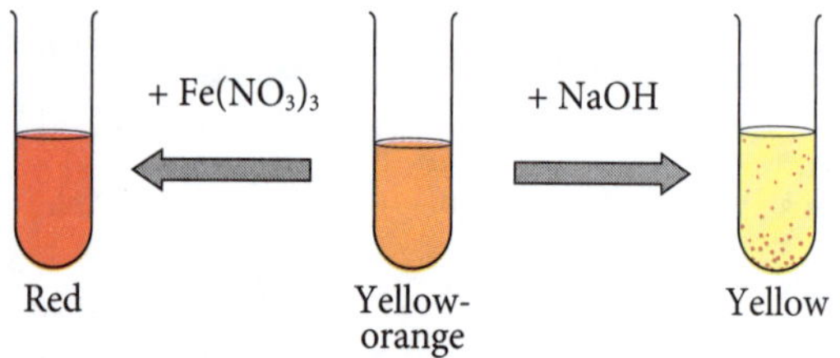

Figure 1.11 Iron (III) thiocyanate equilibrium

Explanation

The equilibrium reaction is:

$$Fe^{3+}(aq) + SCN^{-}(aq) \leftrightharpoons FeSCN^{2+}(aq)$$

yellow colourless blood-red

The addition of iron (III) nitrate crystals increases the concentration of $Fe^{3+}(aq)$ on dissolution of the crystals. This promotes the forward reaction.

The addition of sodium hydroxide decreases the concentration of $Fe^{3+}(aq)$ due to precipitation of $Fe(OH)_3$. This promotes the reverse reaction:

$$Fe^{3+}(aq) + 3OH^{-}(aq) \rightarrow Fe(OH)_3(s)$$

FIRSTHAND INVESTIGATION 3

Combustion of magnesium and steel wool

Aim

to investigate the combustion of magnesium and steel wool in air

Method

1. Hold a 5 cm strip of magnesium with tongs and ignite the tip in the blue flame of a Bunsen burner.
2. Hold the burning magnesium over an evaporating basin so the white solid product falls into the basin. Use safety glasses and do not look directly at the burning magnesium.
3. Tease out a piece of steel wool and hold a small piece with tongs. Hold the steel wool in a blue Bunsen burner flame and observe the reaction. The reaction product forms on the surface of the steel wool.
4. Place the reaction products in separate crucibles and place the crucibles on pipeclay triangles on tripods. Attempt to decompose the reaction products into their component elements by heating strongly with a blue Bunsen burner flame.

Sample observations

The grey magnesium ribbon burns with a bright white light, forming a crumbly white solid. The grey steel wool burns more slowly than magnesium, producing yellow-orange light and some sparks. A dark grey-black solid forms.

No changes were observed when the two reaction products were heated strongly in crucibles. Figure 1.12 shows the burning of steel wool.

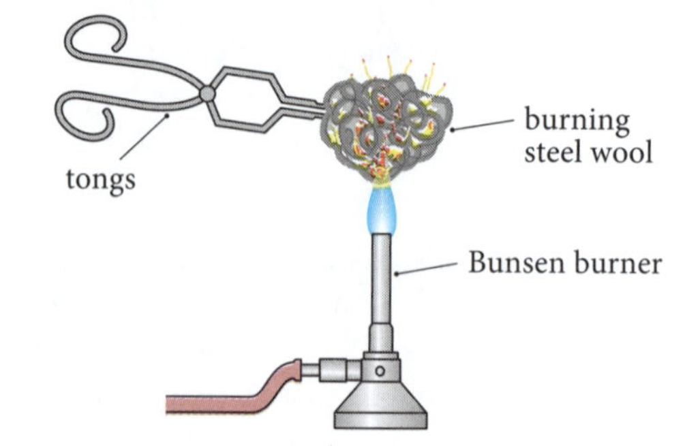

Figure 1.12 Burning steel wool

Explanation

The reactions are:

$2Mg(s) + O_2(g) \rightarrow 2MgO(s)$
grey — white

$3Fe(s) + 2O_2(g) \rightarrow Fe_3O_4(s)$
grey — black

The crucible experiments showed the reactions were not reversible.

Fe_3O_4 is an oxide of iron in which some iron atoms have a +2 oxidation state and others have a +3 oxidation state. Its chemical name is iron (II,III) oxide.

→ KEY QUESTIONS

1. **Distinguish between static and dynamic equilibrium.**
2. **Solid barium carbonate is decomposed in a closed vessel to form solid barium oxide and carbon dioxide gas. Write the equation for the dynamic equilibrium that is achieved.**
3. **A sample of metallic sodium is heated in an atmosphere of chlorine gas. A very rapid reaction occurs in which heat is released and fine crystals of sodium chloride form. Explain whether this reaction is readily reversible.**

Answers ➲ p. 15

3 Non-equilibrium systems

» Students analyse examples of non-equilibrium systems in terms of the effect of entropy and enthalpy, for example: combustion reactions and photosynthesis.

Combustion

→ When hydrocarbons undergo complete combustion in an excess of oxygen, carbon dioxide and water are formed. These reactions are strongly **exothermic** ($\Delta H < 0$). The water formed is initially in the vapour state but as the products cool back down to 25 °C, it condenses back to the liquid state. The equation for the complete combustion of methane is:

$$CH_4(g) + 2O_2(g) \rightarrow CO_2(g) + 2H_2O(l)\ (\Delta H < 0)$$

exothermic: reactions in which heat is liberated to the surroundings

→ The following table lists the standard **enthalpy of formation** ($\Delta_f H^o$) and standard **entropy** (S^o) values for reactants and products.

	$CH_4(g)$	$O_2(g)$	$CO_2(g)$	$H_2O(l)$
$\Delta_f H^o$ (kJ/mol)	–74	0	–394	–286
S^o (J/K/mol)	186	205	214	70

enthalpy of formation: the change in enthalpy when a compound forms from its elements in their standard states

entropy: a measure of the degree of disorder of a system

The tabulated data allows a calculation of ΔH^o and ΔS^o.

$$\Delta H^o = \Sigma\Delta_f H^o(\text{products}) - \Sigma\Delta_f H^o(\text{reactants})$$
$$= (-394) + 2(-286) - 2(-74) - 0 = -818 \text{ kJ/mol}$$

$$\Delta S^o = \Sigma\Delta S^o(\text{products}) - \Sigma\Delta S^o(\text{reactants})$$
$$= 214 + 2(70) - 186 - 2(205) = -242 \text{ J/K/mol}$$
$$= -0.242 \text{ kJ/K/mol}$$

$$\Delta G^o = \Delta H^o - T\Delta S^o = (-818) - 298(-0.242)$$
$$= -746 \text{ kJ/mol} < 0$$

As the **Gibbs free energy** (ΔG^o) change is negative, the combustion reaction is spontaneous under standard conditions.

Gibbs free energy: a measure of the spontaneity of a process

→ The release of heat energy in this exothermic reaction drives the reaction in the forward direction despite the decrease in entropy. The high activation energy for the reverse reaction reduces the chance that products will recombine to form reactants.

→ Hydrogen gas burns explosively in air. This is the basis of the laboratory 'pop' test for hydrogen (see Figure 1.13).

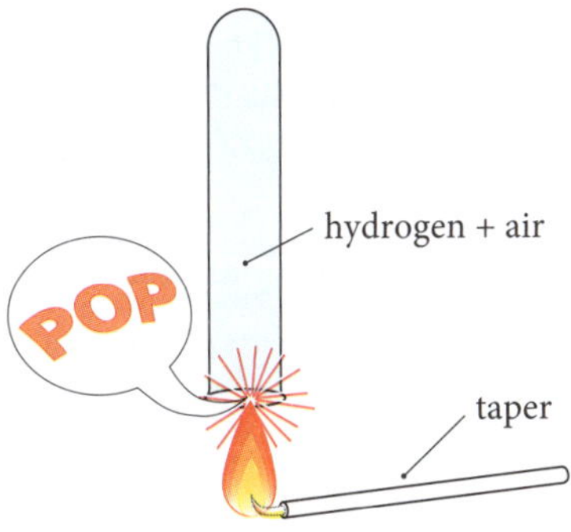

Figure 1.13 Pop test for hydrogen

- The water formed when hydrogen burns in air is initially in the vapour state but as the product cools back down to 25 °C, it condenses back to the liquid state. A film of liquid water can be detected on the inside of the test tube. The equation for the combustion of hydrogen is:

$$H_2(g) + \frac{1}{2}O_2(g) \rightarrow H_2O(l)$$

- The following table lists the standard enthalpies of formation and standard entropies for reactants and product.

	$H_2(g)$	$O_2(g)$	$H_2O(l)$
$\Delta_f H^o$ (kJ/mol)	0	0	-286
S^o (J/K/mol)	131	205	70

The tabulated data allows a calculation of ΔH^o and ΔS^o.

$\Delta H^o = \Sigma\Delta_f H^o(\text{products}) - \Sigma\Delta_f H^o(\text{reactants})$
$= (-286) - 0 - 0 = -286$ kJ/mol (reaction is exothermic)

$\Delta S^o = \Sigma\Delta S^o(\text{products}) - \Sigma\Delta S^o(\text{reactants})$
$= 70 - 131 - \frac{1}{2}(205) = -163.5$ J/K/mol
$= -0.1635$ kJ/K/mol

$\Delta G^o = \Delta H^o - T\Delta S^o = (-286) - 298(-0.1635)$
$= -237 \text{ kJ/mol} < 0$

As the Gibbs free energy change is negative, the combustion reaction is spontaneous under standard conditions.

- When combustion reactions occur in open systems, the products can escape into the surroundings and therefore they cannot recombine to produce reactant molecules.

Photosynthesis

- Photosynthesis occurs in the chloroplasts of green plant cells. Solar energy is absorbed by chlorophyll molecules inside the chloroplasts. Chlorophyll molecules selectively absorb red and violet light in the visible spectrum. This energy is used to convert carbon dioxide and water into glucose and oxygen. The reaction involves the oxidation of water and the reduction of carbon dioxide. Figure 1.14 shows inputs and outputs of substances during photosynthesis.

Oxidation: $2H_2O(l) \rightarrow O_2(g) + 4H^+(aq) + 4e^-$

Reduction: $6CO_2(g) + 24H^+(aq) + 24e^- \rightarrow C_6H_{12}O_6(s) + 6H_2O(l)$

Net reaction: $6CO_2(g) + 6H_2O(l) \rightarrow C_6H_{12}O_6(s) + 6O_2(g)$

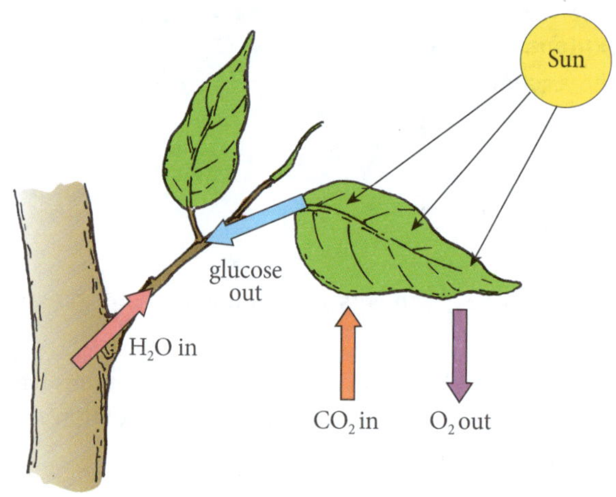

Figure 1.14 Photosynthesis

- The following table lists the standard enthalpies of formation and standard entropies to determine the spontaneity of the photosynthesis reaction.

	$CO_2(g)$	$H_2O(l)$	$C_6H_{12}O_6(s)$	$O_2(g)$
$\Delta_f H^o$ (kJ/mol)	-394	-286	-1273	0
S^o (J/K/mol)	214	70	212	205

The tabulated data allows for a calculation of ΔH^o and ΔS^o.

$\Delta H^o = (-1273) + 6(0) - 6(-394) - 6(-286) = +2807$ kJ/mol (reaction is endothermic)

$\Delta S^o = 212 + 6(205) - 6(214) - 6(70) = -262$ J/K/mol
$= -0.262$ kJ/K/mol

$\Delta G^o = \Delta H^o - T\Delta S^o = (+2807) - 298(-0.272)$
$= +2888 \text{ kJ/mol} > 0$

As the Gibbs free energy change is positive, the photosynthesis reaction is non-spontaneous under standard conditions. This reaction requires solar energy to occur.

- The reaction products are lost from the system and this prevents the reverse reaction occurring.

4 Rate and energy considerations in reversible reactions

» Students investigate the relationship between collision theory and reaction rate in order to analyse chemical equilibrium reactions.

- Energy profile diagrams demonstrate the differences in activation energy for the forward and reverse reactions. Figure 1.15 shows the energy profiles for endothermic and exothermic reactions.

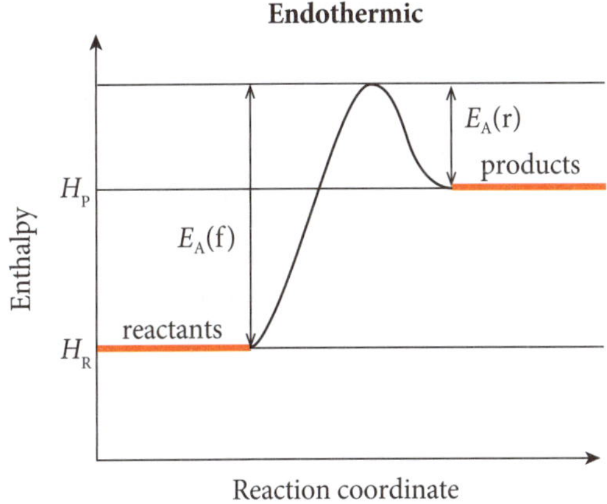

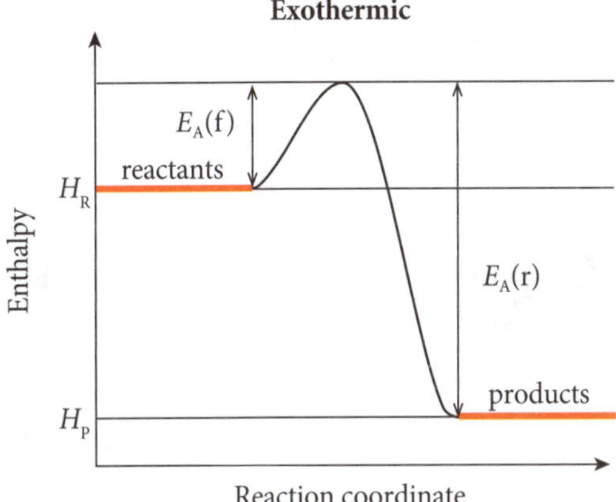

Figure 1.15 Activation energy diagrams

- In an endothermic process the activation energy for the forward reaction ($E_A(f)$) in which reactants turn into products is much greater than the activation energy for the reverse reaction ($E_A(r)$) in which products turn back into reactants. Heating the system increases the kinetic energy of the colliding particles. Therefore the system needs to be heated to increase the forward rate at which reactants are turned into products.
- In an exothermic process the activation energy for the reverse reaction is much greater than for the forward reaction. The forward reaction is therefore more energetically favoured and it will occur initially at a greater rate than the reverse reaction. If the system is closed the heat released will increase the kinetic energies of both the reactants and products, and equilibrium will be achieved faster.
- The rate of a reaction is dependent on the frequency of collision between reactant particles. The greater the concentration of the reacting particles, the greater the collision frequency.
- The rate law for a chemical reaction relates the reaction rate with the concentrations of the reactants for the forward reaction and the concentration of products in the reverse reaction.

Consider the reaction:

$NO(g) + O_3(g) \leftrightharpoons NO_2(g) + O_2(g)$

- Experimental results show that the rate of the forward reaction (R_f) depends on the product of the reactant concentrations (at constant temperature):

  $R_f = k_f[NO][O_3]$

  where k_f is called the *forward rate constant*
- Experimental results show that the rate of the reverse reaction (R_r) depends on the product of the product concentrations (at constant temperature):

  $R_r = k_r[NO_2][O_2]$

  where k_f is called the *reverse rate constant*
- Over time the reactants are consumed and their concentration decreases. As reactants combine to form products, the rate of the forward reaction therefore decreases and the rate of the reverse reaction increases. Eventually a point is reached when the rates of the forward and reverse reactions become equal.

  At equilibrium:

  Rate(forward) = Rate(reverse)

  $k_f[NO][O_3] = k_r[NO_2][O_2]$

  or $k_f/k_r = [NO_2][O_2]/[NO][O_3]$

  The ratio k_f/k_r is a constant called the *equilibrium constant* (K_{eq}). This constant is temperature dependent. Calculations involving the equilibrium constant will be examined in Chapter 3.

- When the forward and reverse rates are equal the system is at equilibrium (see Figure 1.16). Chemical equilibrium is a dynamic process and is recognised by constancy in measurable **macroscopic properties** of the system. The macroscopic properties used depends on the system investigated but may include gas pressure, gas volume, electrical conductivity of a solution and colour.

macroscopic properties: quantities that can be measured (e.g. colour, pressure, concentration)

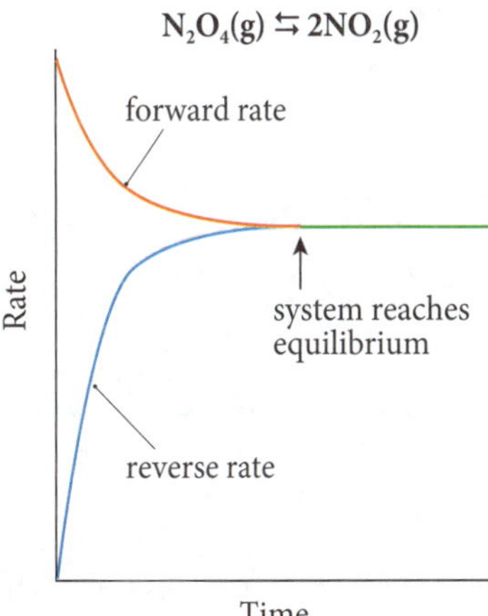

Figure 1.16 Rate of forward and reverse reactions

➔ In the equilibrium between colourless N_2O_4 gas and brown NO_2 gas, the equilibrium state at constant temperature is readily recognised experimentally by a constant colour of the gaseous mixture. The concentrations of each gas in the mixture do not change further once equilibrium is established:

$$N_2O_4(g) \leftrightarrows 2NO_2(g)$$

colourless brown

➔ KEY QUESTIONS

4 **The standard Gibbs free energy change for the complete combustion of methane is negative. Is this reaction spontaneous or non-spontaneous?**

5 **In photosynthesis, identify the substance that is reduced and the substance that is oxidised.**

6 **Use collision theory to explain the factors that lead to a reaction reaching a state of dynamic equilibrium.**

Answers ➲ p. 15

CHAPTER SYLLABUS CHECKLIST

Are you able to answer every question from the syllabus for this chapter? Tick each question as you go through the checklist if you are able to answer it. If you cannot answer a question, turn to the relevant page in the study guide to find the answer. For NESA key word meanings, go to www.educationstandards.nsw.edu.au and search 'key words'.

	FOR A COMPLETE UNDERSTANDING OF THIS TOPIC:	PAGE NO.	✓
1	Can I distinguish between static and dynamic balance?	4	
2	Can I explain the difference between open and closed systems?	5	
3	Can I explain why chemical equilibrium can only be achieved in closed systems?	5	
4	Can I describe examples of experiments that demonstrate the reversibility of reactions in closed systems?	7	
5	Can I calculate the enthalpy change, entropy change and the Gibbs free energy change for combustion reactions?	9	
6	Can I calculate the enthalpy change, entropy change and the Gibbs free energy change for photosynthesis?	10	
7	Can I describe the differences between energy profile diagrams for endothermic and exothermic reactions?	11	
8	Can I compare the rates of the forward and reverse reactions for systems approaching equilibrium?	11	

A QUICK NOTE! All the **dark green panels** with information (like the one below) have the **syllabus inquiry questions** and all the **light green panels** (like the one below) have the **syllabus dot points** for each topic. Note the checklist questions in the table above are based on these.

INQUIRY QUESTION:

What happens when chemical reactions do not go through to completion?

Many chemical reactions proceed to completion and no reactants remain if the stoichiometric ratio is achieved.

1 Modelling static and dynamic equilibrium

» Students model static and dynamic equilibrium and analyse the differences between open and closed systems.

Chemistry Stage 6 Syllabus

Static and Dynamic Equilibrium

Inquiry question: What happens when chemical reactions do not go through to completion?

Students:

- conduct practical investigations to analyse the reversibility of chemical reactions, for example:
  - cobalt(II) chloride hydrated and dehydrated
  - iron(III) nitrate and potassium thiocyanate
  - burning magnesium
  - burning steel wool (ACSCH090)
- model static and dynamic equilibrium and analyse the differences between open and closed systems (ACSCH079, ACSCH091)

HSC EXAM-TYPE QUESTIONS

Now for the real thing! The following questions are modelled on the types of questions you will face in the HSC Examination. Think about it: if you get extensive practice at answering these sorts of questions, you will be more confident in answering them in the actual HSC Examination. It makes sense, doesn't it?

Another advantage for your exam preparation is the format of the answers: they are deliberately structured to give you strategies on how to answer examination questions. This will help you aim for full marks!

- For each objective-response question you will have the correct answer and an explanation, and reasons why the other answers are incorrect.
- For each short-answer question you will have a detailed answer marked with ticks to indicate what part of the question gains which marks and also an examiner's plan (Examiner Maximiser/ EM) to help you get full marks.

When you mark your work, highlight any questions you found difficult and earmark these areas for extra study.

Objective-response questions (1 mark each)

1 **Identify the change that is not readily reversible.**

A Liquid petrol turns into petrol vapour.

B Paraffin wax turns into liquid wax.

C Magnesium carbonate decomposes to form magnesium oxide and carbon dioxide.

D Water vapour condenses to form droplets of water.

2 **Select the true statement concerning the combustion of methane in a Bunsen burner.**

A The system will come to a dynamic equilibrium.

B The reaction is spontaneous.

C The combustion is complete.

D The reaction is endothermic.

3 **Green copper (II) carbonate thermally decomposes to form black copper (II) oxide and carbon dioxide gas, shown in Figure 1.17.**

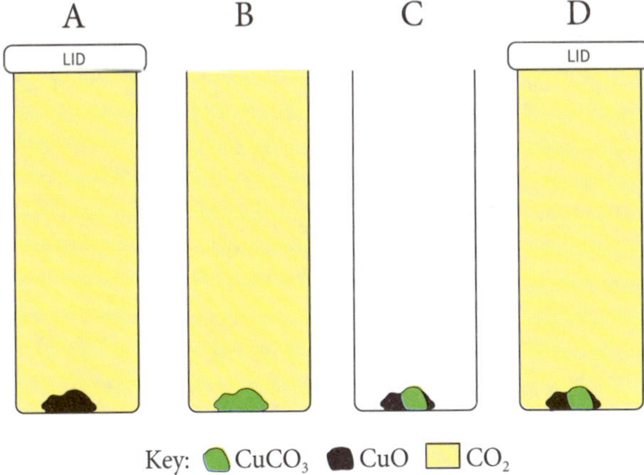

Figure 1.17 Copper (II) carbonate decomposition

Identify which container shows a system in dynamic equilibrium.

A Container A

B Container B

C Container C

D Container D

4 **A student dissolved green copper (II) chloride crystals in a beaker of water to produce a saturated solution. Identify a property of the system which the student could observe that would readily indicate the system had reached a state of dynamic balance.**

A the gas evolution ceases

B the volume of the solution

C the colour of the solution

D the temperature of the solution

5 **Crystals of cobalt (II) carbonate are added to an excess of dilute hydrochloric acid in a 250 mL beaker. A colourless gas is evolved. Select the true statement about the reaction.**

A The system is an example of a static equilibrium.

B A dynamic equilibrium is established.

C The entropy of the system increases.

D The reaction is not spontaneous.

Extended-response questions

6 **Ethanol (C_2H_5OH) is a volatile liquid. Samples of liquid ethanol were placed in separate containers A and B, as shown in Figure 1.18.**

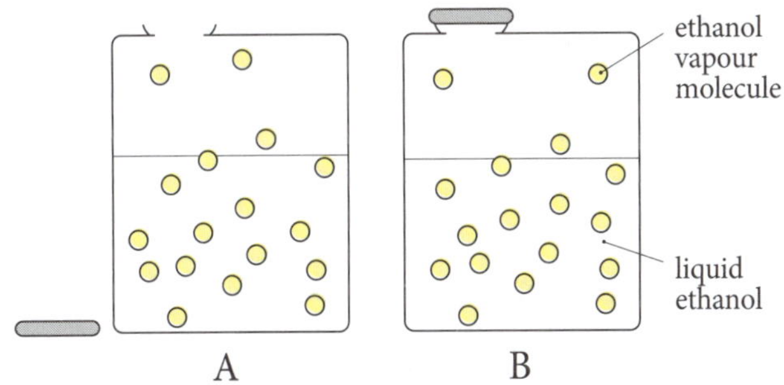

Figure 1.18 Ethanol vaporisation

a **Discuss whether or not an equilibrium between the liquid ethanol and ethanol vapour can be established in each container.** (2 marks)

b **Write the equilibrium equation for this physical change.** (1 mark)

7 A student was supplied with the following chemicals in order to investigate the equilibrium between yellow chromate ions (CrO_4^{2-}) and orange dichromate ions ($Cr_2O_7^{2-}$) in aqueous solution.

Chemicals:
solid potassium chromate (K_2CrO_4)
solid potassium dichromate ($K_2Cr_2O_7$)
1.0 mol/L hydrochloric acid
water

The equilibrium equation is:

$$2CrO_4^{2-}(aq) + 2H^+(aq) \leftrightharpoons Cr_2O_7^{2-}(aq) + H_2O(l)$$

Describe a series of experiments that the student could perform to investigate the chromate/dichromate equilibrium. Predict the results of these experiments. (4 marks)

8 Propane (C_3H_8) gas will undergo complete combustion in an excess of oxygen gas. Under the high temperatures of the reaction, water is formed in the gaseous state.

a Write a balanced equation for the complete combustion of propane. (1 mark)

b Identify and explain the drivers of the forward reaction. (2 marks)

9 Calcium nitrate solid thermally decomposes to form calcium oxide, nitrogen dioxide gas and oxygen gas. The table lists the standard enthalpies of formation and standard entropies for reactants and products.

	$Ca(NO_3)_2(s)$	$CaO(s)$	$NO_2(g)$	$O_2(g)$
$\Delta_f H^\circ$ (kJ/mol)	-938	-635	33	0
S° (J/K/mol)	193	38	240	205

a Write a balanced equation for the thermal decomposition of 1 mole of calcium nitrate. (1 mark)

b Calculate the enthalpy change for the reaction under standard conditions. (1 mark)

c Calculate the entropy change for the reaction under standard conditions. (1 mark)

d Determine whether the reaction is spontaneous under standard conditions. (1 mark)

10 200 mL of water was poured into a flask and a digital pressure gauge was attached (Figure 1.19). The initial temperature was 25 °C. The external air pressure was 100 kPa. The tap was closed and the total gas pressure in the sealed vessel was recorded as a function of time. Table 1.1 shows the results.

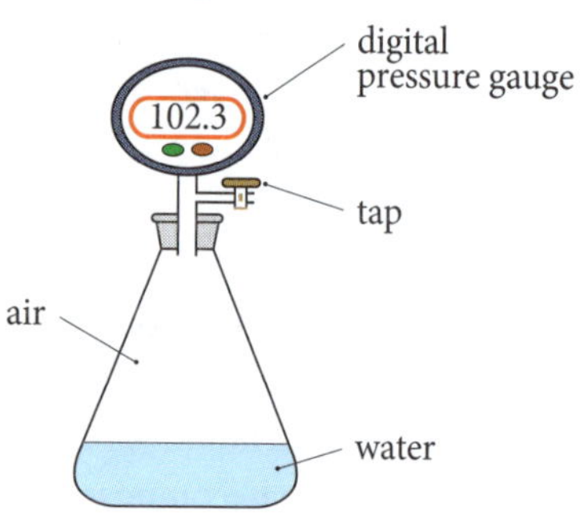

Figure 1.19 Vapour pressure of water

Table 1.1 Results

Time (s)	0	100	200	300	400	500	600	700
Pressure (kPa)	100.0	100.9	101.7	102.3	102.8	103.1	103.2	103.2

a Plot a line graph of the data. (4 marks)

b Determine the time when the system has reached a state of dynamic balance. (1 mark)

c How will the total pressure change if the tap is opened so the system becomes an open system? Explain. (2 marks)

ANSWERS

KEY QUESTIONS

Key questions ➲ p. 9

1 Static equilibrium is the result of no net forces acting on a stationary object. Dynamic equilibrium occurs when no net forces are acting on an object or particle that is in a state of constant motion.

2 $BaCO_3(s) \leftrightarrows BaO(s) + CO_2(g)$

3 This reaction is not readily reversible as the reaction is driven by a large loss of heat to the surroundings and the activation energy for the reverse reaction is very high.

Key questions ➲ p. 12

4 Spontaneous as $\Delta G < 0$.

5 Carbon dioxide is reduced and water is oxidised.

6 The rate of a reaction is dependent on the frequency of collision between reactant particles. The greater concentration of the reacting particles, the greater the collision frequency. Over time the reactants are consumed and their concentration decreases. As reactants combine to form products, the rate of the forward reaction therefore decreases and the rate of the reverse reaction increases. Eventually a point is reached when the rates of the forward and reverse reactions become equal and a dynamic equilibrium is achieved.

HSC EXAM-TYPE QUESTIONS

Objective-response questions

1 **C**. The reaction is not readily reversible as carbon dioxide escapes into the environment. **A**, **B** and **C** are incorrect as these are all physical changes that can be readily reversed.

2 **B**. The system is open and the exothermic change drives this spontaneous process. **A** is incorrect as no equilibrium is achieved in an open system. **C** is incorrect as the combustion is incomplete. **D** is incorrect as this is an exothermic reaction.

3 **D**. Container D is closed and the reactant and two products are all present. **A** is incorrect as no green copper (II) carbonate is present to maintain the equilibrium. **B** is incorrect as no black copper (II) oxide is present. **C** is incorrect as the system is open and the carbon dioxide has escaped.

4 **C**. Equilibrium is achieved when the green colour of the saturated solution no longer changes. **A** is incorrect as no gas is produced in this dissolution process. **B** is incorrect as the volume of the solution is not an indicator of dynamic balance in a solution. **D** is incorrect as the beaker is open and heat changes will not reliably indicate that dynamic balance is achieved.

5 **C**. The escape of gas and the conversion of a solid into aqueous ions will increase the entropy. **A** is incorrect as static equilibrium does not exist in a chemical system. **B** is incorrect as the system is open. **D** is incorrect as the negative enthalpy change and the increase in entropy ensure a spontaneous reaction.

Extended-response questions

6 EM Students need to demonstrate an understanding of the difference between open and closed systems.

a Container B is a closed vessel. The ethanol vapour cannot escape and this prevents the complete evaporation of the ethanol. The ethanol vapour can condense back into liquid ethanol. At a fixed temperature the air in the closed container becomes saturated in ethanol vapour. Thus an equilibrium can be established. ✓

Container A is an open container. The ethanol vapour can escape into the surroundings as evaporation continues. Thus no equilibrium can form in the open container. ✓

b $C_2H_5OH(l) \leftrightarrows C_2H_5OH(g)$ ✓

7 EM Students need to demonstrate skills in experimental design and then show an understanding of dynamic equilibrium by making predictions about the direction in which the equilibria shift.

Method:

a Prepare 0.1 mol/L aqueous solutions of both salts.

b Place about 3 mL of the yellow potassium chromate solution in a test tube. Add drops of HCl and observe the colour change. Then add water and observe the colour change. ✓

c Place about 3 mL of the orange potassium dichromate solution in a test tube. Add water and observe any colour change. Then add HCl and observe the colour change. ✓

Predictions:

a The addition of HCl leads to an increase in the concentration of H^+. This will promote the forward reaction and the solution will become more orange. When water is added the solution becomes more yellow as the reverse reaction is promoted. ✓

b The addition of water will promote the reverse reaction and the solution will become less orange and more yellow-orange. When HCl is added the solution becomes more orange again as the forward reaction is promoted. ✓

8 EM Students need to demonstrate a high-level understanding of the drivers of chemical reactions.

a $C_3H_8(g) + 5O_2(g) \rightarrow 3CO_2(g) + 4H_2O(g)$ ✓

b The release of heat energy in this exothermic reaction drives the reaction in the forward direction. ✓

The high activation energy for the reverse reaction reduces the chance that products will recombine to form reactants. ✓

9 EM Students need to demonstrate a high-level understanding between enthalpy, entropy and Gibbs free energy.

a $Ca(NO_3)_2(s) \rightarrow CaO(s) + 2NO_2(g) + \frac{1}{2}O_2(g)$ ✓

b $\Delta H^o = (-635) + 2(33) + \frac{1}{2}(0) - (-938) = 369$ kJ/mol ✓

c $\Delta S^o = 38 + 2(240) + \frac{1}{2}(205) - (198) = 422.5$ J/K/mol
$= 0.4225$ kJ/K/mol ✓

d $\Delta G^o = \Delta H^o - T\Delta S^o = (369) - 298(0.4225)$
$= +243$ kJ/mol > 0

Reaction is not spontaneous as $\Delta G^o > 0$. ✓

10 EM Students need to demonstrate skills in graph drawing and interpretation of graphical data. Graphs should have titles and the graph should occupy at least 80% of the grid space.

a The graph is shown as Figure A1.1.

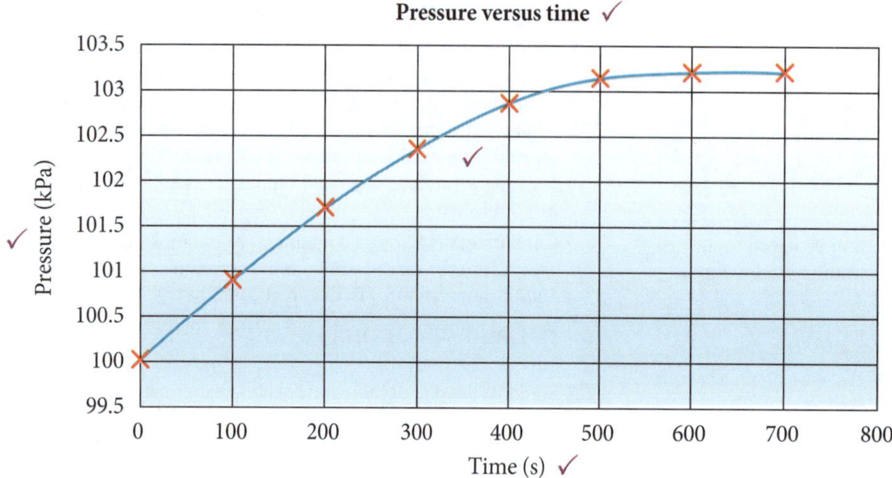

Figure A1.1 Pressure versus time

b ~ 580–600 s ✓

c The total pressure will gradually decrease ✓ as water vapour escapes from the system. ✓

MODULE 5 EQUILIBRIUM REACTIONS

CHAPTER 2 FACTORS THAT AFFECT EQUILIBRIUM

INQUIRY QUESTION:

What factors affect equilibrium and how?

Chemical equilibrium can be achieved in closed systems for many chemical reactions. This chapter investigates a variety of factors that affect the position of the equilibrium. These factors include temperature, concentration, pressure and volume. We will relate these changes to collision theory, activation energy and heats of reaction.

1 Factors causing changes in equilibrium

» Students:

- investigate the effects of temperature, concentration, volume and/or pressure on a system at equilibrium and explain how Le Chatelier's principle can be used to predict such effects.
- explain the overall observations about equilibrium in terms of the collision theory.
- examine how activation energy and heat of reaction affect the position of equilibrium.

Le Chatelier's principle

➔ Equilibrium systems can be affected by changes made to the system. The French chemist Henri Le Chatelier (1850–1936) developed a principle that could be used to predict changes in equilibrium systems. His principle states:

> When a change is made to an equilibrium system, the system moves to counteract the imposed change and restore the system to a new equilibrium.

➔ The new equilibrium position differs from the old equilibrium position. The position of an equilibrium refers to the comparative concentration of reactants and products. An equilibrium in which the product concentration is greater than the reactant concentration is said to lie to the right.

An equilibrium in which the product concentration is less than the reactant concentration is said to lie to the left. Figure 2.1 shows the equilibrium concentrations for two different homogeneous equilibria. The first equilibrium lies to the right and the second equilibrium lies to the left.

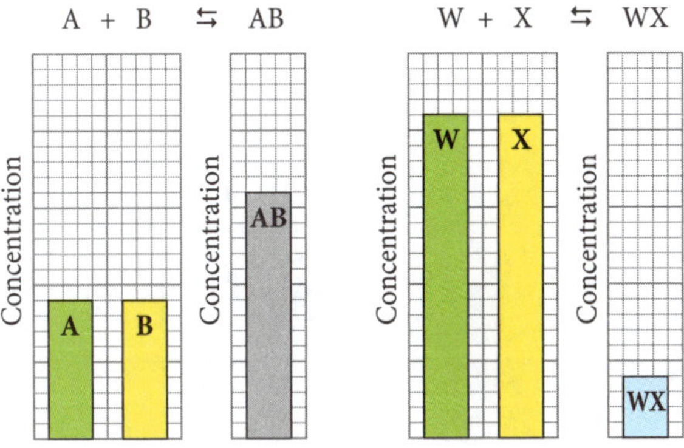

Figure 2.1 Positions of equilibria

➔ Figure 2.2 demonstrates the dynamic nature of these equilibria. In each equilibrium system the reactant and product molecules are in constant random motion and collisions continue to occur between molecules. As the rate of the forward reaction equals the rate of the reverse reaction then the systems remain at equilibrium.

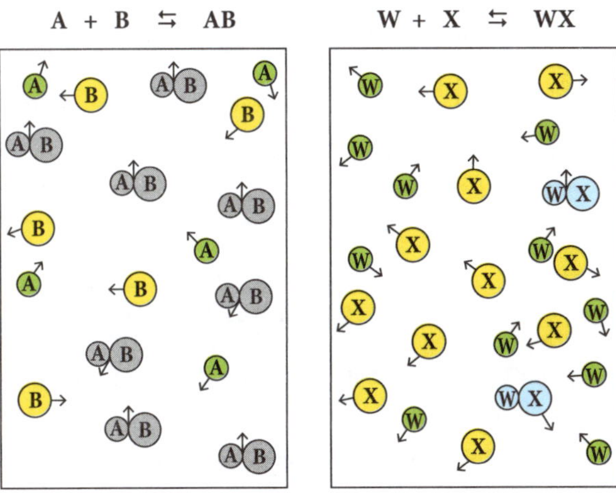

Figure 2.2 Dynamic equilibria

Variables affecting equilibria

- A variety of factors can be investigated and changes can be explained using Le Chatelier's principle.

Temperature changes

- If an equilibrium is exothermic in the forward direction it will be **endothermic** in the reverse direction.
- If the equilibrium is **exothermic** in the forward direction, an increase in temperature causes the equilibrium to shift towards the left (reactants). This shift counteracts the increase in heat content by using up some of the added heat to convert some products into reactants. In the following example colourless hydrogen iodide decomposes into colourless hydrogen gas and violet iodine vapour. The system becomes less violet as the temperature is increased (see Figure 2.3):

$$2HI(g) \leftrightarrows H_2(g) + I_2(g); \Delta H = -48 \text{ kJ/mol}$$

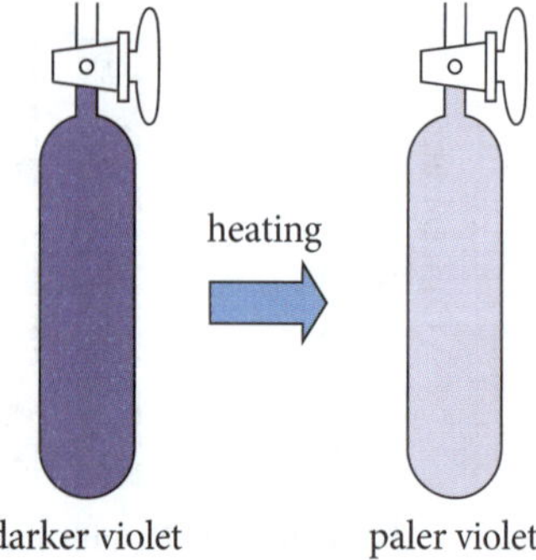

Figure 2.3 Exothermic equilibrium

The increase in temperature increases the kinetic energy of the molecules and therefore a greater proportion of product molecules have sufficient energy to overcome the higher activation energy of the reverse reaction. Figure 2.4 shows the activation energy profile.

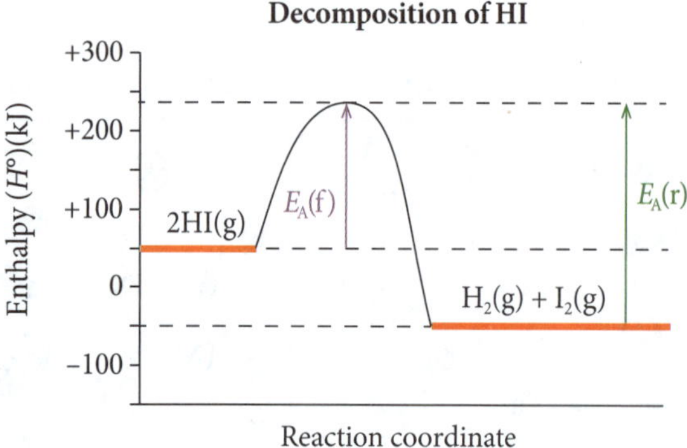

Figure 2.4 Energy profile diagram for the HI decomposition equilibrium

- If an equilibrium is endothermic then an increase in temperature causes the equilibrium to shift to the right (products) to use up some of the added heat. In the following example the concentrations of nitrogen and hydrogen increase in the reaction vessel as the temperature increases:

$$2NH_3(g) \leftrightarrows N_2(g) + 3H_2(g); \Delta H = +92 \text{ kJ/mol}$$

The increase in temperature increases the kinetic energy of the molecules and therefore a greater proportion of reactant molecules have sufficient energy to overcome the high activation energy of the forward reaction.

endothermic: a reaction that absorbs heat from the surroundings ($\Delta H > 0$)

exothermic: a reaction that releases heat to the surroundings ($\Delta H < 0$)

EXAMPLE 1

Use the graphical data in Figure 2.5 to explain the connection between molecular velocities in gaseous systems and the activation energy in an exothermic equilibrium system involving reacting gases.

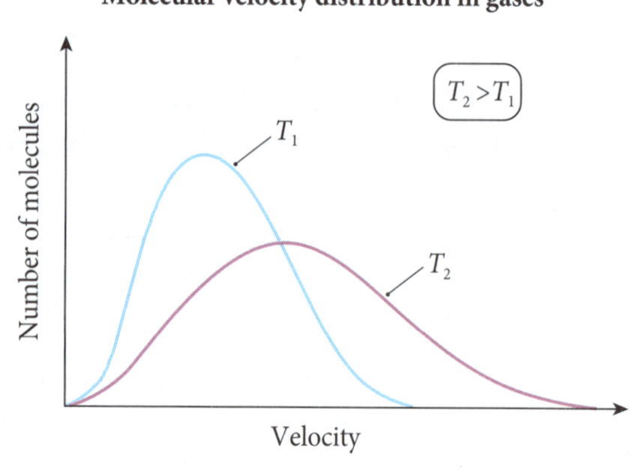

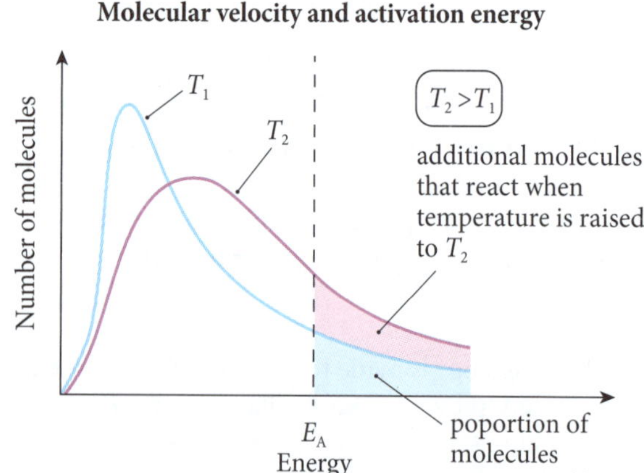

Figure 2.5 Molecular velocity and activation energy

Relate temperature changes to changes in molecular velocity

Answer:

In a system of gas molecules at a fixed temperature, there is a distribution of molecular velocities. Some molecules are moving more rapidly than others at a fixed temperature. As the temperature of the system increases there is an increasing proportion of molecules with higher molecular velocities.

If the system consists of a mixture of reacting gases their kinetic energies increase if the temperature is increased. If the kinetic energy of the molecules exceeds the activation energy on collision then reaction will occur. The activation energy is like a barrier that separates the reacting molecules from their potential products. At the higher temperature (T_2) the number of molecules having an energy greater than E_A has increased compared with the lower temperature (T_1).

In an exothermic equilibrium involving gases the activation energy for the reverse reaction (E_A(r)) is greater than the activation energy for the forward reaction (E_A(f)). As the temperature increases from T_1 to T_2 the rates of the forward and reverse reaction both increase, but the reverse rate increases to a greater extent due to a greater proportion of product molecules having kinetic energies greater than the reverse activation energy. When equilibrium is attained at T_2 there will be a decrease in the concentration of products as a result.

Concentration changes

- Increasing the concentration of one or more reactants (in a homogeneous equilibrium) or decreasing the concentration of the products causes the equilibrium to shift towards the products to counteract the imposed change.
- Decreasing the concentration of reactants results in fewer collisions of reactant particles and so the forward rate decreases. Thus the equilibrium position shifts to the left (reactant side). Increasing the concentration of the products also causes the equilibrium to shift towards the reactants as the more product particles, the greater the collision frequency and the greater the reverse rate.
- In the following example the chromate ion/dichromate ion equilibrium is investigated:

  $2CrO_4^{2-}(aq) + 2H^+(aq) \leftrightharpoons Cr_2O_7^{2-}(aq) + H_2O(l)$

  yellow orange

  - An increase in hydrogen ion concentration (by adding an acid such as HCl) causes the equilibrium to shift to counteract the change. The system becomes more orange as some of the added hydrogen ions react with chromate ions. The equilibrium shifts to the right due to an increase in the forward reaction rate on the addition of hydrogen ions. Eventually a new equilibrium is attained when the rates of the forward and reverse reactions become equal once more.
  - Decreasing the concentration of hydrogen ions (by neutralisation with an alkali such as NaOH) causes the equilibrium to shift to restore some hydrogen ions. In the new equilibrium system the solution is more yellow and less orange (see Figure 2.6).

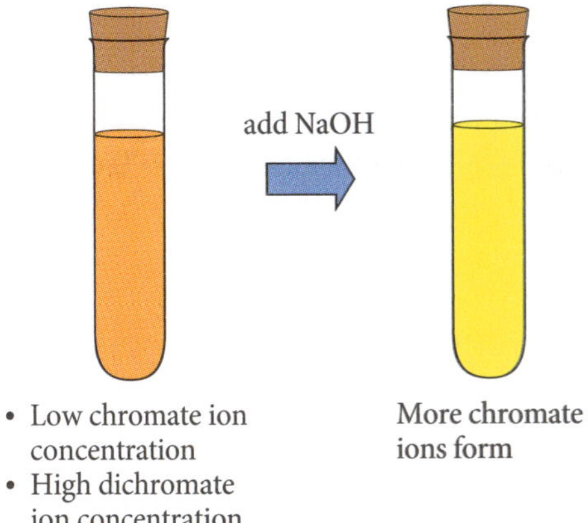

Figure 2.6 Dichromate/chromate equilibrium

- In an aqueous heterogeneous equilibrium, changes in concentration only refer to species in the dissolved state. Changes in the amounts of solids or liquids do not alter the equilibrium position. For example, in a saturated solution of sodium chloride, the equilibrium equation is:

  $NaCl(s) \leftrightharpoons Na^+(aq) + Cl^-(aq)$

  Increases in sodium ion concentration or chloride ion concentration will cause the equilibrium to shift to the left. The addition of more salt crystals has no effect on the equilibrium as the solution is already saturated and no more salt can dissolve.
- In a closed system liquid water is in equilibrium with water vapour. This is a heterogeneous equilibrium. At constant temperature the addition of more liquid water does not affect the equilibrium as the air in the container is saturated in water vapour. The vapour pressure of the water is at a maximum at that temperature. However, the removal of some water vapour will cause the equilibrium to shift to the right to restore the saturation levels:

  $H_2O(l) \leftrightharpoons H_2O(g)$

Gas pressure and volume changes

- In homogeneous gaseous systems changes in pressure (caused by changes in volume of the vessel) may lead to changes in concentrations of reactants or products. In general, an increase in total pressure causes an equilibrium to shift to the side of fewer gas molecules. In this way the system counteracts the change by reducing the number of particles (and thus the pressure). If the number of particles is the same on each side of the equilibrium then a change in pressure has no effect.

EXAMPLE 2

An equilibrium mixture of NO_2 and N_2O_4 is placed in a vessel at a constant temperature. The initial equilibrium mixture is pale brown:

$2NO_2(g) \leftrightharpoons N_2O_4(g)$

brown colourless

The volume of the vessel is suddenly decreased by half at time (t). The following graph (Figure 2.7) shows how the partial pressures of each gas changes over time.

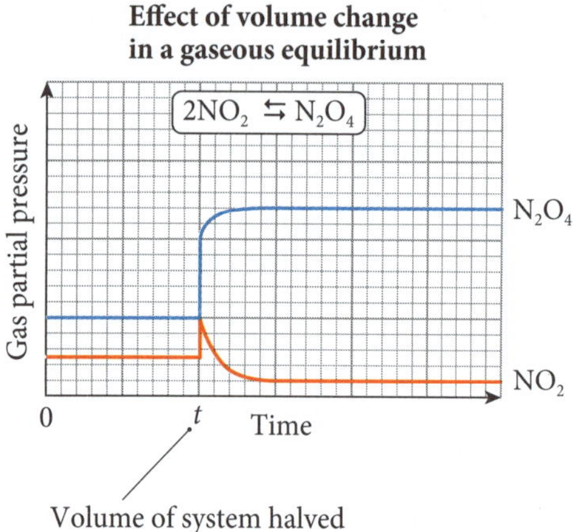

Figure 2.7 Partial pressure changes in gaseous equilibrium

Explain the changes in partial pressures and describe how the colour of the mixture will change over time.

> Determine whether the total pressure increases or decreases as a result of the volume change

Answer:

The total gas pressure suddenly doubles as the volume halves. This is an example of Boyle's law ($PV = k$). Initially the brown colour deepens. This change disturbs the equilibrium and, to counteract the change, the equilibrium shifts to the side of lower number of particles (i.e. more N_2O_4) to reduce the total pressure. As the N_2O_4 is colourless, the colour of the system changes to a paler brown than in the original system.

EXAMPLE 3

Figure 2.8 is a graph of the rate of formation of N_2O_4 as 2 NO_2 molecules react:

$2NO_2(g) \leftrightharpoons N_2O_4(g)$

At time t_1 more NO_2 is injected into the reaction vessel at constant temperature.

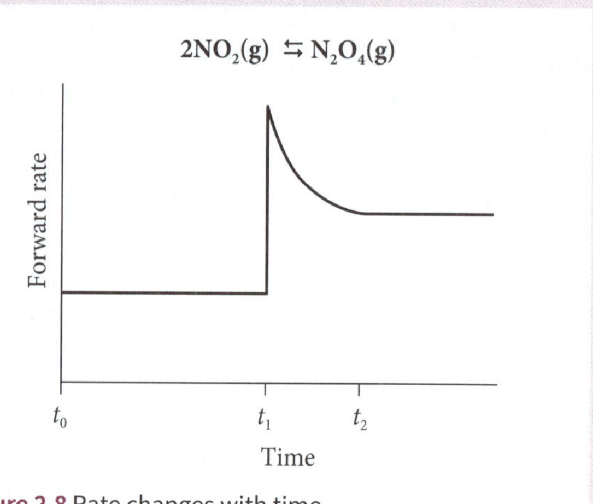

Figure 2.8 Rate changes with time

a **Explain why the rate of the forward reaction is constant between times t_0 and t_1.**

b **Explain why the rate of the forward reaction decreases between times t_1 and t_2.**

c **Explain why the rate of the forward reaction is constant after time t_2.**

> Relate rate changes to the concentration of colliding molecules

Answer:

a The system is at equilibrium and the forward rate and reverse rate are equal.

b The injection of more NO_2 at t_1 causes more collisions between reactant molecules and the forward rate increases. The forward rate now decreases between t_1 and t_2, as the concentration of NO_2 decreases (thus reducing the partial pressure) as N_2O_4 forms.

c At time t_2 the system has reached a new equilibrium. The reverse rate has been increasing between t_1 and t_2 and, at t_2, the reverse and forward rates have become equal and a new equilibrium has been established.

EXAMPLE 4

Carbon dioxide gas dissolves in water to a limited extent to form a solution of carbonic acid (H_2CO_3). The equilibria involved are:

$CO_2(g) \leftrightharpoons CO_2(aq)$ ... (1)

$CO_2(aq) + H_2O(l) \leftrightharpoons H_2CO_3(aq)$ ... (2)

$H_2CO_3(aq) \leftrightharpoons H^+(aq) + HCO_3^-(aq)$ ... (3)

The equilibria are exothermic. Explain the effect of changes in the following variables on the solubility of carbon dioxide in water:

a **partial pressure of carbon dioxide**

b **temperature**

c **addition of sodium hydroxide.**

> The equilibria are interdependent

Answer:

a The solubility of carbon dioxide increases as the partial pressure of gaseous carbon dioxide above the water increases. Le Chatelier's principle predicts that an increase in the reactant concentration (pressure) will cause the equilibrium (1) to shift to the right to counteract the change. Thus more carbon dioxide dissolves. An increase in dissolved carbon dioxide in turn causes equilibrium (2) to shift to the right to form more carbonic acid. An increase in carbonic acid concentration causes equilibrium (3) to shift to the right and thus the acidity of the solution due to hydrogen ions increases.

b The solubility of carbon dioxide decreases with increasing temperature. The reaction equilibria are exothermic. According to Le Chatelier's principle, an increase in temperature causes the reaction equilibria to shift back towards the reactants, to use up some of the added heat. The acidity will also decrease as fewer hydrogen ions will be present at the new equilibrium.

c The presence of OH^- ions causes a decrease in H^+ ions in equilibrium (3) due to a neutralisation reaction. Thus equilibrium (3) shifts to the right to counteract the change. This reduces the carbonic acid concentration. In turn, both equilibrium (2) and equilibrium (1) shift to the right to counteract the change. Carbon dioxide solubility will therefore increase in the presence of an alkali such as hydroxide ions. Thus carbon dioxide is more soluble in alkaline solutions than water.

Effect of catalysts

→ Catalysts have no effect on the *position* of an equilibrium. They increase both the forward and reverse reaction rates to the same extent so that equilibrium is achieved faster by lowering the activation energy. The concentrations of reactants and products at equilibrium does not change. Figure 2.9 shows that equilibrium is achieved at 3.2 hours without the catalyst, but with the catalyst equilibrium is reached at 2.2 hours. Equilibrium is achieved faster due to the lowering of the activation energy in the forward and reverse reactions.

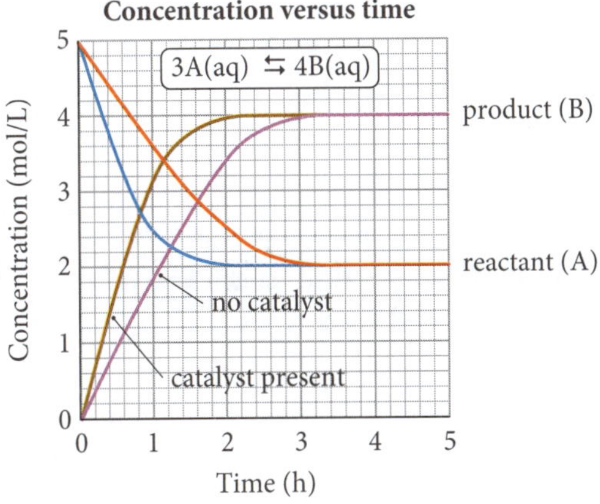

Figure 2.9 Effect of catalysts on equilibrium

EXAMPLE 5

The following graph (Figure 2.10) shows the change in rates for the forward and reverse reactions in which molecules X and Y react to form an XY molecule:

$$X(g) + Y(g) \leftrightarrows XY(g)$$

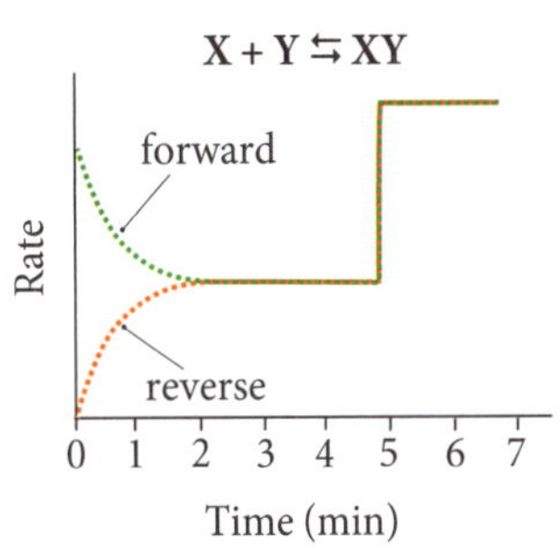

Figure 2.10 Rate versus time

a **Explain at which time the reaction mixture first reaches equilibrium.**

b **Explain whether the change at $t = 5$ minutes is the result of a change in temperature or the addition of a catalyst.**

Consider how the forward and reverse rates change with time

Answer:

a Equilibrium is achieved initially at $t = 2$ minutes as the rate of the forward and reverse reactions are equal.

b At $t = 5$ minutes, both the forward and reverse rates increase equally. The system remains in equilibrium after this time. This can only occur with the addition of a catalyst, which increases the forward and reverse rates equally so that the position of the equilibrium remains the same. A temperature change would lead to a rate change in either the forward or reverse rate.

→ KEY QUESTIONS

1 **The temperature of a system in equilibrium is increased. Explain whether or not this change will ultimately lead to the restoration of the original equilibrium position at the higher temperature.**

2 **A saturated solution of sodium chloride is prepared. Explain how the position of this equilibrium will change if some of the chloride ions are removed by precipitation on the addition of a few drops of concentrated silver nitrate solution.**

3 **An unopened bottle of soda water contains water in which an equilibrium exists between the dissolved carbon dioxide and the gaseous carbon dioxide in the space above the soda water. The bottle is opened and the soda water degases. Explain whether a new carbon dioxide equilibrium will ever be established in this open system.**

Answers ➲ p. 30

2 Investigating disturbances in equilibrium systems

» Students investigate the effects of temperature, concentration, volume and/or pressure on a system at equilibrium and explain how Le Chatelier's principle can be used to predict such effects, for example: heating cobalt (II) chloride hydrate; interaction between nitrogen dioxide and dinitrogen tetroxide; and iron (III) thiocyanate and varying concentration of ions.

Disturbing a heterogeneous equilibrium

➔ Chemists use macroscopic properties to determine if a system has reached equilibrium. When blue crystals of hydrated copper (II) sulfate ($CuSO_4{\cdot}5H_2O$) are added to 5 mL of water in a test tube the crystals slowly dissolve and the solution becomes blue. Shaking the solution increases the rate of dissolution and the blue colour of the solution increases in intensity. The flask is stoppered and allowed to stand. If excess crystals are added to the water, eventually a saturated solution of copper (II) sulfate forms (Figure 2.11).

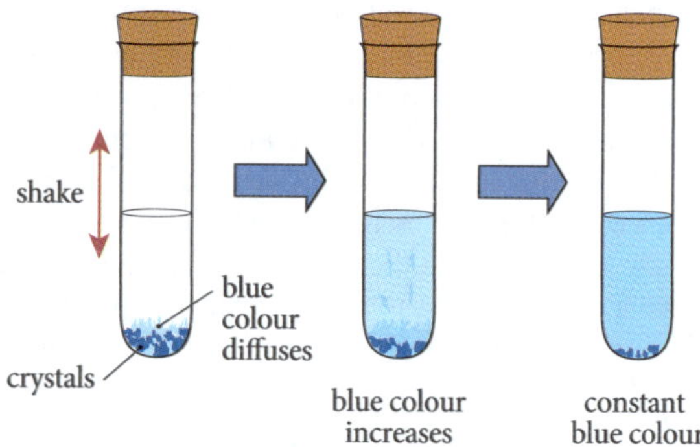

Figure 2.11 Copper (II) sulfate dissolution to form a saturated solution

➔ Equilibrium is reached when there is no further change in measurable macroscopic properties. Colour is a macroscopic property. When the blue colour of the solution becomes constant then equilibrium has been achieved. This can be measured instrumentally using a colourimeter. Electrical conductivity is another macroscopic property that can be used to determine if equilibrium has been attained. The number of ions in solution will rise as the solid dissolves. Eventually the solution becomes saturated in ions and so there will be no further rise in electrical conductivity.

➔ The equation for the dynamic equilibrium in the copper (II) sulfate equilibrium is:

$$CuSO_4{\cdot}5H_2O(s) \leftrightarrows Cu^{2+}(aq) + SO_4^{2-}(aq) + 5H_2O(l)$$

The hydrated copper (II) ion is blue whereas the sulfate ion is colourless.

➔ Various changes can be made to the copper (II) sulfate equilibrium that disturb the equilibrium.

- Change 1: Add additional water
  If additional water is added to the solution more crystals dissolve. If a large amount of water is added then all the crystals dissolve and the solution is no longer saturated. The solution becomes paler blue due to dilution. No equilibrium will then exist.
- Change 2: Evaporate the water
  If the stopper is removed and the water slowly evaporates in the open system more blue crystals form but the solution does not change colour (at constant temperature) as it remains saturated (Figure 2.12). If all the water evaporates then no equilibrium exists.

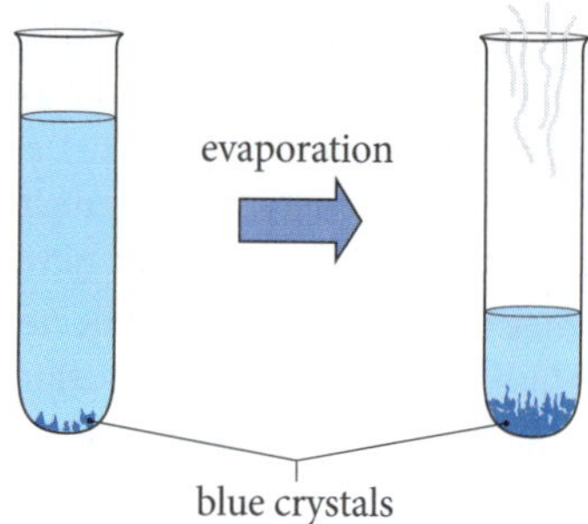

Figure 2.12 Evaporation of water from a saturated solution of copper (II) sulfate

- Change 3: Heat the saturated solution
  If the stoppered test tube is placed in a beaker of warm water (say 30 °C) more crystals dissolve. Most ionic crystals increase in solubility as the temperature increases. In the case of copper (II) sulfate the solubility increases from 22 g/100 g water at 25 °C to 24 g/100 g water at 30 °C. The system remains saturated at the higher temperature as long as some crystals remain.
- Change 4: Adding a common ion
  The addition of drops of a concentrated potassium sulfate causes an increase in the concentration of sulfate ions. Le Chatelier's principle predicts that the equilibrium will shift to the left to counteract the disturbance. More copper (II) sulfate crystals will form. This result is known as the *common ion effect.* The addition of a few drops of a concentrated sodium nitrate solution has no effect on the equilibrium as no common ions have been added.

EXAMPLE 6

Iodine monochloride (ICl) is a brown liquid. It reacts with chlorine gas to form yellow crystals of iodine trichloride (ICl_3). All three species exist in equilibrium in a closed vessel (Figure 2.13):

$$ICl(l) + Cl_2(g) \leftrightarrows ICl_3(s)$$

Explain each of the following observations:

a **If more chlorine is pumped into the vessel the amount of the brown liquid decreases.**

b If the lid of the container is removed the yellow crystals start to disappear.

c A little of the brown liquid is removed (without changing the amount of chlorine) and yet the amount of yellow crystals does not change.

Use Le Chatelier's principle for a heterogeneous equilibrium

d The volume of the vessel is suddenly doubled. More brown liquid forms and the amount of yellow crystals decreases.

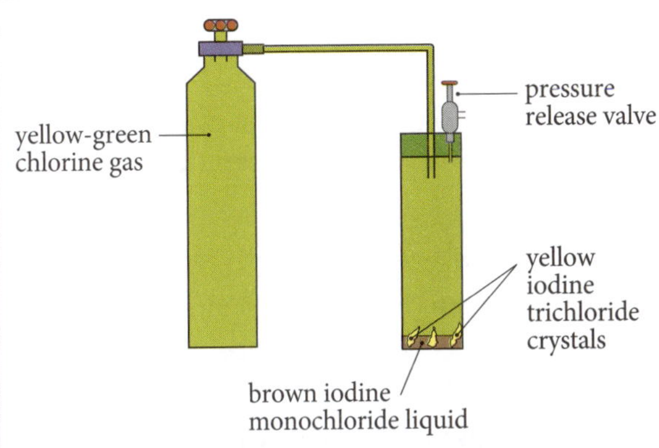

Figure 2.13 Addition of extra chlorine to equilibrium system

Answer:

a According to Le Chatelier's principle, the equilibrium shifts to the right as the partial pressure of chlorine increases. The additional chlorine reacts with the brown liquid and more yellow crystals form.

b The chlorine gas escapes from the open system. The equilibrium is disturbed and the yellow crystals decompose and more brown liquid and chlorine forms. No equilibrium can be established in the open system.

c This is a heterogeneous equilibrium and so removal of some of the brown liquid will not cause a change to the equilibrium as some remains to maintain the equilibrium.

d The doubling of the volume immediately reduces the partial pressure of chlorine gas. Therefore according to Le Chatelier's principle the equilibrium will shift to the left to increase the partial pressure and consequently some ICl_3 decomposes and more ICl forms. The system becomes browner.

Disturbing a homogeneous equilibrium

➔ The Contact process is the major method used in the commercial manufacture of sulfuric acid. One of the steps in this manufacture is the oxidation of sulfur dioxide gas to produce sulfur trioxide.

➔ The reaction equilibrium is homogeneous as all molecules are gases. The reaction is exothermic:

$$2SO_2(g) + O_2(g) \leftrightarrows 2SO_3(g); \Delta H = -99 \text{ kJ/mol}$$

➔ Let us examine the changes to conditions that ensure a greater yield of sulfur trioxide.

- Change 1: Increase the oxygen : sulfur dioxide mole ratio
  The sulfur dioxide is mixed with excess air (in a 1 : 5 mole ratio). This is equivalent to a mole ratio of $O_2 : SO_2$ of approximately 1 : 1, which is twice as great as that required by the stoichiometry of the reaction. The increase in oxygen levels shifts the equilibrium to the right as the rate of the forward reaction increases. More sulfur trioxide will form.
- Change 2: Increase the total pressure
  The gas mixture is maintained at a pressure above atmospheric pressure (between 100–200 kPa) and passes into the catalyst tower. By increasing the total gas pressure the rate of the forward reaction increases as more collisions occur between reactants. As more sulfur trioxide forms the mole ratio of reactants to products decreases.
- Change 3: Adding a catalyst
  The tower contains layers of vanadium (V) oxide catalyst impregnated on the surface of porous silica pellets. The catalyst increases the rate of the forward and reverse reactions and shortens the time to reach equilibrium. In this way the rate can be increased without changing the temperature. The catalyst does not produce a greater yield. Instead it reduces the time to reach equilibrium.
- Change 4: Changing the temperature
  A moderate temperature of 400–550 °C is used rather than a low or high temperature. This temperature range is a compromise temperature. If too low a temperature was used the reactant molecules would have reduced kinetic energies and fewer molecules would have sufficient energy on collision to overcome the activation energy for the forward reaction. Because the reaction is exothermic heat is a product. A higher temperature would add more heat to the system and this would increase the reverse reaction rate as more product molecules would have energies greater than the reverse activation energy. Figure 2.14 shows the effect of temperature on the yield of sulfur trioxide.

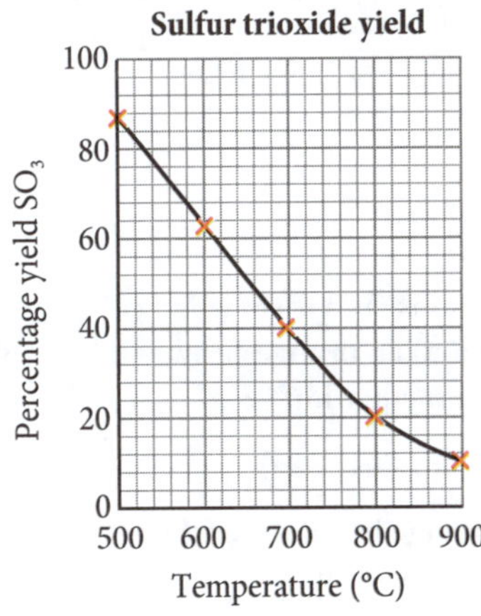

Figure 2.14 Yield of sulfur trioxide versus temperature

EXAMPLE 7

Ammonia can be manufactured industrially using the Haber process, in which hydrogen and nitrogen are reacted over a porous iron catalyst:

$$N_2(g) + 3H_2(g) \leftrightarrows 2NH_3(g)$$

The total gas pressure can be changed by increasing or decreasing the volume of the reaction vessel. Explain how an increase in the volume of the reaction vessel at constant temperature would alter the yield of ammonia.

Examine the reaction stoichiometry and apply Le Chatelier's principle

Answer:

Increasing the volume decreases the total pressure. Figure 2.15 shows how the amount (or partial pressure) of ammonia is affected by increasing the vessel volume.

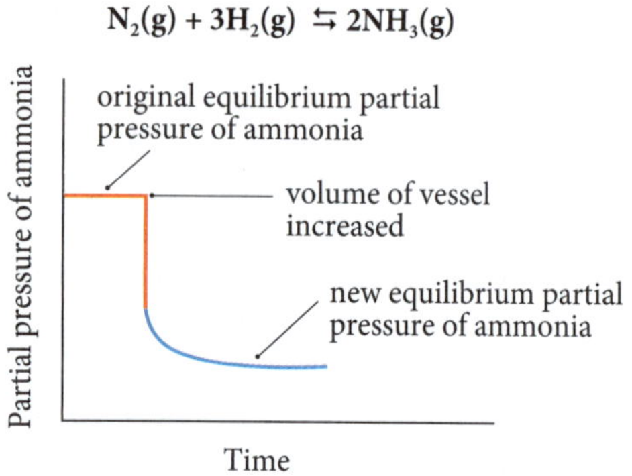

Figure 2.15 The sudden increase in volume to the equilibrium system leads to a shift in the equilibrium to the left to counteract the change

The stoichiometric equation shows that four moles of reactants are converted to two moles of products. Le Chatelier's principle predicts that a decrease in total pressure in the system (by volume increase) will favour the reaction direction that leads to an increase of gas pressure due to the formation of more gas molecules. This is the reverse reaction and consequently the lower the system pressure, the lower the yield of ammonia.

EXAMPLE 8

Red triiodide ions (I_3^-) form when colourless iodide ions are added to an orange-brown iodine solution. The equilibrium equation is:

$$I_2(aq) + I^-(aq) \leftrightarrows I_3^-(aq)$$

Predict the change in colour of the system as the water evaporates at constant temperature.

Compare the number of reactant and product particles

Answer:

The concentration of all species increases as the water evaporates. The equilibrium shifts to the right to counteract the change and reduce the number of particles in the solvent. The right-hand side of the equilibrium has less particles than the left-hand side and thus more triiodide ions form. Therefore the colour of the system becomes less orange-brown and more red as more triiodide ions form.

EXAMPLE 9

Consider the following homogeneous equilibrium in which gaseous XY decomposes to form gaseous X and gaseous Y_2:

$$2XY(g) \leftrightarrows 2X(g) + Y_2(g); \Delta H < 0$$

The following graph (Figure 2.16) shows the change in concentration of XY and Y_2 as a function of time.

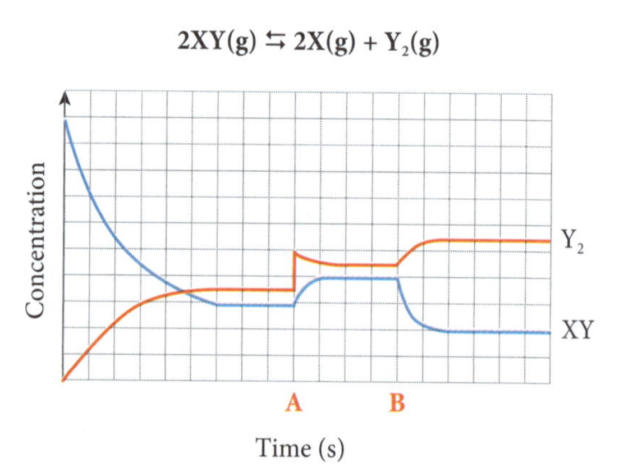

Figure 2.16 Changes in a homogeneous equilibrium

Two changes to the system were made at times A and B that caused a shift in the equilibrium. Explain what changes were made at A and B.

At each time determine which species has increased in concentration

Answer:

At time A: More Y_2 was added to the system. This caused a disturbance that caused the equilibrium to shift to the left (according to Le Chatelier's principle) to counteract the change. Thus more XY formed. A new equilibrium was established.

At time B: The concentration of XY decreased and Y_2 increased. Thus the equilibrium has shifted to the right. As the equilibrium is exothermic ($\Delta H < 0$) the system has been cooled and the removal of heat has caused the equilibrium to move to the right to counteract the change.

FIRSTHAND INVESTIGATION 1

Investigating the hydrated cobalt (II) chloride equilibrium

» Students investigate the effects of temperature on a system at equilibrium and explain how Le Chatelier's principle can be used to predict such effects, for example heating cobalt(II) chloride hydrate.

Aim

to determine the effect of heating hydrated cobalt (II) chloride crystals

Method

1 Place purple cobalt (II) chloride hexahydrate crystals ($CoCl_2{\cdot}6H_2O$) in a test tube and use a test tube holder to heat the tube gently over a blue Bunsen flame till the crystals are dehydrated.
2 Allow the tube to cool back to room temperature in a test tube rack.
3 Add drops of water to the cooled solid and observe the colour change.

Sample results

1 The purple crystals turn blue on heating.
2 The blue solid turns back to red when water is added and additional water produces a red or pink solution, depending on the amount of water added.

Explanation

The equilibrium reaction is:

$$\underset{\text{purple}}{CoCl_2{\cdot}6H_2O(s)} \leftrightharpoons \underset{\text{blue}}{CoCl_2(s)} + 6H_2O(l)$$

Heating the purple crystals promotes the forward reaction as water is removed from the crystals. The water evaporates into the air on continued heating. The equilibrium shifts to the right as water is removed. The reverse reaction is promoted by the addition of liquid water to the blue crystals. The equilibrium shifts back to the left as water is added.

Conclusion

The equilibrium between hydrated cobalt (II) chloride and anhydrous cobalt (II) chloride is reversible.

EXAMPLE 10

Washing soda is the common name of sodium carbonate decahydrate. When placed in an open dish the crystals lose water molecules into the air and a white powder forms. This process is called *efflorescence*. The process can be accelerated in a closed container called a desiccator. The desiccator contains a drying agent (desiccant) called silica gel (SiO_2) impregnated with blue cobalt (II) chloride (see Figure 2.17). The silica gel granules turn pink as water vapour is removed from the air. The following diagram shows an experiment in which 10.00 g of sodium carbonate decahydrate crystals (washing soda crystals) was placed in a Petri dish in a desiccator. The desiccant slowly turned pink as hydrated cobalt (II) ions form.

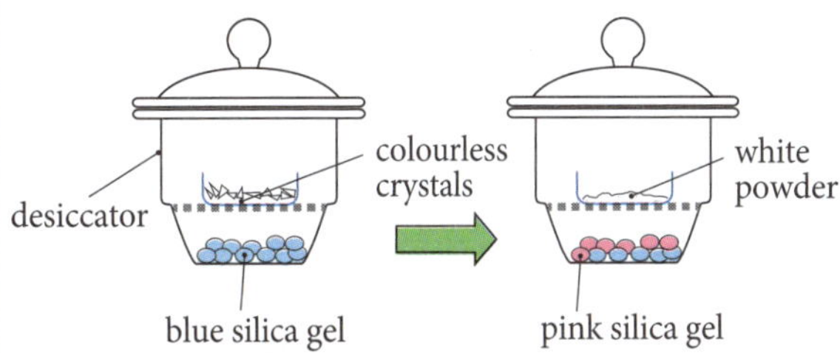

Figure 2.17 Dehydration of washing soda crystals

The white powder which had formed was removed and reweighed. The mass of the white powder was 4.33 g.

a Explain with the aid of an equation why the blue desiccant turns pink.

b The washing soda crystals lose water molecules on dehydration, as shown in the following equation:

$$Na_2CO_3{\cdot}10H_2O(s) \leftrightharpoons Na_2CO_3{\cdot}xH_2O(s) + (10 - x)H_2O(g) \quad (x = \text{integer})$$

i Explain why this equilibrium shifts to the right during the experiment.

ii Use the gravimetric data to determine the formula of the white powder.

Calculate the molar mass of sodium carbonate decahydrate

Answer:

a The blue compound is anhydrous cobalt (II) chloride. The absorption of water vapour by the silica gel causes hydration of the cobalt salt, forming hexaaquacobalt (II) ions, which are pink in the presence of water:

$$CoCl_2(s) + 6H_2O(l) \leftrightharpoons Co(H_2O)_6^{2+}(aq) + 2Cl^-(aq)$$

b i Water vapour molecules are lost from the washing soda due to absorption of water vapour out of the air by the silica gel desiccant. The crystals continue to lose water due to the disturbance of the equilibrium.

ii $M(Na_2CO_3{\cdot}10H_2O) = (2 \times 22.99) + 12.01 + (3 \times 16.00) + (20 \times 1.008) + (10 \times 16.00)$
$= 286.15$ g/mol

$n(Na_2CO_3{\cdot}10H_2O) = m/M = \frac{10.00}{286.15} = 0.03495$ mol

$M(H_2O) = (2 \times 1.008) + 16.00 = 18.016$ g

$m(H_2O)$ lost on drying $= 10.00 - 4.33 = 5.67$ g

$n(H_2O) = m/M = \frac{5.67}{18.016} = 0.3147$ mol

Ratio $n(H_2O)/n(Na_2CO_3{\cdot}10H_2O) = \frac{0.3147}{0.03495} = 9:1$

Therefore nine molecules of water have been lost during dehydration. Sodium carbonate monohydrate has formed.

The formula of white powder = $Na_2CO_3{\cdot}H_2O$

FIRSTHAND INVESTIGATION 2

Investigating the nitrogen dioxide/dinitrogen tetroxide equilibrium

» Students investigate the effects of temperature on a system at equilibrium and explain how Le Chatelier's principle can be used to predict such effects, for example the interaction between nitrogen dioxide and dinitrogen trioxide.

Aim

to determine the effect of temperature changes on the nitrogen dioxide/dinitrogen tetroxide equilibrium

Method

1 Three sealed glass tubes contain brown nitrogen dioxide gas and colourless dinitrogen tetroxide gas.
2 Place the first tube in a beaker of ice and the second tube in a beaker of water at room temperature. Place the third tube placed in a beaker of hot water.
3 Allow the tubes time to reach equilibrium.
4 Record the colour of the gas mixture in each tube.
5 Show that the equilibrium is reversible by moving the tubes into different beakers (see Figure 2.18).

Sample results

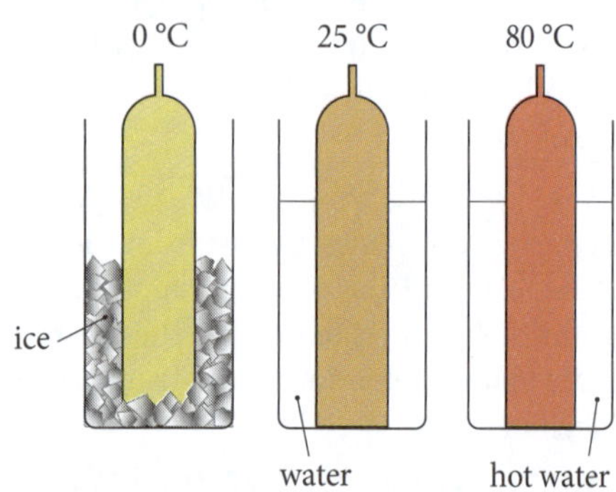

Figure 2.18 Nitrogen dioxide/dinitrogen tetroxide equilibrium

1 The gas tube in the ice turns a pale yellow-brown.
2 The gas tube at room temperature turns medium brown.
3 The gas tube in the hot water turns a deep reddish brown.

Explanation

The equilibrium reaction is:

$2NO_2(g) \rightleftharpoons N_2O_4(g)$

brown colourless

Heating the gaseous equilibrium causes the equilibrium to shift to the left, producing more nitrogen dioxide. Therefore the forward reaction is exothermic. Heating disturbs the equilibrium and, to counteract the change, the system uses some of the added heat to promote the reverse reaction, which has a higher activation energy. Cooling also disturbs the equilibrium and, to counteract the change, the system moves to generate heat. The forward reaction is therefore promoted as it has a lower activation energy.

EXAMPLE 11

Use the supplied data to determine the spontaneity of the nitrogen dioxide/dinitrogen tetroxide equilibrium at 25 °C:

$2NO_2(g) \rightleftharpoons N_2O_4(g)$

Data:

	$\Delta_f H^o$ (kJ/mol)	S^o (J/K/mol)
$NO_2(g)$	33	240
$N_2O_4(g)$	9	304

Calculate the Gibbs free energy change

Answer:

$\Delta H^o = \Sigma\Delta_f H^o(\text{products}) - \Sigma\Delta_f H^o(\text{reactants})$
$= 9 - 2(33) = -57$ kJ/mol

$\Delta S^o = \Sigma\Delta S^o(\text{products}) - \Sigma\Delta S^o(\text{reactants})$
$= 304 - 2(240) = -176$ J/K/mol $= -0.176$ kJ/K/mol

At 25 °C (298 K):

$\Delta G^o = \Delta H^o - T\Delta S^o = (-57) - (298)(-0.176)$
$= -4.6$ kJ/mol < 0

Reaction is spontaneous in the forward direction.

FIRSTHAND INVESTIGATION 3

Investigating the iron (III) thiocyanate equilibrium

» Students investigate the effects of concentration on a system at equilibrium and explain how Le Chatelier's principle can be used to predict such effects, for example iron (III) thiocyanate and varying concentration of ions.

Aim

to determine the effect of concentration changes on the iron (III) thiocyanate equilibrium

Method

1. Prepare 0.010 mol/L (4.04 g/L) solutions of hydrated iron (III) nitrate ($Fe(NO_3)_3 \cdot 9H_2O$) and 0.010 mol/L (0.98 g/L) potassium thiocyanate.
2. Systematically dilute the potassium thiocyanate solution to produce a solution of 0.0010 mol/L concentration. Repeat this 1 : 10 dilution to produce KSCN solutions with concentrations of 0.00010 mol/L and 0.000010 mol/L.
3. Place 10 mL of the iron (III) nitrate solution in each of five beakers. Add 10 mL of water to the first beaker. Into the remaining beakers add 10 mL of each of the four dilute solutions of KSCN. Mix the solutions thoroughly.
4. Compare the colours of the solutions. Iron (III) ions are yellow in water. Iron (III) thiocyanate ions ($FeSCN^{2+}$) are deep red in colour. Use your observations to explain how the concentration of colourless thiocyanate ions has caused changes to the equilibrium.

Sample results

1. The equilibrium is:

$$\underset{\text{yellow}}{Fe^{3+}(aq)} + \underset{\text{colourless}}{SCN^-(aq)} \leftrightarrows \underset{\text{red}}{FeSCN^{2+}(aq)}$$

2. As the KSCN is diluted the equilibrium mixture becomes less red, as there are fewer thiocyanate ions to shift the equilibrium to the right.

Conclusion

The position of the iron (III) thiocyanate equilibrium is changed by altering the concentration of thiocyanate ions. Le Chatelier's principle predicts that a reduction in the thiocyanate ion concentration will cause the equilibrium to shift to the left and reduce the concentration of the red iron (III) thiocyanate ions.

→ KEY QUESTIONS

4 A saturated solution of cobalt (II) chloride is prepared and placed in an open beaker. Over several days some water evaporates. Explain how this disturbance will affect the equilibrium.

....

5 The commercial production of ammonia gas from nitrogen and hydrogen gas is an exothermic equilibrium. Explain why a moderate temperature rather than a high or low temperature is used to synthesise ammonia.

....

6 Explain the effect of the presence of a catalyst on the yield of products in a chemical equilibrium.

Answers ➲ p. 30

CHAPTER SYLLABUS CHECKLIST

Are you able to answer every question from the syllabus for this chapter? Tick each question as you go through the checklist if you are able to answer it. If you cannot answer a question, turn to the relevant page in the study guide to find the answer. For NESA key word meanings, go to www.educationstandards.nsw.edu.au and search 'key words'.

FOR A COMPLETE UNDERSTANDING OF THIS TOPIC:		PAGE NO.	✓
1	Can I state Le Chatelier's principle?	17	
2	Can I explain shifts in equilibria using Le Chatelier's principle?	18	
3	Can I recall the factors that can affect the position of different equilibria?	19	
4	Can I explain why catalysts have no effect on the position of an equilibrium?	21	
5	Can I describe examples of equilibrium shifts in both heterogeneous and homogeneous equilibria?	22	
6	Can I recall firsthand investigations involving equilibrium shifts?	25	

HSC EXAM-TYPE QUESTIONS

Objective-response questions (1 mark each)

1 **One stage in the refining of nickel is the following equilibrium:**

$Ni(s) + 4CO(g) \leftrightharpoons Ni(CO)_4(g); \Delta H = -163$ kJ/mol

Identify which change will shift the equilibrium to the right.

A Lower the temperature.

B Add more nickel (Ni).

C Remove some carbon monoxide (CO).

D Increase the volume of the closed vessel.

2 **100 mL of a saturated solution of sodium sulfate is prepared at 25 °C. The equilibrium equation is:**

$Na_2SO_4(s) \leftrightharpoons 2Na^+(aq) + SO_4^{2-}(aq)$

Identify which change will shift the equilibrium to the right.

A Add more sodium sulfate solid.

B Add liquid water.

C Pour the mixture into a smaller flask.

D Add crystals of potassium chloride.

3 **Methanol vapour forms when carbon monoxide reacts with hydrogen gas, as shown in the following homogeneous equilibrium:**

$CO(g) + 2H_2(g) \leftrightharpoons CH_3OH(g)$

Identify which change will shift the equilibrium to the right.

A Increase the volume of the reaction vessel.

B Remove some carbon monoxide.

C Add some nitrogen gas.

D Increase the total gas pressure by reducing the volume of the reaction vessel.

4 **Identify which of the following changes will increase the solubility of carbon dioxide in water.**

A Acidify the water.

B Increase the temperature of the water.

C Decrease the partial pressure of carbon dioxide gas above the water surface.

D Make the water slightly alkaline by adding a small amount of sodium hydroxide.

5 **The following graph (Figure 2.19) shows changes in concentration versus time for the following equilibrium:**

$2A(g) \leftrightharpoons B(g) + C(g)$

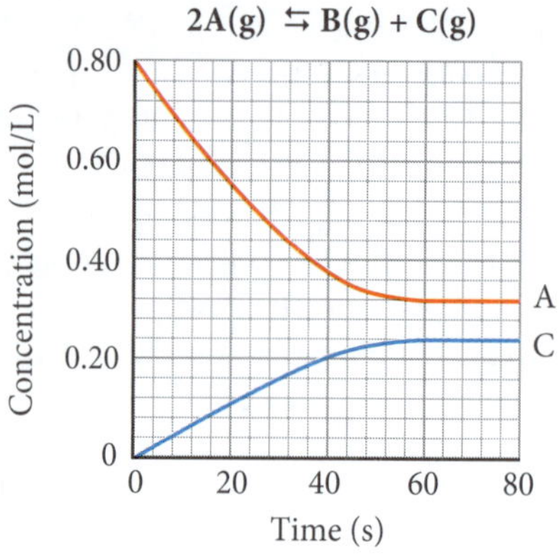

Figure 2.19 Reaction of A

Select the correct statement about this equilibrium.

A The concentration of B at equilibrium is 0.24 mol/L.

B Equilibrium is reached at 40 s.

C 0.32 mol/L is the decrease in the concentration as A decomposes.

D The rate of decomposition of A is greater at 45 s than at 10 s.

Extended-response questions

6 **In an aqueous solution of cobalt (II) chloride the cobalt (II) ions are bonded to 6 water molecules to form the hexaaquacobalt (II) ion ($Co(H_2O)_6^{2+}$), which is pink. If drops of concentrated hydrochloric acid are now added to the pink solution the following equilibrium is established:**

$Co(H_2O)_6^{2+}(aq) + 4Cl^-(aq) \leftrightharpoons CoCl_4^{2-}(aq) + 6H_2O(l)$

The tetrachlorocobalt (II) ion ($CoCl_4^{2-}$) is blue.

At 25 °C the equilibrium mixture is violet.

a **If the reaction mixture is heated the solution turns from violet to blue-violet. Classify this equilibrium as endothermic or exothermic in the forward direction.** (1 mark)

b **Predict and explain the colour change when additional water is added to the violet equilibrium mixture.** (2 marks)

7 **A dynamic equilibrium exists in a saturated solution of sodium hydrogen carbonate.**

a **Write a balanced equilibrium equation for a saturated solution of sodium hydrogen carbonate.** (1 mark)

b **Discuss the effect on this equilibrium of the addition of more water.** (2 marks)

8 The dissociation of nitrosyl chloride into nitric oxide and chlorine takes place according to the following equilibrium:

$$2NOCl(g) \leftrightharpoons 2NO(g) + Cl_2(g)$$

Discuss the effect on the position of this equilibrium by the following changes.

a A catalyst is added. (1 mark)

b The volume of the vessel is halved. (1 mark)

9 The following reaction can be used industrially to generate hydrogen gas from water vapour:

$$H_2O(g) + C(s) \leftrightharpoons CO(g) + H_2(g); \Delta H = +131 \text{ kJ/mol}$$

a Identify the conditions that will increase the yield of hydrogen from a fixed amount of water vapour. (2 marks)

b Explain what happens to the position of this equilibrium if additional solid carbon is added to the equilibrium system. (2 marks)

10 Methanol undergoes decomposition to form carbon monoxide and hydrogen gas in a closed system:

$$CH_3OH(g) \leftrightharpoons CO(g) + 2H_2(g)$$

Figure 2.20 shows how the concentration of methanol vapour changes at different temperatures.

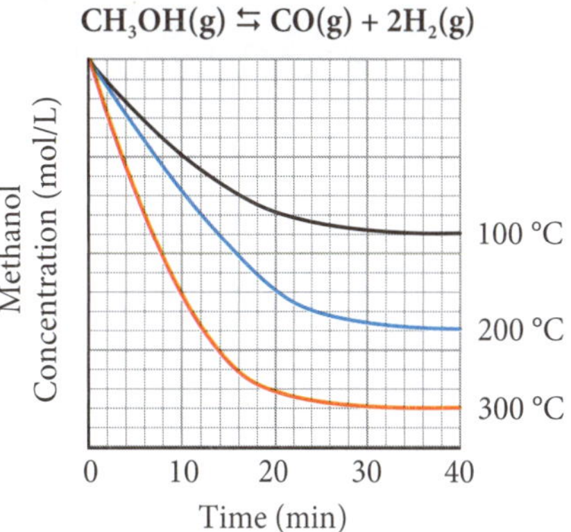

Figure 2.20 Methanol decomposition

a Determine whether the equilibrium is endothermic or exothermic. (1 mark)

b Explain the differences between the initial reaction rates at the different temperatures. (2 marks)

c At what time was equilibrium first achieved at 300 °C? (1 mark)

d How would the equilibrium concentration of methanol change if a catalyst was present? (1 mark)

ANSWERS

KEY QUESTIONS

Key questions ➲ p. 21

1 No. The original equilibrium position is not achieved. A new equilibrium position is created in which different quantities of reactants and products are present.

2 The removal of chloride ions disturbs the equilibrium and some sodium chloride crystals dissolve to increase the chloride ion concentration to counteract the change. A new equilibrium is established as long as sufficient NaCl crystals are present.

3 The loss of gaseous carbon dioxide from the bottle leads to a decrease in the dissolved carbon dioxide concentration. This process of carbon dioxide loss from the bottle continues until the dissolved carbon dioxide comes into equilibrium with the atmospheric carbon dioxide concentration (~403 ppm in 2017). This explains why water in contact with the atmosphere is not neutral as some atmospheric carbon dioxide dissolves to slightly increase acidity.

Key questions ➲ p. 27

4 The loss of water concentrates the dissolved ions and pushes the equilibrium in favour of the solid crystals. Thus more red crystals of cobalt (II) chloride form in order to maintain the heterogeneous equilibrium in which the solution remains saturated.

5 If the temperature is too high the reverse reaction is promoted and the equilibrium shifts in favour of the reactants and the yield of ammonia decreases. If the temperature is too low then the rate of the forward reaction is too low and thus ammonia production is slowed down. Therefore a moderate temperature is selected as a compromise to ensure sufficient yield at a reasonable rate.

6 The catalyst does not alter the yield of products. The catalyst increases the rate of the forward and reverse reactions and so equilibrium is achieved faster.

HSC EXAM-TYPE QUESTIONS

Objective-response questions

1 A. The reaction is exothermic and removal of heat will promote the forward reaction and the equilibrium will shift to the right. **B** is incorrect as this is a heterogeneous equilibrium and the amount of solid will not affect the equilibrium as long as some solid is present. **C** is incorrect as this will shift the equilibrium to the left. **D** is incorrect as this will shift the equilibrium to the left to produce more molecules of gas to raise the pressure.

2 B. The addition of water will reduce the ion concentration and the solution is no longer saturated. More solid sodium sulfate will dissolve to restore saturation equilibrium. **A** is incorrect as the amount of solid added does not affect a heterogeneous equilibrium. **C** is incorrect as the container size cannot alter an equilibrium involving solids and aqueous solutions. **D** is incorrect as the equilibrium has no potassium ions and chloride ions.

3 D. The reduction in volume raises the gas pressure and the equilibrium shifts to the side of fewer molecules (i.e. the products) so as to counteract the change. **A** is incorrect as this will shift the equilibrium to the left to raise the pressure. **B** is incorrect as removal of a reactant will shift the equilibrium to the left to counteract the CO removal. **C** is incorrect as it will not alter the partial pressures of each gas.

4 D. Carbon dioxide forms carbonic acid in water. Adding NaOH will neutralise the carbonic acid and more carbon dioxide will dissolve to counteract the change. **A** is incorrect as extra acid will decrease the solubility as the equilibrium will reverse to counteract the change. **B** is incorrect as gases are less soluble in hot water as the equilibrium is exothermic. **C** is incorrect as less carbon dioxide will dissolve as the equilibrium reverses.

5 A. Products B and C are formed in equal amounts as shown in the equation. Thus the concentration of product B is 0. 24 mol/L. **B** is incorrect as equilibrium is reached at 60 s. **C** is incorrect as the decrease in concentration of product A is 0.80 – 0.32 = 0.48 mol/L. **D** is incorrect as the rate is greater at 10 s than 45 s as the tangent to the curve has a greater slope. The gradient of the tangent is a measure of the rate of the reaction.

Extended-response questions

6 EM Students need to recall the conditions which lead to a saturated solution of a salt. Students need to recall Le Chatelier's principle and in each case identify the disturbance and how it is related to the equilibrium.

- a The forward reaction is endothermic and the final colour of blue-violet indicates that the equilibrium has shifted to the right on heating to produce more blue ions and less pink ions. ✓
- b The mixture will turn more pink as the additional water will cause the equilibrium to shift to the left. ✓

  The extra water molecules collide with the blue ions and increase the reverse rate. ✓

7 EM Students need to recall how heterogeneous equilibria differ from homogeneous equilibria and the factors that can cause equilibrium disturbance.

- a $NaHCO_3(s) \leftrightarrows Na^+(aq) + HCO_3^-(aq)$ ✓
- b The addition of water will cause the concentration of ions to fall below their equilibrium saturation levels. ✓ Therefore the equilibrium will shift to the right to increase the concentration of ions and thus some solid sodium hydrogen carbonate will dissolve. ✓

8 EM Students need to identify what catalysts do in a chemical change. Catalysts do not change the yield of products. Students should revise Boyle's law from the Year 11 course to determine the relationship between gas volume and pressure.

- a No change to the equilibrium position. Catalysts only change the rate at which equilibrium is achieved. ✓
- b The reduction in volume leads to a total pressure increase. The equilibrium will shift to the left to reduce the number of moles of particles and therefore reduce the total pressure. ✓

9 EM Students need to show that this is a heterogeneous equilibrium and that care must be taken in applying Le Chatelier's principle.

- a Higher temperatures, as the equilibrium is endothermic in the forward direction. ✓

  Using low pressure (by increasing the vessel volume) will shift the equilibrium to the right (1 mole reactant gas ⟶ 2 moles product gases). ✓
- b No change to the equilibrium position ✓ as this is a heterogeneous equilibrium that is unaffected by the amount of

solid present. As long as there is sufficient solid to maintain the equilibrium, the equilibrium is not affected by changes in the amount of the solid carbon. ✓

10 EM Students need to show that equilibrium is achieved when the concentration no longer changes. The change in concentration in the first five minutes is linear and therefore the gradient of these lines is a measure of rate.

a Endothermic as more methanol decomposes as the temperature increases and the equilibrium shifts to the right. ✓

b The initial rate is greatest for the highest temperature ✓ as the methanol molecules have much higher kinetic energies. The rate is the slope of the tangent to the curve initially. ✓

c ~30 minutes ✓

d No change to the equilibrium concentration of methanol. The catalyst will only increase the rate of decomposition. ✓

CHAPTER 3 CALCULATING THE EQUILIBRIUM CONSTANT (K_{eq})

INQUIRY QUESTION:

How can the position of equilibrium be described and what does the equilibrium constant represent?

In Chapter 2 the position of an equilibrium was described in terms of favouring the products or reactants. Thus an equilibrium that lies to the right was one in which there was a greater concentration of products than reactants. An equilibrium that lies to the left was one in which there was a greater concentration of reactants than products. This chapter quantifies the position of the equilibrium by calculating the equilibrium constant.

1 Equilibrium constants for homogeneous systems

» Students deduce the equilibrium expression (in terms of K_{eq}) for homogeneous reactions occurring in solution.

The equilibrium position

- In different chemical equilibria the concentration of reactants and products varies considerably. For example, in the precipitation of barium sulfate on the mixing of barium ions and sulfate ions the equilibrium lies very much to the right as barium sulfate is very insoluble in water at 25 °C. The concentration of barium ions and sulfate ions is quite low at equilibrium:

  $Ba^{2+}(aq) + SO_4^{2-}(aq) \leftrightarrows BaSO_4(s)$

  For the equilibrium between nitrogen dioxide, nitrogen and oxygen the equilibrium lies strongly to left at 200 °C. At equilibrium the concentration of NO gas is quite low:

  $N_2(g) + O_2(g) \leftrightarrows 2NO(g)$

- Consider the following homogeneous equilibrium in which reactants A and B combine to form products C and D:

  $A + B \leftrightarrows C + D$

  Figure 3.1 shows the relative concentration of reactants and products if this equilibrium lies to the right.

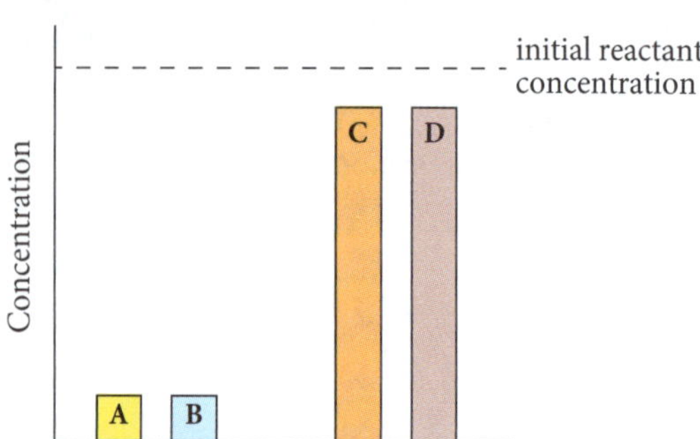

Figure 3.1 Equilibrium lies to the right

- Consider the following homogeneous equilibrium in which reactants W and X combine to form products Y and Z:

  $W + X \leftrightarrows Y + Z$

  Figure 3.2 shows the relative concentration of reactants and products if this equilibrium lies to the left.

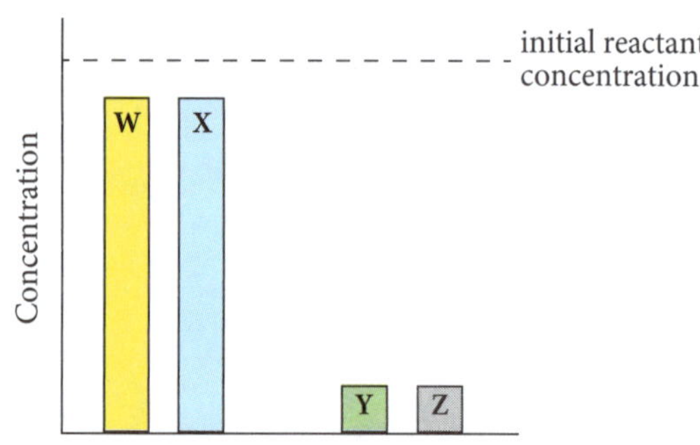

Figure 3.2 Equilibrium lies to the left

- The equilibrium between hydrofluoric acid and its ions in water lies to the left as the acid only partially dissociates in water. For a 0.10 mol/L HF solution at room temperature, about 92% of the acid is in the unionised (HF) form:

  $HF(aq) \leftrightarrows H^+(aq) + F^-(aq)$

The equilibrium constant

- The position or extent of a chemical equilibrium can be expressed quantitatively using the *equilibrium constant* (K_{eq}). If the value of K_{eq} is large then the equilibrium lies to the product side. If K_{eq} is small then the equilibrium lies to the reactant side.

→ Consider the following homogeneous equilibrium in which reactants A and B combine to form products C and D. The lowercase letters are the coefficients that balance the equation:

$aA + bB \leftrightarrows cC + dD$

The equilibrium constant (K_{eq}) for this equilibrium is given by the following formula:

$$K_{eq} = \frac{[C]^c \cdot [D]^d}{[A]^a \cdot [B]^b}$$

→ The square brackets represent the concentration of a species. The concentration of the products is shown on the numerator and the reactant concentrations in the denominator. The concentrations are raised to the power of the coefficients in the balanced equation.

2 Equilibrium constant calculations

» Students perform calculations to find the value of K_{eq} and concentrations of substances within an equilibrium system, and use these values to make predictions on the direction in which a reaction may proceed.

Comparing values for K_{eq}

→ The size of the equilibrium constant (K_{eq}) indicates the position of the equilibrium. Consider the following examples.

- Cadmium ions react with cyanide ions in solution

  $Cd^{2+}(aq) + 4CN^{-}(aq) \leftrightarrows Cd(CN)_4^{2-}(aq)$;

  $K_{eq} = [Cd(CN)_4^{2-}]/[Cd^{2+}][CN^{-}]^4 = 10^{18}$ (at 25 °C)

  The extremely high value of K_{eq} means that this equilibrium lies very strongly to the right and that very few free cadmium ions will be present in the equilibrium mixture.

- Decomposition of sulfur trioxide

  $2SO_3(g) \leftrightarrows 2SO_2(g) + O_2(g)$;

  $K_{eq} = [SO_2]^2[O_2]/[SO_3]^2 = 10^{-12}$ (at 227 °C)

  This very low value of K_{eq} indicates that sulfur trioxide is very stable at elevated temperatures and does not readily decompose into sulfur dioxide and oxygen.

Calculating K_{eq}

→ Graphical data involving equilibria can be used to calculate the equilibrium constant.

Equilibrium 1: Decomposition equilibrium

To determine the value of K_{eq} the equilibrium concentrations of reactants and products have to be measured experimentally.

In this example, a gaseous oxide of element X decomposes to form gaseous X_2 and O_2. Figure 3.3 shows the concentration of two species (XO and O_2) in the gaseous equilibrium:

$2XO(g) \leftrightarrows X_2(g) + O_2(g)$

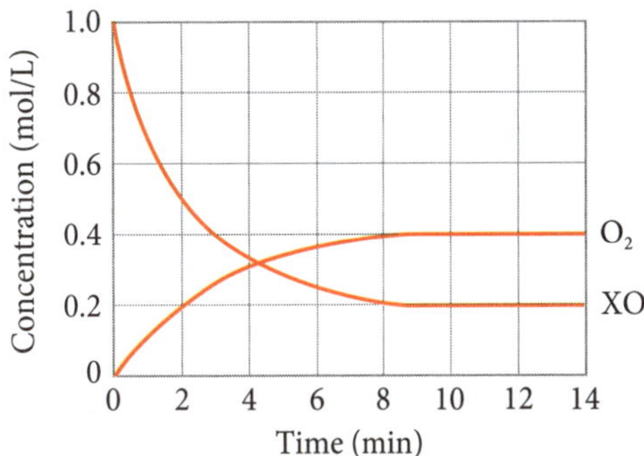

Figure 3.3 Gaseous equilibrium

Figure 3.3 indicates that the concentration of XO was 1.0 mol/L at the start. As XO decomposed, its concentration dropped and the concentration of O_2 increased. Equilibrium was achieved at 10 minutes. Thereafter the concentrations did not change.

The following table is useful for determining the equilibrium concentrations of each species. The graph and the stoichiometry of the reaction can be used to complete the table.

Concentrations (mol/L)	XO	X_2	O_2
Initial	1.0	0	0
Change	−0.8	+0.4	+0.4
Equilibrium	0.2	0.4	0.4

The equilibrium constant can now be calculated:

$K_{eq} = [X_2][O_2]/[XO]^2$

$= \frac{(0.4)(0.4)}{(0.2)^2} = 4$

As $K_{eq} > 1$ then the equilibrium lies to the right.

Equilibrium 2: Oxidation of sulfur dioxide

Sulfur dioxide can be oxidised to form sulfur trioxide according to the following exothermic equilibrium:

$2SO_2(g) + O_2(g) \leftrightarrows 2SO_3(g)$

At a temperature of 450 K (177 °C), 0.60 moles of sulfur dioxide and 0.60 moles of oxygen is placed in a 1 L vessel with a vanadium pentoxide catalyst. The sulfur dioxide is oxidised and equilibrium is slowly attained. Figure 3.4 shows the experimental data.

The equilibrium constant for this equilibrium can be calculated using the graphical data and the reaction stoichiometry.

The number of moles of each gas is divided by the volume of the vessel (1 L) to calculate the concentration.

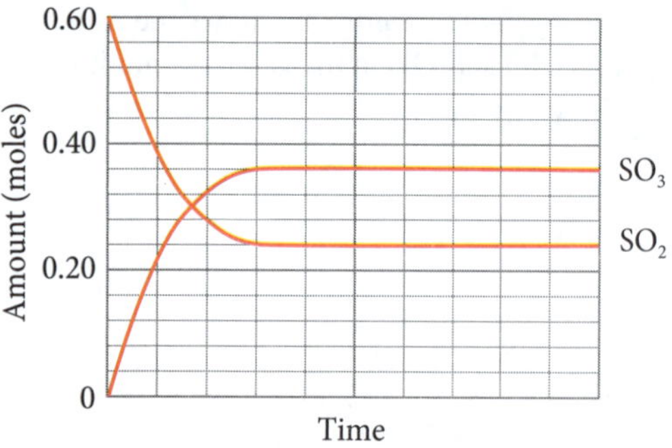

Figure 3.4 Sulfur dioxide–sulfur trioxide equilibrium

Concentration (mol/L)	SO_2	O_2	SO_3
Initial	0.60	0.60	0
Change	–0.36	–0.18	+0.36
Equilibrium	0.24	0.42	0.36

$K_{eq} = [SO_3]^2/[SO_2]^2[O_2]$

$= \dfrac{(0.36)^2}{(0.24)^2(0.42)} = 5.4$

At 450 K the equilibrium lies to the right.

At time X a change was made that disturbed the equilibrium. Eventually a new equilibrium was established, as shown in Figure 3.5.

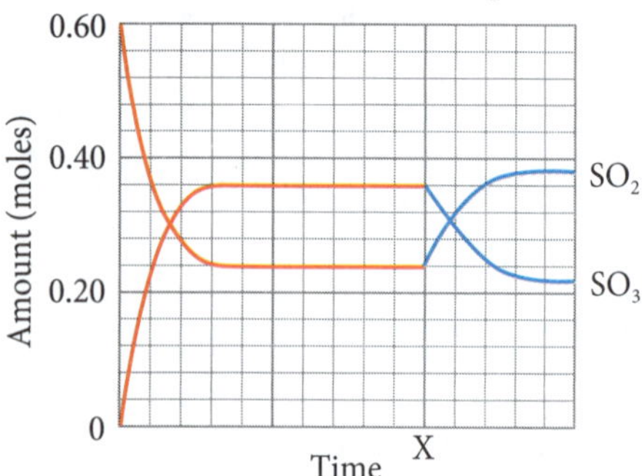

Figure 3.5 Changes to equilibrium

The equilibrium constant for the new equilibrium position attained can be calculated.

At the new equilibrium after change at X:

Concentration (mol/L)	SO_2	O_2	SO_3
Initial	0.24	0.42	0.36
Change	+0.14	+0.07	–0.14
Equilibrium	0.38	0.49	0.22

$K_{eq} = [SO_3]^2/[SO_2]^2[O_2]$

$= \dfrac{(0.22)^2}{(0.38)^2(0.49)} = 0.68$

K_{eq} has decreased. This indicates that the change has caused a shift in the equilibrium to the left (reactants). As this is an exothermic equilibrium, the change is most likely an increase in temperature of the system.

EXAMPLE 1

Acetic acid (CH_3COOH) reacts with ethanol (CH_3CH_2OH) in the presence of an acid catalyst to form ethyl acetate ($CH_3COOCH_2CH_3$):

$$CH_3COOH(l) + CH_3CH_2OH(l) \leftrightarrows CH_3COOCH_2CH_3(l) + H_2O(l)$$

0.10 moles of acetic acid and 0.10 moles of ethanol were mixed with 2 mL of concentrated sulfuric acid catalyst. The mixture was heated in a reflux apparatus (Figure 3.6) for several hours until equilibrium was reached.

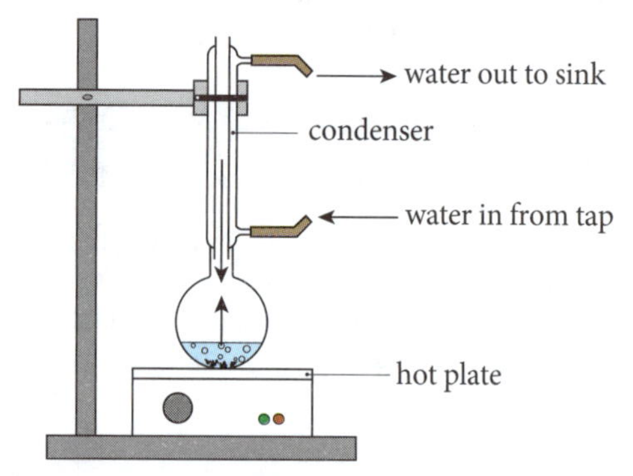

Figure 3.6 Reflux apparatus used to prepare ethyl acetate

The final mixture was analysed for the amount of acetic acid present. The results are tabulated.

Initial amounts (mol)	Equilibrium amounts (mol)
$CH_3COOH = 0.10$ $CH_3CH_2OH = 0.10$	$CH_3COOH = 0.066$
Final volume of reaction mixture = 0.10 L	

Calculate the equilibrium constant for this equilibrium.

Use the reaction stoichiometry to calculate the equilibrium amounts of ethanol and reaction products

Answer:

The equilibrium concentrations are calculated using the reaction stoichiometry.

Initial amounts (mol)	Equilibrium amounts (mol)	Equilibrium concentration (mol/L)
$CH_3COOH = 0.10$	$CH_3COOH = 0.066$	$CH_3COOH = \dfrac{0.066}{0.10} = 0.66$
$CH_3CH_2OH = 0.10$	$CH_3CH_2OH = 0.066$	$CH_3CH_2OH = \dfrac{0.066}{0.10} = 0.66$

Initial amounts (mol)	Equilibrium amounts (mol)	Equilibrium concentration (mol/L)
$CH_3COOCH_2CH_3 = 0$	$CH_3COOCH_2CH_3 = 0.10 - 0.066 = 0.034$	$CH_3COOCH_2CH_3 = \frac{0.034}{0.10} = 0.34$
$H_2O = 0$	$H_2O = 0.10 - 0.066 = 0.034$	$H_2O = \frac{0.034}{0.10} = 0.34$

The expression for the equilibrium constant is:

$$K_{eq} = \frac{(CH_3COOCH_2CH_3)(H_2O)}{(CH_3COOH)(CH_3CH_2OH)}$$

$$K_{eq} = \frac{(0.34)(0.34)}{(0.66)(0.66)} = 0.27$$

EXAMPLE 2

The tribromide ion undergoes dissociation in aqueous solutions to form bromide ions and bromine molecules:

$$Br_3^-(aq) \leftrightarrows Br^-(aq) + Br_2(aq)$$

The equilibrium constant (K_{eq}) is equal to 5.6×10^{-2} at 25 °C.

Calculate the tribromide ion concentration in an equilibrium mixture in which the concentration of bromide ions and bromine were both 0.20 mol/L.

Answer:

Use the balanced equation to write an expression for the equilibrium constant

$$K_{eq} = [Br^-][Br_2]/[Br_3^-]$$

$$5.6 \times 10^{-2} = (0.20)(0.20)/[Br_3^-]$$

$$[Br_3^-] = \frac{(0.20)(0.20)}{(5.6 \times 10^{-2})} = 0.71 \text{ mol/L}$$

→ KEY QUESTIONS

1 **Compare the concentrations of products and reactants in equilibria that lie strongly to the left.**

2 **Write an expression for the equilibrium constant for the following equilibrium:**

$$2NO(g) + O_2(g) \leftrightarrows N_2O_4(g)$$

3 **Hydrogen gas and iodine vapour (I_2) react to form hydrogen iodide gas. At equilibrium the concentration of reactants and products are: $[H_2] = [I_2] = 0.242$ mol/L; $[HI] = 1.666$ mol/L Calculate the value of the equilibrium constant.**

Answers ➲ p. 44

3 Temperature and pressure effects on equilibrium

» Students:

- qualitatively analyse the effect of temperature on the value of K_{eq}.
- perform calculations to find the value of K_{eq} and concentrations of substances within an equilibrium system, and use these values to make predictions on the direction in which a reaction may proceed.

Temperature and pressure changes

→ Consider the following information concerning the effect of temperature on the dinitrogen tetroxide equilibrium.

Dinitrogen tetroxide molecules decompose to form nitrogen dioxide molecules:

$$N_2O_4(g) \leftrightarrows 2NO_2(g)$$

→ The equilibrium constant expression is:

$K_{eq} = [NO_2]^2/[N_2O_4]$

The table shows how K_{eq} varies with temperature.

Temperature (°C)	0	25	100
K_{eq}	5.7×10^{-4}	4.7×10^{-3}	0.50

→ This data shows that at 0 °C and 25 °C the equilibrium lies to the left. As the temperature increases the equilibrium is shifting towards the right. This tells us that the equilibrium is endothermic. At 100 °C the higher K_{eq} value indicates that both the reactant and product are in appreciable concentrations (see Figure 3.7).

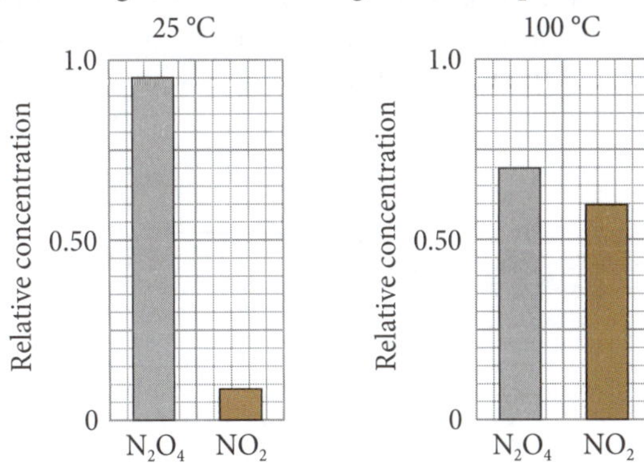

Figure 3.7 Equilibrium shift with increasing temperature

→ If the temperature is held constant and the volume of the vessel is decreased the equilibrium is disturbed due to the increase in total pressure. To counteract the change the equilibrium shifts to the side of lower number of molecules (i.e. N_2O_4) and so the colour of the mixture is less brown when a new equilibrium is attained at the higher pressure. As there is no temperature change K_{eq} has not changed.

Reaction quotient (Q)

- When reactants are mixed together the reaction may be quite rapid or very slow. To gauge how far from equilibrium a reaction is, chemists analyse the reaction mixture and determine the concentration of reactants and products at that time. This data is used to calculate the reaction quotient (Q).
- The reaction quotient has the same form as the equilibrium constant but the concentrations are those measured at any time. If the reaction quotient is less than K_{eq} then the system has not reached equilibrium. The reaction quotient equals K_{eq} once equilibrium has been achieved.
    - Example: The Haber process

      In the industrial preparation of ammonia (called the *Haber process*), nitrogen and hydrogen react over a porous iron catalyst to form ammonia. The reaction is performed at an elevated temperature (about 500 °C):

      $N_2(g) + 3H_2(g) \leftrightarrows 2NH_3(g)$

      At 500 °C the equilibrium constant (K_{eq}) is 0.060. This low value indicates that the equilibrium lies to the left.

      In order to shift the equilibrium to the right the reaction is performed at very high pressures (25–30 MPa). At high pressures the equilibrium shifts to the right to reduce the particle number (four moles of reactants convert to two moles of gaseous products). This illustrates Le Chatelier's principle that the system counteracts the pressure increase by moving to the side that has less particles and therefore the pressure reduces.

      This effect is shown in Figure 3.8, where increased concentrations of ammonia result from reducing the volume of the reaction vessel (and thus raising the pressure), which in turn shifts the equilibrium to the right. The catalyst does not alter the position of the equilibrium. It increases the rate at which equilibrium is reached.

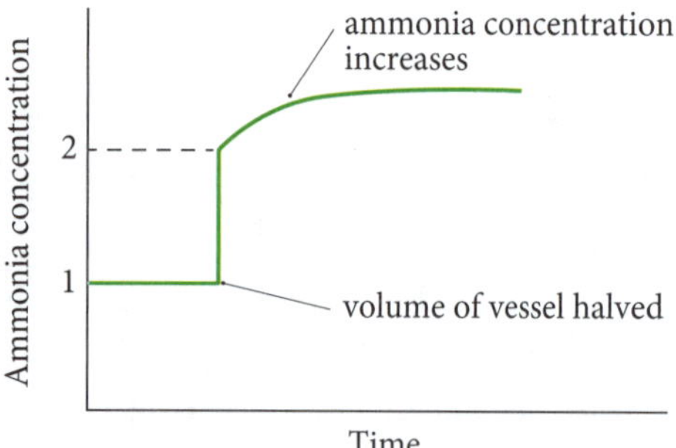

Figure 3.8 Reducing container volume leads to an increase in the concentration of ammonia

If samples of the reaction mixture are analysed at various times, the reaction quotient can be calculated. For the reaction as written, the expression for the reaction quotient is:

$Q = [NH_3]^2/[N_2][H_2]^3$

At a particular time the concentrations of reactants and products are measured and Q is calculated. The data collected and the calculated value of Q is shown in the table.

$[N_2]$ (mol/L)	$[H_2]$ (mol/L)	$[NH_3]$ (mol/L)	Q
0.95	1.60	0.40	0.041

At 500 °C $K_{eq} = 0.060$. Therefore $Q < K_{eq}$. Therefore the system has not reached equilibrium at the time of sampling.

EXAMPLE 3

Figure 3.9 shows the effect of temperature changes on a homogeneous equilibrium in which W and X are in equilibrium with Y and Z in a 1 L vessel:

$W(g) + X(g) \leftrightarrows Y(g) + Z(g)$

The initial concentrations of W and X are 1.0 mol/L.

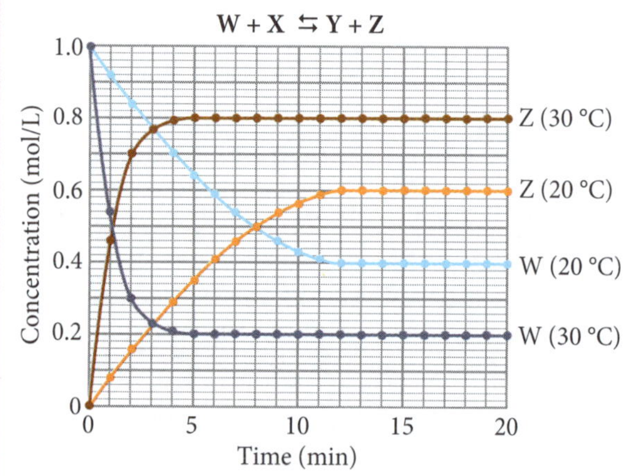

Figure 3.9 Effect of temperature on a homogeneous equilibrium

a **Use the graphical data to calculate the equilibrium constant for the reaction at 20 °C and 30 °C.**

b **Determine whether the equilibrium is endothermic or exothermic in the forward direction.**

c **Explain why equilibrium is achieved faster at the higher temperature.**

Write an expression for K_{eq} and show the data substitution from the graph

Answer:

a Temperature = 20 °C

Concentration (mol/L)	W(g)	X(g)	Y(g)	Z(g)
Initial	1.0	1.0	0	0
Change	−0.6	−0.6	+0.6	+0.6
Equilibrium	0.4	0.4	0.6	0.6

$$K_{eq} = [Y][Z]/[W][X] = \frac{(0.6)(0.6)}{(0.4)(0.4)} = 2.25$$

Temperature = 30 °C

Concentration (mol/L)	W(g)	X(g)	Y(g)	Z(g)
Initial	1.0	1.0	0	0
Change	–0.8	–0.8	+0.8	+0.8
Equilibrium	0.2	0.2	0.8	0.8

$$K_{eq} = [Y][Z]/[W][X] = \frac{(0.8)(0.8)}{(0.2)(0.2)} = 16.0$$

b At 30 °C the equilibrium constant is higher than at 20 °C. Therefore the equilibrium has shifted to the right. Therefore the equilibrium is endothermic in the forward direction.

c At a higher temperature the particles have a higher kinetic energy and a greater proportion of particles can overcome the activation energy barrier. Thus the reaction rate is greater. Equilibrium is therefore attained in a shorter time.

4 Investigating equilibrium constants

» Students conduct an investigation to determine K_{eq} of a chemical equilibrium system, for example, K_{eq} of the iron(III) thiocyanate equilibrium.

➔ The following investigation using secondhand data will assist with an understanding of equilibrium constants.

SECONDARY-SOURCED INVESTIGATION

Equilibrium constant determination

Aim

to determine the equilibrium constant for the iron (III) thiocyanate equilibrium

Background

The iron (III) thiocyanate equilibrium has been qualitatively investigated previously. The equilibrium is:

$Fe^{3+}(aq) + SCN^{-}(aq) \leftrightharpoons FeSCN^{2+}(aq)$

yellow colourless blood-red

The product of the reaction is intensely red. This red ion strongly absorbs blue light (wavelength = 480 nm). If blue light is passed through a solution of these red ions, some of the blue light is absorbed and less is transmitted. The higher the iron (III) thiocyanate concentration, the greater the amount of blue light absorbed and the less transmitted. An instrument called a colourimeter can measure the percentage of light transmitted (Figure 3.10). Pure water is used as a control.

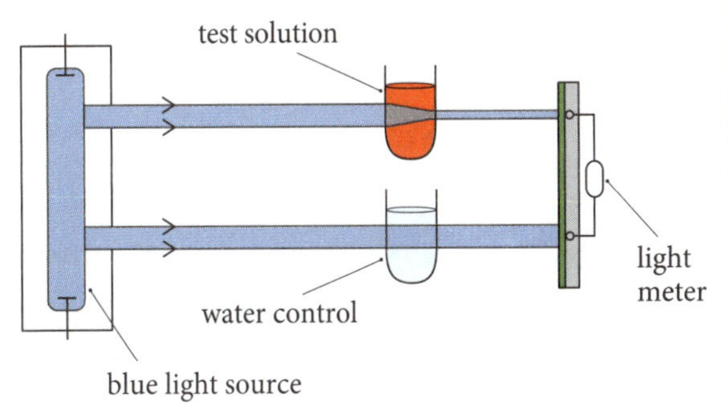

Figure 3.10 Colourimeter

Part A: Establishing a calibration graph

1 Five solutions of known iron (III) thiocyanate concentration (standards) were prepared by using very high iron (III) ion concentrations so that essentially all the thiocyanate ions were converted into red iron (III) thiocyanate ions.

2 Blue light (λ = 480 nm) was passed through each standard and the percentage of light transmitted was measured.

Sample results

Standard	$[FeSCN^{2+}]$ (mol/L) ($\times 10^{-4}$)	Percentage transmission
1	1.00	38
2	0.93	42
3	0.80	51
4	0.70	57
5	0.50	68
6	0 (control)	100

The tabulated data can be graphed, as shown in Figure 3.11.

3 Plot your own calibration graph and compare your graph to the graph shown in Figure 3.11.

Graphical results

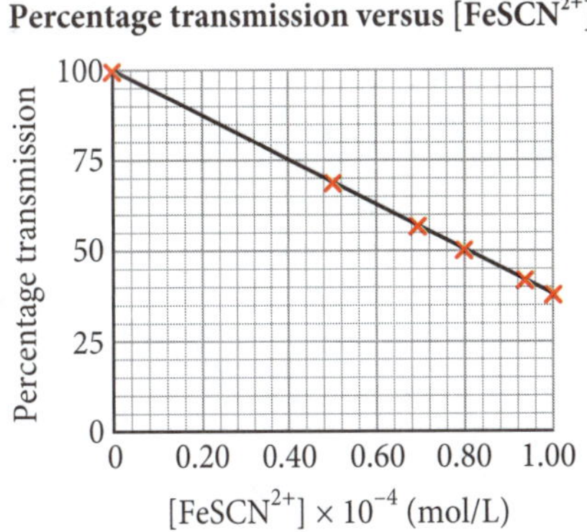

Figure 3.11 Calibration graph

Part B: Determining the equilibrium constant (K_{eq})

1 Standard solutions of $Fe(NO_3)_3$ and KSCN were prepared.

2 Two mixtures (A and B) of these solutions were prepared and the initial $[Fe^{3+}]$ and $[SCN^-]$ are shown in the table.

Tube	Initial $[Fe^{3+}]$ (mol/L) $\times 10^{-4}$	Initial $[SCN^-]$ (mol/L) $\times 10^{-4}$
A	2.28	1.00
B	2.52	2.00

3 The percentage transmission of each solution was measured.

Sample results

Tube	Percentage transmission
A	90
B	80

4 The calibration graph is used to determine the concentrations of A and B.

5 Plot your own calibration graph to determine the concentrations of iron (III) thiocyanate in A and B and compare your results with Figure 3.12.

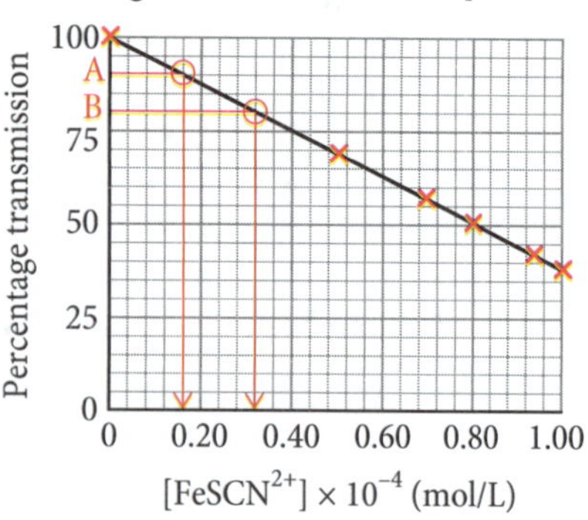

Figure 3.12 Determining concentration of iron (III) thiocyanate ion in solutions A and B

Sample results

Tube	Percentage transmission	$[FeSCN^{2+}]$ (mol/L) $\times 10^{-4}$
A	90	0.16
B	80	0.33

Sample calculations

a The reaction stoichiometry and the equilibrium concentration of $FeSCN^{2+}$ is used to determine the equilibrium concentrations of Fe^{3+} and SCN^- in tubes A and B.

Equilibrium concentrations

Tube	$[Fe^{3+}]$ (mol/L) ($\times 10^{-4}$)	$[SCN^-]$ (mol/L) ($\times 10^{-4}$)	$[FeSCN^{2+}]$ (mol/L) ($\times 10^{-4}$)
A	2.28 – 0.16 = 2.12	1.00 – 0.16 = 0.84	0.16
B	2.52 – 0.33 = 2.19	2.00 – 0.33 = 1.67	0.33

b The tabulated data can now be used to calculate a value of K_{eq} for each solution (A and B). Perform your own calculation using your calculated results and compare your equilibrium constant with the following sample calculation.

Sample calculation

$K_{eq} = [FeSCN^{2+}]/[Fe^{3+}][SCN^-]$

$$\text{A: } K_{eq} = \frac{(0.16 \times 10^{-4})}{(2.12 \times 10^{-4})(0.84 \times 10^{-4})} = 898$$

$$\text{B: } K_{eq} = \frac{(0.33 \times 10^{-4})}{(2.19 \times 10^{-4})(1.67 \times 10^{-4})} = 902$$

By repeating the experiment a minimum of five times, the reliability can be improved. The average value of K_{eq} can then be calculated.

5 Equilibrium constants for a variety of equilibria

» Students explore the use of K_{eq} for different types of chemical reactions, including but not limited to dissociation of ionic solutions, and dissociation of acids and bases.

Dissociation of ionic compounds

- Some ionic compounds completely dissolve in water and produce solutions classified as *electrolytes* as these solutions conduct electricity because the ions are free to move.
- Ionic compounds have different solubilities in water at 25 °C.

  Examples:

  KCl: 36 g/100 g H_2O

  $AgNO_3$: 245 g/100 g H_2O
- Equilibrium is only established when the solution is saturated and some solid is present to maintain the equilibrium. This is achieved for KCl by adding more than 36 g to 100 g of water. For $AgNO_3$ more than 245 g needs to be added to 100 g of water:

  $KCl(s) \leftrightharpoons K^+(aq) + Cl^-(aq)$

  $AgNO_3(s) \leftrightharpoons Ag^+(aq) + NO_3^-(aq)$

- Some ionic compounds have very low solubilities in water and a saturation equilibrium is quickly established.

  Example:

  AgCl: 0.000 19 g/100 g H_2O

  $AgCl(s) \leftrightarrows Ag^+(aq) + Cl^-(aq)$

  In a saturated solution of AgCl the molar concentration of silver ions and chloride ions is only 1.34×10^{-5} mol/L.

- For a solution of an ionic compound the product of the ion concentrations is called the *ionic product* (*IP*).

  Example:

  The equation for the dissociation of calcium chloride in water is:

  $CaCl_2(s) \rightarrow Ca^{2+}(aq) + 2Cl^-(aq)$

  At concentrations *below* saturation, the ionic product is defined as:

  $IP(CaCl_2) = [Ca^{2+}][Cl^-]^2$

  In this example the reaction stoichiometry of 1:2 is used to raise the chloride ion concentration to the power '2'.

  Generally the equation and ionic product for the dissociation of an ionic compound M_xN_y is:

  $M_xN_y(s) \rightarrow xM^{y+}(aq) + yN^{x-}(aq)$

  Ionic product = $[M^{y+}]^x[N^{x-}]^y$

  Example:

  $Al_2(SO_4)_3(s) \rightarrow 2Al^{3+}(aq) + 3SO_4^{2-}(aq)$

  $IP = [Al^{3+}]^2[SO_4^{2-}]^3$

- If increasing amounts of an ionic compound are dissolved in water the ionic product continues to increase. Eventually the ionic product reaches a critical value where a saturation equilibrium is established.

  Example:

  Barium hydroxide is sparingly soluble and forms a saturated solution in water when the ionic product (*IP*) exceeds 3×10^{-4} (mol^3/L^3):

  $Ba(OH)_2(s) \leftrightarrows Ba^{2+}(aq) + 2OH^-(aq)$

  This will occur when 0.72 g of $Ba(OH)_2$ or more is added to 100 g of water at 25 °C.

Dissociation of acids

- Acids can be classified as strong or weak in water solution.
- Hydrochloric acid (HCl) is classified as a **strong acid**. Hydrogen chloride is a gaseous covalent molecular compound. When hydrogen chloride dissolves in water the HCl molecule reacts with a water molecule, forming hydronium ions (H_3O^+) and chloride ions:

  $HCl(g) + H_2O(aq) \rightarrow H_3O^+(aq) + Cl^-(aq)$

  This solution is called *hydrochloric acid*. The **dissociation** process is complete.

  Note: The term **ionisation** is also commonly used to describe the production of ions when acid molecules dissolve in water.

- Hydrofluoric acid (HF) is classified as a **weak acid**. When hydrogen fluoride dissolves in water the molecules do not completely dissociate. An equilibrium is established between the unionised molecules and the ions in solution. The solution is called *hydrofluoric acid*:

  $HF(aq) + H_2O(aq) \leftrightarrows H_3O^+(aq) + F^-(aq)$

strong acid: a strong acid is completely dissociated in water solution

dissociation:

- the process in which ionic solids separate into ions in a solvent
- the process in which a molecular substance interacts with a solvent and produces ions

ionisation: the process in which a molecule forms ions

weak acid: a weak acid is partly dissociated in water solution

- The degree of dissociation (or ionisation) of a weak acid can be calculated by measuring the concentration of hydronium ions formed from a solution of the acid of known concentration. The degree of dissociation represents the fraction or percentage of acid molecules that are ionised. Consider a weak acid, HA. The equilibrium equation for its dissociation is:

  $HA(aq) + H_2O(l) \leftrightarrows H_3O^+(aq) + A^-(aq)$

  Degree of dissociation = $[H_3O^+]/[HA] \times 100/1\%$

- A 0.10 mol/L acetic acid solution has a degree of dissociation of 1.3% at 25 °C, whereas a 0.10 mol/L HCl solution is 100% dissociated because it is a strong acid. The degree of dissociation depends on the initial acid concentration. A 1.0 mol/L acetic acid solution is only 0.4% dissociated. As the solution is diluted the equilibrium shifts to the right and more acetic acid molecules are dissociated.
- Formic acid (HCOOH) is a weak acid. The graph in Figure 3.13 shows how the degree of dissociation changes as the acid is diluted. The equilibrium shifts to the right as more water is added:

  $HCOOH(aq) + H_2O(l) \leftrightarrows HCOO^-(aq) + H_3O^+(aq)$

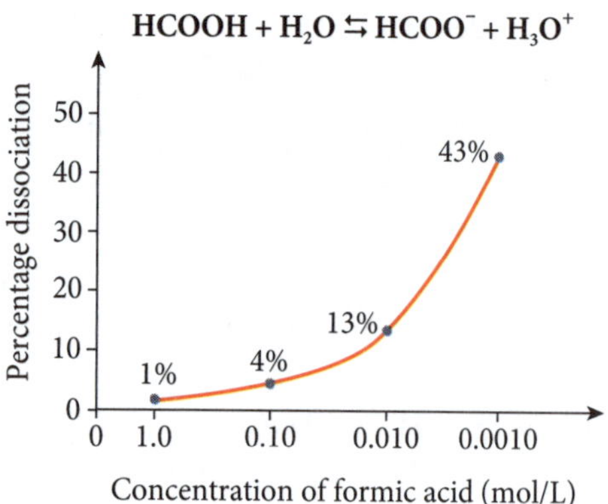

Figure 3.13 Formic acid dissociation increases on dilution

EXAMPLE 4

A 1.0 mol/L solution of acetic acid at 25 °C has an acetate ion concentration of 0.0040 mol/L. Calculate the degree of dissociation of 1.0 mol/L acetic acid.

Use the reaction stoichiometry to determine the hydronium ion concentration

Answer:

At equilibrium, $[H_3O^+] = [CH_3COO^-] = 0.0040$ mol/L

Degree of dissociation $= [H_3O^+]/[CH_3COOH] \times 100/1\%$

$$= \frac{(0.0040)}{(1.0)} \times 100$$

$$= 0.40\%$$

➔ Figure 3.14 is a particle model of a dilute solution of the weak hydrofluoric acid. The diagram shows very few hydronium and fluoride ions at equilibrium.

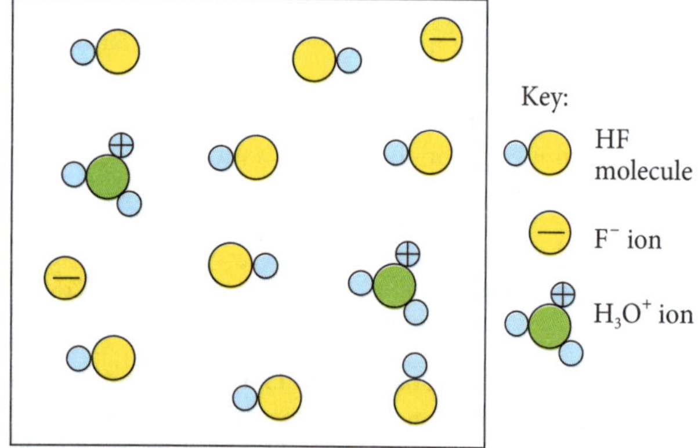

Figure 3.14 Hydrofluoric acid solution

➔ Equilibrium constants can be used to determine the position of weak acid equilibria. Consider the equilibrium for a solution of a weak acid HA:

$$HA(aq) + H_2O(l) \leftrightharpoons H_3O^+(aq) + A^-(aq)$$

The equilibrium constant (K_{eq}) is:

$$K_{eq} = [H_3O^+][A^-]/[HA]$$

A 0.0100 mol/L solution of hydrofluoric acid is prepared. At 25 °C the hydronium ion and fluoride ion concentrations are 0.00253 mol/L and the unionised HF equilibrium concentration at equilibrium is 0.00747 mol/L:

$$K_{eq} = [H_3O^+][F^-]/[HF] = \frac{(0.002\,53)(0.002\,53)}{(0.007\,47)} = 6.76 \times 10^{-4}$$

This equilibrium lies to the left.

➔ The equilibrium constant for weak acid dissociation (ionisation) equilibria is given the symbol K_a. It is called the *acid dissociation constant* or *acid ionisation constant*. Weak acid equilibria will be investigated further in Module 6.

Dissociation of bases

➔ Sodium hydroxide is an ionic compound that dissolves readily in water to produce a solution of sodium ions and hydroxide ions. The solution is said to be alkaline due to the presence of the hydroxide ion. NaOH is classified as a strong base as the dissociation is complete:

$$NaOH(s) \rightarrow Na^+(aq) + OH^-(aq)$$

➔ Ammonia gas dissolves in water and reacts with water to form a weakly alkaline solution. The dissociation process is incomplete and an equilibrium is established:

$$NH_3(aq) + H_2O(l) \leftrightharpoons NH_4^+(aq) + OH^-(aq)$$

Ammonia is classified as a weak base as the formation of ions is incomplete.

➔ The equilibrium constants for weak base solutions can be calculated using experimental data. The equilibrium constant (K_{eq}) for a weak base is given the symbol K_b. For the ammonia equilibrium, the K_b expression is:

$$K_b = [NH_4^+][OH^-]/[NH_3]$$

Note: The term for water is omitted in the K_b expression as its concentration essentially does not change.

At 25 °C the K_b value for the ammonia equilibrium is 1.8×10^{-5}. This low value demonstrates that the equilibrium lies strongly to the left. Figure 3.15 is a particle model of this equilibrium.

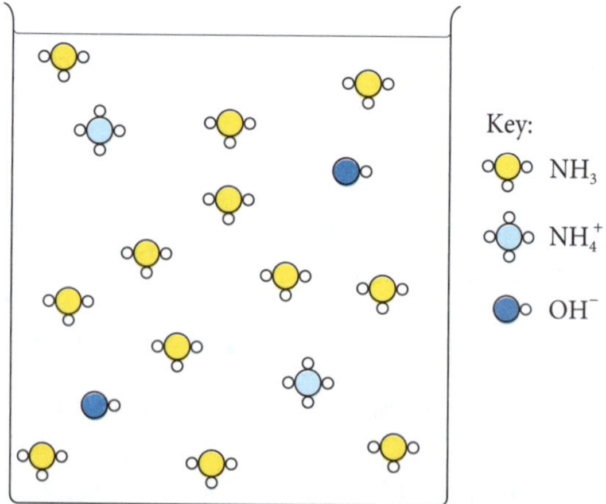

Figure 3.15 Ammonia equilibrium in water solution

→ KEY QUESTIONS

4 Explain how the ionic product (IP) can be used to determine whether the reaction has reached equilibrium at different times during the reaction.

5 Explain how a colourimeter can be used to determine the equilibrium constant for the iron (III) thiocyanate equilibrium.

6 Distinguish between dissociation equilibria involving ionic salts and weak acids.

7 Hypochlorous acid (HOCl) is a weak acid. In a 0.0010 mol/L solution its degree of dissociation is 0.54%. Calculate the hydronium ion concentration in this solution.

Answers ⊃ p. 44

CHAPTER SYLLABUS CHECKLIST

Are you able to answer every question from the syllabus for this chapter? Tick each question as you go through the checklist if you are able to answer it. If you cannot answer a question, turn to the relevant page in the study guide to find the answer. For NESA key word meanings, go to www.educationstandards.nsw.edu.au and search 'key words'.

FOR A COMPLETE UNDERSTANDING OF THIS TOPIC:		PAGE NO.	✓
1	Can I explain how the concentrations of reactants and products vary if equilibria lie to the right or left?	32	
2	Can I write an expression for an equilibrium constant from a balanced equilibrium equation?	33	
3	Can I relate the values of equilibrium constants to the position of the equilibria?	33	
4	Can I use numerical or graphical data to calculate values for equilibrium constants?	34	
5	Can I describe examples of changes in temperature or gas pressure on the equilibrium constant for various equilibria?	35	
6	Can I explain the connection between the reaction quotient (Q) and K_{eq}?	36	
7	Can I describe experiments in which K_{eq} are measured?	37	
8	Can I describe examples of dissociation equilibria of ionic compounds?	38	
9	Can I describe examples of dissociation equilibria for acids and bases?	39	

HSC EXAM-TYPE QUESTIONS

Objective-response questions

(1 mark each)

1 The equilibrium constant for an equilibrium involving gases A, B, C and D is:

$K_{eq} = [A]^2[B]/[C][D]^2$

Identify the equilibrium related to this K_{eq} expression.

A $A(g) + B(g) \leftrightarrows C(g) + D(g)$

B $2A(g) + B(g) \leftrightarrows C(g) + 2D(g)$

C $C(g) + 2D(g) \leftrightarrows 2A(g) + B(g)$

D $C(g) + 2D(s) \leftrightarrows 2A(g) + 2B(g)$

2 Sulfur dioxide gas is oxidised to form sulfur trioxide gas, as shown in the following equilibrium:

$2SO_2(g) + O_2(g) \leftrightarrows 2SO_3(g)$

0.80 mol of sulfur dioxide is placed in a 10 L vessel and 0.40 mol of oxygen is added. The sulfur dioxide is oxidised to form sulfur trioxide. At equilibrium 0.50 mol of sulfur dioxide remains. Calculate the equilibrium constant (K_{eq}) for this equilibrium.

A 14.4

B 24.0

C 11.1

D 0.069

3 Consider the following homogeneous, endothermic equilibrium:

$3F_2(g) + Cl_2(g) \leftrightarrows 2ClF_3(g)$

The system is at equilibrium and the temperature is then raised. The amount of ClF_3 present at the new equilibrium is increased by 0.20 mol. Select the correct answer that identifies the changes in all species.

	F_2	Cl_2	ClF_3
A	Increases by 0.30 mol	Decreases by 0.10 mol	Increases by 0.20 mol
B	Decreases by 0.30 mol	Decreases by 0.10 mol	Increases by 0.20 mol
C	Increases by 0.30 mol	Decreases by 0.10 mol	Decreases by 0.20 mol
D	Increases by 0.30 mol	Increases by 0.10 mol	Decreases by 0.20 mol

4 At 450 °C sulfur dioxide gas was oxidised to form sulfur trioxide gas as shown in the following exothermic equilibrium:

$2SO_2(g) + O_2(g) \leftrightarrows 2SO_3(g)$

The system reaches equilibrium and, at 5 minutes after equilibrium is achieved, a change is made to the system, as shown in Figure 3.16.

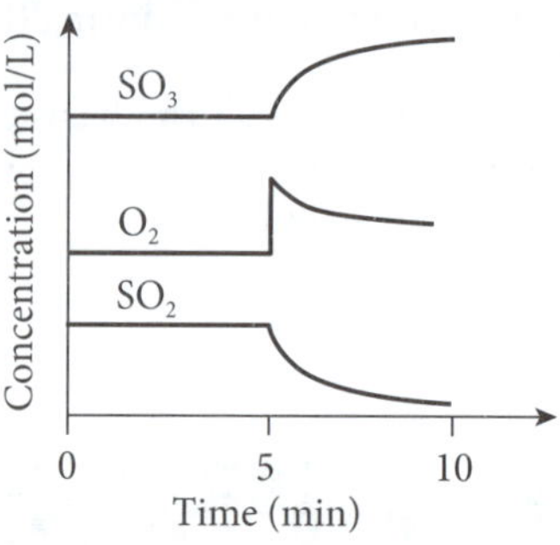

Figure 3.16 Changes to an equilibrium

Identify the change that occurred at 5 minutes.

A The temperature was lowered.

B The volume of the vessel was decreased.

C More oxygen was added to the vessel.

D A catalyst was added to the vessel.

5 A 0.10 mol/L solution of a weak acid HA is 5% dissociated at 25 °C. Calculate the concentration of hydronium ions in the acid solution.

A 0.50 mol/L

B 0.050 mol/L

C 0.0050 mol/L

D 2.0%

Extended-response questions

6 Butyl acetate ($CH_3COOC_4H_9$) can be prepared by reacting butan-1-ol (C_4H_9OH) and excess acetic acid (CH_3COOH) together in the presence of an acid catalyst:

$$CH_3COOH(l) + C_4H_9OH(l) \leftrightarrows CH_3COOC_4H_9(l) + H_2O(l)$$

The following data was collected for an experiment in which butyl acetate was prepared by reacting butan-1-ol with an excess of acetic acid.

Initial $[C_4H_9OH]$ (mol/L)	Initial $[CH_3COOH]$ (mol/L)	Equilibrium $[C_4H_9OH]$ (mol/L)	Equilibrium $[CH_3COOH]$ (mol/L)
0.150	0.450	0.014	0.314

a Write an expression for the equilibrium constant. (1 mark)

b Use the data to calculate a value for the equilibrium constant and state the position of the equilibrium. (3 marks)

7 X and Y react in aqueous solution to form Z, according to the following equilibrium:

$X(aq) + Y(aq) \leftrightarrows 2Z(aq)$

Figure 3.17 shows the progress of the reaction.

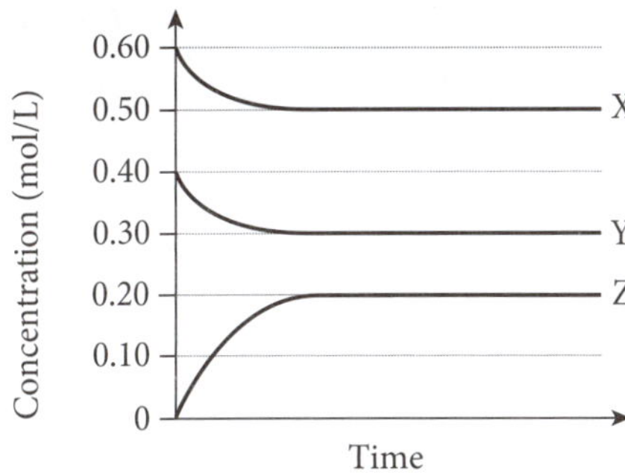

Figure 3.17 Progress of a reaction

Use the graph to calculate a value for the equilibrium constant for this reaction. (3 marks)

8 Colourless nitrous oxide (N_2O) reacts with oxygen gas to form brown nitrogen dioxide gas, according to the following equilibrium:

$$2N_2O(g) + 3O_2(g) \leftrightarrows 4NO_2(g); \Delta H = -32 \text{ kJ}$$

0.27 mol of N_2O was placed in a 5 L vessel with 0.42 mol of oxygen. The system was allowed to reach equilibrium. At equilibrium 0.15 mol of N_2O remained. The temperature of the vessel was monitored over this time.

a Identify whether the temperature of the reaction mixture has increased or decreased as the reaction occurred. (1 mark)

b Describe the colour change during the reaction. (1 mark)

c Write an expression for the equilibrium constant. (1 mark)

d Calculate a value for the equilibrium constant. (4 marks)

9 Soda water is made by dissolving carbon dioxide in water under high pressure. The following exothermic equilibrium is established in a closed soda water bottle:

$$CO_2(g) \leftrightarrows CO_2(aq)$$

A rubber hose is connected to a gas syringe and 80 mL of the soda water is drawn from a refrigerated soda water bottle into the syringe at 5 °C. The rubber hose is then clamped to prevent loss of water or gas (Figure 3.18).

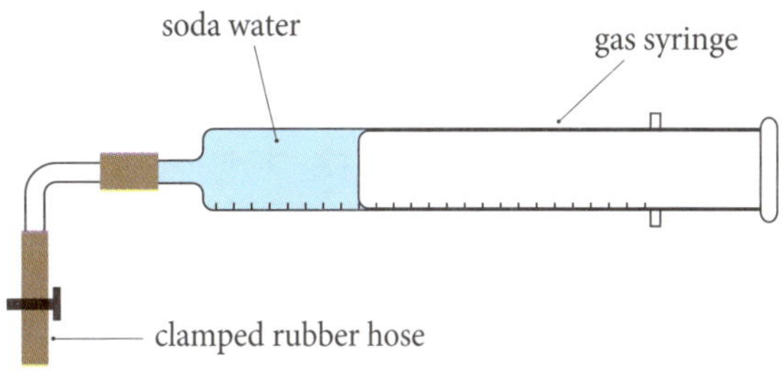

Figure 3.18 Soda water in a gas syringe

The syringe is then allowed to warm from 5 °C to 35 °C. Describe and explain what is observed. (2 marks)

10 Sulfuric acid (H_2SO_4) dissociates in two stages. The first dissociation is complete and leads to the formation of the hydrogen sulfate ion (HSO_4^-). The second stage involves the incomplete dissociation of the hydrogen sulfate ion to form sulfate ions. Write the equations for the two stages. (2 marks)

ANSWERS

KEY QUESTIONS

Key questions ➲ p. 35

1 The reactants are in a much higher concentration if the equilibrium lies strongly to the left or reactants' side.

2 $K_{eq} = [N_2O_4]/[NO]^2[O_2]$

3 $H_2(g) + I_2(g) \rightleftharpoons 2HI(g)$

$$K_{eq} = [HI]^2/[H_2][I_2] = \frac{(1.666)^2}{(0.242)(0.242)} = 47.4$$

Key questions ➲ p. 41

4 When $IP < K_{eq}$, then equilibrium has not been reached. When IP equals K_{eq}, then the system has reached equilibrium.

5 The iron (III) thiocyanate ion absorbs blue light and the greater its concentration the more light is absorbed. A colourimeter can be used to measure the iron (III) thiocyanate concentration by creating a calibration graph using known standards. This graph can then be used to measure the concentrations of unknowns.

6 The dissociation of hydrofluoric acid in water is an example of a molecule partially breaking up into ions. Ionic compounds are already composed of ions and these dissociate into free ions in water. Once the solution is saturated, an equilibrium exists. A saturated solution of KCl is an example.

7 $HOCl(aq) + H_2O(l) \rightleftharpoons H_3O^+(aq) + OCl^-(aq)$
Percentage dissociation = $[H_3O^+]/[HOCl] \times 100$

$$0.54 = \frac{[H_3O^+]}{[0.0010]} \times 100$$

$$[H_3O^+] = 5.4 \times 10^{-6} \text{ mol/L}$$

HSC EXAM-TYPE QUESTIONS

Objective-response questions

1 **C**. In the equilibrium constant the concentrations are raised to the powers shown by the coefficients in the balanced equation. Thus **A**, **B** and **D** are incorrect as the coefficients do not match the K_{eq}.

2 **A**. 0.30 mol of sulfur dioxide react and $\frac{0.30}{2} = 0.15$ mol of oxygen react and 0.30 mol of sulfur trioxide form.

$$[SO_2] = \frac{0.50}{10} = 0.050 \text{ mol/L}$$

$$[SO_3] = \frac{0.30}{10} = 0.030 \text{ mol/L}$$

$$[O_2] = \frac{(0.40 - 0.15)}{10} = 0.025 \text{ mol/L}$$

$$K_{eq} = [SO_3]^2/[SO_2]^2[O_2] = \frac{(0.030)^2}{(0.050)^2(0.025)} = 14.4$$

Thus **B**, **C** and **D** are mathematically incorrect.

3 **B**. The stoichiometry is 3 : 1 : 2 and thus if ClF_3 increases by 0.20 mol then F_2 and Cl_2 decrease by 0.30 mol and 0.10 mol respectively as the equilibrium shifts to the right. **A**, **C** and **D** are incorrect as they state that F_2 increases.

4 **C**. The sudden rise in oxygen concentration at 5 minutes was due to the injection of oxygen. This disturbed the equilibrium and the equilibrium shifted to the right and more sulfur trioxide formed. **A** is incorrect as a temperature change would not produce such a sudden change in oxygen concentration. **B** is incorrect as a volume decrease would increase all the concentrations. **D** is incorrect as a catalyst does not alter the equilibrium position.

5 **C**.
$HA(aq) + H_2O(l) \rightleftharpoons H_3O^+(aq) + A^-(aq)$
Degree of dissociation = $[H_3O^+]/[HA] \times 100/1$
$5 = [H_3O^+]/(0.10) \times 100/1$
$[H_3O^+] = 0.0050$ mol/L
Thus **A**, **B** and **D** are incorrect mathematically.

Extended-response questions

6 EM Students need to use the stoichiometric equation to determine the concentration of products before substitution into the equilibrium constant expression.

a $K_{eq} = [CH_3COOC_4H_9][H_2O]/[C_4H_9OH][CH_3COOH]$ ✓

b Equilibrium concentration of products:
$[CH_3COOC_4H_9] = [H_2O] = 0.150 - 0.014 = 0.136$ mol/L ✓

$$K_{eq} = \frac{[0.136][0.136]}{[0.014][0.314]} = 4.21 \checkmark$$

As $K_{eq} > 1$, the equilibrium lies to the right. ✓

7 EM Students are required to interpret a graph to determine the equilibrium concentrations and then write a correct equilibrium constant expression. Students must show the substitution into this expression to score full marks.

$K_{eq} = [Z]^2/[X][Y]$ ✓

$$= \frac{(0.20)^2}{(0.50)(0.30)} \checkmark$$

$= 0.27$ ✓

8 EM Students must be able to relate the equilibrium equation to observations made in the experiment. The correct use of stoichiometry and creating a correct expression for the equilibrium constant is required to score full marks.

a The temperature increases as the reaction is exothermic. ✓

b The initial reactants are colourless and the product is brown. Therefore the mixture will become browner until equilibrium is achieved. ✓

c $K_{eq} = [NO_2]^4/[N_2O]^2[O_2]^3$ ✓

d $2N_2O(g) + 3O_2(g) \rightleftharpoons 4NO_2(g)$
Mole ratio of gases = 2 : 3 : 4
Amount of N_2O at equilibrium = 0.15 mol ✓
Thus 0.27 – 0.15 = 0.12 mol of N_2O reacted. Thus $\frac{3}{2} \times 0.12$ = 0.18 mol of O_2 reacted.
Therefore the amount of O_2 at equilibrium is 0.42 – 0.18 = 0.24 mol.
Therefore the amount of NO_2 product at equilibrium is 0.24 mol. ✓

$$[N_2O] = \frac{0.15}{5} = 0.030 \text{ mol/L}$$

$$[O_2] = \frac{0.24}{5} = 0.048 \text{ mol/L}$$

$$[NO_2] = \frac{0.24}{5} = 0.048 \text{ mol/L} \checkmark$$

$$K_{eq} = \frac{(0.048)^4}{(0.030)^2(0.048)^3} = 53.3 \checkmark$$

9 EM Students need to understand that heating the carbon dioxide equilibrium leads to a loss of gas from the solution.

The equilibrium is exothermic. As the soda water is warmed the equilibrium shifts to the left and gaseous carbon dioxide bubbles out of solution. ✓

The plunger of the syringe will be pushed outwards by the pressure of the carbon dioxide gas. ✓

10 EM Students must use reversible arrows in the second ionisation. Aqueous states should be included for all species.

$H_2SO_4(aq) + H_2O(l) \rightarrow H_3O^+(aq) + HSO_4^-(aq)$ ✓

$HSO_4^-(aq) + H_2O(l) \leftrightarrows H_3O^+(aq) + SO_4^{2-}(aq)$ ✓

MODULE 5 EQUILIBRIUM REACTIONS

CHAPTER 4 SOLUTION EQUILIBRIA

INQUIRY QUESTION:

How does solubility relate to chemical equilibrium?

When soluble ionic salts are added to water the ions break free of the crystal lattice and dissolve in the water. In the solution the ions are stabilised by ion–dipole attraction with water molecules. When different solutions of ions are mixed, a precipitate may form. Thus when lead (II) nitrate solution is mixed with potassium iodide solution a yellow precipitate of lead (II) iodide forms. This chapter investigates solution equilibria.

1 Dissolution equilibria

» Students describe and analyse the processes involved in the dissolution of ionic compounds in water.

→ **Dissolution** is the process in which solutes dissolve in water and form a solution. In the case of an ionic solute such as nickel (II) chloride crystals, the surface of the crystal undergoes collisions with water molecules. Some collisions lead to ions breaking free of the crystal lattice. Positive nickel (II) ions become solvated (hydrated) by the water solvent as the negative ends of the water molecule are attracted to the nickel (II) ions. This is an ion–dipole attraction. The chloride ions are solvated by water molecules because the positive end of the water dipole is attracted to the negative chloride ions.

Example:

$NiCl_2(s) \rightarrow Ni^{2+}(aq) + 2Cl^-(aq)$

This example is illustrated in Figure 4.1, where cations and anions are stabilised by interactions with the water dipole.

→ If more solute is added to the water, a point is reached when no more will dissolve and a saturated solution forms. The saturated solution is an example of a heterogeneous equilibrium. This is a dynamic equilibrium in which ions **dissociate** from the crystal at the same rate as ions **associate** together and form a crystal:

$NiCl_2(s) \leftrightharpoons Ni^{2+}(aq) + 2Cl^-(aq)$

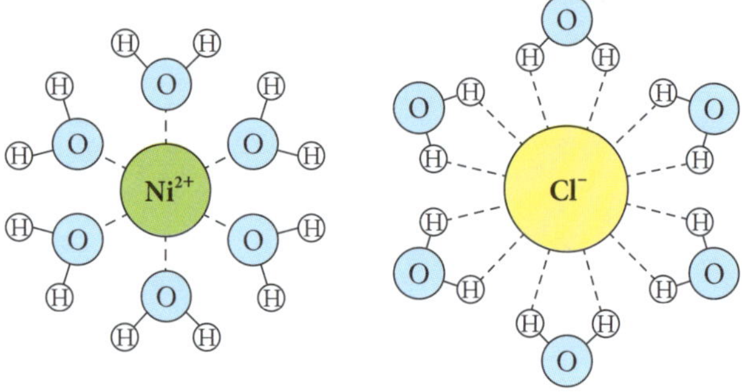

Figure 4.1 Hydration of nickel (II) ions and chloride ions

Thus the shape of the crystals changes over time but the total mass of crystals remains constant. Figure 4.2 shows this heterogeneous equilibrium.

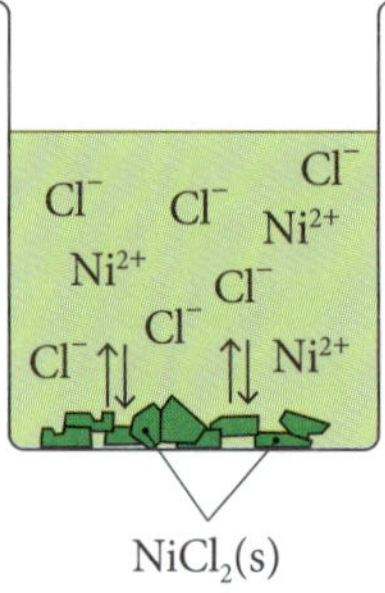

Figure 4.2 Heterogeneous dissolution equilibrium of $NiCl_2$

dissolution: the process in which a solute dissolves in a solvent to form a solution

dissociation: the breakdown (dissolution) of an ionic crystal in water to produce separate cations and anions in solution

association: the attraction of oppositely charged ions in solution leading to the formation of an ionic solid

→ Ionic compounds have a wide range of solubilities in water. 39 g of NH_4Cl is the maximum amount that will dissolve in 100 g of water at 25 °C. If 40 g of NH_4Cl is added to 100 g of water 1 g remains undissolved and remains as crystals on the base of the beaker. The solution is saturated and the dissolved ions and the ions in the crystal lattice are in equilibrium.

EXAMPLE 1

5.0 g of potassium bromide was added to 5.0 mL (5.0 g) of water in a test tube at 25 °C. The test tube was stoppered and shaken at a uniform rate for 1 minute. The mass of the KBr dissolved was measured. The experiment was repeated with different times of agitation. The results are shown in Figure 4.3.

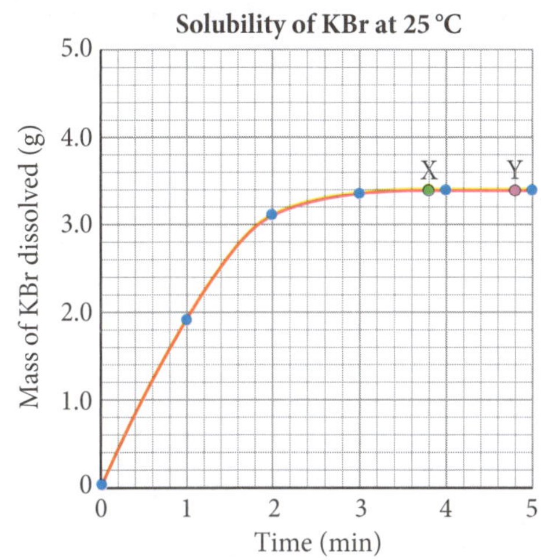

Figure 4.3 Solubility of KBr in water

a **Explain the procedure used to measure the amount of KBr that dissolved after 1 minute.**

b **Determine when the solution first becomes saturated.**

c **Write an equation for the equilibrium in the saturated solution.**

Identify the equipment used in the experiment

d **Determine the solubility of KBr in water in the units of g/100 g water.**

e **Compare the masses of crystals of KBr that remain undissolved at time X and time Y.**

Answer:

a At 1 minute the mixture is filtered through a pre-weighed filter paper. The filter paper and crystals are dried in a desiccator or drying oven until constant mass is achieved. The mass of KBr crystals is obtained by subtracting the mass of the filter paper. The mass of KBr that has dissolved is obtained by subtracting the mass of KBr from 5.0 g.

b At 3.4 minutes there is no further change in the mass that dissolves. This is the time when the solution first becomes saturated.

c $KBr(s) \leftrightarrows K^+(aq) + Br^-(aq)$

d 3.4 g of KBr dissolves in 5.0 g of water.

Therefore solubility = $\left(\frac{3.4}{5.0}\right) \times 20 = 68$ g/100 g water

e At X and Y, 3.4 g of KBr have dissolved.

Therefore the mass of undissolved crystals = 5.0 – 3.4
= 1.6 g

- Dissolution equilibria are temperature dependent. Figure 4.4 compares the solubility of ammonium chloride and sodium chloride as a function of temperature. Both salts increase in solubility with temperature. Both equilibria are endothermic. To create a saturated solution of ammonium chloride, at any temperature between 20 °C and 80 °C, requires a greater mass than is required for sodium chloride.

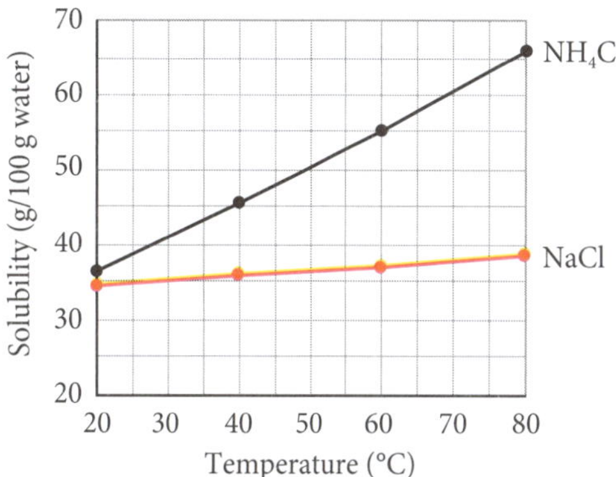

Figure 4.4 Solubility versus temperature

- The Year 11 course investigated exothermic and endothermic dissolution of ionic compounds in water. When sodium hydroxide crystals undergo dissolution in water the temperature of the water rises. This is an example of an exothermic dissolution. When ammonium nitrate crystals undergo dissolution in water the temperature of the water decreases. This is an example of an endothermic dissolution.
- The process of dissolution can be analysed in terms of the following three processes.
  - *Step 1:* To overcome the ionic bonds in the crystal lattice the ions absorb energy from water molecules as water molecules collide with the lattice. This is an endothermic step.
  - *Step 2:* In order for ions to be solvated the hydrogen bonds between water molecules must be broken. The attraction of free ions for water dipoles helps the water molecules to separate from one another. This is an endothermic step.
  - *Step 3:* Energy is released as ion–dipole forces stabilise the ions in the solution during the solvation (hydration) process. The ions are less free to move about and their kinetic energies decrease. This is an exothermic step.
- The sum of the enthalpy changes for these three steps determines whether the dissolution is exothermic or endothermic.
  - For an *exothermic dissolution*, the energy released in step 3 is greater than the sum of the energy absorbed for steps 1 and 2.
  - For an *endothermic dissolution*, the energy released in step 3 is less than the sum of the energy absorbed for steps 1 and 2.

  Figure 4.5 summarises these ideas.

$NaOH(s) \rightarrow Na^+(aq) + OH^-(aq); \Delta H < 0$

$NH_4NO_3(s) \rightarrow NH_4^+(aq) + NO_3^-(aq); \Delta H > 0$

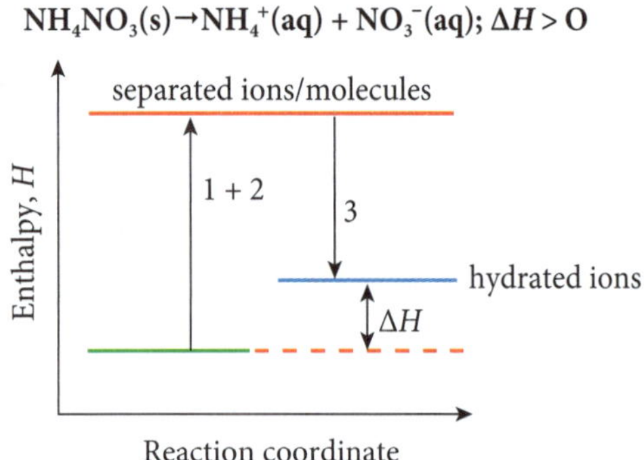

Figure 4.5 Exothermic and endothermic dissolutions in water

EXAMPLE 2

Calcium iodide crystals are dissolved in water at 25 °C.

a **Write an equation for the dissolution of calcium iodide in water.**

b **Use the following standard thermodynamic data to calculate the enthalpy change, entropy change and Gibbs free energy change for the dissolution reaction. Comment on the drivers of this reaction.**

	$CaI_2(s)$	$Ca^{2+}(aq)$	$I^-(aq)$
$\Delta_f H^o$ (kJ/mol)	-533	-543	-57
S^o (J/K/mol)	142	-56	106
$\Delta_f G^o$ (kJ/mol)	-529	-553	-52

Use the stoichiometry in these calculations

Answer:

a $CaI_2(s) \rightarrow Ca^{2+}(aq) + 2I^-(aq)$

b Enthalpy change:

$\Delta H^o = \Sigma\Delta_f H^o$ products $- \Sigma\Delta_f H^o$ reactants $= (-543) + 2(-57) - (-533) = -124$ kJ/mol

This is an exothermic reaction and the heat loss to the environment drives the reaction forward.

Entropy change:

$\Delta S^o = \Sigma S^o$ products $- \Sigma S^o$ reactants $= (-56) + 2(106) - (142) = +14$ J/K/mol

The increase in entropy as ions become freed from the crystal also drives the reaction forward.

Gibbs free energy change:

$\Delta G^o = \Sigma\Delta_f G^o$ products $- \Sigma\Delta_f G^o$ reactants $= (-553) + 2(-52) - (-529) = -128$ kJ/mol < 0

The negative Gibbs free energy change indicates the reaction is spontaneous.

2 Solubility of polar molecular compounds

» Students investigate the use of solubility equilibria used by Aboriginal and Torres Strait Islander peoples when removing toxicity from foods; for example, toxins in cycad fruit.

➔ Molecular compounds also vary considerably in their water solubility. **Polar** functional groups (e.g. OH, COOH, NH_2) improve solubility, particularly if they can hydrogen-bond with water molecules. A simple sugar such as table sugar (sucrose, $C_{12}H_{22}O_{11}$) has many polar –OH groups and has a solubility of 210 g/100 g water at 25 °C.

polar molecule: a molecule with a permanent dipole due to the asymmetric arrangement of functional groups

➔ The functional groups illustrated in the organic molecules in Figure 4.6 can hydrogen-bond with water which improves solubility. As the molar mass of such compounds increases due to an increase in the number of carbon atoms, the solubility decreases until these compounds become insoluble. The following table uses organic compounds called *alkanols* to illustrate this observation.

Alcohol	CH_3OH	C_2H_5OH	C_3H_7OH	C_4H_9OH	$C_5H_{11}OH$
Solubility (mol/L)	miscible*	miscible*	miscible*	0.11	0.030

* soluble in all proportions

methanol   formic acid   methanamine

Figure 4.6 Polar molecules

➔ Figure 4.7 shows how water molecules interact with a methanol molecule by hydrogen bonding. The highly electropositive hydrogen atoms are attracted to the non-bonding electron lone pairs of the oxygen atoms.

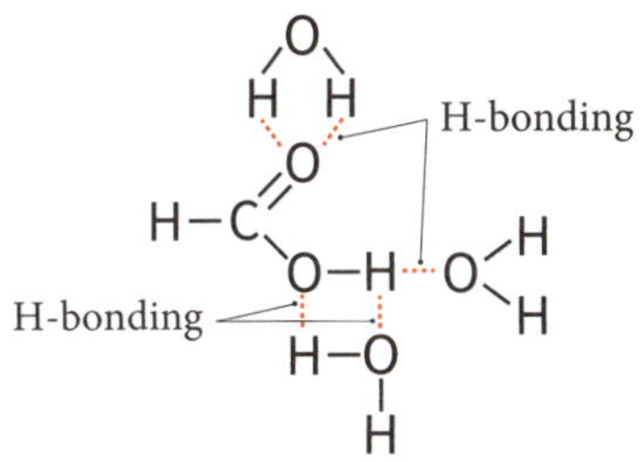

H-bonding

H-bonding

Figure 4.7 Hydrogen bonding between methanol and water

➔ A carcinogenic and neurotoxic compound called *cycasin* is present in the orange-coloured nuts of the fruit of cycad plants. The structure of cycasin in Figure 4.8 shows that it has many polar groups (e.g. hydroxyl (OH)), which makes it slightly water soluble. Such nuts cannot be eaten unless the cycasin is removed.

carcinogen: a substance that causes cancer

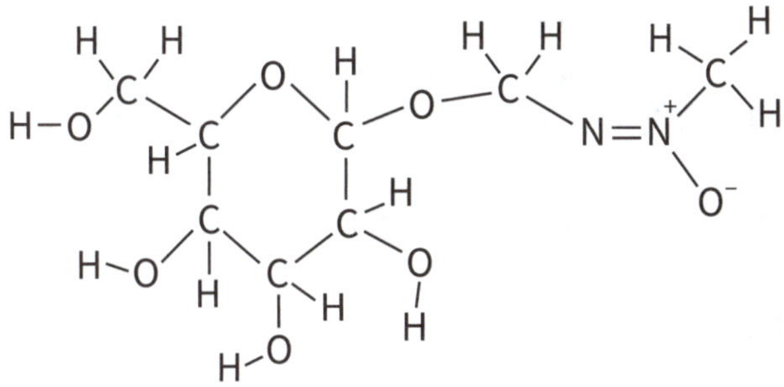

Figure 4.8 Cycasin

➔ Indigenous Australians have used cycad nuts as a major food source for thousands of years. Responding to knowledge that the nuts were toxic, Indigenous people used trial and error methods to find ways to make the nuts edible. Aged nuts were found to contain much less toxin than fresh nuts and caused limited illness. Another approach to make the nuts edible was to remove the shell and cut open the kernels and repeatedly leach the toxin out of the flesh using running water until the toxin was removed. Crushing the flesh led to an increase in the surface area and the rate of dissolution of the toxin into the water was increased. The dried flesh was then powdered to make a type of flour, which could be used to make bread.

➔ The toxicity of these cycad nuts in humans is due to microbes in the intestine breaking the cycasin into methylazomethanol, which is then converted in the liver to a carcinogenic compound (see Figure 4.9).

3 Solubility rules

» Students conduct an investigation to determine solubility rules, and predict and analyse the composition of substances when two ionic solutions are mixed, for example: potassium chloride and silver nitrate; potassium iodide and lead nitrate; and sodium sulfate and barium nitrate.

➔ Precipitation reactions can be used to investigate the solubility of salts in water. Solubility rules can be formulated based on these investigations. Table 4.1 summarises the solubility rules for some common salts based on these experiments.

Table 4.1 Solubility rules

Anion	Solubility rule
NO_3^-	All nitrates are soluble
Cl^-	Most chlorides are soluble, except AgCl and $PbCl_2$ ($PbCl_2$ is soluble in hot water)
Br^-	Most bromides are soluble, except AgBr and $PbBr_2$
I^-	Most iodides are soluble, except AgI and PbI_2
SO_4^{2-}	Most sulfates are soluble, except $BaSO_4$, $PbSO_4$ and $CaSO_4$ ($CaSO_4$ is slightly soluble in dilute solutions)
CO_3^{2-}	Most carbonates are insoluble, except Group 1 carbonates
OH^-	Most hydroxides are insoluble, except Group 1 hydroxides and $Ba(OH)_2$; $Ca(OH)_2$ is slightly soluble in dilute solutions

➔ Figure 4.10 shows the formation of a yellow precipitate of lead (II) iodide when colourless lead (II) nitrate solution is mixed with colourless potassium iodide solution. Table 4.1 predicts the formation of a precipitate of PbI_2, whereas the potassium ions and nitrate ion remain in solution as all nitrates are soluble:

$$Pb^{2+}(aq) + 2I^-(aq) \rightarrow PbI_2(s)$$

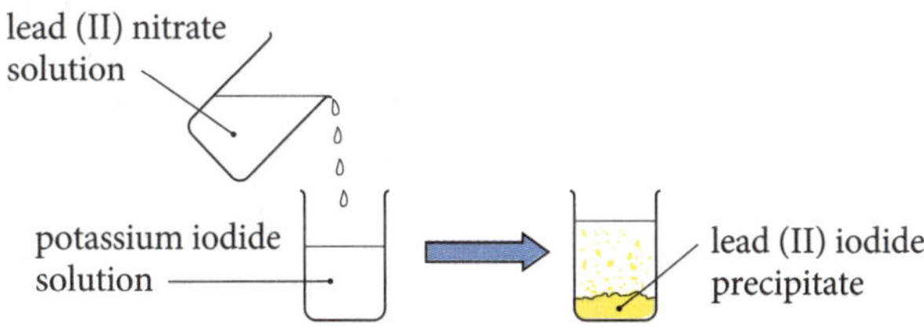

Figure 4.10 Precipitation of yellow lead (II) iodide

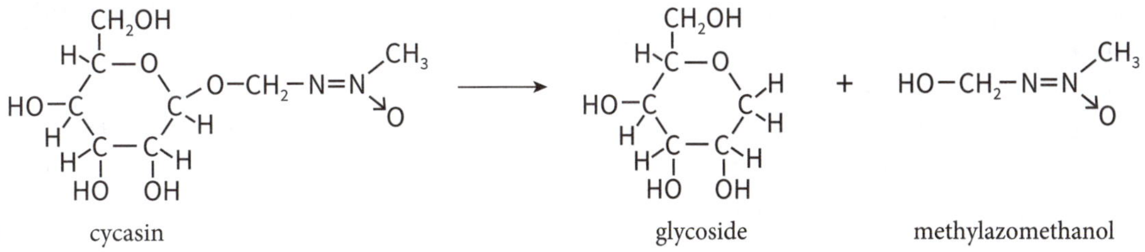

cycasin → glycoside + methylazomethanol

Figure 4.9 Cycasin conversion to methylazomethanol

EXAMPLE 3

The following table shows the results of experiments involving the addition of silver nitrate and barium nitrate solutions to solutions containing carbonate, chloride and sulfate ions as their potassium salts.

Anion	Silver nitrate solution	Barium nitrate solution
CO_3^{2-}	(A) yellow precipitate	(D) white precipitate
Cl^-	(B) white precipitate	(E) no precipitate
SO_4^{2-}	(C) no precipitate	(F) dense white precipitate

Write net ionic equations for the precipitation reactions.

Net ionic equations show no spectator ions

Answer:

The experimental results are consistent with the solubility rules. Potassium ions do not form any insoluble salts.

(A): $2Ag^+(aq) + CO_3^{2-}(aq) \rightarrow Ag_2CO_3(s)$

(B): $Ag^+(aq) + Cl^-(aq) \rightarrow AgCl(s)$

(D): $Ba^{2+}(aq) + CO_3^{2-}(aq) \rightarrow BaCO_3(s)$

(F): $Ba^{2+}(aq) + SO_4^{2-}(aq) \rightarrow BaSO_4(s)$

No precipitate forms in experiment (C) as silver sulfate is soluble. No precipitate forms in experiment (E) as barium chloride is soluble.

EXAMPLE 4

The following table shows precipitation tests performed on cations present in solution as their nitrate salts. Figure 4.11 shows an example of a precipitation test for copper (II) ions using sodium hydroxide solution.

Cation	Hydrochloric acid	Sulfuric acid	Sodium hydroxide solution
Pb^{2+}	(A) white precipitate	(E) white precipitate	(I) white precipitate
Cu^{2+}	(B) no reaction	(F) no reaction	(J) blue precipitate
Fe^{3+}	(C) no reaction	(G) no reaction	(K) orange-brown precipitate
Ca^{2+}	(D) no reaction	(H) white precipitate	(L) faint white precipitate

Write net ionic equations for the precipitation reactions.

Review the table of solubilities to determine which ion interactions will produce precipitation

Answer:

The experimental results are consistent with the solubility rules.

(A): $Pb^{2+}(aq) + 2Cl^-(aq) \rightarrow PbCl_2(s)$

(E): $Pb^{2+}(aq) + SO_4^{2-}(aq) \rightarrow PbSO_4(s)$

(H): $Ca^{2+}(aq) + SO_4^{2-}(aq) \rightarrow CaSO_4(s)$

(I): $Pb^{2+}(aq) + 2OH^-(aq) \rightarrow Pb(OH)_2(s)$

(J): $Cu^{2+}(aq) + 2OH^-(aq) \rightarrow Cu(OH)_2(s)$

(K): $Fe^{3+}(aq) + 3OH^-(aq) \rightarrow Fe(OH)_3(s)$

(L): $Ca^{2+}(aq) + 2OH^-(aq) \rightarrow Ca(OH)_2(s)$

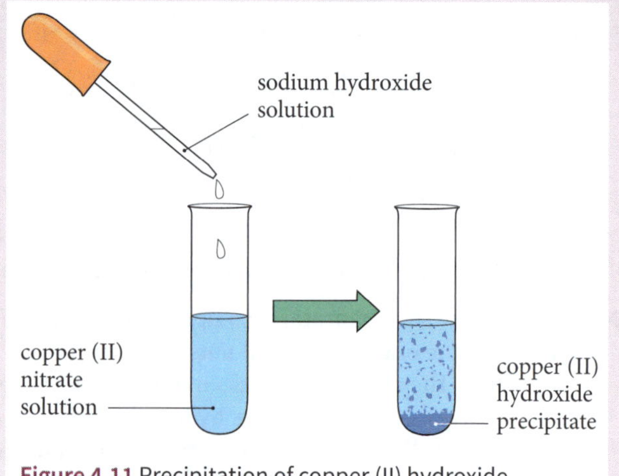

Figure 4.11 Precipitation of copper (II) hydroxide (experiment J)

FIRSTHAND INVESTIGATION

Investigating solubility using precipitation reactions

Aim

to investigate solubility rules using precipitation reactions

Materials

Plastic acetate sheet or spot plate

1.0 mol/L solutions of the following salts in dropper bottles:

- Group (A): silver nitrate; copper (II) nitrate; barium nitrate; lead nitrate
- Group (B): potassium chloride; potassium carbonate; potassium sulfate; potassium iodide

Method

1 The reactions are performed by mixing drops of a selected solution from Group (A) with a selected solution from Group (B) on an acetate plastic sheet or spot plate (see Figure 4.12).

2 Record the colour of any precipitate that forms.

Sample results

The following table shows the results of these experiments.

	KCl	K_2CO_3	K_2SO_4	KI
$AgNO_3$	white precipitate	yellow precipitate	no precipitate*	cream precipitate
$Zn(NO_3)_2$	no precipitate	white precipitate	no precipitate	no precipitate
$Ba(NO_3)_2$	no precipitate	white precipitate	white precipitate	no precipitate
$Pb(NO_3)_2$	white precipitate	white precipitate	white precipitate	yellow precipitate

* A faint white precipitate may form depending on concentration of the solutions.

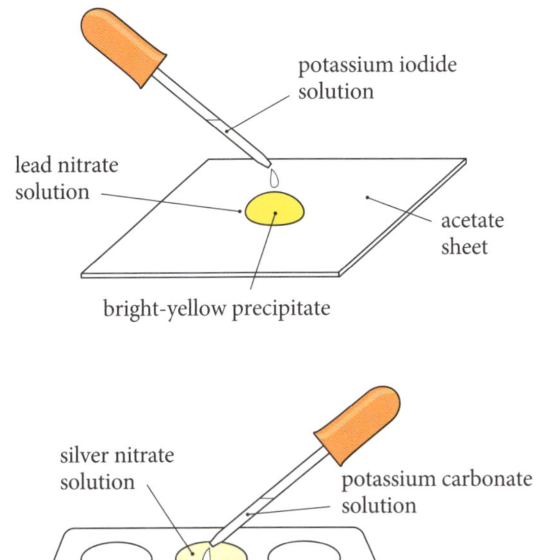

Figure 4.12 Testing for precipitation

Analysis

1 Compare these results with the solubility generalisation rules in the table of solubilities.
2 For all reactions that produced precipitates, write net ionic equations.

Sample answers

1 The results were consistent with the table of solubilities.
2 Net ionic equations:
- $AgNO_3$
  $Ag^+(aq) + Cl^-(aq) \rightarrow AgCl(s)$
  $2Ag^+(aq) + CO_3^{2-}(aq) \rightarrow Ag_2CO_3(s)$
  $Ag^+(aq) + I^-(aq) \rightarrow AgI(s)$
- $Zn(NO_3)_2$
  $Zn^{2+}(aq) + CO_3^{2-}(aq) \rightarrow ZnCO_3(s)$
- $Ba(NO_3)_2$
  $Ba^{2+}(aq) + CO_3^{2-}(aq) \rightarrow BaCO_3(s)$
  $Ba^{2+}(aq) + SO_4^{2-}(aq) \rightarrow BaSO_4(s)$
- $Pb(NO_3)_2$
  $Pb^{2+}(aq) + 2Cl^-(aq) \rightarrow PbCl_2(s)$
  $Pb^{2+}(aq) + CO_3^{2-}(aq) \rightarrow PbCO_3(s)$
  $Pb^{2+}(aq) + SO_4^{2-}(aq) \rightarrow PbSO_4(s)$
  $Pb^{2+}(aq) + 2I^-(aq) \rightarrow PbI_2(s)$

Conclusion

The experimental results agreed with the predictions made using the table of solubilities.

→ KEY QUESTIONS

1 **Write an ionic equation for the dissolution equilibrium in a saturated solution of ammonium sulfate.**

2 **Potassium bromide solution is added to a solution of lead (II) nitrate. Write a net ionic equation for the precipitation reaction.**

3 **Calcium nitrate solution is mixed with sodium chloride solution. Use the solubility rules to predict whether a precipitation reaction will occur.**

4 **Barium chloride solution is mixed with silver sulfate solution. Write the whole formula equation and net ionic equations for the reaction.**

Answers ➲ p. 56

4 Solubility product constant

» Students:
- derive equilibrium expressions for saturated solutions in terms of K_{sp} and calculate the solubility of an ionic substance from its K_{sp} value.
- predict the formation of a precipitate given the standard reference values for K_{sp}.

→ Ionic solids undergo dissolution in water to varying extents. As more solid dissolves, the concentration of cations and anions increases. The solution remains **unsaturated** until the ionic product reaches a critical value when saturation equilibrium is achieved. In the **saturated** solution, the association of ions to form crystals is balanced by dissociation. Figure 4.13 shows models of unsaturated and saturated solutions.

unsaturated solution: a solution which contains less than the amount of solute than is required to saturate it at a fixed temperature

saturated solution: a solution which contains the maximum amount of solute based on its solubility at a fixed temperature

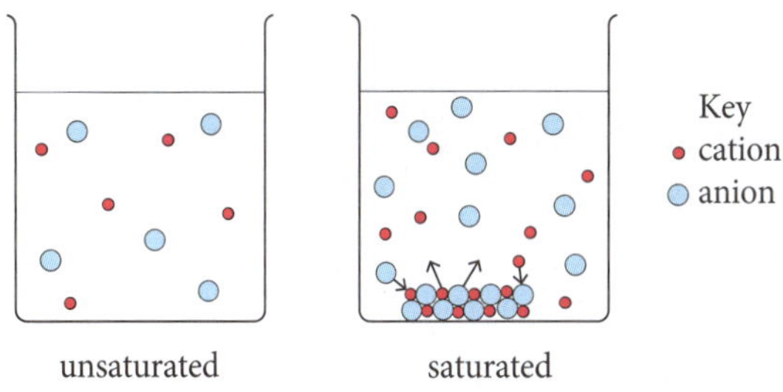

Figure 4.13 Unsaturated and saturated solutions

- For any saturated solution of an ionic solid M_xN_y, the dissolution equilibrium equation is:

  $M_xN_y(s) \leftrightarrows xM^{y+}(aq) + yN^{x-}(aq)$

  The ionic product is:

  $[M^{y+}]^x[N^{x-}]^y$
- Once a saturated solution is established the ionic product (*IP*) is equal to the equilibrium constant (K_{eq}). The equilibrium constant expression for this solubility equilibrium is called the *solubility product constant*, symbol K_{sp}.

$$K_{sp} = [M^{y+}]^x[N^{x-}]^y$$

- This equilibrium constant expression does not include the solid ionic compound in equilibrium with its ions.
- The smaller the value of K_{sp}, the more insoluble the ionic compound in water.
- Table 4.2 shows that lead (II) iodide is the most insoluble of the lead halides.

Table 4.2 Lead halide solubility

Solubility equilibrium	K_{sp} expression	K_{sp}
$PbF_2(s) \leftrightarrows Pb^{2+}(aq) + 2F^-(aq)$	$[Pb^{2+}][F^-]^2$	2.7×10^{-8}
$PbCl_2(s) \leftrightarrows Pb^{2+}(aq) + 2Cl^-(aq)$	$[Pb^{2+}][Cl^-]^2$	1.7×10^{-5}
$PbBr_2(s) \leftrightarrows Pb^{2+}(aq) + 2Br^-(aq)$	$[Pb^{2+}][Br^-]^2$	6.6×10^{-6}
$PbI_2(s) \leftrightarrows Pb^{2+}(aq) + 2I^-(aq)$	$[Pb^{2+}][I^-]^2$	9.8×10^{-9}

- Solubility product constant data can be used to determine the solubility of ionic salts in water.

EXAMPLE 5

The K_{sp} value for magnesium hydroxide is 5.61×10^{-12}.

a Write the equation for the solubility equilibrium for magnesium hydroxide.

b Write the K_{sp} expression for magnesium hydroxide.

c Calculate the solubility of magnesium hydroxide at 25 °C.

Ensure that the dissociation equation is correctly balanced

Answer:

a $Mg(OH)_2(s) \leftrightarrows Mg^{2+}(aq) + 2OH^-(aq)$

b $K_{sp} = [Mg^{2+}][OH^-]^2$

c Let x mol of $Mg(OH)_2$ dissolve per litre.

At equilibrium: $[Mg^{2+}] = x$; $[OH^-] = 2x$

$K_{sp} = [Mg^{2+}][OH^-]^2 = (x)(2x)^2 = 4x^3 = 5.61 \times 10^{-12}$

Solve for x.

$x = 1.12 \times 10^{-4}$ mol/L

Solubility of $Mg(OH)_2 = 1.12 \times 10^{-4}$ mol/L

EXAMPLE 6

Barium sulfate is an insoluble ionic compound. Its solubility in water at 20 °C is 0.00025 g/100 g water. Calculate the K_{sp} for barium sulfate (1 g water = 1 mL).

Convert the mass of water to litres

Answer:

$n(BaSO_4) = m/M = \dfrac{0.00025}{(137.3 + 32.07 + (4 \times 16.00))}$

$= 1.07 \times 10^{-6}$ mol per 100 mL water

Solubility $= 1.07 \times 10^{-6}/0.100$

$= 1.07 \times 10^{-5}$ mol/L

$BaSO_4(s) \leftrightarrows Ba^{2+}(aq) + SO_4^{2-}(aq)$

$[Ba^{2+}] = [SO_4^{2-}] = 1.07 \times 10^{-5}$ mol/L

$K_{sp} = [Ba^{2+}][SO_4^{2-}]$

$= (1.07 \times 10^{-5})(1.07 \times 10^{-5}) = 1.14 \times 10^{-10}$

EXAMPLE 7

20 mL of 0.010 mol/L calcium nitrate is added to 20 mL of 0.010 mol/L sodium sulfate. Determine whether precipitation of calcium sulfate will occur: (K_{sp} ($CaSO_4$) = 4.93×10^{-5})

Determine the number of moles of calcium ions and sulfate ions

Answer:

$n(Ca^{2+}) = n(Ca(NO_3)_2) = cV = (0.010)(0.020)$

$= 2.0 \times 10^{-4}$ mol

$n(SO_4^{2-}) = n(Na_2SO_4) = cV = (0.010)(0.020)$

$= 2.0 \times 10^{-4}$ mol

Volume on mixing = 20 + 20 = 40 mL = 0.040 L

$[Ca^{2+}] = n/V = \dfrac{(2.0 \times 10^{-4})}{0.040} = 5.0 \times 10^{-3}$ mol/L

$[SO_4^{2-}] = n/V = \dfrac{(2.0 \times 10^{-4})}{0.040} = 5.0 \times 10^{-3}$ mol/L

$Ca^{2+}(aq) + SO_4^{2-}(aq) \rightarrow CaSO_4(s)$

Calculate the ionic product (*IP*).

$IP = [Ca^{2+}][SO_4^{2-}] = (5.0 \times 10^{-3})(5.0 \times 10^{-3}) = 2.5 \times 10^{-5}$

$IP < K_{sp}$

Therefore no precipitate will form.

EXAMPLE 8

Calculate the minimum concentration of silver nitrate that will cause precipitation of silver bromide when 25 mL of the silver nitrate solution is added to 25 mL of 0.020 mol/L sodium bromide: (K_{sp} (AgBr) = 5.35×10^{-13})

Consider the change in concentration when solutions are mixed

Answer:

$Ag^+(aq) + Br^-(aq) \rightarrow AgBr(s)$

Final volume = 25 + 25 = 50 mL = 0.050 L

On dilution, $[Br^-] = 0.010$ mol/L

$K_{sp} = [Ag^+][Br^-] = 5.35 \times 10^{-13}$

$[Ag^+][0.010] = 5.35 \times 10^{-13}$

$[Ag^+] = 5.35 \times 10^{-11}$ mol/L

Before dilution, $c(AgNO_3) = 2 \times 5.35 \times 10^{-11}$

$= 1.07 \times 10^{-10}$ mol/L

(minimum concentration)

→ KEY QUESTIONS

5 Write the K_{sp} expression for cobalt (II) hydroxide.

6 The K_{sp} expression for the carbonate of metal X is $K_{sp} = [X^{3+}]^2[CO_3^{2-}]^3$. Write the equation for the dissolution equilibrium.

7 The K_{sp} value for lead (II) hydroxide is 8×10^{-16}. Calculate the solubility of lead (II) hydroxide in water at 25 °C.

Answers ➲ p. 56

CHAPTER SYLLABUS CHECKLIST

Are you able to answer every question from the syllabus for this chapter? Tick each question as you go through the checklist if you are able to answer it. If you cannot answer a question, turn to the relevant page in the study guide to find the answer. For NESA key word meanings, go to www.educationstandards.nsw.edu.au and search 'key words'.

FOR A COMPLETE UNDERSTANDING OF THIS TOPIC:		PAGE NO.	✓
1	Can I describe and analyse the processes involved in the dissolution of ionic compounds in water?	46	
2	Can I write equations for the equilibria in saturated solutions of ionic compounds in water?	46	
3	Can I explain how polar groups in molecular compounds assist their solubility in water?	48	
4	Can I describe experiments to determine the solubility rules of ionic compounds in water?	49	
5	Can I write net ionic equations for precipitation reactions?	49	
6	Can I derive equilibrium expressions for saturated solutions in terms of K_{sp}?	51	
7	Can I calculate the solubility of ionic compounds from their K_{sp} values?	52	
8	Can I predict the formation of a precipitate given the standard reference values for K_{sp}?	52	

HSC EXAM-TYPE QUESTIONS

Objective-response questions

(1 mark each)

1 Identify two electrolyte solutions that will form a precipitate on mixing.

A calcium nitrate + barium chloride

B calcium chloride + sodium carbonate

C silver nitrate + potassium nitrate

D sodium chloride + copper (II) nitrate

2 Potassium sulfate has a solubility of 12 g of salt per 100 g of water at 25 °C. 20 g of potassium sulfate crystals and 200 g of water are shaken together and allowed to stand in a stoppered flask. Select the statement that best describes the resulting system.

A The solution is saturated.

B A heterogeneous equilibrium is formed.

C An unsaturated solution forms.

D A homogeneous equilibrium is formed.

3 The K_{sp} expression for the solubility equilibrium involving iron (III) hydroxide in water is:

A $[Fe^{3+}][OH^-]$

B $[Fe^{3+}]^2[OH^-]^3$

C $[Fe^{3+}][OH^-]^3$

D $[Fe^{3+}][OH^-]^3/[Fe(OH)_3]$

4 The K_{sp} for calcium fluoride is 3.2×10^{-11}. Determine the concentration of calcium ions in a saturated aqueous solution.

A 2.0×10^{-4} mol/L

B 5.7×10^{-6} mol/L

C 3.2×10^{-4} mol/L

D 4.0×10^{-4} mol/L

5 Identify the salt that has the highest concentration of carbonate ions in its saturated aqueous solution.

A $PbCO_3$ ($K_{sp} = 7.4 \times 10^{-14}$)

B $SrCO_3$ ($K_{sp} = 5.4 \times 10^{-10}$)

C $CaCO_3$ ($K_{sp} = 3.36 \times 10^{-9}$)

D $MgCO_3$ ($K_{sp} = 2.0 \times 10^{-5}$)

Extended-response questions

6 An ionic compound XY_2 dissolves in water to form a divalent cation and a univalent anion. Figure 4.14 shows its solubility as a function of temperature.

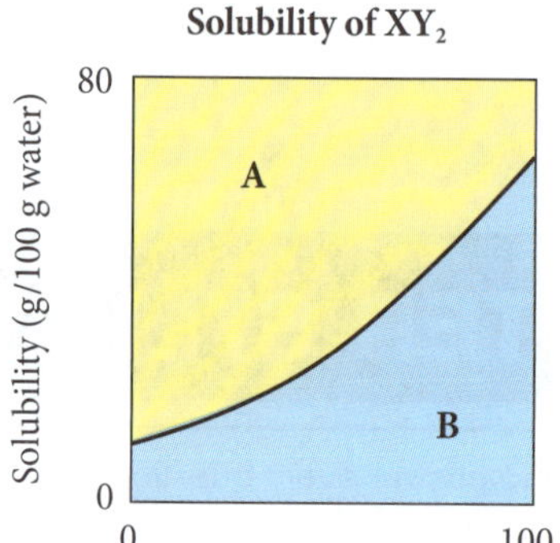

Figure 4.14 Solubility versus temperature

a Write the equation for the dissolution equilibrium. (1 mark)

b Two regions (A and B) are shown on the graph. Identify each region. (2 marks)

c Determine whether this dissolution is endothermic or exothermic. (1 mark)

7 a A test tube contains a solution of a sodium salt. Explain the chemical test(s) that a student could perform to determine whether the solution contains either chloride ions or sulfate ions. Include a relevant equation. (3 marks)

b A second test tube contains a solution of a nitrate salt. Explain the chemical test(s) that a student could perform to determine whether the solution contains either potassium ions or copper (II) ions. Include a relevant equation. (3 marks)

8 The K_{sp} value for silver cyanide (AgCN) is 2.2×10^{-16}.

a Write the equation for the solubility equilibrium for silver cyanide. (1 mark)

b Write the K_{sp} expression for silver cyanide. (1 mark)

c Calculate the solubility of silver cyanide at 25 °C. (3 marks)

9 The following table shows the solubility of aluminium sulfate in water as a function of temperature.

Temperature (°C)	20	30	40	50	60	70
Solubility (g/100 g H_2O)	36.4	40.4	45.8	52.2	59.2	66.2

a Plot a line graph of this data. (4 marks)

b Determine the solubility of aluminium sulfate at 25 °C. (1 mark)

c Write an equation for the solubility equilibrium that exists in a saturated solution of aluminium sulfate. (1 mark)

10 Potassium hydroxide solution is added dropwise to a blue solution of copper (II) nitrate until all the blue colour of the solution disappears and a blue precipitate of copper (II) hydroxide forms.

a Write an equation for the precipitation reaction. (1 mark)

b The precipitate was filtered out of the mixture and added to 100 mL (100 g) of water at 25 °C. The mixture was allowed to come to equilibrium.

Determine the solubility of the copper (II) hydroxide (K_{sp} for copper (II) hydroxide is 4.8×10^{-20}). (3 marks)

ANSWERS

KEY QUESTIONS

Key questions ➲ p. 51

1 $(NH_4)_2SO_4(s) \leftrightarrows 2NH_4^+(aq) + SO_4^{2-}(aq)$

2 $Pb^{2+}(aq) + 2Br^-(aq) \rightarrow PbBr_2(s)$

3 No precipitation occurs. Calcium chloride and sodium nitrate are both soluble.

4 $BaCl_2(aq) + Ag_2SO_4(aq) \rightarrow BaSO_4(s) + 2AgCl(s)$
$Ba^{2+}(aq) + SO_4^{2-}(aq) \rightarrow BaSO_4(s)$; $Ag^+(aq) + Cl^-(aq) \rightarrow AgCl(s)$

Key questions ➲ p. 53

5 $K_{sp} = [Co^{2+}][OH^-]^2$

6 $X_2(CO_3)_3(s) \leftrightarrows 2X^{3+}(aq) + 3CO_3^{2-}(aq)$

7 $Pb(OH)_2(s) \leftrightarrows Pb^{2+}(aq) + 2OH^-(aq)$
Let x mol of $Pb(OH)_2$ dissolve per litre.
At equilibrium: $[Pb^{2+}] = x$; $[OH^-] = 2x$
$K_{sp} = [Pb^{2+}][OH^-]^2 = (x)(2x)^2 = 4x^3 = 8 \times 10^{-16}$
Solve for x.
$x = 5.8 \times 10^{-6}$ mol/L
Solubility of $Pb(OH)_2 = 5.8 \times 10^{-6}$ mol/L

HSC EXAM-TYPE QUESTIONS

Objective-response questions

1 **B.** Calcium carbonate is insoluble. **A** is incorrect as both calcium chloride and barium nitrate are soluble. **C** is incorrect as both salts are nitrates. **D** is incorrect as sodium nitrate and copper (II) chloride are both soluble.

2 **C.** 20 g/200 g = 10 g/100 g, which is less than 12 g/100 g, thus the solution is unsaturated. **A** is incorrect as 12 g/100 g is required to produce saturation. **B** is incorrect as the solution is homogeneous and no equilibrium has formed. **D** is incorrect as no equilibrium forms.

3 **C.** The formula of iron (III) hydroxide is $Fe(OH)_3$. Therefore the K_{sp} expression will be $[Fe^{3+}][OH^-]^3$. **A** is incorrect as the hydroxide concentration has not been cubed. **B** is incorrect as the iron (III) ion concentration should not be squared. **D** is incorrect as no solids appear in the K_{sp} expression.

4 **A.** $CaF_2(s) \leftrightarrows Ca^{2+}(aq) + 2F^-(aq)$
$K_{sp} = [Ca^{2+}][F^-]^2 = 3.2 \times 10^{-11}$
Let x = concentration of calcium ions in the saturated solution.
$[x][2x]^2 = 4x^3 = 3.2 \times 10^{-11}$
$x = 2.0 \times 10^{-4}$ mol/L
Therefore **B, C** and **D** are mathematically incorrect.

5 **D.** This is the greatest K_{sp} value and so magnesium carbonate is the least insoluble. **A, B** and **C** are incorrect as the K_{sp} values are all smaller than that for magnesium carbonate.

Extended-response questions

6 EM Students need to interpret the zones of the graph in terms of unsaturated and saturated solutions; students need to relate increasing solubility with temperature in terms of the enthalpy change.
- a $XY_2(s) \leftrightarrows X^{2+}(aq) + 2Y^-(aq)$ ✓
- b A = saturated solution ✓
  B = unsaturated solution ✓
- c Endothermic (adding heat increases solubility) ✓

7 EM Students are required to recall and apply solubility rules to predict whether precipitation occurs; students must be able to write net ionic equations to score full marks.
- a Add drops of barium nitrate solution to a sample of the solution. If sulfate ions are present a white precipitate of barium sulfate will form. ✓
  If chloride ions are present there will be no precipitation. ✓
  $Ba^{2+}(aq) + SO_4^{2-}(aq) \rightarrow BaSO_4(s)$ ✓
- b Add drops of sodium hydroxide solution to a sample of the solution. If copper (II) ions are present a blue precipitate of copper (II) hydroxide will form. ✓
  If potassium ions are present there will be no precipitation. ✓
  $Cu^{2+}(aq) + 2OH^-(aq) \rightarrow Cu(OH)_2(s)$ ✓

8 EM Students need to examine the solubility equilibrium equation to create a K_{sp} expression; full marks are awarded if the steps of the calculation are shown.
- a $AgCN(s) \leftrightarrows Ag^+(aq) + CN^-(aq)$ ✓
- b $K_{sp} = [Ag^+][CN^-]$ ✓
- c Let x mol of AgCN dissolve per litre.
  At equilibrium: $[Ag^+] = x$; $[CN^-] = x$ ✓
  $K_{sp} = [Ag^+][CN^-] = (x)(x) = x^2 = 2.2 \times 10^{-16}$ ✓
  Solve for x.
  $x = 1.5 \times 10^{-8}$ mol/L
  Solubility of AgCN = 1.5×10^{-8} mol/L ✓

9 EM Students need to show skills in graphing and interpreting the graphical data. The graph must occupy at least 80% of the grid space and it must have a title. The line of best fit should be drawn.
- a Figure A4.1 represents the graph.

Solubility of aluminium sulfate versus temperature ✓

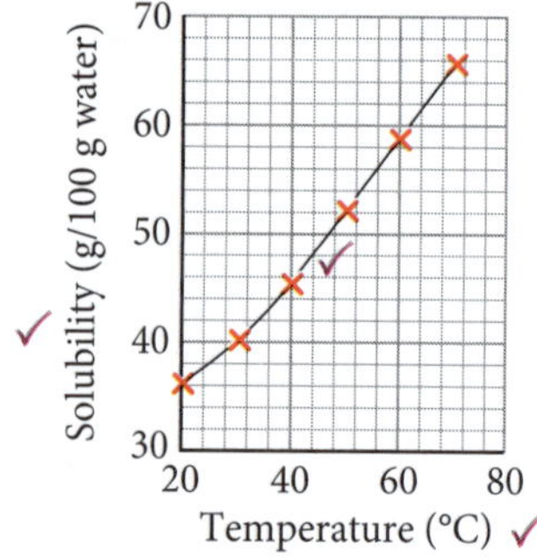

Figure A4.1 Solubility of aluminium sulfate

- b 38 g/100 g water ✓
- c $Al_2(SO_4)_3(s) \leftrightarrows 2Al^{3+}(aq) + 3SO_4^{2-}(aq)$ ✓

10 EM Students are required to predict the nature of the precipitate and to write a correct K_{sp} expression; all calculation steps must be shown for full marks.
- a $2KOH(aq) + Cu(NO_3)_2(aq) \rightarrow Cu(OH)_2(s) + 2KNO_3(aq)$ ✓
- b $K_{sp} = [Cu^{2+}][OH^-]^2$ ✓
  $Cu(OH)_2(s) \leftrightarrows Cu^{2+}(aq) + 2OH^-(aq)$
  Let x moles of $Cu(OH)_2$ dissolve per litre.
  At equilibrium: $[Cu^{2+}] = x$; $[OH^-] = 2x$
  $K_{sp} = (x)(2x)^2 = 4x^3 = 4.8 \times 10^{-20}$ ✓
  Solve for x.
  $x = 2.3 \times 10^{-7}$ mol/L
  Solubility of $Cu(OH)_2 = 2.3 \times 10^{-7}$ mol/L ✓

MODULE 6 ACID–BASE REACTIONS

CHAPTER 5 PROPERTIES OF ACIDS AND BASES

INQUIRY QUESTION:

What is an acid and what is a base?

In the Year 11 course we classified compounds in a variety of ways. Compounds could be classified as ionic, covalent molecular or covalent networks. Another way of classifying substances is to group them according to their ability to change the colour of a natural dye or indicator. This method was used to classify acids and bases. This method of classification was also related to other observable properties of these substances, such as their ability to react with materials such as active metals or minerals such as calcite. This chapter investigates the changing concepts of acids and bases.

1 Inorganic acids and bases

» Students investigate the correct IUPAC nomenclature and properties of common inorganic acids and bases.

Common acidic and basic substances

➔ Chemical substances can be classified as *acidic*, *basic* or *neutral*. This classification is usually based on the concentration of hydrogen ions (H^+) or hydronium ions (H_3O^+) produced when the substance is dissolved in water. Hydronium ions form when the hydrogen ion bonds to a water molecule by sharing one of the electron pairs of the oxygen atom. This type of bond is a type of covalent bond called a **coordinate bond**. Coordinate bonds in structural formulae are shown using an arrow, illustrated in Figure 5.1:

$$H^+(aq) + H_2O(l) \rightarrow H_3O^+(aq)$$

coordinate bond: a type of covalent bond in which one atom provides the shared electron bonding pair

➔ The concentration of hydrogen ions can be determined by the change in colour of natural or synthetic dyes called indicators or the use of electronic probes such as pH electrodes.

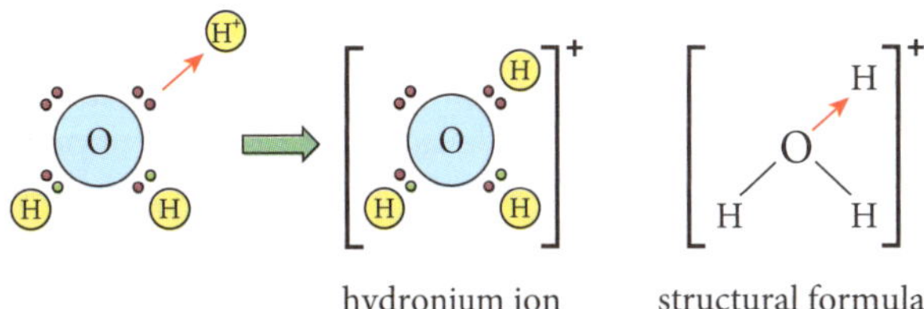

Figure 5.1 Formation of the hydronium ion

➔ Soluble acidic substances *increase* the hydrogen ion concentration above that found in pure water (1.0×10^{-7} mol/L at 25 °C).

➔ Soluble basic substances *decrease* the hydrogen ion concentration below that found in pure water. This is caused by the neutralisation of some of the hydrogen ions by ions such as hydroxide ions (OH^-) or carbonate ions (CO_3^{2-}). Solutions containing hydroxide ions are said to be **alkaline solutions**. Compounds such as NaOH and KOH are called **alkalis**. Alkalis are a subgroup of **bases**.

➔ Soluble neutral substances do not change the concentration of hydrogen ions normally present in pure water.

alkaline solution: solutions containing hydroxide ions
alkalis: a group of bases that contain hydroxide ions
base: substances that neutralise acids

➔ Table 5.1 lists some common acidic, basic and neutral substances. Figure 5.2 shows examples of common acidic and basic substances and their relative degrees of acidity or basicity.

Table 5.1 Common acidic, basic and neutral substances

Acidic substances	Basic substances	Neutral substances
vinegar	drain and oven cleaners	pure water
carbonated soft drinks	ammonia cleanser	sugar solution
lemon juice	baking soda	pure alcohol solution
car battery acid	washing soda	sodium chloride solution

Figure 5.2 Common acidic and basic substances

- Litmus is a water-soluble purple dye mixture extracted from a lichen such as *Rocella tinctoria*. The dye solution can be made blue or red by adding very small amounts of weak bases or weak acids to the purple extract. Red and blue litmus papers can then be prepared from the red or blue extracts by absorbing the extract onto absorbent paper strips. The strips are allowed to dry. These litmus papers can be used to determine whether unknown solutions are acidic or alkaline. Acidic solutions turn blue litmus red, and alkaline solutions turn red litmus blue, as shown in Figure 5.3.

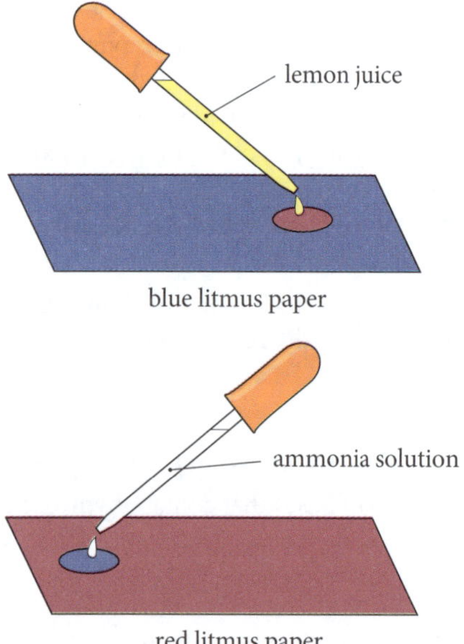

Figure 5.3 Using litmus paper to test for acidity or alkalinity

- Not all acidic and basic substances are water soluble. Although inorganic acids are soluble in water many organic acids (e.g. dodecanoic acid, $C_{11}H_{23}COOH$) are not water soluble yet they are classified as acids because they can be neutralised by aqueous solutions of strong bases such as sodium hydroxide.
- Basic oxides such as copper (II) oxide and iron (III) oxide are also insoluble in water. These insoluble bases can be neutralised by aqueous solutions of strong acids such as sulfuric acid.
- When bases react with acids the properties of each are destroyed or neutralised. The solution which is formed contains a salt of the acid and base. Salts are ionic compounds. Soluble salts can be recovered from the solution by evaporation and crystallisation.

Laboratory acids and bases

- In a chemistry laboratory there are a variety of common acidic and basic substances. Concentrated acids are quite dangerous and cannot be used by students. Trained laboratory staff dilute these acids to produce solutions with typical concentrations of 2.0 mol/L and 1.0 mol/L (Figure 5.4). Most bases (except ammonia gas) are solids. Solid bases are dissolved in water to produce solutions with similar concentrations to the diluted acids. Dilute ammonia solutions are prepared from concentrated ammonia solution.

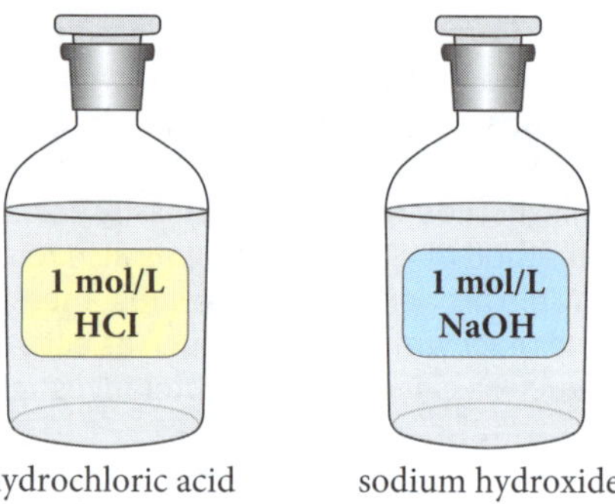

Figure 5.4 Dilute laboratory acid and base

Common acids

- HCl—hydrochloric acid
- HNO_3—nitric acid
- H_2SO_4—sulfuric acid
- CH_3COOH—acetic acid (ethanoic acid) (present in vinegar)
- H_3PO_4—phosphoric acid

Acetic acid and phosphoric acid are classified as weak acids.

Common bases

- NaOH—sodium hydroxide
- KOH—potassium hydroxide
- $Ca(OH)_2$—calcium hydroxide
- $NH_3(aq)$—ammonia solution
- Na_2CO_3—sodium carbonate
- $NaHCO_3$—sodium hydrogen carbonate (baking soda)

Ammonia, sodium carbonate and sodium hydrogen carbonate are classified as weak bases.

Properties of acids and bases

- Table 5.2 compares the properties of acidic and basic solutions. The taste test applies only to very dilute solutions or weak acids and bases so that no damage is done to the tongue. For example, lemon juice and a solution of sodium hydrogen carbonate (baking soda) can be safely tested.

Table 5.2 Properties of acidic and basic solutions

Acidic solutions	Basic (alkaline) solutions
Dilute acid solutions taste sour	Dilute alkaline solutions taste bitter
Aqueous acid solutions turn blue litmus red	Alkaline solutions turn red litmus blue
Neutralise bases	Neutralise acids
React with and dissolve active metals with the release of a gas (*)	Generally do not react with active metals (#)
React with carbonates with the evolution of carbon dioxide gas	Do not react with carbonates
Acid solutions are **electrolytes**	Solutions of soluble bases are electrolytes

* Dilute HCl and H_2SO_4 liberate hydrogen gas on reaction with active metals (metal + acid → salt + hydrogen).
* Dilute HNO_3 liberates nitric oxide (NO) gas rather than hydrogen when it reacts with many metals.
# A few metals such as zinc and aluminium will react with strong bases to form salts and hydrogen gas.

electrolyte: a substance that on dissolution in water produces a solution that conducts electricity due to the formation of mobile ions

2 Indicators

» Students conduct an investigation to demonstrate the preparation and use of indicators as illustrators of the characteristics and properties of acids and bases and their reversible reactions.

Acid–base indicators

- Acid–base indicators are natural or synthetic chemical dyes that change colour as the acidity of the surroundings change. Indicators are organic molecules that can change colour as the concentration of acid or base changes.
- Litmus is a natural indicator. The colour of litmus varies in different solutions. In neutral solutions the litmus is purple. In acid solutions it is red and in alkaline solutions it is blue.
- Industrial chemists have manufactured synthetic indicators. Litmus (a natural indicator) and some common synthetic indicators used in the school laboratory are listed in Table 5.3.

Table 5.3 Common acid–base indicators

Indicator	High acidity	Low acidity	Neutral	Low alkalinity	High alkalinity
Methyl orange	red	orange to yellow	yellow	yellow	yellow
Litmus	red	red	purple	blue	blue
Bromothymol blue	yellow	yellow	green	blue	blue
Phenolphthalein	colourless	colourless	colourless	pink	crimson

- Acidity in water solutions is caused by hydronium ions whereas alkalinity is caused by hydroxide ions. Approximate ranges:
  - High acidity: $[H_3O^+] > 10^{-3}$ mol/L; low acidity: $[H_3O^+] = 10^{-4} - 10^{-7}$ mol/L
  - Neutral: $[H_3O^+] = [OH^-] = 10^{-7}$ mol/L
  - High alkalinity: $[OH^-] > 10^{-3}$ mol/L; low alkalinity: $[OH^-] = 10^{-4} - 10^{-7}$ mol/L
- The pH scale is another measure of acidity. The scale is shown in Figure 5.5.

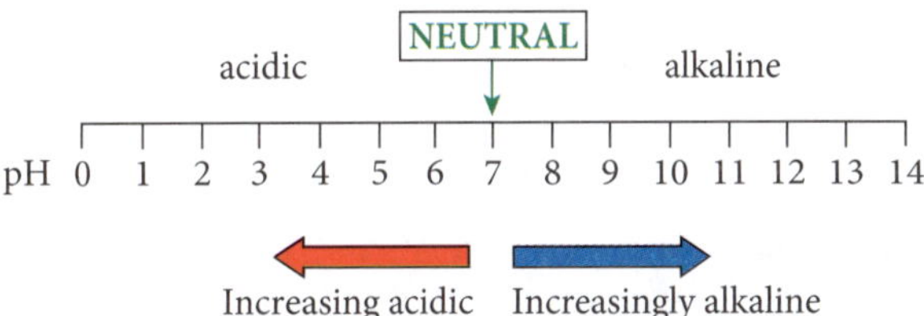

Figure 5.5 A pH scale

- If the pH of a solution is less than 7 then the solution is acidic.
- The lower the pH below 7 then the more acidic is the solution.
- If the pH is greater than 7 then the solution is alkaline.
- The greater the pH above 7 the more alkaline is the solution.

Calculations for pH are investigated in Chapter 6.

- Rain can become acidic due to gaseous air pollutants such as sulfur dioxide, sulfur trioxide and nitrogen dioxide. Such rain is called **acid rain**. These pollutants produce acids on dissolving in water. Sulfur dioxide produces sulphurous acid (H_2SO_3). Sulfur trioxide forms sulfuric acid (H_2SO_4) and nitrogen dioxide forms a mixture of nitrous acid (HNO_2) and nitric acid (HNO_3).

acid rain: rain that has a pH less than 5

➔ Universal indicator (UI) is a convenient mixed indicator. It produces a rainbow of colours that can be associated with different levels of acidity and alkalinity and the pH scale. The actual colours vary between commercial brands so students should use the colour card provided. Table 5.4 shows the colours of universal indicator in solutions of varying degrees of acidity or alkalinity.

EXAMPLE 1

A sample of milk was tested with three indicators. The results are shown below. What conclusion can be made about the acidity of milk?

- **methyl orange: yellow**
- **bromothymol blue: yellow-green**
- **phenolphthalein: colourless**

Relate the colours to Table 5.3

Answer:

The milk is very weakly acidic. This is best shown by the colour of the bromothymol blue indicator. The pH of milk is approximately 6.6.

EXAMPLE 2

A synthetic indicator called phenol red has the following colours and pH ranges:

yellow: pH < 6.8; orange: pH 6.8–8.4; red: pH > 8.4

a **Pure water was tested with phenol red indicator. Predict the colour of this indicator in pure water.**

b **Drops of a colourless solution were added to the water and the mixture tested with a phenol red indicator. The indicator turns yellow. Identify which of the following substances would produce this colour change with phenol red?**

ammonia solution; baking soda solution; lemon juice

Review the acid–base properties of common household substances

Answer:

a Pure water is neutral and has a pH of 7. Phenol red is orange at that pH level.

b Yellow indicates the pH is less than 6.8. Therefore the unknown solution would be lemon juice, which is acidic.

➔ The colour changes of indicators are reversible. For example, if bromothymol blue is added to water the colour of the indicator is green. If drops of hydrochloric acid are added, the indicator turns yellow. If drops of sodium hydroxide solution are now added the indicator reverts to a green colour and then blue if more NaOH solution is added. Finally if drops of HCl are now added to the blue solution, the indicator turns green when the base is neutralised (see Figure 5.6).

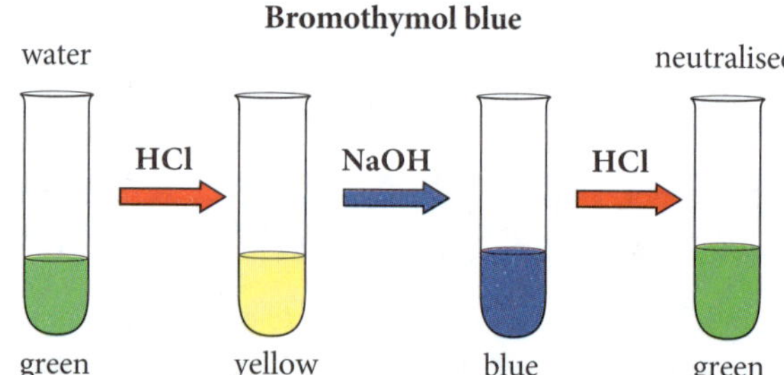

Figure 5.6 Bromothymol blue colour changes are reversible

FIRSTHAND INVESTIGATION 1

Preparing and testing a natural indicator

» Students conduct an investigation to demonstrate the preparation and use of indicators as illustrators of the characteristics and properties of acids and bases and their reversible reactions.

Aims

1 to prepare a natural indicator using red (or purple) cabbage leaves and to determine the colour of the indicator in different pH solutions

2 to determine the approximate pH of common household chemicals

Table 5.4 Universal indicator colours

Acidity/basicity level	Strongly acidic pH 0–3	Moderately acidic pH 4–5	Weakly acidic pH 6	Neutral pH 7	Weakly alkaline pH 8	Moderately alkaline pH 9–10	Strongly alkaline pH 11–14
UI colour	red	orange	yellow	green	greenish-blue	bluish-violet	violet

Method

1 Red cabbages vary in colour from red to purple. Strip some leaves from the cabbage and shred or grind them in a mortar. Add the shredded leaves to a beaker and just cover them with water (see Figure 5.7). Add heat to extract some dye into the water. Heat for 10 minutes without boiling. Allow the extract to cool and decant it into a clean beaker. The cabbage extract will be a purplish colour but this will vary depending on the variety or kind of cabbage.

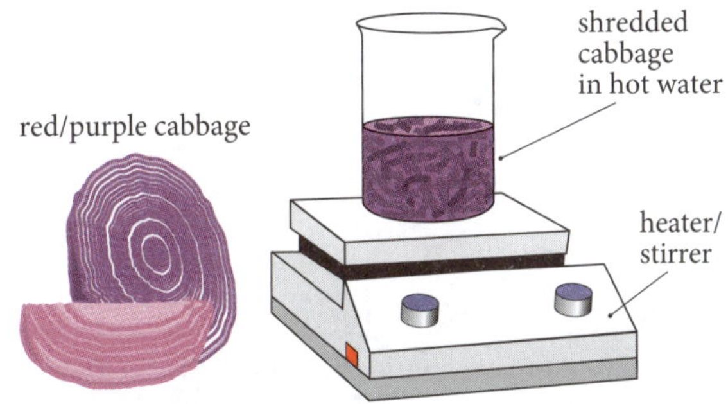

Figure 5.7 Cabbage extract

2 You are provided with two solutions of known pH:
Solution A = 0.010 mol/L HCl (pH = 2)
Solution B = 0.010 mol/L NaOH (pH = 12)

- Dilute solution A to obtain solutions with pH levels of 4 and 6
- Dilute solution B to obtain solutions with pH levels of 10 and 8

Prepare the dilutions as follows:

- Solution A1 (pH 4): 10 mL of solution A + 190 mL water
- Solution A2 (pH 6): 10 mL of solution A1 + 190 mL water
- Solution B1 (pH 10): 10 mL of solution B + 190 mL water
- Solution B2 (pH 8): 10 mL of solution B1 + 190 mL water

If a pH electrode is available you can check the accuracy of your dilutions.

3 Place 5 mL of each acid and base solution in six separate test tubes.

4 Add 2 mL of the cabbage indicator to each tube and record the colours that the indicator turns, comparing the colours with the original extract.

5 Place the cabbage indicator in a small beaker and test its pH with a calibrated pH meter. If the pH is less than 7, then add drops of sodium hydrogen carbonate solution until the pH is 7. Record the colour of the cabbage indicator at pH 7.

6 Some common household chemicals can be tested. Test your indicator with milk of magnesia, potato juice, lemon juice and ammonia cleaner

Sample results

1 Figure 5.8 shows typical results for the change of colour of the cabbage indicator at different pH levels.

Red cabbage indicator

	red	pink–purple	violet	blue–violet	blue	green	yellow
pH	2	4	6	7	8	10	12

Figure 5.8 Cabbage indicator colours as a function of pH

2 The following table shows the results of experiments in which common household substances were tested with the cabbage extract.

Solutions	Milk of magnesia	Potato juice	Lemon juice	Ammonia cleaner solution
Colour of cabbage extract	green	pale violet	reddish-purple	green-yellow
Estimated pH	~10	~5–6	~2–3	~11–12

Analysis

- Milk of magnesia (an antacid medication used for acid reflux) is weakly alkaline.
- Potato juice is very weakly acidic.
- Lemon juice is moderately acidic.
- Ammonia cleaner solution is strongly alkaline.

Conclusion

The cabbage indicator is a useful indicator as it changes colour in solutions of different pH. The colours were determined in known solutions of acids and bases and these results were applied to common household chemicals to determine their acidity or alkalinity.

EXAMPLE 3

Flower petals were shredded and the dye pigment was extracted into water. The extract was yellow in colour. The following table shows the effect of adding drops of HCl and NaOH solution to the yellow extract. Figure 5.9 shows the colour of the flower petal indicator in acidic, alkaline and neutral solutions.

Test	Observation
Hydrochloric acid added	Turned orange
Sodium hydroxide solution added	Turned green
Water added	Remained yellow

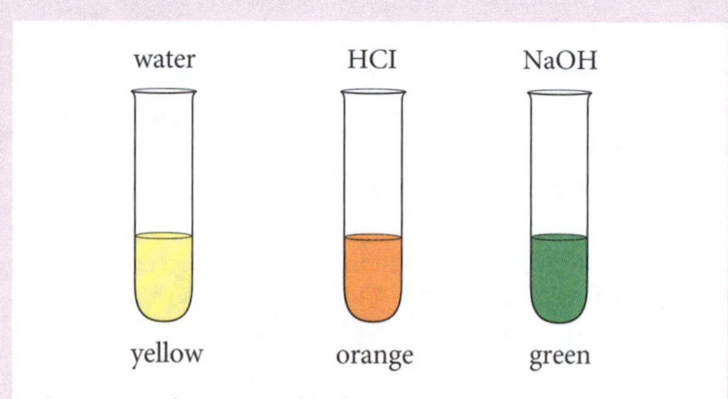

Figure 5.9 Flower petal indicator

a Identify the colour of the extract in acidic and alkaline solutions.

b Predict the colour of the extract in lemon-flavoured soft drink.

c Predict the colour of the extract in washing soda (sodium carbonate) solution.

d Predict the colour of the extract in sugar solution.

Revise the classification of household substances

Answer:

a Orange in acidic solution; green in alkaline solution

b Orange or orange-yellow because lemon-flavoured soft drink contains weak carbonic acid (H_2CO_3) and lemon juice, which contains weak citric acid

c Yellow-green because washing soda solution is moderately alkaline

d Yellow because a sugar solution is neutral

→ KEY QUESTIONS

1 Classify the following substances as acidic, basic or neutral:
- a baking soda
- b solution of alcohol in water
- c orange juice

2 State the colour of red litmus papers placed in the following solution:
- a salt water
- b lemonade
- c ammonia cleanser

3 Classify the following compounds as acids or bases:
- a $Ba(OH)_2$
- b HNO_3
- c CH_3COOH

4 State the colour of phenolphthalein indicator in:
- a sulfuric acid
- b sodium hydroxide solution
- c pure water

Answers ➲ p. 72

3 Reactions of acids with bases, carbonates and metals

» Students predict the products of acid reactions and write balanced equations to represent acids and bases, acids and carbonates, and acids and metals.

Acid–base reactions

→ When an acidic solution reacts with a base, such as a metal oxide or metal hydroxide, a **neutralisation** reaction occurs. The products of such acid–base reactions are water and a **salt**.

ACID + BASE → SALT + WATER

neutralisation: the reaction between an acid and a base to produce a new compound that usually has neither acidic or basic properties

salt: the ionic compound formed when a base is neutralised by an acid

Acid–base reaction 1

- Magnesium oxide solid + dilute sulfuric acid

  Magnesium oxide is basic and is neutralised by the sulfuric acid to form magnesium sulfate and water. The white magnesium oxide solid gradually dissolves and the magnesium sulfate product is water soluble. The sulfuric acid solution contains acidic hydronium ions that react with the magnesium oxide. The sulfate ions are spectator ions.

  Whole formula equation:

  $$MgO(s) + H_2SO_4(aq) \rightarrow MgSO_4(aq) + H_2O(l)$$

  Net ionic equation:

  $$MgO(s) + 2H_3O^+(aq) \rightarrow Mg^{2+}(aq) + 2H_2O(l)$$

Acid–base reaction 2

- Iron (III) hydroxide solid + dilute hydrochloric acid

  The iron (III) hydroxide is basic and is neutralised by the hydrochloric acid to form iron (III) chloride and water. The brown iron (III) hydroxide solid gradually dissolves and a solution of yellow iron (III) chloride forms. The hydrochloric acid solution contains acidic hydronium ions that neutralise the iron (III) hydroxide. Figure 5.10 shows the neutralisation reaction of iron (III) hydroxide and hydrochloric acid.

  Whole formula equation:

  $$Fe(OH)_3(s) + 3HCl(aq) \rightarrow FeCl_3(aq) + 3H_2O(l)$$

  Net ionic equation:

  $$Fe(OH)_3(s) + 3H_3O^+(aq) \rightarrow Fe^{3+}(aq) + 6H_2O(l)$$

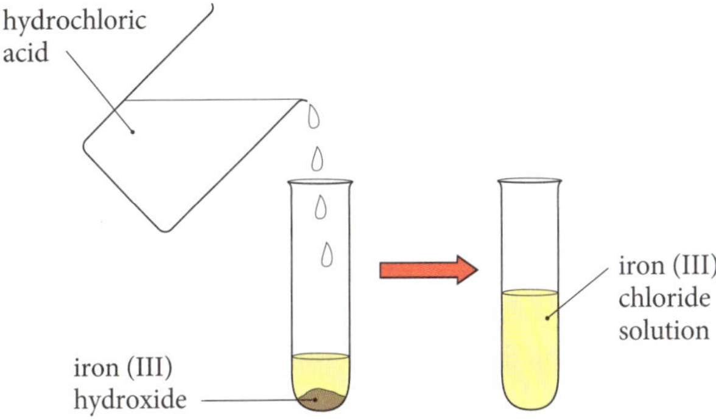

Figure 5.10 Hydrochloric acid neutralises iron (III) hydroxide

Acid–carbonate reactions

➔ When an acidic solution reacts with a metal carbonate, a neutralisation reaction occurs. The products of such acid–carbonate reactions are carbon dioxide, water and a salt.

ACID + CARBONATE → SALT + WATER + CARBON DIOXIDE

Acid–base reaction 3

- Nickel (II) carbonate solid + dilute nitric acid
  The nickel (II) carbonate is neutralised by the dilute nitric acid to form nickel (II) nitrate, carbon dioxide and water. The green solid gradually dissolves and a solution of green nickel (II) nitrate forms. Figure 5.11 shows the neutralisation reaction of nickel (II) carbonate and nitric acid. Note the effervescence as carbon dioxide gas is evolved.

  Whole formula equation:

  $$NiCO_3(s) + 2HNO_3(aq) \rightarrow Ni(NO_3)_2(aq) + H_2O(l) + CO_2(g)$$

  Net ionic equation:

  $$NiCO_3(s) + 2H_3O^+(aq) \rightarrow Ni^{2+}(aq) + CO_2(g) + 3H_2O(l)$$

  The nitrate ions in the nitric acid are spectator ions. Spectator ions are not involved in the chemical reaction and are not shown in net ionic equations.

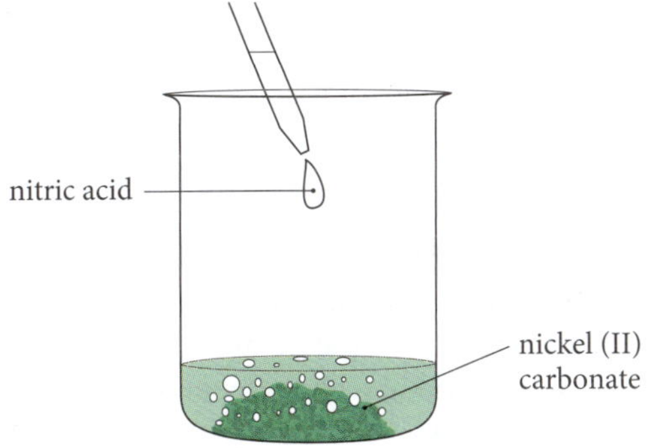

Figure 5.11 Dilute nitric acid neutralises nickel (II) carbonate

EXAMPLE 4

Cobalt (II) carbonate is a red solid that is insoluble in water. It dissolves in dilute sulfuric acid to produce a reddish-pink solution of a salt, water and carbon dioxide gas.

a **Identify the salt formed.**

b **Explain how the salt can be recovered from the solution.**

c **Identify a test for carbon dioxide and describe the observations for a positive test.**

d **Write a whole formula equation for the acid–carbonate reaction.**

e **Write the net ionic equation.**

Eliminate the spectator ions to create the net ionic equation

Answer:

a Cobalt (II) sulfate. Salts of sulfuric acid are called *sulfates*.

b Evaporate the solution slowly in an evaporating basin to allow large crystals to form. These crystals are hydrated cobalt (II) sulfate ($CoSO_4 \cdot 7H_2O$).

c The limewater test is used. Limewater is a colourless solution of calcium hydroxide. As carbon dioxide bubbles through the limewater, the limewater turns milky-white as a precipitate of calcium carbonate forms.

d $CoCO_3(s) + H_2SO_4(aq) \rightarrow CoSO_4(aq) + H_2O(l) + CO_2(g)$

e $CoCO_3(s) + 2H_3O^+(aq) \rightarrow Co^{2+}(aq) + 3H_2O(l) + CO_2(g)$

(The sulfate ions are spectator ions and are not shown in the net ionic equation.)

Acid–metal reactions

➔ The reaction of active metals with dilute acids is an example of an oxidation–reduction reaction rather than a neutralisation reaction.

➔ Dilute hydrochloric acid and dilute sulfuric acid react with active metals and hydrogen gas is released. Bubbles of colourless hydrogen gas are observed on the metal's surface. The reactive metals will dissolve in the excess acid to form solutions of the metal salt. The hydrogen ion (or hydronium ion) is the **oxidant** in these examples. The metal is the **reductant**.

oxidant: a substance that accepts electrons from another substance called the reductant

reductant: a substance that donates electrons to another substance called the oxidant

Acid–metal reaction

- Magnesium + dilute sulfuric acid

  Whole formula redox equation:

  $$Mg(s) + H_2SO_4(aq) \rightarrow FeSO_4(aq) + H_2(g)$$

  Net ionic equation:

  $$Mg(s) + 2H_3O^+(aq) \rightarrow Mg^{2+}(aq) + H_2(g) + 2H_2O(l)$$

  The hydrogen gas can be collected in an inverted test tube and identified using the 'pop' test. A mixture of hydrogen and air will explode when ignited with a flame from a wax taper.

- In general, dilute nitric acid reacts with metals and nitric oxide gas rather than hydrogen gas is released. In these cases the nitrate ion is the oxidant rather than the hydronium ion. Dilute nitric acid will even react with copper whereas dilute HCl and dilute H_2SO_4 will not.

  Whole formula redox equation:

  $$3Cu(s) + 8HNO_3(aq) \rightarrow 3Cu(NO_3)_2(aq) + 2NO(g) + 4H_2O(l)$$

  Net ionic equation:

  $$3Cu(s) + 8H_3O^+(aq) + 2NO_3^-(aq) \rightarrow 3Cu^{2+}(aq) + 2NO(g) + 12H_2O(l)$$

4 Applications of neutralisation reactions

» Students investigate applications of neutralisation reactions in everyday life and industrial processes.

- Many neutralisation processes occur in our everyday lives as well as in chemical industries.

Neutralising excess stomach acidity

- Our stomach contains gastric juices, which are required in the digestive process. Hydrochloric acid is a major component of gastric juice.
- Some people suffer from gastric reflux in which gastric juices rise up the oesophagus and irritate it, causing them pain.
- Neutralisation can be used to reduce the symptoms of acid reflux. Commercial tablets or suspensions called *antacids* are used to neutralise excess acid. These antacids contain weak bases such as magnesium hydroxide, aluminium hydroxide and sodium hydrogen carbonate:

  $$Mg(OH)_2(s) + 2H_3O^+(aq) \rightarrow Mg^{2+}(aq) + 4H_2O(l)$$

  $$Al(OH)_3(s) + 3H_3O^+(aq) \rightarrow Al^{3+}(aq) + 6H_2O(l)$$

  $$NaHCO_3(s) + H_3O^+(aq) \rightarrow Na^+(aq) + 2H_2O(l) + CO_2(g)$$

- Effervescent antacid tablets contain sodium hydrogen carbonate and citric acid. When these tablets are added to water the weak acid and base undergo dissolution and an acid–carbonate neutralisation reaction occurs with the release of carbon dioxide. Such products are often used to relieve indigestion symptoms:

  $$3NaHCO_3(aq) + C_3H_5O(COOH)_3(aq) \rightarrow Na_3C_3H_5O(COO)_3(aq) + 3H_2O(l) + 3CO_2(g)$$

Adjusting swimming pool acidity

- Sodium hypochlorite (NaOCl) is added to swimming pools to kill dangerous microbes such as bacteria. The hypochlorite ion reacts with water to produce unstable hypochlorous acid (HOCl) and hydroxide ions. It is the HOCl that kills the microbes:

  $$OCl^-(aq) + H_2O(l) \rightleftharpoons HOCl(aq) + OH^-(aq)$$

- As a result of the reaction, the pool becomes alkaline due to the production of hydroxide ions (OH^-). The hydroxide ions irritate the skin and eyes and so they need to be neutralised. Hydrochloric acid is then added in small amounts to return the water to near neutrality:

  $$H_3O^+(aq) + OH^-(aq) \rightarrow 2H_2O(l)$$

- Phenol red indicator is commonly used in swimming pool test kits (see Figure 5.12). When the indicator turns orange-pink in a sample of the pool water then the pool water is near neutrality.

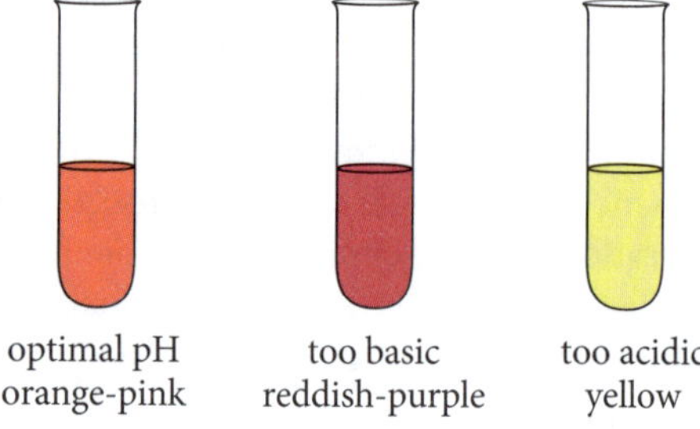

Figure 5.12 Testing swimming pool water with phenol red indicator

Adjusting soil acidity

- Optimal plant growth depends on the acidity or pH of soils. Camellias, for example, grow best in moderately acidic soils (pH = 4.5; $[H^+] = 3.16 \times 10^{-5}$ mol/L) whereas chrysanthemums grow best in mildly acidic soils (pH = 6; $[H^+] = 1.0 \times 10^{-6}$ mol/L). Soil pH can be monitored and adjusted to produce optimal conditions for different flower or vegetable crops. Indicators can be used as part of this monitoring and adjustment process.

Increasing soil acidity

- This can be achieved in a number of ways, including adding decaying organic leaf matter containing organic acids, or adding weakly acidic compounds such as ammonium nitrate or ammonium sulfate, which are found in many commercial fertilisers. These substances dissolve in the soil water to increase the soil acidity.

Decreasing soil acidity

- This can be achieved by adding weakly basic substances that neutralise excess soil acidity. Adding powdered limestone ($CaCO_3$) to an acidic soil reduces the acidity of the soil water as calcium carbonate is a weak base and a neutralisation reaction occurs:

$$CaCO_3(s) + 2H_3O^+(aq) \rightarrow Ca^{2+}(aq) + 3H_2O(l) + CO_2(g)$$

A universal indicator can be used to monitor the acidity of soil water during these adjustments.

- The following method can be used to monitor soil pH:
  1. Saturate the collected soil sample with distilled water.
  2. Add some white barium sulfate powder to the surface of the wet soil. Allow the soil water to soak into the barium sulfate.
  3. Add drops of universal indicator to the wet barium sulfate and note the colour observed (Figure 5.13).
  4. Use a universal indicator colour chart to determine the acidity (pH) of the soil water.

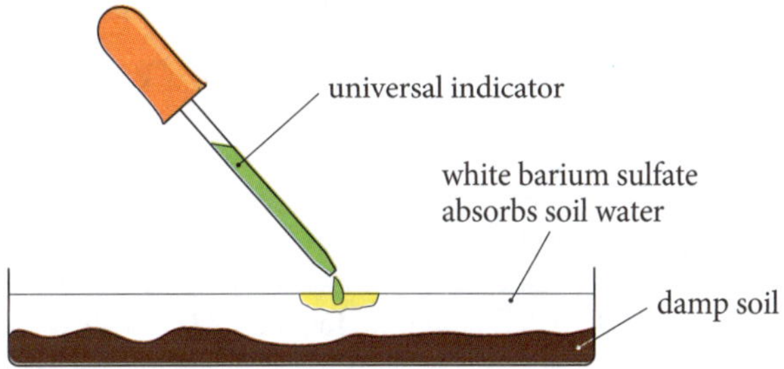

Figure 5.13 Determining soil acidity

The neutral, insoluble barium sulfate absorbs the soil water and provides a white coloured background to observe changes in the indicator colour. If the indicator turns orange then the pH is about 5 ($[H^+] = 1.0 \times 10^{-5}$ mol/L). If the indicator turns yellow then the pH is about 6 ($[H^+] = 1.0 \times 10^{-6}$ mol/L).

Monitoring the basicity of commercial window cleaner

- The concentration of hydroxide ions in a commercial ammonia window cleaner can be determined by reaction with a standard hydrochloric acid solution using methyl orange indicator. These tests are performed by industrial chemists to ensure the quality of the product.

EXAMPLE 5

A window cleaner contains dissolved ammonia. An industrial analytical chemist dilutes 25.00 mL of the window cleaner until its total volume is 250.00 mL. Using methyl orange indicator, 25.00 mL of the diluted cleaner was found to be neutralised by 37.20 mL of 0.200 mol/L HCl.

a **Write the neutralisation equation.**

b **Calculate the ammonia concentration in the diluted solution.**

c **Calculate the concentration of ammonia in the original window cleaner.**

Ensure the dilution factor is taken into account

Answer:

a Neutralisation equation:

$$NH_3(aq) + HCl(aq) \rightarrow NH_4Cl(aq)$$

b $n(HCl) = cV = (0.200)(0.03720) = 0.00744$ mol

$n(NH_3) = n(HCl) = 0.00744$ mol

Diluted cleaner: $c(NH_3) = n/V = \dfrac{0.00744}{0.0250} = 0.298$ mol/L

c Dilution equation: $c_1V_1 = c_2V_2$

$c_1(0.0250) = (0.298)(0.250)$

$c_1 = 2.98$ mol/L

In the original cleaner the ammonia concentration is 2.98 mol/L.

Neutralisation in the Solvay process

- The Solvay process is an industrial process used to manufacture sodium carbonate. Sodium carbonate is used in the manufacture of glass, soaps and detergents.
- One step in the Solvay process is the reaction of carbonic acid (H_2CO_3) with an ammoniated solution of salt water.
  - The carbonic acid forms when carbon dioxide dissolves in the salt water which is saturated with ammonia.
  - The neutralisation reaction occurs at very cold temperatures and high salt concentration to promote the crystallisation of sodium hydrogen carbonate.

    The neutralisation reaction involves the carbonic acid neutralising the ammonia:

    $$H_2CO_3(aq) + NH_3(aq) \leftrightharpoons HCO_3^-(aq) + NH_4^+(aq)$$
  - The crystallisation reaction is $Na^+(aq) + HCO_3^-(aq) \leftrightharpoons NaHCO_3(s)$. The sodium carbonate is formed by the thermal decomposition of sodium hydrogen carbonate.

KEY QUESTIONS

5. **Write a balanced equation for the reaction of dilute nitric acid and solid sodium hydrogen carbonate. Explain the test that can be used to identify the gaseous product.**
6. **Zinc dissolves in dilute nitric acid and nitric oxide is liberated. Write a net ionic equation for zinc reacting with dilute nitric acid.**
7. **Barium sulfate and universal indicator are used to test the acidity of the wet soil. State reasons as to why barium sulfate is used in this experiment.**
8. **Explain why NaOH is not suitable as an antacid to treat excess stomach acidity.**

Answers ➲ p. 72

5 Enthalpy of neutralisation

» Students conduct a practical investigation to measure the enthalpy of neutralisation.

Comparing enthalpies of neutralisation

➔ Neutralisation reactions are exothermic. The amount of heat liberated depends on whether the acid or base is strong or weak and their concentrations. Table 5.5 shows the enthalpies of neutralisation of some acids and bases. Strong acid–strong base neutralisation (HCl/NaOH) produces the greatest amount of heat. The neutralisation between weak acids and bases (CH_3COOH/NH_3) produces the least amount of heat.

Table 5.5 Enthalpies of neutralisation

Neutralisation reaction (using 1 mol/L solutions at 25 °C)	ΔH (kJ/mol)
$HCl(aq) + NaOH(aq) \rightarrow NaCl(aq) + H_2O(l)$	-57.1
$CH_3COOH(aq) + NaOH(aq) \rightarrow NaCH_3COO(aq) + H_2O(l)$	-56.1
$HCl(aq) + NH_3(aq) \rightarrow NH_4Cl(aq)$	-53.4
$CH_3COOH(aq) + NH_3(aq) \rightarrow NH_4CH_3COO(aq)$	-50.4

➔ The amount of heat liberated in a neutralisation reaction can be measured using a polystyrene calorimeter, which is an excellent heat insulator. The heat liberated by the neutralisation reaction is absorbed by the solution, which rises in temperature.

FIRSTHAND INVESTIGATION 2

Enthalpy of neutralisation

» Students conduct a practical investigation to measure the enthalpy of neutralisation.

Aim

to measure the enthalpy of neutralisation when nitric acid neutralises sodium hydroxide solution

Apparatus

Figure 5.14 illustrates the apparatus used in this investigation.

Method

1 Allow the acid and base solutions to come to room temperature.
2 Use a pipette to deliver 25.0 mL (= 25.8 g = 0.0258 kg) of 1.00 mol/L nitric acid to the calorimeter.
3 Measure the initial temperature of the acid.
4 Use a pipette to deliver 25.0 mL (= 26.0 g = 0.0260 kg) of 1.00 mol/L sodium hydroxide to the acid in the calorimeter.
5 Stir and record the highest temperature reached.
6 Repeat the experiment to improve reliability.

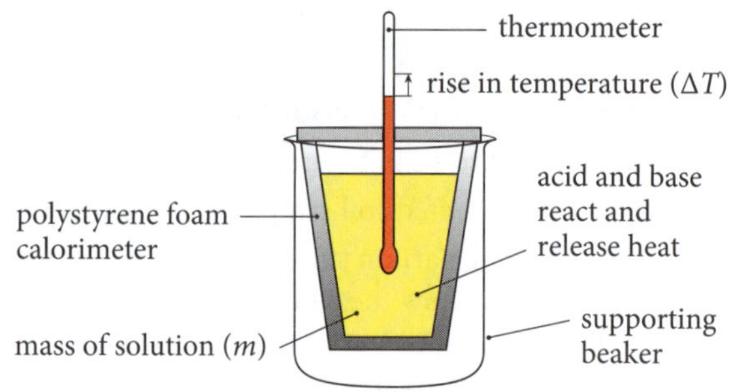

Figure 5.14 Neutralisation calorimetry experiment

Sample results

Initial temperature (°C)	24.0
Final temperature (°C)	30.4

(Assume c = specific heat capacity = 4.18×10^3 J/kg/K)

Sample calculation

In this calculation we use the calorimetry formula ($q = mc\Delta T$) to calculate the heat liberated. The enthalpy change is calculated using the formula $\Delta H = -q/n$.

$\Delta T = 30.4 - 24.0 = 6.4$ °C

$m = 0.0258 + 0.0260 = 0.0518$ kg

$q = mc\Delta T = (0.0518)(4.18 \times 10^3)(6.4) = 1386$ J = 1.386 kJ

$n(HNO_3) = cV = (1.00)(0.0250) = 0.0250$ mol

$n(NaOH) = cV = (1.00)(0.0250) = 0.0250$ mol

$HNO_3(aq) + NaOH(aq) \rightarrow NaNO_3(aq) + H_2O(l)$

$$\Delta H = -q/n = \frac{-1.386}{0.0250} = -55.4 \text{ kJ/mol}$$

(Compare this result with the literature value for the standard heat of neutralisation of nitric acid and sodium hydroxide: −57.1 kJ/mol; the experimental result is 2.5% low. (This suggests that the experimental method is valid within the limits of experimental error.)

6 Overview of acid–base theories

» Students explore the changes in definitions and models of an acid and a base over time to explain the limitations of each model, including but not limited to Arrhenius's theory and Brønsted–Lowry theory.

The history of acids and bases

➔ Acids were originally defined in terms of their properties. They had a sour taste and were **corrosive** towards metals.

They also changed the colour of vegetable dyes such as litmus. Soluble bases tasted bitter and changed the colour of litmus. Various theories gradually emerged to try and explain these observations.

corrosive: a substance that damages other substances with which it comes into contact; in the case of metals the corrosive substance attacks the metal surface and gradually wears it away

Antoine Lavoisier (1776)

→ Lavoisier observed that oxides of non-metals generally produced acidic solutions when dissolved in water. Lavoisier believed that the presence of oxygen in non-metallic oxides (e.g. sulfur trioxide) caused the acidity in water. His theory was known as the *oxygen theory* of acidity:

$SO_3(g) + H_2O(l) \rightarrow H_2SO_4(aq)$

Lavoisier's theory could not explain why metal oxides were not acidic.

Humphry Davy (1810)

→ Davy observed that hydrogen chloride gas when dissolved in water produced an acidic solution (called hydrochloric acid). As HCl did not contain oxygen atoms then Lavoisier's theory did not apply to all acidic substances. He showed that hydrochloric acid could dissolve metals such as zinc and hydrogen gas was released. He believed that the zinc atoms replaced the hydrogen atoms in the acid. Davy concluded that *replaceable hydrogen atoms* were the cause of acidity in the hydrochloric acid solution:

$Zn(s) + 2HCl(aq) \rightarrow ZnCl_2(aq) + H_2(g)$

Davy's theory could not explain why other hydrogen compounds were not acidic.

Svante Arrhenius (1884)

→ Arrhenius observed that acidic solutions conducted electricity in a similar way to salt solutions. As a result of these observations, he proposed that acids released *hydrogen ions* when they dissolved in water. Litmus indicator turns red in acidic solutions.

- As alkaline solutions also conduct electricity Arrhenius also proposed that alkalinity was due to the presence of *hydroxide ions* in solution. Litmus turns blue in alkaline solutions.
- Arrhenius explained how acid solutions neutralised alkaline solutions in terms of the interaction of the aqueous hydrogen ions and hydroxide ions. The whole formula equation for the neutralisation of sodium hydroxide solution by hydrochloric acid is:

  $NaOH(aq) + HCl(aq) \rightarrow NaCl(aq) + H_2O(l)$

- However, Arrhenius's explanation of the neutralisation is shown by the net ionic equation:

  $OH^-(aq) + H^+(aq) \rightarrow H_2O(l)$

- The water produced is neutral. The sodium and chloride ions are neutral spectator ions (see Figure 5.15).

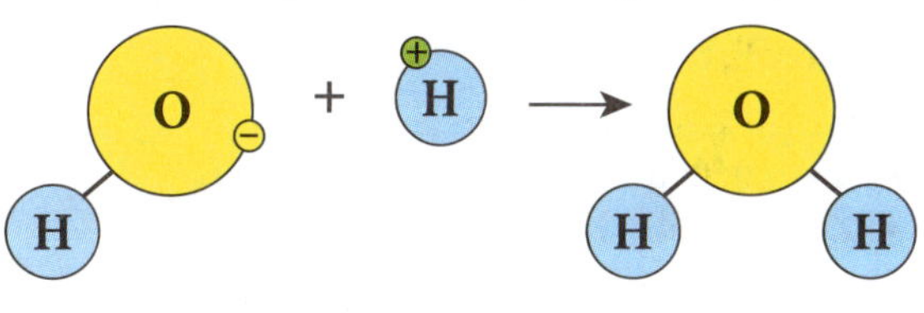

Figure 5.15 Neutralisation explained using Arrhenius's theory

- Arrhenius's theory also explained the difference between **strong acids**, such as hydrochloric acid (HCl), and **weak acids**, such as hydrofluoric acid (HF), as being due to their different degrees of dissociation in water.

strong acid: an acid that completely dissociates in water
weak acid: an acid that only partially dissociates in water

- A weak acid (HA) dissociates according to the following equilibrium:

  $HA(aq) \leftrightharpoons H^+(aq) + A^-(aq)$

- The degree of dissociation is defined as:

  Percentage dissociation = $[H^+]/[HA] \times 100/1\%$

  Thus a 1.0 mol/L solution of a weak acid that dissociates to produce a hydrogen ion concentration of 0.060 mol/L has a degree of dissociation of 6.0%:

  $$\text{Percentage dissociation} = \frac{(0.060)}{(1.0)} \times \frac{100}{1}\% = 6.0\%$$

- When hydrogen chloride molecules dissolve in water the molecules completely dissociate and ions are formed. The solution is called *hydrochloric acid*:

  $HCl(aq) \rightarrow H^+(aq) + Cl^-(aq)$

- However, when hydrogen fluoride molecules dissolve in water the molecules do not completely dissociate. An equilibrium is established between the unionised molecules and the ions in solution. The solution is called *hydrofluoric acid*:

  $HF(aq) \leftrightharpoons H^+(aq) + F^-(aq)$

- Bases can also be classified as *strong* or *weak*. **Strong bases** such as NaOH and KOH are ionic compounds that completely dissociate into free ions in water to produce high concentrations of hydroxide ions. **Weak bases** such as $NH_3(aq)$ incompletely dissociate in water and the concentration of hydroxide ions is low.

strong base: a base which completely dissociates in water
weak base: a base that only partially dissociates in water

- The previous discussion illustrates the important role of the water in Arrhenius's theory. Arrhenius's theory could not explain the basicity of oxides and carbonates. It could not explain why not all salts were neutral when dissolved in water.

EXAMPLE 6

Two acid solutions are prepared:

- **X—0.010 mol/L nitric acid**
- **Z—2.0 mol/L hydrofluoric acid**

Identify the statement that is true concerning these two solutions.

A X is a weak acid and more dilute than acid Z.
B X is a strong acid and more dilute than acid Z.
C X is a weak acid and less concentrated than acid Z.
D Z is a strong acid and more concentrated than acid X.

Distinguish between the terms strong and concentrated as well as weak and dilute

Answer:

B is correct. Nitric acid (X) is a strong acid and hydrofluoric acid (Y) is a weak acid. The 0.010 mol/L solution is more dilute than the 2.0 mol/L solution as X has a much lower concentration than Z.

Johannes Brønsted and Thomas Lowry (1923)

- Johannes Brønsted and Thomas Lowry (1923) independently developed a new theory of acids, which has since become known as the Brønsted–Lowry (B/L) theory. Their theory is more general than Arrhenius's theory and includes substances that neutralise acids but which are not necessarily alkalis. These substances are called *bases*. Their theory also applies to any solvent, whereas Arrhenius's theory is applicable only to water. The theory also applies to gaseous acid–base reactions.
- The definitions of an acid and base in this theory are:

B/L acid: a proton donor
B/L base: a proton acceptor

- Hydrogen ions (H^+) are protons as the valence electron of the hydrogen atom has been removed. These hydrogen ions associate with water to form hydronium ions (H_3O^+).
- Hydrochloric acid (HCl) is classified as a monoprotic acid as it can donate one proton to a base whereas H_2SO_4 is a diprotic acid as it can donate two protons to a base.

EXAMPLE 7

Figure 5.16 shows the reaction between hydrogen chloride gas and ammonia gas when test tubes containing fuming concentrated hydrochloric acid and fuming ammonia solution are placed next to each other. A white smoke consisting of fine crystals of ammonium chloride forms above the test tubes.

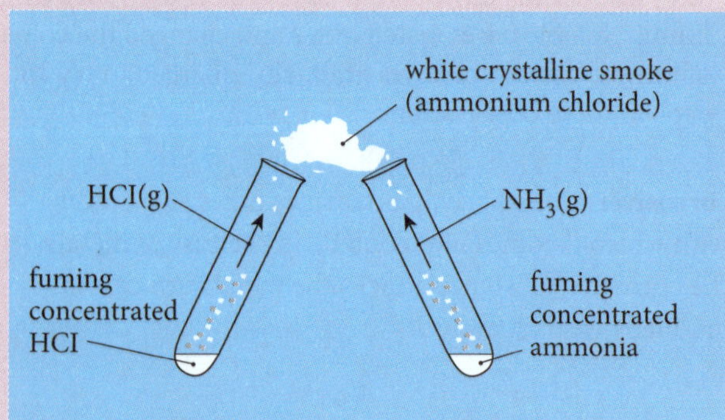

Figure 5.16 Reaction between hydrogen chloride gas and ammonia gas

Write an equation for the reaction and explain how this reaction is an example of the Brønsted–Lowry (B/L) theory of acids and bases.

Identify the proton donor

Answer:

$NH_3(g) + HCl(g) \rightarrow NH_4Cl(s)$

Ammonium chloride is an ionic crystal containing ammonium ions (NH_4^+) and chloride ions (Cl^-).

The hydrogen chloride molecule has donated a proton (H^+) to the ammonia molecule to form the ammonium ion. Thus the hydrogen chloride molecule is a Brønsted–Lowry acid (proton donor) and ammonia is a Brønsted–Lowry base (proton acceptor).

The B/L theory provided explanations about the basicity of oxides and carbonates and why not all salts are neutral in water.

Gilbert Lewis (1923)

- Gilbert Lewis (1923) proposed the electron pair acceptor/electron pair donor theory of acids and bases. In the Lewis theory, when hydrogen ions and hydroxide ions undergo neutralisation, the H^+ ion accepts a pair of non-bonding electrons from the oxygen atom of the OH^- ion to form a covalent bond in the water molecule formed.

Lewis acid: an electron-pair acceptor
Lewis base: an electron-pair donor

- All cations are Lewis acids since they are able to accept electrons. Species such as water and ammonia, which have non-bonding electron pairs, can donate these electron pairs to the cations to form special covalent bonds called *coordinate bonds*.
  Example:

  $Cu^{2+}(aq) + 4NH_3(aq) \rightarrow Cu(NH_3)_4^{2+}(aq)$

  The copper (II) ion is the Lewis acid and the ammonia is the Lewis base (see Figure 5.17).

  In the Lewis theory, hydrogen ions are not necessarily involved and the presence or absence of a solvent is not relevant.

tetraammine copper (II) ion

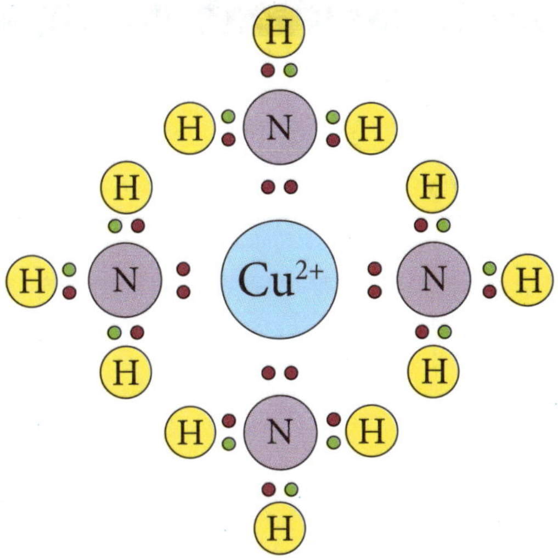

H:N:H (with H above N) ; H:N: Cu²⁺ :N:H ; H:N:H (with H below N)

Figure 5.17 Ammonia molecules donate electron pairs to the copper (II) ion

→ Arrhenius's theory and the Brønsted–Lowry theory are particularly relevant to the requirements of the HSC (Year 12) Chemistry course. Neutralisation reactions, for example, were explained using Arrhenius's theory of hydrogen ions and hydroxide ions. The Brønsted–Lowry theory will be used in Chapter 6 to explain areas of acid–base chemistry that Arrhenius's theory cannot explain.

→ KEY QUESTIONS

9 **Compare the heat released when 1 mole of acetic acid is neutralised by:**
  a **1 mole of NaOH.**
  b **1 mole of NH_3.**

10 **Compare the Lavoisier and Davy theories of acids.**

11 **Explain Arrhenius's explanation of neutralisation of an acid and a base in water solution.**

Answers ⊃ p. 72

CHAPTER SYLLABUS CHECKLIST

Are you able to answer every question from the syllabus for this chapter? Tick each question as you go through the checklist if you are able to answer it. If you cannot answer a question, turn to the relevant page in the study guide to find the answer. For NESA key word meanings, go to www.educationstandards.nsw.edu.au and search 'key words'.

	FOR A COMPLETE UNDERSTANDING OF THIS TOPIC:	PAGE NO.	✓
1	Can I state examples of common acidic and basic substances?	57	
2	Can I describe the colour change of litmus in acidic and basic solutions?	58	
3	Can I name and write the formulas of common laboratory acids and bases?	58	
4	Can I recall the properties that distinguish acids and bases?	59	
5	Can I describe the colour changes of common laboratory indicators, including universal indicator, in acidic, basic or neutral solutions?	59	
6	Can I describe the procedure to prepare and test a natural indicator?	61	
7	Can I write equations for acid–base reactions?	62	
8	Can I write equations for acid–carbonate reactions?	63	
9	Can I write equations for acid–metal reactions?	64	
10	Can I recall examples of applications of neutralisation reactions in everyday life and industrial processes?	64	
11	Can I recall the procedure to measure the enthalpy of neutralisation?	66	
12	Can I describe the changes in definitions and models of acids and bases over time and explain the limitations of each model?	67	

HSC EXAM-TYPE QUESTIONS

Objective-response questions (1 mark each)

1 A 2.0 mol/L solution of ammonia was tested with four different indicators. Select the answer that correctly shows the colour of the indicator in the ammonia solution.

	Indicator	Colour
A	methyl orange	red
B	litmus	red-purple
C	bromothymol blue	yellow-green
D	phenolphthalein	crimson

2 Select the substance which is basic.

A sugar solution
B baking soda
C orange juice
D lemonade

3 Select the true statement.

A Carbon dioxide is released when zinc reacts with dilute nitric acid.
B The reaction of copper (II) oxide with sulfuric acid is an example of a redox reaction.
C The reaction of magnesium with hydrochloric acid is a neutralisation reaction.
D Barium nitrate is formed when nitric acid reacts with barium hydroxide.

4 50.0 mL (0.0516 kg) of 1.00 mol/L HNO_3 is placed in a polystyrene foam calorimeter. The initial temperature was 22.5 °C. 50.0 mL (0.0520 kg) of 1.00 mol/L NaOH at the same initial temperature was added to the calorimeter and the mixture stirred. Determine the final temperature of the solution.

Data: $c = 4.18 \times 10^3$ J/kg/K; $\Delta H = -57.1$ kJ/mol

A 15.9 °C
B 35.7 °C
C 29.1 °C
D 6.6 °C

5 Select the correct statement about the theories of acids.

A Arrhenius proposed that acids released hydrogen ions when they dissolved in water.
B Lavoisier proposed that the presence of oxygen in metallic oxides caused acidity in water.
C Brønsted–Lowry theory defines an acid as a proton acceptor.
D Davy proposed that all metallic hydrides were acidic.

Extended-response questions

6 Pineapples grow best in weakly acidic soils. A chemist collected some soil from a pineapple farm to determine whether the soil was weakly acidic. She placed a soil sample in a Petri dish and moistened it. She applied some powdered barium sulfate on the soil and allowed soil water to be absorbed. She added drops of phenolphthalein indicator and found that the indicator remained colourless.

a Her supervisor criticised her choice of indicator. Explain. (1 mark)
b What would be a good indicator to use? Explain. (2 marks)

7 100 mL of 0.100 mol/L sodium carbonate solution can be neutralised by hydrochloric acid in two steps.

a Write the two equations for the step-wise neutralisation of sodium carbonate solution with HCl. (2 marks)
b Calculate the mass of sodium chloride that will form if excess hydrochloric acid is used. (3 marks)

8 200 mL of 0.0100 mol/L barium hydroxide is neutralised using 0.0150 mol/L sulfuric acid solution.

a Write a balanced equation for this reaction. (1 mark)
b Calculate the volume of sulfuric acid required. (3 marks)

9 Write net ionic equations for the following neutralisation reactions:

a solid cobalt (II) oxide + dilute hydrochloric acid (1 mark)
b solid manganese (II) hydroxide + dilute sulfuric acid (1 mark)
c potassium hydroxide solution + acetic acid solution (1 mark)

10 A metal can be defined operationally or conceptually.

Operational definition: a metal is lustrous and conducts electricity.

Conceptual definition: a metal consists of positive ions in a lattice where mobile electrons exist.

The following statements are definitions of a soluble base. Classify each statement as an operational or conceptual definition.

a Soluble bases taste bitter. (1 mark)
b Soluble bases release hydroxide ions in water. (1 mark)
c Soluble bases turn red litmus blue. (1 mark)
d Hydroxide ions in soluble bases react with hydrogen ions in acids to form water. (1 mark)

ANSWERS

KEY QUESTIONS

Key questions ➲ p. 62

1 a basic b neutral c acidic

2 a stays red b stays red c turns blue

3 a base b acid c acid

4 a colourless b crimson c colourless

Key questions ➲ p. 65

5 $NaHCO_3(s) + HNO_3(aq) \rightarrow NaNO_3(aq) + H_2O(l) + CO_2(g)$

The limewater test is used to identify carbon dioxide. The carbon dioxide turns the clear limewater solution a milky white due to the formation of white calcium carbonate.

6 $3Zn(s) + 8H_3O^+(aq) + 2NO_3^-(aq) \rightarrow 3Zn^{2+}(aq) + 2NO(g) + 12H_2O(l)$

7 Barium sulfate is white and insoluble in water. It is also neutral.

8 NaOH is such a strong base that it is caustic to human tissue, causing severe ulceration.

Key questions ➲ p. 69

9 More heat is released when NaOH is neutralised than when ammonia is neutralised.

10 Lavoisier theory: Oxygen atoms present in non-metal oxides caused acidity; Davy theory: Replaceable hydrogen atoms in compounds cause acidity.

11 Arrhenius explained neutralisation in terms of the reaction of hydrogen ions and hydroxide ions to form water.

HSC EXAM-TYPE QUESTIONS

Objective-response questions

1 **D.** Phenolphthalein is crimson in basic solutions. **A** is incorrect as methyl orange is yellow in basic solutions. **B** is incorrect as litmus is blue in basic solutions. **C** is incorrect as bromothymol blue is blue in basic solutions.

2 **B.** Baking soda is a weak base. **A** is incorrect as sugar solution is neutral. **C** and **D** are incorrect as they are acidic.

3 **D.** The salt formed when nitric acid reacts with a metal hydroxide is a nitrate salt. **A** is incorrect as hydrogen gas is evolved. **B** is incorrect as this is an acid–base reaction. **C** is incorrect as this is a redox reaction.

4 **C.**

$n(HNO_3) = cV = (1.00)(0.0500) = 0.0500$ mol

$n(NaOH) = cV = (1.00)((0.0500) = 0.0500$ mol

$HNO_3(aq) + NaOH(aq) \rightarrow NaNO_3(aq) + H_2O(l)$

For an exothermic reaction: $\Delta H = -q/n$

$\therefore q = -n\Delta H = -(0.0500)(-57.1) = 2.855 \text{ kJ} = 2855 \text{ J}$

$q = mc\Delta T$

$2855 = (0.0516 + 0.0520)(4.18 \times 10^3)(\Delta T)$

$\Delta T = 6.6$ °C

As the neutralisation is exothermic, the final temperature will be higher than the initial temperature.

Final temperature = 22.5 + 6.6 = 29.1 °C

Thus **A**, **B** and **D** are incorrect.

5 **A.** Arrhenius explained acidity in terms of hydrogen ions in water. **B** is incorrect as the oxygen theory only applied to non-metal oxides. **C** is incorrect as the B/L theory states that acids are proton donors. **D** is incorrect as Davy's theory applies to replaceable hydrogen when acids reacted with metals.

Extended-response questions

6 EM Students need to explain the relationship between weak acidity and the colour of the phenolphthalein indicator.

a Phenolphthalein is a not a suitable indicator to detect acidity as it is colourless at pH values less than 8. ✓

b Universal indicator is yellow in weakly acidic solutions. ✓ In solutions of greater acidity it is orange or red. ✓

7 EM Students are required to show that sodium carbonate is neutralised in two steps and that $NaHCO_3$ is the intermediate. Full marks are awarded only if all the steps of the mole/mass calculation are shown.

a $Na_2CO_3(aq) + HCl(aq) \rightarrow NaHCO_3(aq) + NaCl(aq)$ ✓

$NaHCO_3(aq) + HCl(aq) \rightarrow NaCl(aq) + H_2O(l) + CO_2(g)$ ✓

b Net equation:

$Na_2CO_3(aq) + 2HCl(aq) \rightarrow 2NaCl(aq) + H_2O(l) + CO_2(g)$ ✓

$n(Na_2CO_3) = cV = (0.100)(0.100) = 0.0100$ mol

$n(NaCl) = 0.0200$ mol ✓

$m(NaCl) = n.M = (0.0200)(22.99 + 35.45) = 1.17$ g ✓

8 EM Students are required to demonstrate the reaction stoichiometry in the calculation. Students must show states in the equation.

a $Ba(OH)_2(aq) + H_2SO_4(aq) \rightarrow BaSO_4(s) + 2H_2O(l)$ ✓

b $n(Ba(OH)_2) = cV = (0.0100)(0.200) = 0.00200$ mol ✓

Stoichiometry: $Ba(OH)_2 : H_2SO_4 = 1 : 1$

$n(H_2SO_4) = 0.00200$ mol ✓

$V = n/c = \dfrac{(0.00200)}{(0.0150)} = 0.133 \text{ L} = 133 \text{ mL}$ ✓

9 EM Students are required to write balanced equations that must include states to obtain full marks. Students need to correctly identify spectator ions that must be eliminated in the net ionic equation.

a $CoO(s) + 2H_3O^+(aq) \rightarrow Co^{2+}(aq) + 3H_2O(l)$ ✓

b $Mn(OH)_2(s) + 2H_3O^+(aq) \rightarrow Mn^{2+}(aq) + 4H_2O(l)$ ✓

c $OH^-(aq) + CH_3COOH(aq) \rightarrow H_2O(l) + CH_3COO^-(aq)$ ✓

10 EM Students need to use the supplied definitions for metals and then apply them to definitions of acids and bases. Operational definitions are based on measurable macroscopic properties. Conceptual definitions are based on theories.

a operational ✓

b conceptual ✓

c operational ✓

d conceptual ✓

CHAPTER 6 USING BRØNSTED–LOWRY THEORY

MODULE 6 ACID–BASE REACTIONS

INQUIRY QUESTION:

What is the role of water in solutions of acids and bases?

Svante Arrhenius explained that the sourness and acidity of citrus juices was due to the presence of hydrogen ions in the juice. Some citrus juices such as lemons are more sour than others, such as orange juice. This chapter quantifies the degree of acidity in terms of a function called *pH*. The degree of alkalinity of basic solutions can also be quantified in terms of the pOH function. We will investigate the applications of the Brønsted–Lowry theory.

1 Investigating the pH of solutions

» Students conduct a practical investigation to measure the pH of a range of acids and bases.

- The concentrations of acids can be compared using the pH scale. The pH scale is a logarithmic scale based on powers of ten. A change of 1 unit on the pH scale equals a tenfold change in hydrogen ion concentration $[H^+]$ or the hydronium ion concentration $[H_3O^+]$.
- The pH of a solution is defined as:

$$\mathbf{pH = -log_{10}[H^+]}$$

- As hydrogen ions react with water to form hydronium ions, the pH formula can also be written as:

$$\mathbf{pH = -log_{10}[H_3O^+]}$$

EXAMPLE 1

a A 0.000 100 mol/L solution of HCl is prepared. Calculate its pH.

b A sulfuric acid solution is prepared with the same concentration as the HCl solution. Calculate its pH.

Express your answer to two decimal places as the concentration is expressed to three significant figures

Answer:

a As HCl is a strong acid (i.e. completely dissociated), the concentration of H^+ (or H_3O^+) is 0.000 100 mol/L:

$$HCl(aq) + H_2O(l) \rightarrow H_3O^+(aq) + Cl^-(aq)$$

$$pH = -log_{10}[0.000\,100] = 4.00$$

b Sulfuric acid is a strong diprotic acid that dissociates completely in dilute solution:

$$H_2SO_4(aq) + 2H_2O(l) \rightarrow 2H_3O^+(aq) + SO_4^{2-}(aq)$$

Therefore $[H_3O^+] = 2 \times [H_2SO_4] = 2(0.000\,100)$
$= 0.000\,200$ mol/L

$$pH = -log_{10}[0.000\,200] = 3.70$$

- The pH scale for most aqueous solutions extends from 0 to 14. The lower the pH, the more acidic the solution. The pH of pure water is 7 at 25 °C. The greater the pH above 7, the more alkaline the solution.
- The pH scale can be related to the colours of universal indicator (UI). Figure 6.1 shows the typical colours of universal indicator as a function of pH. The actual colours vary between commercial brands so students should use the colour card provided.

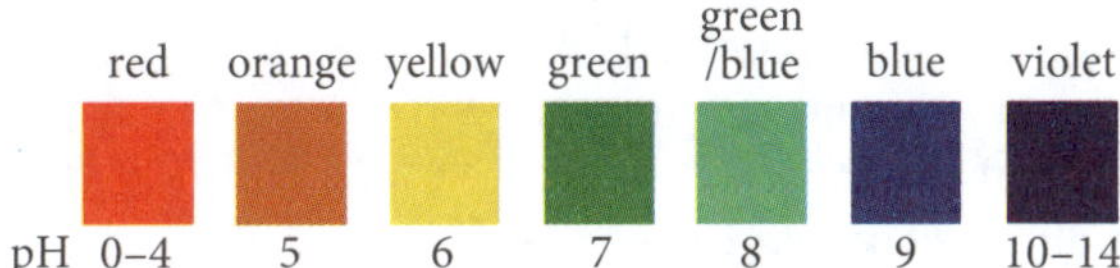

Figure 6.1 Universal indicator colour chart

- A calibrated pH glass electrode and meter can be used to measure more accurately the pH of solutions (see Figure 6.2). The pH electrode is a galvanic cell whose voltage is

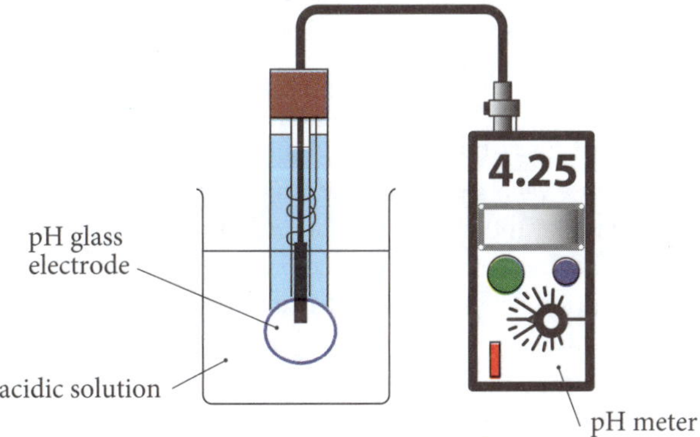

Figure 6.2 pH glass electrode and meter

dependent on the concentration of hydrogen ions. The pH electrode must be calibrated with solutions of known pH. These solutions are known as *buffers*. Typically a pH 4.0 buffer and a pH 7.0 buffer are used to establish a **calibration** between voltage and pH for acid solutions. A pH 10.0 buffer and a pH 7.0 buffer are used to calibrate the pH electrode for alkaline solutions. Values of pH measured by a correctly calibrated pH meter are accurate to within 0.01 pH units.

calibration: a procedure used to check or change the graduations of a measuring instrument

FIRSTHAND INVESTIGATION 1

Investigating the pH of solutions

» Students conduct a practical investigation to measure the pH of a range of acids and bases.

Aim

to use universal indicator (UI) and a universal indicator pH chart to estimate the pH of HCl and NaOH solutions at different degrees of dilution

Chemicals/equipment

0.010 mol/L hydrochloric acid (pH = 2.0)
0.010 mol/L sodium hydroxide (pH = 12)
0.010 mol/L solutions of other acids and bases (e.g. H_2SO_4, CH_3COOH, $Ca(OH)_2$, Na_2CO_3)
Universal indicator solution or pH papers
Universal indicator colour chart
10 mL measuring cylinders
100 mL measuring cylinders
250 mL beakers
Calibrated pH electrode

Method

Part 1: HCl and NaOH solutions

1 Place 1 mL of 0.010 mol/L HCl (A1) and 0.010 mol/L NaOH (B1) in separate, clean beakers.
2 Use the dropper bottle to add 3 to 4 drops of universal indicator (or pieces of universal indicator paper) to each solution.
3 Record the indicator colour and use the pH colour chart to estimate the pH of the solution.
4 Dilute each solution by adding 99 mL of water to make 100 mL of solution. Add more drops of indicator and record the colour of each diluted solution (A2 and B2). Estimate the pH of the diluted solutions.
5 Continue the dilution by removing 1 mL of the dilute solutions from the previous step and dilute them with a further 99 mL of water. Add more indicator and record colours and pH of the diluted solutions (A3 and B3).
6 If a calibrated pH electrode is available then measure the pH of each solution and compare these values with your estimates based on the UI colours.

Part 2: Other acid and base solutions

Repeat the procedure for Part 1 using other acids and bases supplied by your teacher.

Sample results for Part 1

Solution	A1	A2	A3	B1	B2	B3
UI colour	red	orange-red	yellow	violet	blue-violet	green-blue
pH	2	~4	~6	12	~10	~8

Conclusion

In Part 1, as the HCl solution is diluted, the pH rises. As the NaOH solution is diluted the pH decreases. The calibrated pH electrode gives much more accurate pH values.

Similar results should be obtained with the solutions used in Part 2, although the actual pH values will differ from the HCl and NaOH experiments.

➔ KEY QUESTIONS

1 **Identify the colour of universal indicator in solutions with a pH of 6 and 9.**

2 **Explain why a pH electrode and meter must be calibrated prior to use.**

3 **Calculate the pH of an HCl solution with a concentration of 2.50×10^{-4} mol/L.**

Answers ➲ p. 88

2 pH and pOH calculations

» Students:

- calculate pH, pOH, hydrogen ion concentration ($[H^+]$) and hydroxide ion concentration ($[OH^-]$) for a range of solutions.
- calculate the pH of the resultant solution when solutions of acids and/or bases are diluted or mixed.

Water ionisation equilibrium

➔ The pH scale can be understood in terms of the **auto-ionisation** of water, which is illustrated in Figure 6.3. The extent of this ionisation is very low:

$$2H_2O(l) \leftrightarrows H_3O^+(aq) + OH^-(aq).$$

auto-ionisation: a reaction between two water molecules in which hydronium ions and hydroxide ions form by proton transfer

base + acid ⇌ conjugate acid + conjugate base

Figure 6.3 Auto-ionisation of water

- At 25 °C in pure water, hydrogen ions (or hydronium ions) and hydroxide ions exist in equal concentrations. These concentrations have been experimentally measured and are equal to 1.0×10^{-7} mol/L.
- The ionic product for this equilibrium is given the symbol K_w. K_w is called the *water ionisation constant.*

$$K_w = [H^+][OH^-]$$

The value of K_w can be calculated:

$K_w = (1.0 \times 10^{-7})(1.0 \times 10^{-7}) = 1.0 \times 10^{-14}$

This equilibrium constant is listed in the Data Sheet on the inside cover of this book.

➔ When acids are added to water, the hydrogen ion concentration increases and the water ionisation equilibrium is disturbed. The equilibrium shifts to the left and consequently the concentration of the hydroxide ions decreases. The value of K_w remains constant.

EXAMPLE 2

20.0 mL of 0.25 mol/L HCl is added to 80.0 mL of water to form 100.0 mL of solution. Calculate hydrogen ion and hydroxide ion concentrations in this solution.

Use Le Chatelier's principle to predict the shift in the water equilibrium

Answer:

Calculate the number of moles of HCl:

$n(\text{HCl}) = cV = (0.25)(0.0200) = 0.00500$ mol

Calculate the diluted concentration:

$$c(\text{HCl}) = n/V = \frac{(0.0030)}{(0.100)} = 0.0500 \text{ mol/L}$$

As the acid is fully ionised, $c(H^+) = c(\text{HCl})$.
Thus $[H^+] = 0.0500$ mol/L

The higher hydrogen ion concentration than in pure water will shift the water equilibrium so as to reduce the hydroxide ion concentration.

Use the K_w expression to calculate the OH^- concentration:

$K_w = [H^+][OH^-] = (0.0500)[OH^-] = 1.0 \times 10^{-14}$

$$[OH^-] = \frac{1.0 \times 10^{-14}}{0.0500} = 2.0 \times 10^{-13} \text{ mol/L}$$

Thus the hydroxide ion concentration has decreased from 1.0×10^{-7} mol/L to 2.0×10^{-13} mol/L.

➔ The value of K_w is temperature dependent. The graph in Figure 6.4 shows that K_w increases as temperature increases.

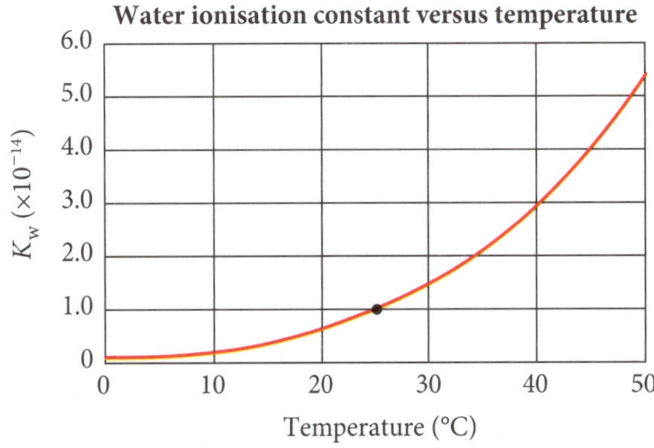

Figure 6.4 Graph of K_w versus temperature

EXAMPLE 3

The table lists the K_w values at three temperatures.

Temperature	10	25	40
K_w	3.0×10^{-15}	1.0×10^{-14}	2.9×10^{-14}

a **Calculate the pH of water at each temperature.**
b **Explain whether the water remains neutral as the temperature increases.**
c **Determine whether the water ionisation equilibrium is endothermic or exothermic.**

Answer:

Equal amounts of hydrogen ions and hydroxide ions form on the ionisation of water

a $K_w = [H^+][OH^-]$
As $[H^+] = [OH^-]$
$\therefore K_w = [H^+]^2$
At 10 °C: $[H^+]^2 = 3.0 \times 10^{-15}$
$\therefore [H^+] = 5.5 \times 10^{-8}$ mol/L
$\therefore \text{pH} = -\log_{10}(5.5 \times 10^{-8}) = 7.3$
At 25 °C: $[H^+]^2 = 1.0 \times 10^{-14}$
$\therefore [H^+] = 1.0 \times 10^{-7}$ mol/L
$\therefore \text{pH} = -\log_{10}(1.0 \times 10^{-7}) = 7.0$
At 40 °C: $[H^+]^2 = 2.9 \times 10^{-14}$
$\therefore [H^+] = 1.7 \times 10^{-7}$ mol/L
$\therefore \text{pH} = -\log_{10}(1.7 \times 10^{-7}) = 6.8$

b Although the pH has decreased as temperature increased, the water remains neutral as equal amounts of hydrogen ions and hydroxide ions are present.

c The concentration of hydrogen ions and hydroxide ions increases as temperature increases. Therefore the position of the water ionisation equilibrium is shifting to the right. Thus the equilibrium is endothermic.

pH and pOH calculations

- The pH formula can be used to calculate the pH of acidic solutions of known hydrogen ion concentration.

EXAMPLE 4

A solution of nitric acid has a concentration of 0.000450 mol/L. Determine the pH of this solution.

Nitric acid is a fully dissociated strong acid

Answer:

The hydronium ion concentration will be equal to the HNO_3 concentration as this acid is fully dissociated:

$$HNO_3(aq) + H_2O(l) \longrightarrow H_3O^+(aq) + NO_3^-(aq)$$

$$pH = -\log_{10}[H^+] = -\log_{10}[H_3O^+]$$
$$= -\log_{10}[0.000\,450] = 3.35$$

- In alkaline solutions, the high OH^- concentration causes the water ionisation equilibrium to shift to the left and lower the $[H^+]$. In alkaline solutions the hydrogen ion concentration is less than 1×10^{-7} mol/L and thus the pH is greater than 7. For example, in a 0.10 mol/L NaOH solution the pH is 13.
- The pH of alkaline solutions can be calculated using the water ionisation constant (K_w).

EXAMPLE 5

Calculate the pH of a 0.00250 mol/L potassium hydroxide solution.

Potassium hydroxide is a fully dissociated base

Answer:

Use K_w to calculate the hydrogen ion concentration:

$$K_w = [H^+][OH^-] = 1.0 \times 10^{-14}$$

The concentration of hydroxide ions will equal the potassium hydroxide concentration as KOH is completely dissociated:

$$[H^+](0.002\,50) = 1.0 \times 10^{-14}$$
$$[H^+] = 4.00 \times 10^{-12} \text{ mol/L}$$
$$pH = -\log_{10}(4.00 \times 10^{-12}) = 11.4$$

- The alkalinity of a solution can also be expressed by another logarithmic function called pOH.

pOH is defined as:

$$\mathbf{pOH = -\log_{10}[OH^-]}$$

The smaller the values of pOH, the more alkaline the solution.

EXAMPLE 6

Calculate the pOH of a 0.00250 mol/L barium hydroxide solution.

First determine the hydroxide ion concentration

Answer:

Barium hydroxide dissociates in water according to the equation:

$$Ba(OH)_2(aq) \longrightarrow Ba^{2+}(aq) + 2OH^-(aq)$$

Stoichiometry: $Ba(OH)_2(aq) : OH^-(aq) = 1:2$

$\therefore\ [OH^-] = 2 \times 0.002\,50 = 0.005\,00$ mol/L

$pOH = -\log_{10}[OH^-] = -\log_{10}[0.005\,00] = 2.30$

- The mathematical relationship between pH and pOH is given by the following equation:

$$\mathbf{pH + pOH = 14}$$

Figure 6.5 shows the relationship between pH and pOH.

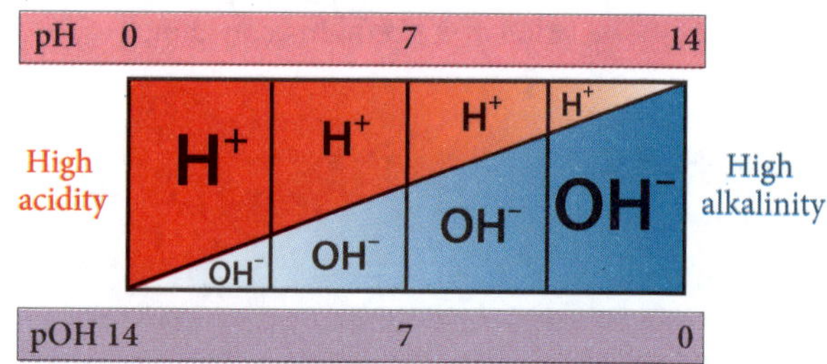

Figure 6.5 pH and pOH

- Table 6.1 shows how the pOH and pH changes as a sodium hydroxide solution is diluted.

Table 6.1 pOH and pH values for diluted NaOH solutions

[NaOH] (mol/L)	pOH	pH
0.100	1.00	13.00
0.0200	1.70	12.30
0.001 50	2.82	11.18
0.000 500	3.30	10.70

EXAMPLE 7

An alkaline solution has been prepared by dilution of 100 mL of 0.0010 mol/L KOH with sufficient water until the total volume was 600 mL.

a Calculate the pOH of the original solution.

b Calculate the pOH and pH of the diluted solution.

Use the dilution formula to determine the diluted concentration of the KOH solution

Answer:

a The concentration of OH^- ions equals the KOH concentration as KOH is a monoprotic strong base:

$pOH = -\log_{10}(0.001\,00) = 3.00$

b Calculate the concentration of the diluted KOH solution:

$c_1V_1 = c_2V_2$

$(0.001\,00)(0.100) = c_2(0.600)$

$c_2 = [OH^-] = 1.67 \times 10^{-4}$ mol/L

$pOH = -\log_{10}(1.67 \times 10^{-4}) = 3.78$

$pH = 14 - pOH = 10.22$

EXAMPLE 8

An acidic solution has been prepared by dilution of 200 mL of 0.0100 mol/L HCl with sufficient water until the total volume was 2.00 L. Calculate the pH and pOH of the diluted solution.

Ensure volume units are the same

Answer:

$n(HCl) = c.V = (0.010)(0.200) = 0.0020$ mol

After dilution, the new concentration of the HCl is:

$$c = n/V = \frac{0.0020}{2.00} = 0.0010 \text{ mol/L}$$

Thus $[H^+] = 0.0010$ mol/L

$pH = -\log_{10}(0.0010) = 3.00$

$pOH = 14 - pH = 14 - 3.00 = 11.00$

➔ Common substances have a wide range of pH. Table 6.2 shows some examples. The pH values vary with temperature and concentration.

Table 6.2 Typical pH of common substances

Substance	pH	Substance	pH
lemon juice	2.0	seawater	7.9
lemonade	2.5	baking soda solution	8.5
milk	6.6	ammonia cleanser	11.5

Neutralisation and pH calculations

➔ When solutions of acids and bases are mixed, a neutralisation reaction occurs. If a strong acid and a strong base are used then the pH of the final solution will be 7 when the solutions are mixed in the correct mole ratio.

➔ If the acid and base solutions are not mixed in the correct ratios then the final mixture will be either acidic or alkaline.

EXAMPLE 9

Calculate the final pH of a solution formed by mixing 200 mL of 0.100 mol/L HCl and 300 mL of 0.0550 mol/L NaOH.

Determine whether the acid or the base is a limiting reagent and which is in excess

Answer:

$HCl(aq) + NaOH(aq) \rightarrow NaCl(aq) + H_2O(l)$

$n(HCl) = cV = (0.100)(0.200) = 0.0200$ mol

$n(NaOH) = cV = (0.0550)(0.300) = 0.0165$ mol

Reaction stoichiometry = 1 : 1

Thus the HCl is in excess and all the NaOH will be neutralised. NaOH is the **limiting reagent**.

Excess HCl: $n(HCl) = 0.0200 - 0.0165 = 0.003\,50$ mol

Total volume = 200 + 300 = 500 mL = 0.500 L

$$c(HCl) = n/V = \frac{0.003\,50}{0.500} = 0.00700 \text{ mol/L}$$

$[H^+] = 0.007\,00$ mol/L

$pH = -\log_{10}(0.007\,00) = 2.15$

limiting reagent: a substance in a chemical reaction that is totally consumed when the chemical reaction is complete

➔ KEY QUESTIONS

4 **Calculate the pOH of a 0.005 50 mol/L solution of calcium hydroxide.**

5 **200 mL of 0.001 00 mol/L NaOH is diluted with water until the total volume is 750 mL. Calculate the pOH and pH of the diluted solution.**

6 **Calculate the final pH of a solution formed by mixing 250 mL of 0.100 mol/L HNO_3 and 250 mL of 0.150 mol/L KOH.**

Answers ➲ p. 88

3 Strength of acids and bases

» Students:

- conduct an investigation to demonstrate the use of pH to indicate the differences between the strength of acids and bases.
- write ionic equations to represent the dissociation of acids and bases in water, conjugate acid/base pairs in solution and amphiprotic nature of some salts; for example, sodium hydrogen carbonate and potassium dihydrogen phosphate.
- construct models and/or animations to communicate the differences between strong, weak, concentrated and dilute acids and bases.

Brønsted–Lowry theory of acids and bases

- The Brønsted–Lowry theory established the importance of the water solvent in understanding the properties of acids and bases. Acids dissociate in water solution due to their interaction with water molecules. The acidic hydrogen ion (or proton) is donated by the acid to a water molecule to form the strongly acidic hydronium ion. In a hydrofluoric acid solution the HF molecule is a Brønsted–Lowry acid, and water is a Brønsted–Lowry base as it accepts the proton. Weak acids such as HF are incompletely dissociated:

$$HF(aq) + H_2O(l) \leftrightarrows H_3O^+(aq) + F^-(aq)$$

- Figure 6.6 illustrates proton donation and acceptance in the forward and reverse reaction.

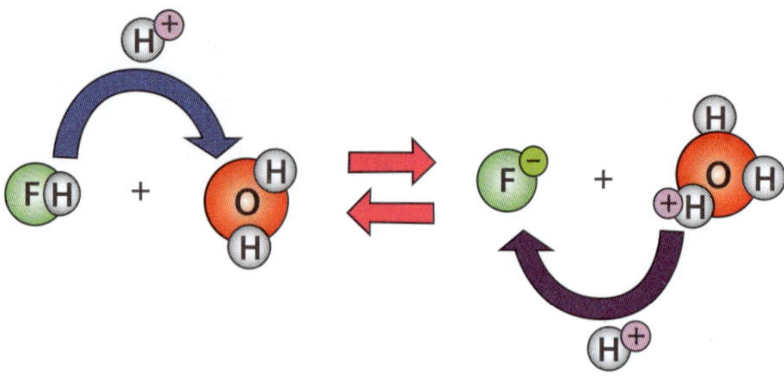

Figure 6.6 Proton transfer equilibrium for hydrofluoric acid solution

- As this reaction is reversible the hydronium ion and the fluoride ion also act as acids and bases. For the reverse reaction the hydronium ion is the acid as it donates a proton to the base (F^-). In the Brønsted–Lowry theory, each acid has a corresponding base called the **conjugate** base. Each Brønsted–Lowry base has a corresponding conjugate acid. In the following equation the conjugate base of acetic acid is the acetate ion, and the conjugate acid of water is the hydronium ion:

$$CH_3COOH\ (aq) + H_2O(l) \leftrightarrows CH_3COO^-(aq) + H_3O^+(aq)$$

conjugate: refers to an acid and its paired base that differ from each other by a proton

Table 6.3 shows some common Brønsted–Lowry acid/base pairs.

Table 6.3 Brønsted–Lowry acid/base pairs

Brønsted–Lowry acid	Conjugate base	Brønsted–Lowry base	Conjugate acid
HNO_3	NO_3^-	OH^-	H_2O
H_3O^+	H_2O	S^{2-}	HS^-
HNO_2	NO_2^-	CO_3^{2-}	HCO_3^-
CH_3COOH	CH_3COO^-	NH_3	NH_4^+
H_2CO_3	HCO_3^-	HS^-	H_2S

- A strong Brønsted–Lowry acid such as nitric acid (HNO_3) has a weak conjugate base (NO_3^-). The conjugate acids of strong bases such as the hydroxide ion (OH^-) are also very weak.
- When strong acid and strong bases react, the products are weak conjugate acids and bases. Consequently such equilibria lie so far to the right that the reactions essentially go to completion. The hydronium ion is a strong acid and the hydroxide ion is a strong base. Water is produced when these ions react:

$$H_3O^+(aq) + OH^-(aq) \rightarrow H_2O(l) + H_2O(l)$$

- When weak acids and weak bases react an equilibrium is established, as illustrated in the following equation. The reaction does not proceed to completion:

$$NH_3(aq) + H_2CO_3(aq) \leftrightarrows NH_4^+(aq) + HCO_3^-(aq)$$

- In the Brønsted–Lowry theory acids and bases can be classified as *monoprotic*, *diprotic* and *triprotic*:
  - Monoprotic acids—donate 1 proton (e.g. HCl)
  - Diprotic acids—donate 2 protons (e.g. H_2SO_4)
  - Triprotic acids—donate 3 protons (e.g. H_3PO_4)
  - Monoprotic bases—accept 1 proton (e.g. NH_3)
  - Diprotic bases—accept 2 protons (e.g. CO_3^{2-})
  - Triprotic bases—accept 3 protons (e.g. PO_4^{3-})

Figure 6.7 uses structural formulae to model how hydrochloric acid and sulfuric acid solutions form when HCl and H_2SO_4 molecules are dissolved in water to form acidic solutions.

Examples:

- Sulfuric acid—a diprotic acid
  Sulfuric acid dissociates in water in two steps. The first step is complete and the second step involves an equilibrium reaction:

$$H_2SO_4(aq) + H_2O(l) \rightarrow HSO_4^-(aq) + H_3O^+(aq)$$
$$HSO_4^-(aq) + H_2O(l) \leftrightarrows SO_4^{2-}(aq) + H_3O^+(aq)$$

  In a 0.10 mol/L sulfuric acid solution the second equilibrium is only 35% complete. If sulfuric acid is diluted sufficiently the second equilibrium shifts strongly to the right. If a strong base such as OH^- ions is added to the sulfuric acid solution the second step of the dissociation is driven completely to the right:

$$H_2SO_4(aq) + 2OH^-(aq) \rightarrow SO_4^{2-}(aq) + 2H_2O(l)$$

- Phosphoric acid—a triprotic acid
  Phosphoric acid is a weak acid that dissociates in water in three steps:

$$H_3PO_4(aq) + H_2O(l) \leftrightarrows H_2PO_4^-(aq) + H_3O^+(aq)$$
$$H_2PO_4^-(aq) + H_2O(l) \leftrightarrows HPO_4^{2-}(aq) + H_3O^+(aq)$$
$$HPO_4^{2-}(aq) + H_2O(l) \leftrightarrows PO_4^{3-}(aq) + H_3O^+(aq)$$

- Carbonate ion—a diprotic base
  The carbonate ion is a weak base that reacts with water to form alkaline solutions. The water acts as a Brønsted–Lowry acid in each equilibrium:

$$CO_3^{2-}(aq) + H_2O(l) \leftrightarrows HCO_3^-(aq) + OH^-(aq)$$
$$HCO_3^-(aq) + H_2O(l) \leftrightarrows H_2CO_3(aq) + OH^-(aq)$$

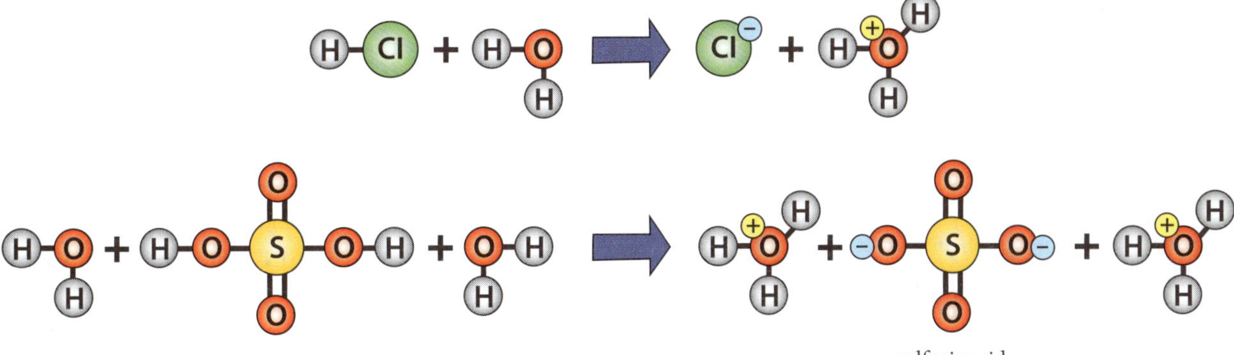

Figure 6.7 Formation of hydrochloric acid and sulfuric acid

If a strong acid such as the hydronium ion is reacted with carbonate ions, two protons are accepted in the neutralisation process. The carbonic acid that forms in the reaction decomposes to form carbon dioxide and water. The net reaction is:

$$CO_3^{2-}(aq) + 2H_3O^+(aq) \rightarrow CO_2(g) + 2H_2O(l)$$

➔ In the Brønsted–Lowry (B/L) theory, some species are **amphiprotic**. This means that they can behave as either a Brønsted–Lowry acid or a Brønsted–Lowry base, depending on the other chemical substances present. Consider the following two examples, one of which is illustrated using Figures 6.8 and 6.9.

> **amphiprotic:** chemical substances that can behave as either a Brønsted–Lowry acid or a Brønsted–Lowry base in different chemical environments

- Hydrogen carbonate ion
  - i HCO_3^- is a B/L acid:
    $HCO_3^-(aq) + OH^-(aq) \rightarrow CO_3^{2-}(aq) + H_2O(l)$

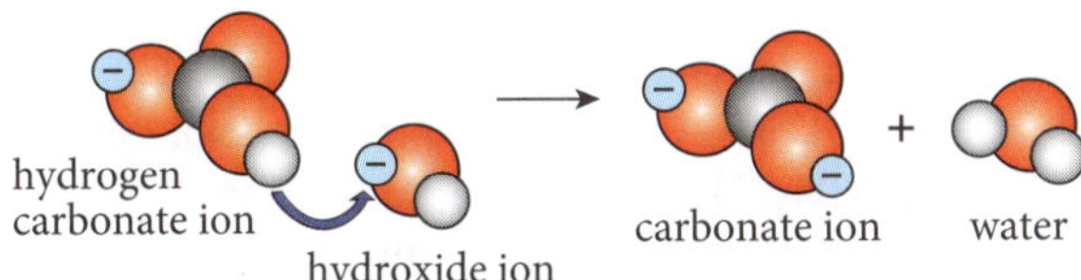

Figure 6.8 Hydrogen carbonate ion acts as a B/L acid

  - ii HCO_3^- is a B/L base:
    $HCO_3^-(aq) + H_3O^+(aq) \rightarrow H_2CO_3(aq) + H_2O(l)$

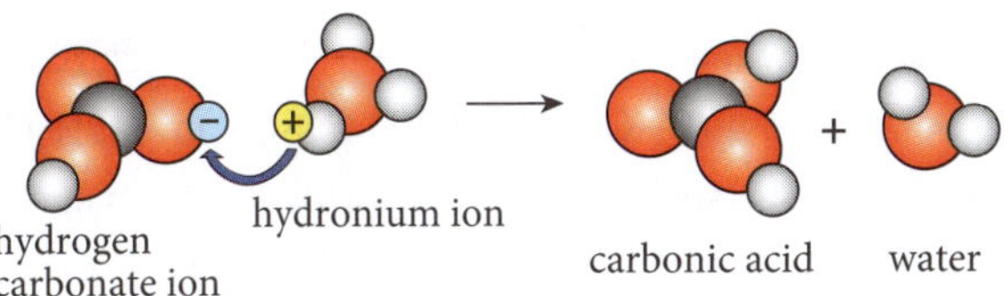

Figure 6.9 Hydrogen carbonate ion acts as a B/L base

These amphiprotic properties of the hydrogen carbonate ion can be demonstrated in the laboratory. Sodium hydrogen carbonate solutions react with strong base solutions such as NaOH and strong acid solutions such as HCl.

- Dihydrogen phosphate ion
  - i $H_2PO_4^-$ is a B/L acid:
    $H_2PO_4^-(aq) + OH^-(aq) \rightarrow HPO_4^{2-}(aq) + H_2O(l)$
  - ii $H_2PO_4^-$ is a B/L base:
    $H_2PO_4^-(aq) + H_3O^+(aq) \rightarrow H_3PO_4(aq) + H_2O(l)$

    These amphiprotic properties of the dihydrogen phosphate ion can also be demonstrated in the laboratory. Potassium dihydrogen phosphate solutions react with strong base solutions such as NaOH and strong acid solutions such as HCl.

EXAMPLE 10

Solutions of potassium hydrogen oxalate exhibit both acidic and basic properties. The hydrogen oxalate ion is amphiprotic. Write whole formula equations and net ionic equations to demonstrate that it is amphiprotic.

Use strong acids and bases to demonstrate amphiprotism

Answer:

- **$HC_2O_4^-$: a B/L base**

  Whole formula equation:

  $$KHC_2O_4(aq) + HCl(aq) \rightarrow H_2C_2O_4(aq) + KCl(aq)$$

  Net ionic equation:

  $$HC_2O_4^-(aq) + H_3O^+(aq) \rightarrow H_2C_2O_4(aq) + H_2O(l)$$

  Note that the hydronium ion is the proton donor in the net ionic equation.

- **$HC_2O_4^-$: a B/L acid**

  Whole formula equation:

  $$KHC_2O_4(aq) + NaOH(aq) \rightarrow KNaC_2O_4(aq) + H_2O(l)$$

  Net ionic equation:

  $$HC_2O_4^-(aq) + OH^-(aq) \rightarrow C_2O_4^{2-}(aq) + H_2O(l)$$

  Note that the hydroxide ion is the proton acceptor in the net ionic equation.

EXAMPLE 11

The production of sodium hydrogen carbonate involves the following steps:

1. **Production of ammonia gas by heating a mixture of calcium hydroxide solution and ammonium chloride.**
2. **The ammonia gas is dissolved and saturates a concentrated solution of sodium chloride.**
3. **Carbon dioxide gas is used to saturate the ammoniated NaCl solution produced in step 2. The carbonic acid that is produced reacts with the ammonia to form ammonium ions and hydrogen carbonate ions. Sodium hydrogen carbonate crystallises as the mixture is cooled and sodium ions and hydrogen carbonate ions associate.**

Write equations for the reactions described and explain how the Brønsted–Lowry (B/L) theory of acids and bases is applicable to the chemistry involved.

Identify which substances are proton donors or acceptors

Answer:

In step 1 the ammonium ion acts as a B/L acid and the hydroxide ion is a B/L base. The ammonia formed is the conjugate base of the ammonium ion and water is the conjugate acid of the hydroxide ion:

$$NH_4^+(aq) + OH^-(aq) \rightarrow NH_3(g) + H_2O(l)$$

In step 3 the carbonic acid is a B/L acid and the dissolved ammonia is a B/L base. The hydrogen carbonate ion is the conjugate base of carbonic acid and ammonium ions are conjugate acids of ammonia:

$$H_2CO_3(aq) + NH_3(aq) \leftrightharpoons HCO_3^-(aq) + NH_4^+(aq)$$

→ KEY QUESTIONS

7. **Write the formula of the conjugate base of carbonic acid.**
8. **Write the formula of the conjugate acid of the hydrogen sulfide ion.**
9. **The hydrogen phosphate ion is amphiprotic. Write net ionic equations to demonstrate its amphiprotism.**

Answers ⮌ p. 88

Hydrolysis of salts

→ Water is a common amphiprotic substance. Water can act as a proton donor or a proton acceptor, as shown in the following equations involving solutions of salts and illustrated in Figure 6.10.

- Water as a proton donor
  If the salt sodium hydrosulfide (NaHS) is dissolved in water the solution becomes alkaline due to the formation of hydroxide ions. The HS^- ion acts as a Brønsted–Lowry base:

  $$HS^-(aq) + H_2O(l) \leftrightharpoons H_2S(aq) + OH^-(aq)$$

- Water as a proton acceptor
  If the salt sodium hydrogen sulfate ($NaHSO_4$) is dissolved in water the solution becomes acidic due to the formation of hydronium ions. The HSO_4^- ion acts as a Brønsted–Lowry acid:

  $$HSO_4^-(aq) + H_2O(l) \leftrightharpoons SO_4^{2-}(aq) + H_3O^+(aq)$$

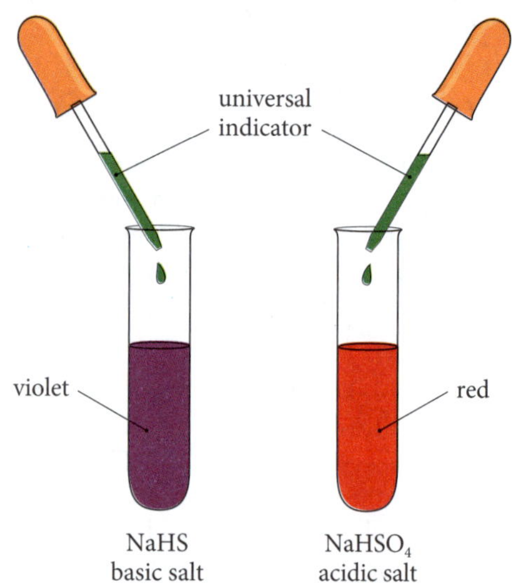

Figure 6.10 Hydrolysis of salts

→ The previous examples are examples of the **hydrolysis** of salts.

hydrolysis: the reaction of a salt with water that results in a change in pH

→ Salts can be classified as *neutral salts*, *acidic salts* or *basic salts*:

- Neutral salts—do not hydrolyse (e.g. NaCl)
- Acidic salts—hydrolyse to produce acidic solutions (e.g. NH_4Cl)
- Basic salts—hydrolyse to produce alkaline (basic) solutions (e.g. NaF)

→ Neutral salts (such as sodium chloride) are the salts formed when a strong acid neutralises a strong base:

$$HCl(aq) + NaOH(aq) \rightarrow NaCl(aq) + H_2O(l)$$

→ Acidic salts (such as copper (II) sulfate) are the salts formed when a strong acid neutralises a weak base:

$$H_2SO_4(aq) + Cu(OH)_2(s) \rightarrow CuSO_4(aq) + 2H_2O(l)$$

→ Basic salts (such as sodium fluoride) are the salts formed when a weak acid neutralises a strong base:

$$HF(aq) + NaOH(aq) \rightarrow NaF(aq) + H_2O(l)$$

→ The Brønsted–Lowry theory explains the pH changes involved in salt hydrolysis, which is described below and illustrated in Figure 6.11.

- $CuSO_4$ solution has a pH < 7
  The hydrated copper (II) ion ($Cu(H_2O)_4^{2+}$) is a Brønsted–Lowry acid:
  $$Cu(H_2O)_4^{2+}(aq) + H_2O(l) \leftrightharpoons Cu(H_2O)_3OH^+(aq) + H_3O^+(aq)$$
  $Cu(H_2O)_3OH^+$ is a weak base whereas the hydronium ion that forms makes the solution acidic as hydronium ions are strong acids. Other hydrated transition metal salts (e.g. $CoCl_2{\cdot}6H_2O$) also form acidic solutions in water.

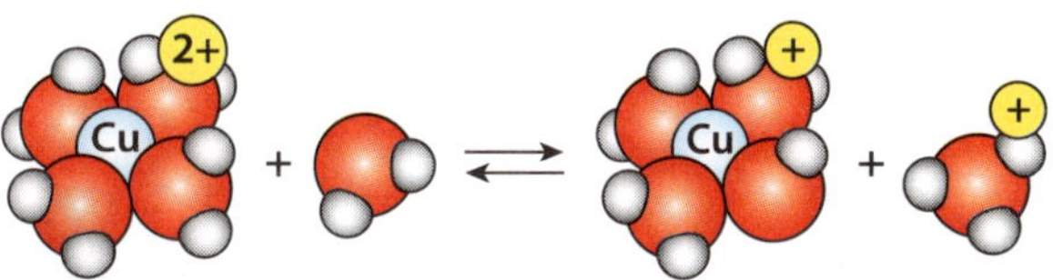

Figure 6.11 Hydrolysis of $Cu(H_2O)_4^{2+}$

- NaF solution has a pH > 7
  The F^- ion is a Brønsted–Lowry base:
  $$F^-(aq) + H_2O(l) \leftrightharpoons HF(aq) + OH^-(aq)$$
  HF is a weak acid whereas the hydroxide ion (a strong base) that forms makes the solution slightly alkaline.

EXAMPLE 12

Ammonium nitrate crystals are dissolved in water to form a 0.10 mol/L solution. The pH of the solution was 5.1.

a **Write an ionic equation to explain why ammonium nitrate is classified as an acidic salt.**

b **A few drops of the following solutions were added (in separate experiments) to the ammonium nitrate solution.**

i **2.0 mol/L HCl**

ii **2.0 mol/L NH_4Cl**

Explain how each change will affect the position of the acid–base equilibrium.

Identify the ions that will cause a shift in the equilibrium

Answer:

a The ammonium ion is a weak Brønsted–Lowry acid. It reacts with water (the Brønsted–Lowry base) to form ammonia (weak conjugate base) and hydronium ions (strong Brønsted–Lowry acid).

Thus the solution at equilibrium is acidic with a pH of 5.1:

$$NH_4^+(aq) + H_2O(l) \leftrightharpoons NH_3(aq) + H_3O^+(aq)$$

b **i** The HCl solution contains a high concentration of hydronium ions. The increase in hydronium ion concentration will increase the reverse rate of the NH_4^+/H_2O equilibrium. The equilibrium will shift to the left to counteract the change and more ammonium ions will form.

ii The addition of NH_4Cl solution will lead to an increase in the concentration of ammonium ions. This will increase the rate of the forward reaction in the NH_4^+/H_2O equilibrium. The equilibrium will shift to the right to counteract the change and more ammonia and hydronium ions will form.

Indicators and Brønsted–Lowry theory

→ The colour change of acid–base indicators can be explained using the Brønsted–Lowry theory. Acid–base indicators exist in two forms, the unionised Brønsted–Lowry acid (HIn) and its conjugate base (In^-). The ionisation equilibrium for the Brønsted–Lowry acid (HIn) is shown in the equation $HIn(aq) \leftrightharpoons H^+(aq) + In^-(aq)$.

→ The HIn indicator molecule has a different colour to the In^- ion. Table 6.4 shows the colours of these acid/conjugate base pairs for common indicators.

Table 6.4 Acid–base indicators

Indicator	HIn	In^-	pH range
methyl orange	red	yellow	3.1–4.4
bromothymol blue	yellow	blue	6.0–7.6
phenolphthalein	colourless	crimson	8.3–10.0

Figure 6.12 shows the structures of the acid and conjugate base forms of methyl orange.

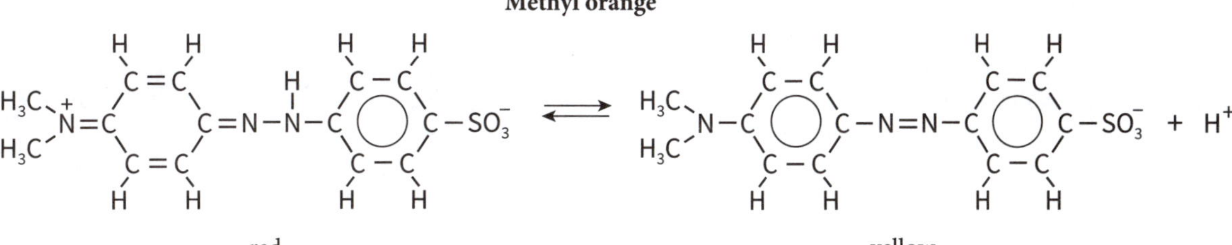

Figure 6.12 Acid and conjugate base structures of methyl orange

- ➔ As a solution becomes increasingly acidic the indicator equilibrium shifts more to the left and more HIn molecules form, according to Le Chatelier's principle. Thus the colour of the indicator becomes increasingly the colour of the unionised acid molecule.
- ➔ As a solution becomes increasingly alkaline the indicator equilibrium shifts more to the right as hydrogen ions become neutralised and more In^- ions form, according to Le Chatelier's principle. Thus the colour of the indicator becomes increasingly the colour of the conjugate base (In^-).
- ➔ The transition from the acid form to the conjugate base form is pH dependent. In the case of methyl orange if the pH is less than 3.1 then the red B/L acid molecule predominates. At pH levels above 4.4 the equilibrium shifts strongly to the right and the yellow In^- ion predominates. At pH levels between 3.1 and 4.4, the colour is orange as the B/L molecule and the In^- ion are both in appreciable concentrations.
- ➔ Table 6.4 shows that different indicators change colour at different concentrations of hydrogen ions or hydroxide ions in aqueous solution. This change in colour occurs over a narrow range of hydrogen ion or hydroxide ion concentration. Outside this range the indicator does not change colour anymore.
  - *Methyl orange* is a useful indicator to detect the difference between solutions that are strongly acidic and weakly acidic. In neutral water the indicator is yellow but as the acidity increases it changes to *orange* (between pH 3.1 and 4.4) and then red below pH 3.1.
  - *Litmus* is often used in its red form or blue form. Red litmus paper will turn blue in alkaline solutions. Blue litmus paper will turn red in acidic solutions. Between pH 5.0 and 8.0 the indicator appears *purple* as both the red form and blue form are present in significant amounts. Red and blue litmus papers do not change colour in pure water.
  - *Bromothymol blue* detects small changes in acidity or alkalinity near neutrality. In solutions with pH levels between 6.0 and 7.6 it is *green*, but the presence of small amounts of acid will turn it yellow. Small quantities of base will turn it blue.
  - *Phenolphthalein* is a useful indicator to detect the difference between solutions that are strongly alkaline and weakly alkaline. In neutral water the indicator is colourless but as alkalinity increases it changes to *pink* (between pH 8.3 and 10.0) and then to crimson above pH 10.0.
- ➔ Figure 6.13 shows the colour of these indicators as a function of pH.

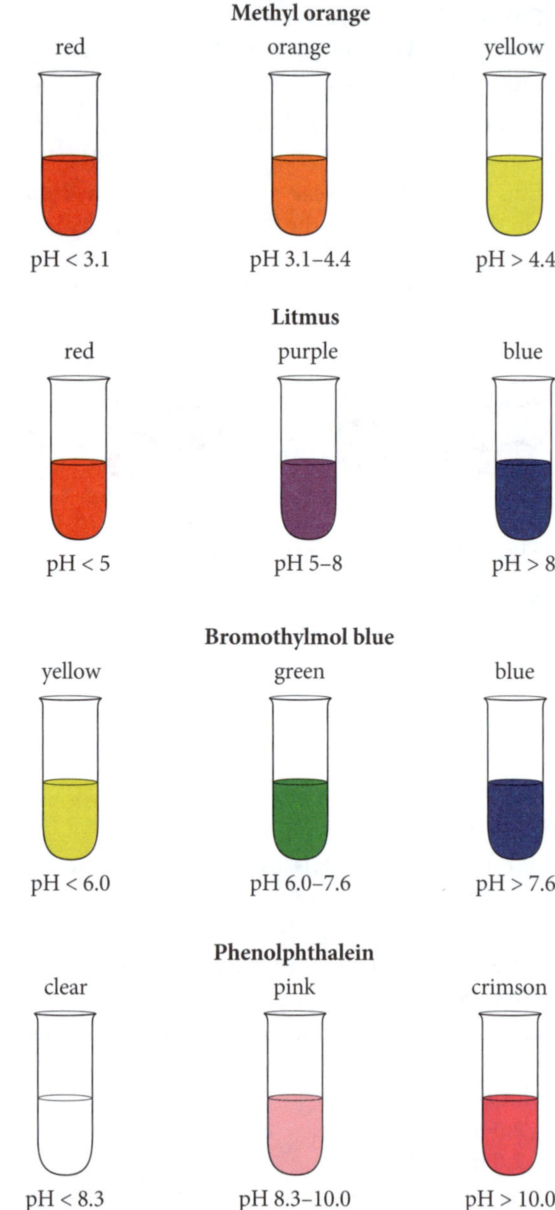

Figure 6.13 Indicator colours

EXAMPLE 13

Use Table 6.3 and Figure 6.13 to identify the colour of each indicator in solutions of the following pH:

a 4.0
b 6.5
c 10.5

Use the pH range for each indicator to determine its colour

Answer:

pH	Methyl orange	Litmus	Bromothymol blue	Phenolphthalein
4.0	orange	red	yellow	colourless
6.5	yellow	purple	green	colourless
10.5	yellow	blue	blue	crimson

EXAMPLE 14

Acetic acid solution A has a concentration of 5.0 mol/L. A series of dilutions with distilled water was made and shown in a Dilutions table. Solutions B to G are formed.

Dilutions

Dilution	Solution formed	Dilution	Solution formed
20 mL of A diluted to 100 mL	B	10 mL of D diluted to 100 mL	E
10 mL of B diluted to 100 mL	C	10 mL of E diluted to 100 mL	F
10 mL of C diluted to 100 mL	D	10 mL of F diluted to 100 mL	G

Four indicators were used to test the pH of each solution. The colours of the indicators were recorded in an indicator table.

Indicator colours

Indicator	A	B	C	D	E	F	G
thymol blue	orange	orange	yellow	yellow	yellow	yellow	yellow
methyl yellow	red	red	red	orange	orange	yellow	yellow
bromocresol green	yellow	yellow	yellow	yellow	green	green	green
methyl red	red	red	red	red	red	red	orange

The pH ranges of these four indicators were tabulated.

pH ranges

Indicator	pH range	Low pH colour	Intermediate pH colour	High pH colour
thymol blue	1.2–2.8	red (pH < 1.2)	orange (pH 1.2–2.8)	yellow (pH > 2.8)
methyl yellow	2.9–4.0	red (pH < 2.9)	orange (pH 2.9–4.0)	yellow (pH > 4.0)
bromocresol green	3.8–5.4	yellow (pH < 3.8)	green (pH 3.8–5.4)	blue (pH > 5.4)
methyl red	4.4–6.2	red (pH < 4.4)	orange (pH 4.4–6.2)	yellow (pH > 6.2)

Estimate the pH ranges of solutions A to G and describe the trend in pH on dilution.

Answer:

Thymol blue is orange in the pH range 1.2 to 2.8

The table summarises the results.

Solution	Thymol blue pH	Methyl yellow pH	Bromocresol green pH	Methyl red pH	Solution pH range
A	1.2–2.8	< 2.9	< 3.8	< 4.4	1.2–2.8
B	1.2–2.8	< 2.9	< 3.8	< 4.4	1.2–2.8
C	> 2.8	< 2.9	< 3.8	< 4.4	2.8–2.9
D	> 2.8	2.9–4.0	< 3.8	< 4.4	2.9–3.8
E	> 2.8	2.9–4.0	3.8–5.4	< 4.4	3.8–4.0
F	> 2.8	> 4.0	3.8–5.4	< 4.4	4.0–4.4
G	> 2.8	> 4.0	3.8–5.4	4.4–6.2	4.4–5.4

An assessment of the data shows that the pH of the acetic acid solution increases on dilution.

FIRSTHAND INVESTIGATION 2

Strengths of acids and bases

» Students conduct an investigation to demonstrate the use of pH to indicate the differences between the strength of acids and bases.

Aim

to use pH measurements to determine whether acids and bases are strong or weak

Chemicals/equipment

0.010 mol/L hydrochloric acid
0.010 mol/L acetic acid
0.010 mol/L citric acid
0.010 mol/L sodium hydroxide
0.010 mol/L ammonia solution
0.010 mol/L sodium hydrogen carbonate solution
Calibrated pH electrode and meter

Method

1. Place 50 mL of the HCl in a clean beaker and measure its pH using the calibrated pH electrode.
2. Rinse the electrode with distilled water and measure the pH of the acetic acid and citric acid solutions.
3. Recalibrate the pH electrode.
4. Repeat the procedure using the basic solutions.

Sample results

Acidic solution	Hydrochloric acid	Acetic acid	Citric acid
pH	2.0	3.4	2.6

Basic solution	Sodium hydroxide solution	Ammonia solution	Sodium hydrogen carbonate solution
pH	12.0	10.6	8.2

Figure 6.14 shows the structure of citric acid.

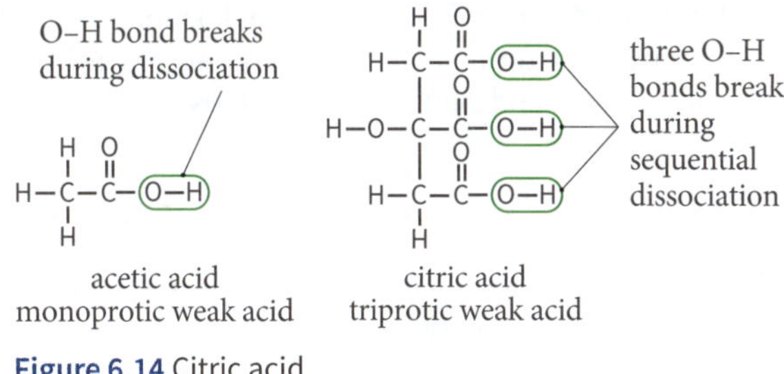

Figure 6.14 Citric acid

Analysis

The 0.010 mol/L HCl solution has a measured pH of 2.0 as expected, based on the pH calculation for a strong acid:

$$pH = -\log_{10}(0.010) = 2.0$$

The acetic acid and citric acid solutions have the same molarity but higher pH levels as they are weak acids and incompletely dissociated.

The 0.010 mol/L NaOH solution has a measured pH of 12.0 as expected, based on the pH calculation for a strong base:

$$pOH = -\log_{10}(0.010) = 2.0$$
$$pH = 14 - pOH = 14 - 2.0 = 12.0$$

The ammonia and sodium hydrogen carbonate solutions have the same molarity but lower pH levels as they are weak bases.

Conclusion

Weak acid and weak base solutions have different pH levels than strong acid and strong base solutions of the same molarity.

→ KEY QUESTIONS

10 Use an equation to explain why potassium sulfite (K_2SO_3) is a basic salt.

11 Explain why phenolphthalein is a suitable indicator to distinguish between weakly basic solutions and strongly basic solutions.

12 A 0.10 mol/L of $FeCl_3$ has a pH of 2.0. The acidity is due to the hexaaquairon (III) ion ($Fe(H_2O)_6^{2+}$), which is a Brønsted–Lowry acid. Use an equation to demonstrate the acidic behaviour of this ion in water.

Answers ➲ p. 88

FIRSTHAND INVESTIGATION 3

Modelling acids and bases in solution

» Students construct models and/or animations to communicate the differences between strong, weak, concentrated and dilute acids and bases.

Aim

to use models to distinguish between strong and weak acids and bases and also to distinguish concentrated and dilute acid and base solutions

Equipment

Molecular model kits
Plastic trays

Method

A Strong versus weak acids

1 Use the model kit to make five HCl molecules and five HF molecules. Use the plastic connectors to join the atoms together.

2 Use two plastic trays to represent beakers. Imagine that each tray contains water.

3 Into the first tray place the HCl molecules and break their plastic connectors to simulate the formation of separate hydrogen ions and chloride ions. This tray now represents the strong acid that is completely dissociated.

4 Into the second tray place the five HF molecules but only break the connector of one molecule to simulate the formation of a hydrogen ion and a fluoride ion. Leave the other molecules intact. This tray now represents the weak acid that is only partially dissociated.

Figure 6.15 models strong and weak acid solutions.

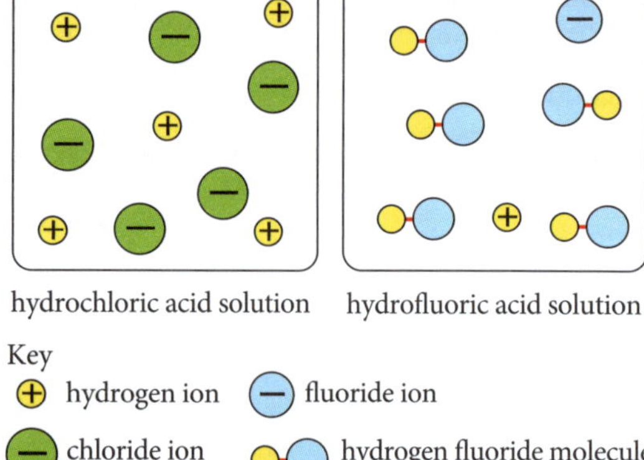

Figure 6.15 Strong and weak acid solutions

B Strong versus weak bases

1 Use the model kit to make five NaOH ionic particles and five ammonia (NH_3) molecules and five water molecules (H_2O). Use the plastic connectors to join the atoms together.

2 Use two plastic trays to represent beakers. Imagine that each tray contains water.

3 Into the first tray place the five NaOH ionic particles and break the plastic connectors between the Na^+ and OH^- ions to simulate the formation of separate sodium ions and hydroxide ions. This tray now represents the strong base that is completely dissociated.

4 Into the second tray place the five NH_3 molecules and the five H_2O molecules. Break the connector of one H_2O molecule to produce separate H^+ and OH^- ion particles. Join the H^+ ion to one of the NH_3 molecules to form an ammonium ion (NH_4^+). Leave the remainder of ammonia molecules and water molecules in the tray. This tray now represents the weak base that is only partially dissociated.

Figure 6.16 models strong and weak base solutions.

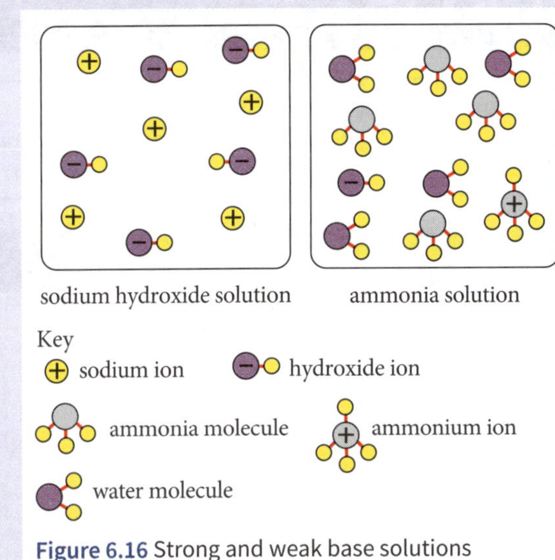

Figure 6.16 Strong and weak base solutions

C Concentrated versus dilute acid and base solutions

1. Into the first tray place six hydrogen ion particles and six chloride ion particles. This represents the concentrated solution of hydrochloric acid.
2. Into the second tray place two hydrogen ion particles and two chloride ion particles. This represents the dilute solution of hydrochloric acid.
3. Repeat steps 1 and 2 with NaOH, in which sodium ion particles and hydroxide ion particles are used.

Figure 6.17 models concentrated and dilute acid and base solutions.

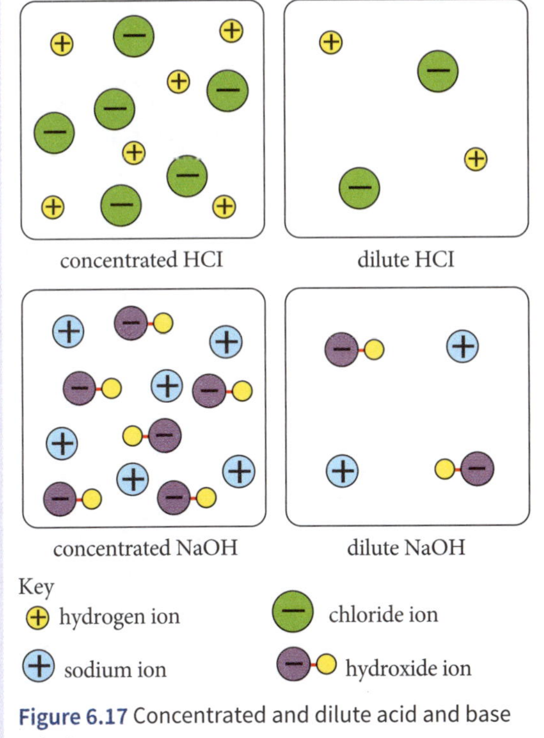

Figure 6.17 Concentrated and dilute acid and base solutions

Conclusion

Strong acid and strong base solutions contain only ions formed from the complete dissociation of the acid or base. Weak acid and weak base solutions contain a mixture of molecules and ions. Concentrated solutions of strong acids or strong bases contain a large number of ions in a given volume of water, whereas dilute solutions contain a much smaller number of ions in a given volume of water.

These simplified models do not show how the ions are solvated by the water molecules due to ion–dipole attractions.

➔ Further information about strong and weak acids and bases can be found in various online videos. Many of these videos show animations of acid–base reactions. To start your investigation, enter the following title in a search bar: strengths and weaknesses of acids and bases

CHAPTER SYLLABUS CHECKLIST

Are you able to answer every question from the syllabus for this chapter? Tick each question as you go through the checklist if you are able to answer it. If you cannot answer a question, turn to the relevant page in the study guide to find the answer. For NESA key word meanings, go to www.educationstandards.nsw.edu.au and search 'key words'.

FOR A COMPLETE UNDERSTANDING OF THIS TOPIC:		PAGE NO.	✓
1	Can I calculate the pH given the hydrogen ion concentration?	73	
2	Can I relate pH to the colours of different indicators?	73	
3	Can I describe an investigation to determine the pH of different solutions?	74	
4	Can I relate K_w to the water ionisation equilibrium?	75	
5	Can I calculate the pOH given the hydroxide ion concentration?	76	
6	Can I calculate the hydrogen ion or hydroxide ion concentration using the formula pH + pOH = 14?	76	
7	Can I perform calculations involving neutralisation reactions where one reagent is limiting?	77	
8	Can I describe examples of the Brønsted–Lowry theory of acids and bases?	78	
9	Can I explain why some chemical species can be amphiprotic?	79	
10	Can I use equations to explain why salts can be classified as acid or basic salts?	80	
11	Can I use the Brønsted–Lowry theory to explain the colour changes of indicators?	82	
12	Can I describe an investigation to demonstrate the use of pH to indicate the differences between the strength of acids and bases?	83	
13	Can I construct models to explain the differences between strong, weak, concentrated and dilute acids and bases?	84	

HSC EXAM-TYPE QUESTIONS

Objective-response questions (1 mark each)

1 Identify which acid is not a weak acid.

A carbonic acid

B hydrofluoric acid

C acetic acid

D nitric acid

2 The conjugate base of carbonic acid is:

A carbonate ion

B hydrogen carbonate ion

C hydronium ion

D hydroxide ion

3 Calculate the pH of a 0.00200 mol/L solution of nitric acid.

A 2.00

B 2.70

C 11.30

D 12.00

4 Identify the household product that has the highest pH.

A vinegar

B ammonia window cleaner

C milk

D apple juice

5 Acid–base indicators are weak Brønsted–Lowry acids. Identify the colour of the conjugate base of methyl orange.

A yellow

B red

C crimson

D blue

Extended-response questions

6 An aqueous solution has a pH of 9.5.

a Use the water ionisation constant to calculate the hydroxide ion concentration of this solution. (3 marks)

b What colour will bromothymol blue and phenolphthalein turn when separately added to samples of this solution? (2 marks)

7 a Calculate the pH of a sulfuric acid solution prepared by mixing 5.00 mL of 0.0050 mol/L sulfuric acid and 995.0 mL water. (3 marks)

b What colour will methyl orange indicator turn in this diluted solution? (1 mark)

8 Universal indicator was added to water, sodium chloride solution, sodium hydrogen sulfate solution and sodium carbonate solution. The indicator colour results are shown in Figure 6.18.

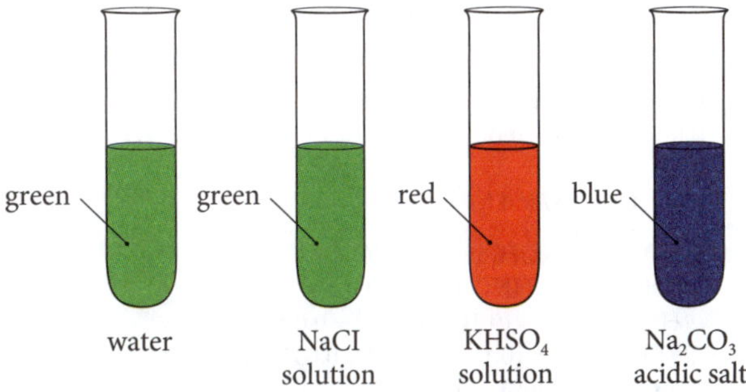

Figure 6.18 Universal indicator experiments

a Explain why the indicator was tested in pure water. (1 mark)

b Use equations to explain the indicator colour in each salt solution. (5 marks)

9 a Explain why a solution of potassium nitrite (KNO_2) turns universal indicator blue. Use an appropriate equation in your answer. (2 marks)

b Zinc sulfate solution is acidic. The tetraaquazinc ion ($Zn(H_2O)_4^{2+}$) is a Brønsted–Lowry acid. Use this information to write an equation to show the acidity of zinc sulfate in water. (2 marks)

10 a Calculate the final pH of a solution formed by mixing 250 mL of 0.002 00 mol/L H_2SO_4 and 350 mL of 0.001 50 mol/L KOH. (5 marks)

b Identify the colour of bromothymol blue and phenolphthalein in the final solution. (2 marks)

ANSWERS

KEY QUESTIONS

Key questions ⊃ p. 74

1 pH 6 = yellow; pH 9 = blue

2 The calibration procedure ensures more accurate readings. Following its use, it should be rinsed with distilled water to remove any adhering solution from a previous experiment.

3 $pH = -\log_{10}[2.50 \times 10^{-4}] = 3.60$

Key questions ⊃ p. 77

4 Calcium hydroxide dissociates to produce two hydroxide ions:

$$pOH = -\log_{10}[2 \times 0.005\,50] = 1.96$$

5 $c_1V_1 = c_2V_2$
$(0.001\,00)(0.200) = c_2(0.750)$
$c_2 = 2.67 \times 10^{-3}$ mol/L = $[OH^-]$
pOH = 2.57
pH = 14 − 2.57 = 11.43

6 $HNO_3(aq) + KOH(aq) \rightarrow KNO_3(aq) + H_2O(l)$
Stoichiometry: $HNO_3(aq) : KOH(aq) = 1 : 1$
$n(HNO_3) = cV = (0.100)(0.250) = 0.0250$ mol
$n(KOH) = cV = (0.150)(0.250) = 0.0375$ mol
Thus HNO_3 is limiting.
Excess KOH: $n(KOH) = 0.0375 - 0.0250 = 0.0125$ mol
$c(KOH) = [OH^-] = n/V = \frac{0.0125}{0.500} = 0.0250$ mol/L
pOH = 1.60
pH = 14 − 1.60 = 12.4

Key questions ⊃ p. 80

7 HCO_3^-

8 H_2S

9 $HPO_4^{2-}(aq) + H_3O^+(aq) \rightarrow H_2PO_4^-(aq) + H_2O(l)$
$HPO_4^{2-}(aq) + OH^-(aq) \rightarrow PO_4^{3-}(aq) + H_2O(l)$

Key questions ⊃ p. 84

10 The sulfite ion is a Brønsted–Lowry base:

$$SO_3^{2-}(aq) + H_2O(l) \leftrightharpoons HSO_3^-(aq) + OH^-(aq)$$

The presence of hydroxide ions makes the solution alkaline.

11 Between pH 8.3 and 10 the indicator is pink and above pH 10 it is crimson. Thus weakly basic solutions can be distinguished from strongly basic solutions.

12 $Fe(H_2O)_6^{2+}(aq) + H_2O(l) \leftrightharpoons Fe(H_2O)_5OH^+(aq) + H_3O^+(aq)$
The equilibrium solution contains hydronium ions, which makes the $FeCl_3$ solution acidic.

HSC EXAM-TYPE QUESTIONS

Objective-response questions

1 **D.** Nitric acid is fully dissociated. **A**, **B** and **C** are incorrect as they are all weak acids.

2 **B.** Subtract a proton from H_2CO_3 to obtain the conjugate base. **A** is incorrect as two protons have been removed. **C** is incorrect as the hydronium ion is the conjugate acid of water. **D** is incorrect as the hydroxide ion is the conjugate base of water.

3 **B.** Nitric acid is a strong acid and fully dissociated.
$[H^+] = 0.002\,00$ mol/L
$pH = -\log[H^+] = -\log(0.002\,00) = 2.70$
Thus **A**, **C** and **D** are mathematically incorrect.

4 **B.** Ammonia is a base whereas vinegar, milk and apple juice are all acidic. Thus **A**, **C** and **D** are incorrect.

5 **A.** Methyl orange turns yellow in solutions with a pH greater than 4.4. Yellow is the colour of its conjugate base. **B** is incorrect as red is the colour in its unionised acid form. **C** is incorrect as crimson is the colour of phenolphthalein's conjugate base. **D** is incorrect as blue is the colour of bromothymol blue's conjugate base.

Extended-response questions

6 EM Students need to demonstrate how the water ionisation constant is related to pH. Students also need to recall indicator colours at different values of pH.

a $[H^+] = 10^{-pH} = 10^{-9.5} = 3.162 \times 10^{-10}$ mol/L ✓
$K_w = [H^+][OH^-] = 1 \times 10^{-14}$
$(3.162 \times 10^{-10})[OH^-] = 1 \times 10^{-14}$ ✓
$[OH^-] = \frac{(1 \times 10^{-14})}{(3.162 \times 10^{-10})} = 3.16 \times 10^{-5}$ mol/L ✓

b Blue as bromothymol turns blue above pH 7.6; ✓ pink as phenolphthalein is pink between pH 8.3 and 10.0 ✓

7 EM Students need to determine the total volume (in litres) of the final solution, as well as taking into account that sulfuric acid is diprotic. Students need to demonstrate an understanding of an indicator's colour within its pH range.

a $n(H_2SO_4) = c.V = (0.005\,00)(0.005) = 2.50 \times 10^{-5}$ mol ✓
New total volume of solution = 5.00 + 995.0 = 1000 mL = 1.000 L
On dilution, the new concentration is:
$c = n/V = \frac{2.50 \times 10^{-5}}{1.000} = 2.50 \times 10^{-5}$ mol/L
Sulfuric acid is fully dissociated in dilute solutions. The acid is diprotic.
$[H^+] = 2 \times 2.50 \times 10^{-5} = 5.00 \times 10^{-5}$ mol/L ✓
$pH = -\log(5.00 \times 10^{-5}) = 4.30$ ✓

b Methyl orange is orange in the range 3.1–4.4. The indicator will be orange-yellow as the pH is close to the transition zone. ✓

8 EM Students are required to show an understanding of controls in firsthand investigations. Students must demonstrate a good understanding of salt hydrolysis, using the Brønsted–Lowry theory.

a Water is the control. ✓

b NaCl—neutral salt; no hydrolysis as sodium ions and chloride ions are too weak to cause hydrolysis ✓
$KHSO_4$—acidic salt; the HSO_4^- ion is a B–L acid ✓
$HSO_4^-(aq) + H_2O(l) \leftrightharpoons SO_4^{2-}(aq) + H_3O^+(aq)$ ✓
Na_2CO_3—basic salt; the CO_3^{2-} ion is a B–L base ✓
$CO_3^{2-}(aq) + H_2O(l) \leftrightharpoons HCO_3^-(aq) + OH^-(aq)$ ✓

9 EM Students are required to use the Brønsted–Lowry theory to explain hydrolysis of salts. Students need to determine whether or not the cation or anion undergoes hydrolysis.

a Potassium nitrite is a basic salt and hydrolyses to produce an alkaline solution, which turns the indicator blue. ✓
$NO_2^-(aq) + H_2O(l) \leftrightharpoons HNO_2(aq) + OH^-(aq)$ ✓

b $Zn(H_2O)_4^{2+}(aq) + H_2O(l) \leftrightharpoons Zn(H_2O)_3(OH)^+(aq) + H_3O^+(aq)$ ✓
The hydronium ions that form in the hydrolysis equilibrium make the solution acidic. ✓

10 EM Students are required to demonstrate that an understanding of reaction stoichiometry and the diprotic nature of sulfuric acid is essential in the calculation. The steps of the calculation must be fully expressed and in a logical order.

a $H_2SO_4(aq) + 2KOH(aq) \rightarrow K_2SO_4(aq) + 2H_2O(l)$ ✓
$n(H_2SO_4) = c.V = (0.002\,00)(0.250) = 0.000\,500$ mol
$n(KOH) = c.V = (0.001\,50)(0.350) = 0.000\,525$ mol ✓
Reaction stoichiometry: $H_2SO_4 : KOH = 1 : 2$
Thus the sulfuric acid is in excess and all the potassium hydroxide will be neutralised.
Excess H_2SO_4: $n(H_2SO_4) = 0.000\,500 - \frac{1}{2}(0.000\,525)$
$= 0.000\,237\,5$ mol ✓
Total volume = 250 + 350 = 600 mL = 0.600 L
$c(H_2SO_4) = n/V = \frac{0.000\,237\,5}{0.600} = 0.000\,396$ mol/L ✓
Sulfuric acid is diprotic.
$[H^+] = 0.000\,792$ mol/L
$pH = -\log_{10}(0.000\,792) = 3.10$ ✓

b bromothymol blue—yellow; ✓ phenolphthalein—colourless ✓

MODULE 6 ACID–BASE REACTIONS

CHAPTER 7 QUANTITATIVE ANALYSIS

INQUIRY QUESTION:

How are solutions of acids and bases analysed?

Volumetric analysis is a common laboratory procedure that chemists use to determine the concentration of acids and bases in an aqueous solution. As these solutions are colourless, acid–base indicators such as methyl orange and phenolphthalein are used to determine when the stoichiometric quantities have reacted. This chapter investigates quantitative analysis of acids and bases.

1 Titrations

» Students conduct practical investigations to analyse the concentration of an unknown acid or base by titration.

Volumetric analysis

- Volumetric analysis is a quantitative laboratory technique used to determine the concentration of a solution by reacting it with a **standard solution**. For example, the concentration of an alkaline solution can be determined by reacting it with a primary standard solution of an acid.
- The first step in acid–base volumetric analysis involves the preparation of a solution of known concentrations. This solution is prepared using a **primary standard**. A primary standard is a stable, pure solid of known composition. The primary standard must also be readily soluble in water.
- Anhydrous sodium carbonate (Na_2CO_3) is a common primary standard base. Sodium hydroxide and potassium hydroxide are unsuitable as they absorb water from the atmosphere. These bases can be classified as *secondary standards* once their exact concentrations are determined.

standard solution: a solution of accurately known concentration

primary standard: a pure solid of known composition, which is used to prepare a standard solution

Preparation of a primary standard base

- Anhydrous sodium carbonate is one of the most commonly used primary standard bases. It is stable, pure and readily dried so that it can be weighed easily.
- To prepare 500.0 mL of a 0.050 00 mol/L solution of sodium carbonate the following procedure is used:
  - Dry the sodium carbonate in a low temperature drying oven and then cool the solid in a desiccator. This procedure ensures the solid is anhydrous.
  - Weigh out 2.650 g ($n = m/M = \frac{2.650}{105.99} = 0.025\,00$ mol) of the anhydrous solid and transfer it quantitatively to a 500 mL volumetric flask (which has been pre-rinsed with distilled water) with the aid of water from a wash bottle.
  - Add distilled water to the flask until the base of the meniscus is on the engraved mark in the volumetric flask. Stopper the flask and mix thoroughly.

  Figure 7.1 shows the steps in the preparation of the standard solution.
- Once the standard base is prepared then the concentration of a solution of an acid can be determined by a procedure called a **titration**.

titration: a laboratory method of quantitative chemical analysis that is used to determine the concentration of a solution

Steps of a titration

- A known volume (25.00 mL) of base is pipetted into a clean conical flask. This is repeated for a second flask, which will act as a control.
- The pipette is cleaned prior to use with rinses of water, followed by rinses with the standard base solution. Then the pipette is filled with the standard base solution.
- The acid solution is used to fill the burette that has previously been cleaned with water, and then rinsed with the acid solution.
- Figure 7.2 shows the pipette being filled and the filled burette with the solution filling the space below the tap.

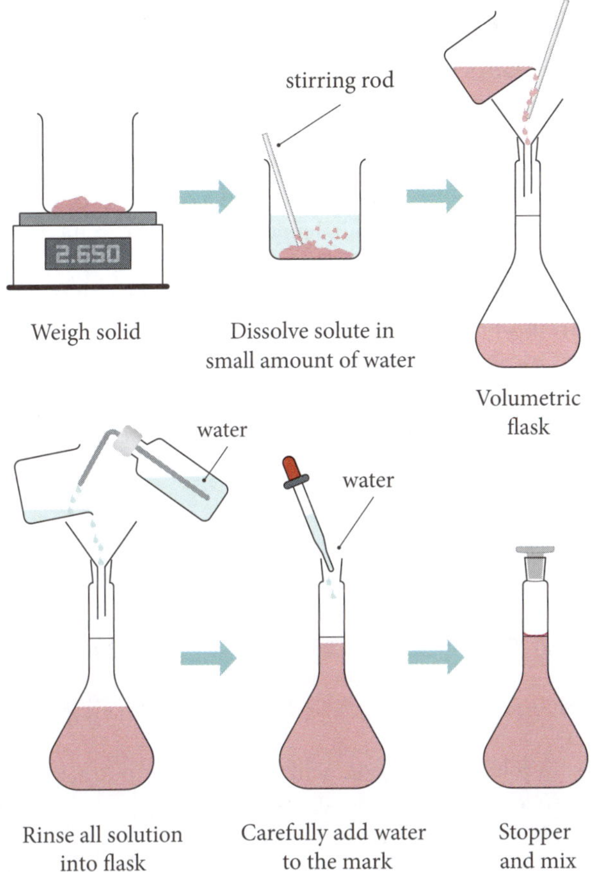

Figure 7.1 Volumetric flask

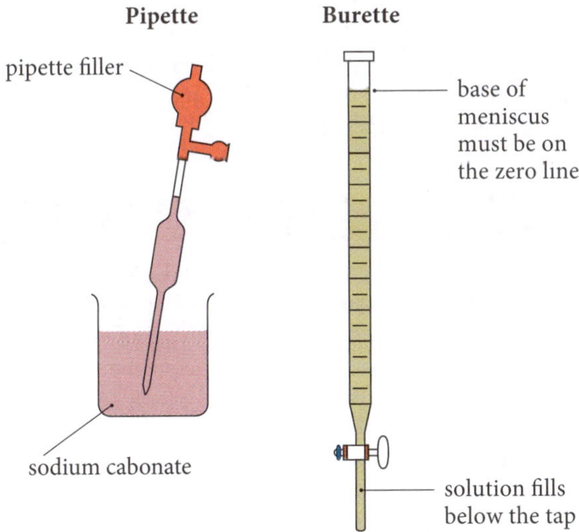

Figure 7.2 Pipette and burette

- Three to four drops of the selected indicator are added to the flask (as well as the control flask) and acid is then run into the flask from the burette. During the acid addition the flask is constantly agitated to ensure mixing and a wash bottle of water is used occasionally to wash solution down the walls of the flask into the bulk of the liquid.
- A change in indicator colour signals the end point of the titration. If the indicator has been selected correctly then the **end point** will coincide with the **equivalence point**. At the equivalence point the amount of acid added is just sufficient to exactly neutralise the base in the flask. For methyl orange the colour change is from yellow to orange.
- The volume of acid **titrant** delivered from the burette is called the **titre**. Normally the first run is quickly executed and is called the *rough titre*. This rough titration establishes the approximate volume at the end point. This volume is not included in subsequent calculations. The next few titrations are done more accurately. With considerable practice students may be able to obtain three titres that agree within 0.10–0.20 mL.
- Figure 7.3 shows the apparatus used in the titration.

end point: the volume of titrant that leads to the first permanent colour change in the indicator
equivalence point: the volume of titrant that is required to exactly neutralise the acid or base in the flask based on the stoichiometry of the reaction
titrant: the solution in the burette which is used to titrate the solution in the flask
titre: the volume of titrant added to achieve an end point

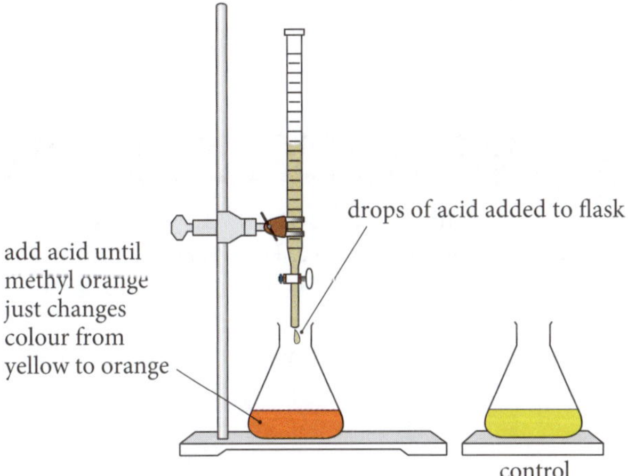

Figure 7.3 Titration

- Titrations are repeated until at least three titres agree to within 0.05 mL. An average titre is calculated but any rough titration is not included in the averaging. The concentration of the 'unknown' solution can now be calculated. This acid solution is now a **secondary standard**.

secondary standard: a solution whose concentration has been determined by titration with a primary standard or other standardised solution

EXAMPLE 1

25.00 mL of a standard 0.050 00 mol/L Na_2CO_3 solution was titrated with a solution of HCl of unknown concentration using methyl orange indicator. The table shows the titres.

Run	Volume HCl (mL)
1	22.30
2	21.85
3	21.80
4	21.90

a Calculate the average titre.

b Calculate the concentration of the HCl.

Consider whether the titre for run 1 should be included in the averaging

Answer:

a The first run is a rough titration and it is not included in the averaging. It is an outlier.

$$\text{Average titre} = \frac{(21.85 + 21.80 + 21.90)}{3} = 21.85 \text{ mL}$$

b The neutralisation equation is:

$$Na_2CO_3(aq) + 2HCl(aq) \rightarrow 2NaCl(aq) + H_2O(l) + CO_2(g)$$

$$n(Na_2CO_3) = c.V = (0.050\,00)(0.025\,00) = 1.250 \times 10^{-3} \text{ mol}$$

Reaction stoichiometry: $Na_2CO_3 : HCl = 1 : 2$

$$n(HCl) = 2 \times 1.250 \times 10^{-3} = 2.500 \times 10^{-3} \text{ mol}$$

$$c(HCl) = n/V = \frac{2.500 \times 10^{-3}}{0.021\,85} = 0.1140 \text{ mol/L}$$

Preparation of a primary standard acid

- Oxalic acid dihydrate ($(COOH)_2{\cdot}2H_2O$) and potassium hydrogen phthalate ($KH(C_8H_4O_4)$) are common primary standard acids. Their structures are shown in Figure 7.4. These primary standards can be used to standardise alkaline solutions of strong bases such as NaOH and KOH.

oxalic acid

potassium hydrogen phthalate

Figure 7.4 Oxalic acid and potassium hydrogen phthalate

- The same method is used is used to prepare a primary standard acid as was used to prepare a primary standard base. Volumetric flasks come in a variety of sizes (e.g. 50 mL, 100 mL, 250 mL, 500 mL) and the requirements of the experiment determine which size to use.

EXAMPLE 2

A standard solution of oxalic acid dihydrate was prepared by dissolving 4.30 g of crystals in sufficient water to make 250 mL of solution in a volumetric flask. This standard solution was used to fill a burette. 25.00 mL of a sodium hydroxide solution was pipetted into a conical flask and four drops of phenolphthalein was added. The oxalic acid solution was added from the burette until the indicator just turned from pink to colourless. The average titre for four titrations was 19.45 mL.

a Calculate the concentration of the oxalic acid solution.

b Calculate the concentration of the sodium hydroxide solution.

Calculate the concentration of the oxalic acid standard

Answer:

a $M(COOH)_2{\cdot}2H_2O = (2 \times 12.01) + (6 \times 16.00) + (6 \times 1.008) = 126.068 \text{ g/mol}$

$$n(COOH)_2{\cdot}2H_2O = m/M = \frac{4.30}{126.068} = 0.034\,11 \text{ mol}$$

$$c(COOH)_2{\cdot}2H_2O = n/V = \frac{0.034\,11}{0.250} = 0.136 \text{ mol/L}$$

b In solution the oxalic acid exists as a diprotic acid ($(COOH)_2$). It is a weak diprotic acid.

$$(COOH)_2(aq) + 2NaOH(aq) \rightarrow Na_2(COO)_2(aq) + 2H_2O(l)$$

Stoichiometry: $(COOH)_2 : NaOH = 1 : 2$

$$n((COOH)_2) = cV = (0.136)(0.019\,45) = 0.002\,645 \text{ mol}$$

$$n(NaOH) = 2 \times 0.002\,645 = 0.005\,29 \text{ mol}$$

$$c(NaOH) = n/V = \frac{0.005\,29}{0.0250} = 0.212 \text{ mol/L}$$

The sodium hydroxide solution is now a secondary standard and it can be used to titrate acids (e.g. acetic acid) of unknown concentration.

Back titrations

- In the analysis of minerals and some mixtures, a direct titration cannot be used. Instead a titration procedure called a **back titration** is used. Consider a sample of limestone, which contains calcium carbonate. The procedure involves two steps. In the first step a known excess of acid is added to the weighed sample of the limestone (an insoluble base). The acid neutralises the calcium carbonate, leaving an excess of acid. The amount of unreacted excess acid is determined by titration with a standardised base solution such as NaOH. With methyl orange indicator the end point will occur when the indicator changes from red to orange.
- Figure 7.5 shows the steps of the back titration.

back titration: a volumetric procedure in which a known excess of a standard reagent is added to the sample to be analysed and the excess reagent is titrated with a standard solution

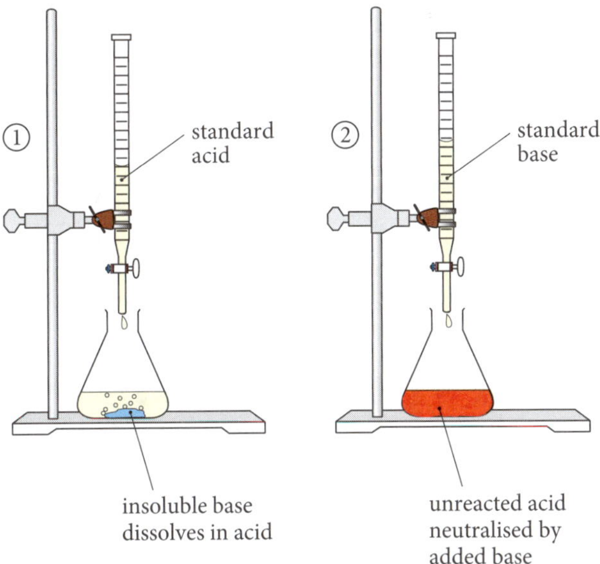

Figure 7.5 Back titration steps used to analyse an insoluble base

EXAMPLE 3

Dolomite is a mixed carbonate mineral. Its chemical formula is $CaMg(CO_3)_2$. An impure sample of dolomite was analysed using a back titration. 0.125 g of impure dolomite was weighed and placed in a conical flask. 40.00 mL of standard 0.100 mol/L hydrochloric acid was added from a burette and the mineral allowed to completely dissolve. The remaining acidic solution was then titrated using 0.100 mol/L sodium hydroxide. The end point occurred when 14.45 mL of NaOH was added.

a **Write a balanced equation for the reaction of dolomite with hydrochloric acid.**

b **Write a balanced equation for the reaction of sodium hydroxide with hydrochloric acid.**

c **Calculate the percentage by weight of pure $CaMg(CO_3)_2$ in the impure dolomite.**

Calcium and magnesium ions form separate chlorides in the reaction with hydrochloric acid

Answer:

a $CaMg(CO_3)_2(s) + 4HCl(aq) \rightarrow CaCl_2(aq) + MgCl_2(aq) + 2H_2O(l) + 2CO_2(g)$

b $NaOH(aq) + HCl(aq) \rightarrow NaCl(aq) + H_2O(l)$

c Calculate the number of moles of NaOH added in the titration:
$n(NaOH) = cV = (0.100)(0.01445) = 0.001\,445$ mol
Calculate the number of moles of unreacted HCl:
Reaction stoichiometry in the titration = 1 : 1
$n(HCl)$ unreacted = 0.001 445 mol
Calculate the initial number of moles of HCl added:
$n(HCl)$ initial $= cV = (0.100)(0.04000) = 0.004\,000$ mol
Calculate the moles of HCl that reacted with the dolomite:
$n(HCl)$ reacted = 0.004 000 − 0.001 445 = 0.002 555 mol
Calculate the moles of $CaMg(CO_3)_2$ present in the mineral sample:
Reaction stoichiometry: $CaMg(CO_3)_2$: HCl = 1 : 4

$$n(CaMg(CO_3)_2 = \frac{0.002\,555}{4} = 0.000\,638\,8 \text{ mol}$$

Calculate the mass of $CaMg(CO_3)_2$:

$$m = nM = (0.000\,638\,8)(40.08 + 24.31 + (2 \times 12.01) + (6 \times 16.00)) = (0.000\,638\,8)(184.41) = 0.118 \text{ g}$$

Calculate the percentage by weight of dolomite in the sample:

$$\text{Percentage } CaMg(CO_3)_2 = \frac{0.118}{0.125} \times 100 = 94.4\%\text{w/w}$$

➔ KEY QUESTIONS

1 **Explain why potassium hydroxide crystals are not suitable to prepare a primary standard solution.**

2 **Explain how a pipette is prepared in order to deliver 25.00 mL of a standard sodium carbonate solution to a conical flask.**

3 **Explain the purpose of conducting a rough titration in the titrimetric analysis of an acidic solution.**

Answers ➲ p. 109

2 Titration graphs

» Students:
- investigate titration curves and conductivity graphs to analyse data to indicate characteristic reaction profiles; for example, strong acid/strong base, strong acid/weak base, weak acid/strong base.
- model neutralisation of strong and weak acids and bases using a variety of media.

Reaction profiles

Indicator selection

➔ To select the most appropriate indicator for the titration, the pH at the equivalence point must be matched to the pH range in which the indicator changes colour.

➔ The end point of a titration occurs when the indicator just changes colour. The indicator must change colour when the graph of pH versus volume of titrant shows its greatest rate of change. This is the steep inflexion region of the graph, which coincides with the equivalence point of the titration. The following pH graphs show the inflexion region for three different titrations. The pH is measured using a calibrated pH glass electrode.

Strong acid/strong base titrations

- When strong bases are titrated with strong acids the pH is 7 at the equivalence point.
- The inflexion in the pH versus volume of titrant graph is very long (pH ~11 down to pH ~3), and consequently *any* of the following indicators may be used as they change colour in this pH inflexion zone:
  - methyl orange (pH 3.1–4.4)
  - bromothymol blue (pH 6.0–7.6)
  - phenolphthalein (pH 8.3–10).
- Figure 7.6 shows the pH graph of a titration in which a strong acid (HCl) is titrated with a strong base (NaOH). The equivalence point is at the centre of the inflexion in the graph. In this case it occurs at a pH of 7.0. The pH ranges of the three indicators are also shown. These ranges are all located in the steep inflexion zone and so all are suitable for strong acid/strong base titrations.

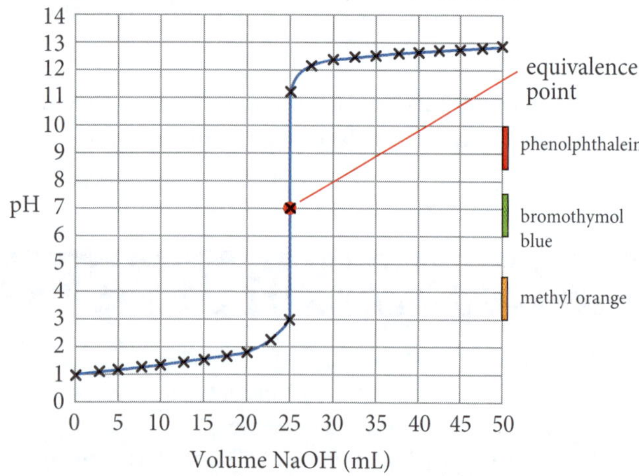

Figure 7.6 Strong acid/strong base titration graph

Strong acid/weak base titrations

- When hydrochloric acid is used to titrate an ammonia solution the ammonium ion is formed. The ammonium ion is a weak conjugate acid and the solution at the equivalence point is weakly acidic rather than neutral.
- Methyl orange is commonly selected as its pH range (3.1–4.4) matches the lower steep section of the inflexion in the pH graph. Bromothymol blue can also be used whereas phenolphthalein starts to change colour too early in the titration and cannot be used. Figure 7.7 shows how the pH changes as ammonia solution (a weak base) is neutralised by hydrochloric acid.

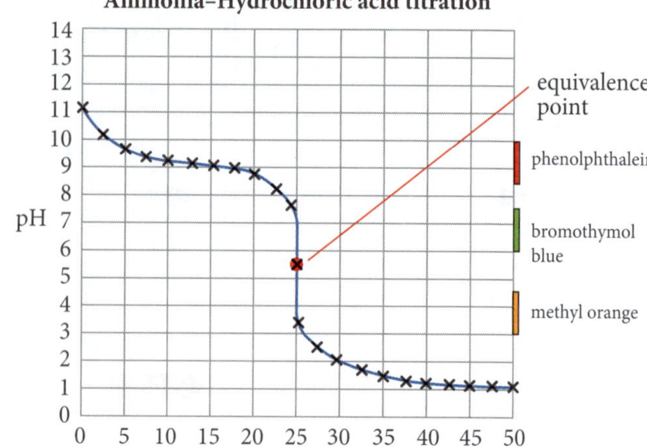

Figure 7.7 Strong acid/weak base titration graph

Strong base/weak acid titrations

- When acetic acid is titrated with NaOH the conjugate base of the acid that is formed at the equivalence point is a weak conjugate base and the solution formed at the equivalence point is alkaline. Phenolphthalein is commonly selected as its pH range (8.3–10) matches the steep inflexion in the pH curve. Figure 7.8 shows how the pH changes as the acetic acid solution (a weak acid) is neutralised by sodium hydroxide (a strong base). Bromothymol blue can also be used but methyl orange cannot be used as it changes colour too early in the titration.

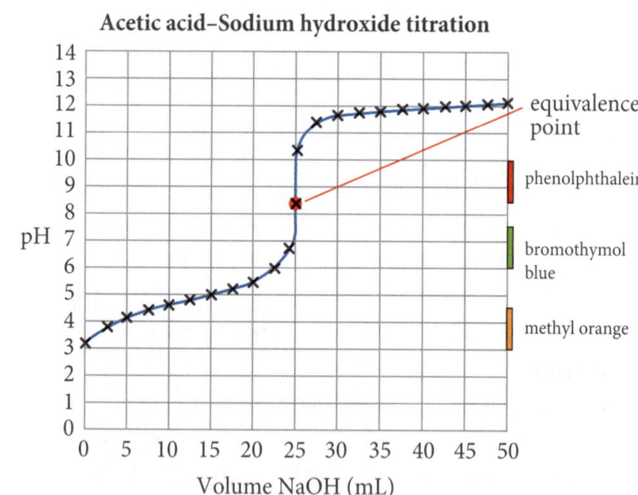

Figure 7.8 Strong base/weak acid titration graph

- Titrations of weak acids with weak bases are not performed as there is no rapid pH change at the equivalence point. Thus there is no discernible end point to the titration. The pH graph of the titration of acetic acid solution with ammonia solution is shown in Figure 7.9.

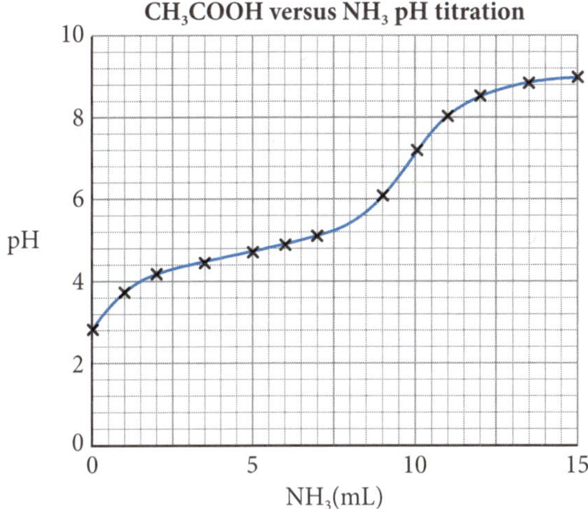

Figure 7.9 pH titration of acetic acid and ammonia

- Ammonia solutions decrease in concentration with time because ammonia escapes into the air each time the container is opened. To determine the concentration of an ammonia solution a back titration can be used.

EXAMPLE 4

A back titration was performed to determine the concentration of an ammonia solution. 50.00 mL of standard 0.100 mol/L sulfuric acid was added to 10.00 mL of the ammonia solution. The excess sulfuric acid was then back titrated with a standard 0.100 mol/L sodium carbonate solution, using methyl orange indicator. The end point occurred at 11.75 mL of sodium carbonate. Calculate the concentration of the ammonia solution.

Calculate the number of moles of unreacted (excess) sulfuric acid

Answer:

Write the equation for the reaction of sulfuric acid and sodium carbonate:

$$H_2SO_4(aq) + Na_2CO_3(aq) \rightarrow Na_2SO_4(aq) + H_2O(l) + CO_2(g)$$

Calculate the number of moles of unreacted sulfuric acid:

Reaction stoichiometry = 1 : 1

$n(Na_2CO_3) = cV = (0.100)(0.01175) = 0.001175$ mol

$n(H_2SO_4)$ unreacted = 0.001175 mol

Calculate the initial number of moles of sulfuric acid added to the ammonia solution:

$n(H_2SO_4)$ initial $= cV = (0.100)(0.0500) = 0.005000$ mol

Calculate the moles of sulfuric acid that reacted with the ammonia:

$n(H_2SO_4)$ reacted $= 0.005000 - 0.001175 = 0.003825$ mol

Write the equation for the reaction of sulfuric acid and ammonia solution:

$$H_2SO_4(aq) + 2NH_3(aq) \rightarrow (NH_4)_2SO_4(aq)$$

Calculate the concentration of the ammonia solution:

Reaction stoichiometry = $H_2SO_4 : NH_3 = 1 : 2$

$n(NH_3) = 2(0.0003825) = 0.0007650$ mol

$$c(NH_3) = n/V = \frac{(0.0007650)}{(0.01000)} = 0.0765 \text{ mol/L}$$

Conductivity titrations

- Electrical conductivity of a solution can be used to determine the equivalence point in acid–base titrations. Figure 7.10 shows the equipment used.

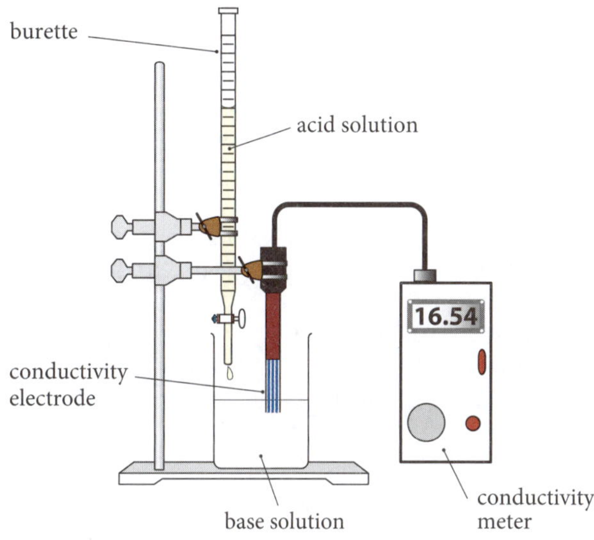

Figure 7.10 Conductivity titration

- In a titration of a strong acid (e.g. HCl) and a strong base (e.g. NaOH) the conductivity decreases as the acid is added to the base solution. This is due to the neutralisation of the hydroxide ions by the hydrogen ions to form water. Water is a non-conductor. As a salt such as NaCl is also formed, the conductivity at the equivalence point is different and lower than the conductivity of the original base solution.

EXAMPLE 5

25.00 mL of 0.0100 mol/L NaOH is pipetted into a beaker and a conductivity electrode was inserted. A burette was filled with a dilute HCl solution. A conductivity titration was performed. The results are tabulated.

a **Plot a graph of this data.**

b **Determine the concentration of the HCl.**

HCl volume (mL)	0	5.0	10.0	15.0	16.0	17.0	18.0	19.0	20.0
Conductivity units	23.0	18.8	14.2	9.7	8.8	8.5	10.0	12.5	14.5

Answer:

Plot conductivity on the vertical axis and select an appropriate scale

a Figure 7.11 shows the conductivity graph.

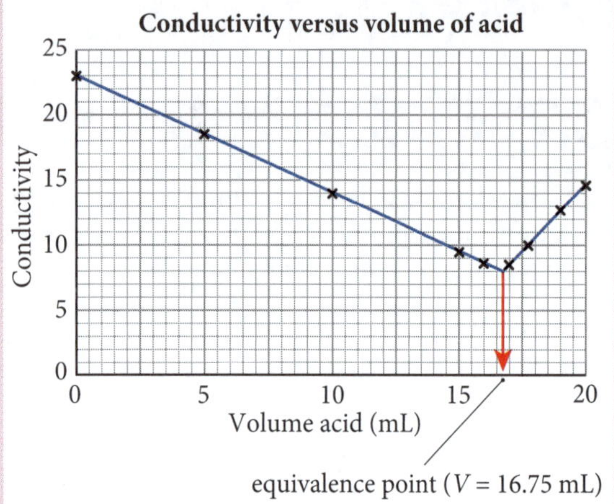

Figure 7.11 Conductivity graph

b $HCl(aq) + NaOH(aq) \rightarrow NaCl(aq) + H_2O(l)$

$n(NaOH) = cV = (0.0100)(0.0250) = 0.000\,250$ mol

Stoichiometry = 1 : 1

$n(HCl) = 0.000\,250$ mol

Titre = 16.75 mL

$c(HCl) = n/V = \dfrac{(0.000\,250)}{(0.01675)} = 0.0149$ mol/L

➔ Conductivity measurements can also be used where an insoluble salt precipitates during the titration. For example, when sulfuric acid is used to titrate a barium hydroxide solution the conductivity of the solution drops to zero at the equivalence point.

➔ As sulfuric acid is added to a barium hydroxide solution, water and a precipitate form and the conductivity drops. When all ions are removed at the equivalence point the conductivity of the solution drops to zero. Any excess acid will then increase the conductivity again:

$$Ba(OH)_2(aq) + H_2SO_4(aq) \rightarrow BaSO_4(s) + 2H_2O(l)$$

Figure 7.12 is a sketch of how the conductivity changes as the sulfuric acid is added. When excess acid is added the conductivity rises again.

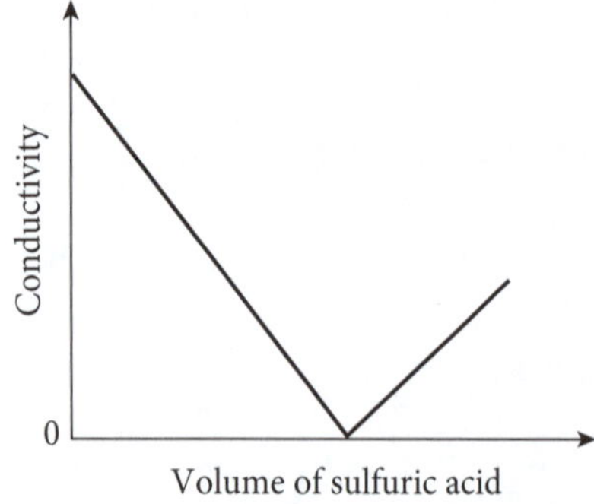

Figure 7.12 Conductivity changes during the precipitation reaction

Modelling neutralisation

➔ The neutralisation reactions involving acids and bases can be modelled using a variety of media.

➔ One method is to investigate demonstrations using the internet. Online videos involving neutralisation are quite common. Go to the online website and type in such key words as:

- acid base neutralisation
- dissociation strong and weak acids
- how are strong and weak acids different

You will soon find other videos that deal with neutralisation.

➔ Neutralisation in titrations can also be investigated using a spreadsheet program. The following activity illustrates the use of this program to construct graphs.

FIRSTHAND INVESTIGATION 1

Modelling a titration

» Students model neutralisation of strong and weak acids and bases using a variety of media.

Aim

to use a spreadsheet application to chart pH versus volume of titrant for a titration involving a strong acid (hydrobromic acid) and an ammonia solution

Method

1 The supplied data is tabulated in the spreadsheet and the charting program is used to create a chart for analysis.

2 Copy the following data into the spreadsheet application. 20.0 mL of 0.100 mol/L HBr is pipetted into a flask and titrated with the ammonia solution. The pH of the mixture is measured throughout the titration.

$NH_3(aq)$ (mL)	0	2.00	4.00	8.00	10.00	10.50	10.70	11.20	11.30	11.40
pH	1.00	1.10	1.20	1.75	2.20	2.60	2.80	3.60	3.90	4.30

$NH_3(aq)$ (mL)	11.50	11.55	11.60	12.00	14.00	15.00	16.00	17.00	18.00	20.00
pH	5.35	6.50	7.60	8.60	8.95	9.01	9.05	9.10	9.17	9.25

3 For the spreadsheet file, select the following menu items: Insert Chart/Scatter/With Smooth Lines and Markers.

4 Insert axes, titles and units, and add a title for the graph.

5 Determine the equivalence point for this titration by determining the midpoint of the inflexion region. Use the titre value at this point to calculate the concentration of the ammonia solution.

Sample answer

Figure 7.13 shows the pH graph of the titration of an ammonia solution with hydrobromic acid (HBr).

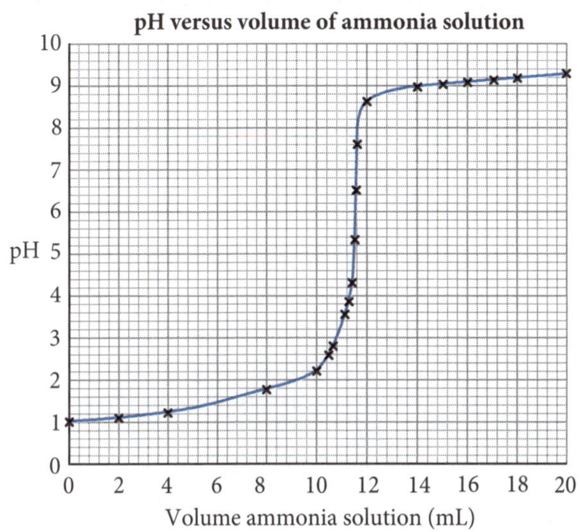

Figure 7.13 pH titration

$NH_3(aq) + HBr(aq) \rightarrow NH_3Br(aq)$

Equivalence point titre (from graph) = 11.55 mL NH_3 (at pH 6.50)

$n(HBr) = cV = (0.100)(0.020) = 0.00200$ mol

$n(NH_3) = 0.00200$ mol

$c(NH_3) = \frac{0.00200}{0.01155} = 0.173$ mol/L

→ KEY QUESTIONS

4 **Distinguish between the end point and equivalence point of a titration.**

5 **Explain why indicators cannot be used in the titration of a weak base and a weak acid.**

6 **Explain how the equivalence point of a conductivity titration can be determined.**

Answers ➲ p. 109

3 Acid dissociation constants

» Students calculate and apply the dissociation constant (K_a) and pK_a (p$K_a = -\log_{10}(K_a)$) to determine the difference between strong and weak acids.

K_a calculations

→ The equilibrium constant for the ionisation of a weak acid is given the symbol K_a and is called the acid **dissociation** constant or the acid **ionisation** constant.

dissociation:
- the process in which ionic solids separate into ions in a solvent
- the process in which a molecular substance interacts with a solvent and produces ions

ionisation: the process in which atoms or molecules are converted into ions

→ Consider the weak acid HA. The following equilibrium equation shows the formation of ions in water. This is a Brønsted–Lowry acid–base reaction in which hydronium ions (the conjugate acid of water) are formed:

$$HA(aq) + H_2O(l) \leftrightarrows H_3O^+(aq) + A^-(aq)$$

The weak acid HA is a molecule that has been partially converted to ions by its reaction with the water solvent.

→ The equilibrium constant for the HA equilibrium is given the symbol K_a. The K_a expression for this acid is:

$$K_a = \frac{[H_3O^+]\cdot[A^-]}{[HA]}$$

Note that no term for the water concentration is present in the K_a expression as the concentration of the water solvent is constant.

→ The lower the value of K_a, the weaker the acid. Table 7.1 shows some K_a values for various weak acids.

Table 7.1 K_a values

Weak acid	Equilibrium	K_a
Hydrofluoric acid	$HF(aq) + H_2O(l) \leftrightarrows H_3O^+(aq) + F^-(aq)$	6.8×10^{-4}
Formic acid	$HCOOH(aq) + H_2O(l) \leftrightarrows H_3O^+(aq) + HCOO^-(aq)$	1.8×10^{-4}
Acetic acid	$CH_3COOH(aq) + H_2O(l) \leftrightarrows H_3O^+(aq) + CH_3COO^-(aq)$	1.7×10^{-5}

→ Acids such as HCl and HF are classified as *monoprotic acids* as their molecules can only dissociate to produce one hydrogen ion. Diprotic and triprotic weak acids dissociate in several steps. Each step has its own value of K_a. For a diprotic acid the acid ionisation constants are called K_{a1} and K_{a2}. Consider the following examples.

- Sulfurous acid—a diprotic acid

  The two steps in the dissociation are:

  $H_2SO_3(aq) + H_2O(l) \leftrightarrows H_3O^+(aq) + HSO_3^-(aq)$; $K_{a1} = 1.38 \times 10^{-2}$

  $HSO_3^-(aq) + H_2O(l) \leftrightarrows H_3O^+(aq) + SO_3^{2-}(aq)$; $K_{a2} = 6.46 \times 10^{-8}$

  The lower value of K_{a2} shows that the second dissociation reaction lies less to the product side of the equilibrium.

- Phosphoric acid—a triprotic acid
  The three steps in the dissociation are:
  $H_3PO_4(aq) + H_2O(l) \leftrightarrows H_3O^+(aq) + H_2PO_4^-(aq)$;
  $K_{a1} = 7.1 \times 10^{-3}$
  $H_2PO_4^{2-}(aq) + H_2O(l) \leftrightarrows H_3O^+(aq) + HPO_4^{2-}(aq)$;
  $K_{a2} = 6.3 \times 10^{-8}$
  $HPO_4^{2-}(aq) + H_2O(l) \leftrightarrows H_3O^+(aq) + PO_4^{3-}(aq)$;
  $K_{a3} = 4.2 \times 10^{-13}$
  The decreasing values of the dissociation constants show that each subsequent dissociation reaction lies less to the product side of the equilibrium.

EXAMPLE 6

Carbonic acid (H_2CO_3) is a weak diprotic acid. The acid dissociates in two stages:

$H_2CO_3(aq) + H_2O(l) \leftrightarrows H_3O^+(aq) + HCO_3^-(aq)$;
$K_{a1} = 4.5 \times 10^{-7}$

$HCO_3^-(aq) + H_2O(l) \leftrightarrows H_3O^+(aq) + CO_3^{2-}(aq)$;
$K_{a2} = 4.7 \times 10^{-11}$

A 0.0100 mol/L solution of carbonic acid was prepared.

a Comment on the validity of the following statements.

Statement 1: The pH of the carbonic acid can be calculated using only the K_{a1} data.

Statement 2: The concentration of H_2CO_3 at equilibrium is approximately 0.0100 mol/L.

b Assuming the statements are valid, calculate the pH of the carbonic acid solution.

Consider the magnitudes of K_{a1} and K_{a2}

Answer:

a Statement 1: The value of K_{a2} is very much smaller than K_{a1} and so the increase in the hydronium ion concentration from the second dissociation equilibrium is negligible compared with the first equilibrium.

Statement 2: The value of K_{a1} is very small and the equilibrium lies strongly to the left. Therefore the change in the concentration of H_2CO_3 will be effectively negligible.

b $K_{a1} = 4.5 \times 10^{-7} = [H_3O^+][HCO_3^-]/[H_2CO_3]$
$[H_3O^+] = [HCO_3^-]$
$4.5 \times 10^{-7} = [H_3O^+]^2/(0.0100)$
$[H_3O^+] = 6.32 \times 10^{-5}$ mol/L
$pH = -\log_{10}(6.32 \times 10^{-5}) = 4.20$

➔ The acid dissociation constant can be related to the degree of dissociation of a weak acid. Consider the next example.

EXAMPLE 7

A 0.100 mol/L solution of formic acid (methanoic acid) was analysed and the concentration of hydronium ions was found to be 4.18×10^{-3} mol/L at 25 °C. Calculate the:

a degree of dissociation
b K_a
c pH of the 0.100 mol/L formic acid solution
d pH of the formic acid solution when it is diluted to 0.0100 mol/L.

The degree of dissociation is expressed as a percentage

Answer:

$HCOOH(aq) + H_2O(l) \leftrightarrows H_3O^+(aq) + HCOO^-(aq)$

a Percentage dissociation
$= [H_3O^+]/[HCOOH] \times 100$
$= \frac{(4.18 \times 10^{-3})}{(0.100)} \times 100$
$= 4.18\%$

b $K_a = [H_3O^+][HCOO^-]/[HCOOH]$
Hydronium ions and formate ions ($HCOO^-$) have the same concentration according to the balanced equation.
Equilibrium concentration of HCOOH
$= 0.100 - \frac{(4.18)}{100}(0.100)$
$= 0.100 - 4.18 \times 10^{-3}$
$= 0.09582$ mol/L
$K_a = \frac{(4.18 \times 10^{-3})(4.18 \times 10^{-3})}{(0.09582)} = 1.82 \times 10^{-4}$

c $pH = -\log_{10}[H_3O^+] = -\log_{10}(4.18 \times 10^{-3}) = 2.38$

d Initial [HCOOH] = 0.0100 mol/L
Let x mol/L HCOOH dissociate, as noted in the table.

Concentrations (mol/L)	HCOOH	H_3O^+	$HCOO^-$
Initial	0.0100	0	0
Change	$-x$	$+x$	$+x$
Equilibrium	$0.0100 - x$	x	x

$K_a = [H_3O^+][HCOO^-]/[HCOOH] = 1.82 \times 10^{-4}$
$(x)(x)/(0.0100 - x) = 1.82 \times 10^{-4}$
Solve for x.
$x = 1.26 \times 10^{-3}$ mol/L
$pH = -\log_{10}[H_3O^+] = -\log_{10}(1.26 \times 10^{-3}) = 2.89$
Therefore the pH has increased from 2.38 to 2.89 on the dilution of formic acid.
The addition of more water has shifted the dissociation equilibrium to the right so that more molecules of HCOOH dissociate but the concentration of all species has decreased due to the large increase in volume of the solution.
Figure 7.14 shows that the degree of dissociation of formic acid increases on dilution.

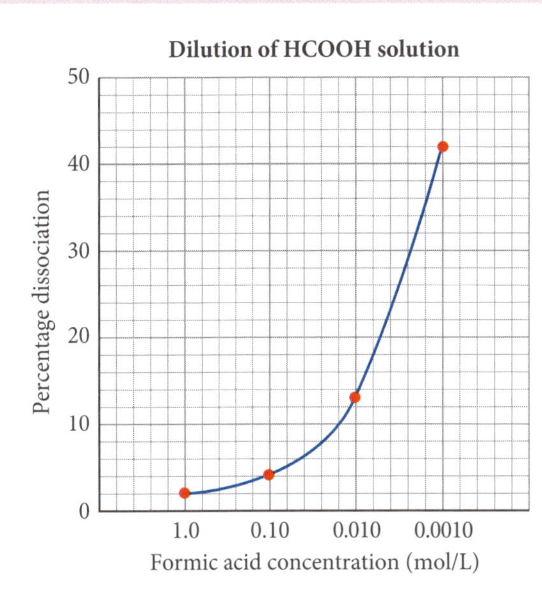

Figure 7.14 Degree of dissociation of formic acid on dilution

EXAMPLE 8

Glycolic acid ($CH_2(OH)COOH$) is a weak organic monoprotic acid that can be extracted from sugarcane and unripe grapes. Its systematic name is 2-hydroxyethanoic acid. The K_a of glycolic acid is 1.48×10^{-4}. The degree of dissociation is 3.8%.

a **Write the equation for the dissociation equilibrium of glycolic acid in water in which hydronium ions and glycolate anions ($CH_2(OH)COO^-$) form.**

b **Calculate the pH of a 0.10 mol/L glycolic acid solution.**

c **Sodium glycolate ($NaCH_2(OH)COO$) is a soluble salt. How will the pH of the 0.10 mol/L solution of glycolic acid change as small amounts of sodium glycolate are dissolved in the solution?**

Figure 7.15 shows the structural formula of glycolic acid.

$$H-O-\underset{H}{\overset{H}{|}}C-\overset{O}{\overset{\|}{C}}-O-H$$

acidic hydrogen

Figure 7.15 Glycolic acid

Determine the equilibrium concentration of glycolic acid

Answer:

a $CH_2(OH)COOH(aq) + H_2O(l) \leftrightharpoons H_3O^+(aq) + CH_2(OH)\ COO^-(aq)$

b As glycolic acid dissociates equal concentrations of hydronium ions and glycolate ions ($CH_2(OH)COO^-$) result. Let this concentration = x mol/L.

The concentration of glycolic acid decreases. The

decrease in concentration = $\frac{(3.8)}{100} \times 0.10$ mol/L

= 3.8×10^{-3} mol/L $CH_2(OH)COOH$

Equilibrium concentration of $CH_2(OH)COOH$

= $0.10 - 3.8 \times 10^{-3} = 0.0962$ mol/L

$K_a = [H_3O^+][CH_2(OH)COO^-]/[CH_2(OH)COOH]$

$= (x)(x)/(0.0962) = 1.48 \times 10^{-4}$

Solve for x.

$x = [H_3O^+] = 3.8 \times 10^{-3}$ mol/L

$pH = -\log_{10}(H_3O^+) = -\log_{10}(3.8 \times 10^{-3}) = 2.4$

c Glycolate ions are conjugate bases. The addition of glycolate ions to the glycolic acid solution causes the equilibrium to shift to the left to counteract the change. The hydronium ion concentration decreases and the acidity will decrease. Thus the pH will increase.

pK_a calculations

➔ Consider the K_a expression for the following weak acid equilibrium:

$HA(aq) + H_2O(l) \leftrightharpoons H_3O^+(aq) + A^-(aq)$

$K_a = [H_3O^+][A^-]/[HA]$

The pK_a function is defined as follows:

$$\mathbf{pK_a = -\log_{10}K_a}$$

The larger the value of pK_a, the weaker the acid. Hydrocyanic acid is much weaker than hydrofluoric acid. Examples are:

- hydrofluoric acid (HF): pK_a = 3.17
- butanoic acid (C_3H_7COOH): pK_a = 4.82
- hydrocyanic acid (HCN): pK_a = 9.21

EXAMPLE 9

Nitrous acid (HNO_2) is a weak acid. Its pK_a is 3.15. A 0.10 mol/L solution of sodium nitrite ($NaNO_2$) is prepared. The solution is tested with bromothymol blue indicator.

a **Write a net ionic equation for the hydrolysis reaction of nitrite ions in water.**

b **Calculate the pH of the sodium nitrite solution.**

c **Identify the colour of bromothymol blue in the sodium nitrite solution.**

Relate the K_a and K_w formulas to the hydrolysis reaction

Answer:

a $NO_2^-(aq) + H_2O(l) \leftrightharpoons HNO_2(aq) + OH^-(aq)$

b The equilibrium constant (K_{eq}) for the hydrolysis equilibrium is:

$K_{eq} = [HNO_2][OH^-]/[NO_2^-]$

The value for K_{eq} can be determined from K_a and K_w. Review Chapter 6 for examples of the use of K_w.

Multiply the K_{eq} expression by $[H^+]/[H^+]$ (i.e. 1)

$K_{eq} = [HNO_2][OH^-]/[NO_2^-] \times [H^+]/[H^+]$
$= [HNO_2]/[NO_2^-][H^+] \times [H^+][OH^-]$
$= 1/K_a \times K_w = K_w/K_a$
$= (1.0 \times 10^{-14})/10^{-3.15}$
$= \frac{(1.0 \times 10^{-14})}{7.08 \times 10^{-4}}$
$= 1.41 \times 10^{-11}$

Let x mol/L of NO_2^- hydrolyse, as shown in the table.

Concentrations (mol/L)	NO_2^-	HNO_2	OH^-
Initial	0.10	0	1×10^{-7}
Change	$-x$	$+x$	$+x$
Equilibrium	$0.10 - x$	x	$1 \times 10^{-7} + x$

$1.41 \times 10^{-11} = [HNO_2][OH^-]/[NO_2^-]$
$= (x)(1 \times 10^{-7} + x)/(0.10 - x)$

Solve for x.

$x = 1.14 \times 10^{-6}$ mol/L

$[OH^-] = 1 \times 10^{-7} + x = 1 \times 10^{-7} + 1.14 \times 10^{-6}$
$= 1.24 \times 10^{-6}$ mol/L

$pOH = -\log_{10}[OH^-] = -\log_{10}(1.24 \times 10^{-6}) = 5.9$

$pH = 14 - pOH = 14 - 5.9 = 8.1$

c The indicator will turn blue as the pH > 7.6.

EXAMPLE 10

50.0 mL of 0.0100 mol/L CH_3COOH was pipetted into a beaker and 50.0 mL of 0.005 00 mol/L NaOH solution was added to the beaker. The mixture was stirred and a pH electrode was used to measure the pH of the solution.

a **Write a balanced whole formula equation and a net ionic equation for the neutralisation reaction.**

b **Calculate the concentrations of acetic acid molecules and acetate ions in the solution following neutralisation.**

Calculate the final volume of the solution

c **Calculate the pH of the solution ($K_a(CH_3COOH) = 1.74 \times 10^{-5}$).**

Answer:

a $CH_3COOH(aq) + NaOH(aq) \rightarrow NaCH_3COO(aq) + H_2O(l)$

$CH_3COOH(aq) + OH^-(aq) \rightarrow CH_3COO^-(aq) + H_2O(l)$

b Initial: $n(CH_3COOH) = cV = (0.0100)(0.0500) = 5.00 \times 10^{-4}$ mol

Initial: $n(OH^-) = cV = (0.00500)(0.0500) = 2.50 \times 10^{-4}$ mol

Stoichiometry: $CH_3COOH : OH^- : CH_3COO^- = 1:1:1$

The OH^- is limiting.

Final:

$n(CH_3COOH) = 5.00 \times 10^{-4} - 2.50 \times 10^{-4}$
$= 2.50 \times 10^{-4}$ mol

$n(CH_3COO^-) = 2.50 \times 10^{-4}$ mol

Final volume = 50.0 + 50.0 = 100.0 mL = 0.100 L

Final concentrations:

$[CH_3COOH] = n/V = \frac{(2.50 \times 10^{-4})}{0.100}$
$= 2.50 \times 10^{-3}$ mol/L

$[CH_3COO^-] = n/V = \frac{(2.50 \times 10^{-4})}{0.100} = 2.50 \times 10^{-3}$ mol/L

c $CH_3COOH(aq) + H_2O(l) \leftrightharpoons H_3O^+(aq) + CH_3COO^-(aq)$

$K_a = [H_3O^+][CH_3COO^-]/[CH_3COOH] = 1.74 \times 10^{-5}$

$[H_3O^+][2.50 \times 10^{-3}]/[2.50 \times 10^{-3}] = 1.74 \times 10^{-5}$

$\therefore [H_3O^+] = 1.74 \times 10^{-5}$

$pH = -\log_{10}[H_3O^+] = -\log_{10}(1.74 \times 10^{-5}) = 4.76$

→ Further information about strong and weak acids and their degree of dissociation can be found in various online videos. Enter the following title in the search bar: dissociation strong and weak acids

→ KEY QUESTIONS

7 **Write the K_a expression for the dissociation of chlorous acid ($HClO_2$), which is a monoprotic weak acid.**

8 **Selenous acid (H_2SeO_3) is a weak diprotic acid that dissociates in two steps. Write equations for each step.**

9 **A 0.200 mol/L solution of a weak acid (HX) is 0.025% dissociated. Calculate K_a for this acid.**

Answers ➲ p. 109

4 Applications of acid–base analysis

» Students explore acid–base analysis techniques that are applied in industries, by Aboriginal and Torres Strait Islander peoples and using digital probes and instruments.

Techniques in industries

Analysis of a commercial drain cleaner

→ Titrations can be used to determine the concentration of sodium hydroxide in a liquid drain cleaner. Industrial chemists perform such titrations to ensure quality control of this product.

EXAMPLE 11

An industrial chemist dilutes 25.00 mL of the drain cleaner until its total volume is 100.00 mL. Using phenolphthalein indicator, 25.00 mL of the diluted drain cleaner was found to be neutralised by 26.50 mL of 0.300 mol/L HCl. Calculate the NaOH concentration in the undiluted drain cleaner.

Take the dilution factor into account

Answer:

Write the equation for the neutralisation reaction:

$NaOH(aq) + HCl(aq) \rightarrow NaCl(aq) + H_2O(l)$

Reaction stoichiometry = 1 : 1

Calculate the number of moles of HCl:

$n(HCl) = c.V = (0.300)(0.0265)$

$= 7.95 \times 10^{-3}$ mol

$= n(NaOH)$ in diluted solution.

Calculate the NaOH concentration in the diluted solution:

$c(NaOH)$ in diluted solution $= n/V = \dfrac{7.95 \times 10^{-3}}{0.025}$

$= 0.318$ mol/L

Calculate the NaOH concentration in the undiluted drain cleaner:

$c_1V_1 = c_2V_2$

$c_1(0.025) = (0.318)(0.100)$

$c_1 = \dfrac{(0.318)(0.100)}{(0.025)} = 1.27$ mol/L

Thus original NaOH concentration in the drain cleaner = 1.27 mol/L.

Acid content of wine

- In the wine industry the acidity of wine can be determined by titration.
- The three main acids in wine are tartaric acid, malic acid and citric acid. These acids are weak acids. The levels of these acids alter the sourness of the wine. For example, a sweet red wine has less acidity than a dry white wine.
- For white wine a sample of the wine is pipetted into a flask and phenolphthalein indicator is added. The wine is then titrated with a standardised solution of sodium hydroxide and the end point determined when the indicator turns pale pink. The titre is then recorded.
- This method is not suitable for red wines as the indicator colour cannot be seen. An alternative method is to perform a pH titration using a pH meter and glass electrode. This method can be used for all types of wine.
- Figure 7.16 shows a pH electrode and pH meter being used in a pH titration in which the acidity of the solution in the beaker is determined by titration with standardised NaOH solution. The electrode is made of glass with a very thin glass membrane (~0.1 mm) at the base. The electrode measures the potential difference across the glass membrane using an Ag/AgCl internal reference electrode and this potential difference is proportional to the concentration of hydrogen ions in the solution. The glass electrode is connected to a pH meter, which records the pH of the solution during the titration. The beaker stands on a magnetic stirrer and a Teflon-coated magnet is placed in the base of the beaker. When the stirrer is turned on, the magnet rotates and the solution is stirred.

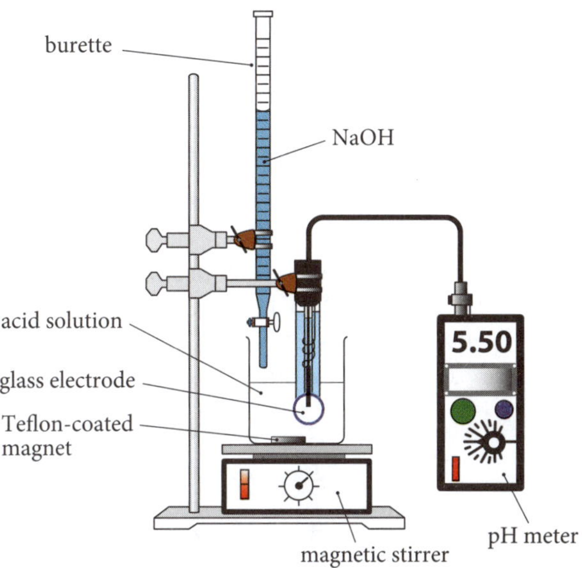

Figure 7.16 pH titration

- If the acidity of wine is to be measured, a known volume of wine is pipetted into the beaker and the glass electrode is inserted. The pH is monitored during the addition of NaOH from the burette. At the equivalence point the pH is about 8.2 as this is a weak acid–strong base titration. The titre is recorded and the total acidity of the wine can be calculated. The total acidity is a measure of the concentration of unionised acid molecules and the hydronium ion concentration.
- Tartaric acid $C_2H_4O_2(COOH)_2$ is a major organic acid present in grapes. Its structural formula is shown in Figure 7.17.

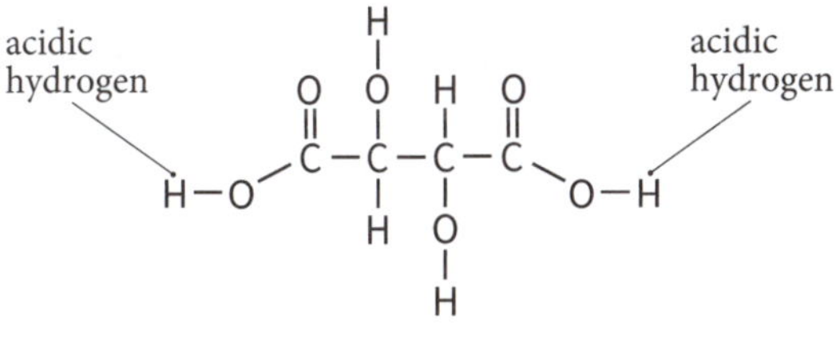

acidic hydrogen — H–O–C(=O)–CH(OH)–CH(OH)–C(=O)–O–H — acidic hydrogen

Figure 7.17 Tartaric acid

- Tartaric acid is a weak diprotic acid. In solution it dissociates in two steps:

$C_2H_4O_2(COOH)_2(aq) + H_2O(l) \leftrightharpoons C_2H_4O_2(COOH)(COO)^-(aq) + H_3O^+(aq)$

$C_2H_4O_2(COOH)(COO)^-(aq) + H_2O(l) \leftrightharpoons C_2H_4O_2(COO)_2^{2-}(aq) + H_3O^+(aq)$

- When a solution of tartaric acid is titrated with sodium hydroxide solution the salt called *sodium tartrate* forms. The reaction stoichiometry is 1 : 2, as shown in the following equation:

$$C_2H_4O_2(COOH)_2(aq) + 2NaOH(aq) \rightarrow Na_2C_2H_4O_2(COO)_2(aq) + 2H_2O(l)$$

- At the equivalence point of this titration the pH is about 8.2 as the tartrate ion acts as a Brønsted–Lowry base in water. Hydroxide ions form and make the solution alkaline:

$$C_2H_4O_2(COO)_2^{2-}(aq) + H_2O(l) \leftrightharpoons C_2H_4O_2(COOH)(COO)^-(aq) + OH^-(aq)$$

Analysis of Vitamin C

- Analytical chemists can use titrations to measure the ascorbic acid content in Vitamin C tablets or products containing Vitamin C.

  Figure 7.18 shows the structural formula of ascorbic acid.

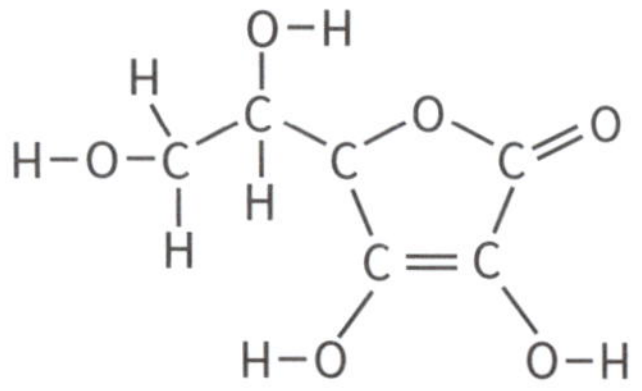

Figure 7.18 Structure of ascorbic acid (Vitamin C)

- Ascorbic acid is a diprotic acid due to the acidity of the H atoms of the hydroxyl (OH) functional groups on the C=C double bond. However, ascorbic acid behaves as a monoprotic acid because the second hydrogen does not ionise until the pH is very high (11). Thus the molecular formula of this acid can be written as $HC_6H_7O_6$.

EXAMPLE 12

0.600 g of a powdered Vitamin C tablet was dissolved in 25 mL of water in a conical flask. A 0.100 mol/L solution of NaOH was placed in a burette and phenolphthalein indicator was used to determine the end point when sodium hydroxide neutralised the Vitamin C solution. The titre at the end point was 27.90 mL. Calculate the ascorbic acid content as a per cent by weight in the Vitamin C tablet:

$$HC_6H_7O_6(aq) + NaOH(aq) \rightarrow NaC_6H_7O_6(aq) + H_2O(l)$$

Answer:

Use the equation to determine the reaction stoichiometry

Determine the reaction stoichiometry from the balanced equation.

Reaction stoichiometry: acid : base = 1 : 1

Determine the number of moles of NaOH:

$$n(NaOH) = cV = (0.100)(0.02790) = 0.00279 \text{ mol}$$

Determine the number of moles of ascorbic acid:

$$n(HC_6H_7O_6) = n(NaOH) = 0.00279 \text{ mol}$$

Determine the mass of ascorbic acid:

$$M(HC_6H_7O_6) = (6 \times 12.01) + (8 \times 1.008) + (6 \times 16.00) = 176.124 \text{ g/mol}$$

$$m = nM = (0.00279)(176.124) = 0.491 \text{ g}$$

Calculate the percentage by weight of ascorbic acid in the Vitamin C tablet:

$$\text{Percentage ascorbic acid} = \frac{0.491}{0.600} \times \frac{100}{1} = 81.8\%\text{w/w}$$

Investigating Indigenous use of acidic and basic materials

- Chemists have investigated the composition and use of natural acidic and basic substances by Indigenous Australians. The tart flavour of the edible fruits of common native plants is due to a wide variety of acids, including citric acid, malic acid, tartaric acid and ascorbic acid. Some examples of Indigenous Australians' use of natural acidic and basic materials, and the subsequent investigations by analytical chemists are summarised here.
  - Aboriginal people in northern Australia have long known of the health benefits of the Kakadu plum (*Terminalia ferdinandiana*) (Figure 7.19). This native plum was known to assist in reducing the symptoms of colds as well as treating skin sores. The ascorbic acid content of natural bush fruits has been determined by analytical chemists using titrimetric procedures. Kakadu plums have been analysed and shown to contain more than ten times the ascorbic acid content of oranges. Native quandong fruits have twice the ascorbic acid content of oranges. Ascorbic acid is called Vitamin C. It is essential for human health in order to avoid a disease called scurvy.

Figure 7.19 Kakadu plum

  - Indigenous Australians discovered that the Emu bush (*Eremophila longifolia*) had excellent medical properties. Biochemists have analysed extracts of this plant and found that it inhibits oral bacteria relating to tooth decay. These oral bacteria interact with sugars in food and produce acids that promote tooth decay. The Emu bush extracts were found to inhibit acid production by these bacteria.
  - Oil extracted from emus has been used by Indigenous peoples as a healing agent applied to the skin.

Chemists have analysed the emu oil and determined that essential fatty acids such as oleic, linoleic and palmitic acids are present in large amounts. These fatty acids are known to increase the body's anti-inflammatory process to help fight off infection.

- Indigenous Australians have used wood charcoal for oral health and the prevention of tooth decay for thousands of years. Chemical investigations have revealed the importance of the removal of alkalis from the charcoal before oral use, as the alkalis in the wood ash irritate the gums. Chemists have discovered that repeated washings of the charcoal with water removes the alkali. This washing procedure was likely to have been used by Aboriginals before applying the charcoal to their teeth.
- Indigenous Australians mixed dried parts of the *Dubisia hopwoodij* plant with alkaline wood ash from *Acacia salicina*, which had the highest concentration of alkalis in its wood ash. The alkali assisted the release of an alkaloid compound. The mixture was chewed and the alkaloid acted on the brain to reduce hunger and thirst on long treks through Australian deserts.

FIRSTHAND INVESTIGATION 2

Analysis of a household weak acid

» Students conduct a chemical analysis of a common household substance for its acidity or basicity; for example, soft drink, wine, juice or medicine.

Aim

to determine the concentration of acetic acid in white vinegar using standardised sodium hydroxide solution

Chemicals

Standardised NaOH solution (0.100 mol/L)—standardise against a primary standard acid

Diluted white vinegar (25.00 mL of vinegar accurately diluted to 250.0 mL)

Phenolphthalein

Method

1 Pipette 25.00 mL of 0.100 mol/L NaOH into a clean conical flask.
2 Add four drops of phenolphthalein.
3 Fill the burette with the diluted vinegar solution.
4 Titrate the NaOH solution until the indicator just turns from pink to colourless. Repeat the titration at least three times and determine the average titre.

Sample results

Run	Volume diluted vinegar (mL)
1	38.65
2	38.30
3	38.35
4	38.25

Sample calculation

Average titre (omitting run 1 as an outlier) = 38.30 mL

The neutralisation equation is:

$$CH_3COOH\,(aq) + NaOH(aq) \rightarrow NaCH_3COO(aq) + H_2O(l)$$

$n(NaOH) = cV = 0.100(25.00 \times 10^{-3}) = 2.50 \times 10^{-3}$ mol

Stoichiometry = 1 : 1

$n(CH_3COOH) = 2.50 \times 10^{-3}$ mol

$$c(CH_3COOH) = n/V = \frac{(2.50 \times 10^{-3})}{(38.30 \times 10^{-3})} = 0.0653 \text{ mol/L (diluted vinegar)}$$

Calculate the concentration of acetic acid in the undiluted vinegar:

$c_1V_1 = c_2V_2$

$c_1(0.025\,00) = (0.0653)(0.2500)$

$c_1 = 0.653$ mol/L

Concentration of acetic acid in undiluted vinegar = 10 × 0.0653 = 0.653 mol/L

Extension: Titration of white wine

The acidity of a sample of white wine can also be determined by titration. A 10 mL sample of wine is pipetted into a flask and phenolphthalein indicator is added. The wine is then titrated with a standardised solution of 0.10 mol/L sodium hydroxide and the end point determined when the indicator turns pale pink. The titre is recorded and the total acidity of the wine can be calculated. In the wine industry, the titratable acidity (TA) is a measure of the tartaric acid concentration in the wine. At the equivalence point (pH ~8.2) the added NaOH neutralises the tartaric acid. The acid : base stoichiometry is 1 : 2.

$$C_2H_4O_2(COOH)_2(aq) + 2NaOH(aq) \rightarrow Na_2C_2H_4O_2(COO)_2(aq) + 2H_2O(l)$$

→ KEY QUESTIONS

10 **An industrial chemist analyses an ammonia window cleaner to determine the percentage by weight of ammonia in the cleaner. A titration is performed. Identify a suitable titrant and indicator that could be used.**

11 **Red wine is titrated to determine its total acidity. Explain why a pH electrode and meter are used rather than using the change in colour of an indicator to determine the equivalence point.**

12 **The acetic acid concentration in household vinegar is determined using a titration with standardised potassium hydroxide. Identify the indicator used and write a balanced equation for the neutralisation.**

Answers ➲ p. 109

5 Buffers

» Students:
- conduct a practical investigation to prepare a buffer and demonstrate its properties.
- describe the importance of buffers in natural systems.

Properties of buffers

- Our bodily fluids must be maintained in a narrow pH range in order for the biochemical processes to occur at an optimal rate. Buffers are also used in a chemical laboratory to calibrate pH meters and also to provide a constant pH environment for a chemical process.
- **Buffer** solutions contain either a weak Brønsted–Lowry acid and its conjugate base, or a weak Brønsted–Lowry base and its conjugate acid. The greatest buffering effect is obtained if the buffer solution contains equimolar concentrations of the weak acid and its conjugate base, or the weak Brønsted–Lowry base and its conjugate acid.

buffer: solutions that resist large changes in pH on the addition of small quantities of strong acids or bases

- Consider a 0.10 mol/L solution of hydrofluoric acid. The following equilibrium exists in this solution. As the acid is a weak acid few ions exist in the solution:

  $HF(aq) + H_2O(l) \leftrightarrows F^-(aq) + H_3O^+(aq)$

  If fluoride ions (in the form of sodium fluoride) are added to this equilibrium, the equilibrium is disturbed and shifts to the left. If sufficient fluoride ions are added until the concentration of fluoride ions equals the HF concentration, the solution is still acidic, with a pH of 3.2. This solution of HF and F^- is called an *acidic buffer*.
- If a solution of ammonia and its conjugate acid (ammonium ions) are prepared such that the Brønsted–Lowry base and its conjugate acid are in equal concentration (e.g. 0.10 mol/L), the pH of the solution is 9.3. The solution of NH_3 and NH_4^+ is called a *basic buffer*:

  $NH_3(aq) + H_2O(l) \leftrightarrows NH_4^+(aq) + OH^-(aq)$

EXAMPLE 13

An acidic buffer solution is prepared containing equimolar amounts of acetic acid and sodium acetate. The following equilibrium is established:

$CH_3COOH(aq) + H_2O(l) \leftrightarrows CH_3COO^-(aq) + H_3O^+(aq)$

What is the effect on the pH of the buffer of the addition of small amounts of:

a NaOH?

b HCl?

Consider the neutralisation reactions as the reason why the pH changes by only a small amount

Answer:

a There is a very small rise in pH.
The small rise in pH is due to the neutralisation of hydroxide ions due to the high concentration of acetic acid molecules in the buffer. The neutralisation reaction proceeds to completion as the hydroxide ion is a strong base:

$$CH_3COOH(aq) + OH^-(aq) \rightarrow CH_3COO^-(aq) + H_2O(l)$$

Hydroxide ions (strong bases) are neutralised by reaction with acetic acid molecules leading to the production of the weak base (acetate ions). The pH does not stay constant but rises by a very small amount. The increase in acetate ion concentration affects the buffer equilibrium (see the following equation), which will shift to the left to counteract this change according to Le Chatelier's principle. This leads to a decrease in hydronium ion concentration and the pH rises slightly:

$$CH_3COOH(aq) + H_2O(l) \leftrightarrows CH_3COO^-(aq) + H_3O^+(aq)$$

b The small decrease in pH is caused by the neutralisation of the added hydronium ions due to the high concentration of acetate ions in the buffer. The neutralisation reaction proceeds to completion because the hydronium ion is a strong acid:

$$CH_3COO^-(aq) + H_3O^+(aq) \rightarrow CH_3COOH(aq) + H_2O(l)$$

Hydronium ions (strong acids) are neutralised by reaction with acetate ions, leading to the production of the weak acid (acetic acid). The pH does not stay constant but decreases by a very small amount. The increase in acetic acid concentration affects the buffer equilibrium (see the following equation), which will shift to the right to counteract this change according to Le Chatelier's principle. This leads to an increase in hydronium ion concentration and the pH decreases slightly:

$$CH_3COOH(aq) + H_2O(l) \leftrightarrows CH_3COO^-(aq) + H_3O^+(aq)$$

Buffers in natural systems

- The pH of our blood must remain in the range 7.35 to 7.45. Buffers are used to maintain blood pH.

  One important buffer system in the blood is the carbonic acid/hydrogen carbonate buffer. The relevant equations are:

  1 $CO_2(g) + H_2O(l) \leftrightarrows H_2CO_3(aq)$

  2 $H_2CO_3(aq) + H_2O(l) \leftrightarrows HCO_3^-(aq) + H_3O^+(aq)$
- By adjusting the speed and depth of breathing, the brain and lungs are able to regulate the blood pH minute by minute:
  - Carbon dioxide is an acidic oxide and, as carbon dioxide accumulates in the blood, the pH of the blood

decreases. Note that acidity increases as equilibria 1 and 2 shift to the right.

- The brain regulates the amount of carbon dioxide that is exhaled by controlling the speed and depth of breathing. The amount of carbon dioxide exhaled, and consequently the pH of the blood, increases as breathing becomes faster and deeper.
- The kidneys are also able to affect blood pH by altering the balance of the HCO_3^- ions it secretes. Increasing the concentration of HCO_3^- ions shifts equilibrium 2 to the left, whereas decreasing the level of HCO_3^- ions shifts the equilibrium to the right. Because the kidneys make these adjustments more slowly than the lungs do, this pH compensation generally takes several days.

➔ This simple buffering system is not the only one involved in keeping the blood plasma pH at 7.35–7.45. Another system involving haemoglobin is also involved. In this system the haemoglobin/oxyhaemoglobin (abbreviated HHb^+/HbO_2) molecules form a buffer:

$$HHb^+(aq) + H_2O(l) + O_2(aq) \rightleftharpoons HbO_2(aq) + H_3O^+(aq)$$

A rise in oxyhaemoglobin levels due to oxygen absorption by the haemoglobin in the blood is associated with an increase in acidity as the haemoglobin equilibrium shifts to the right (Le Chatelier's principle). At this point the carbonic acid/hydrogen carbonate equilibrium buffer system becomes involved to reduce the acidity while oxygen is delivered to the cells.

Preparation of laboratory buffers

➔ Buffer solutions with a wide variety of pH levels can be made in the laboratory. They can be used to calibrate pH meters. Various syntheses of organic compounds must be performed at specific pH levels and buffers provide such an environment. Electrophoresis, which is a technology used to separate mixtures of molecules such as amino acids, must be performed in a buffer solution.

➔ A basic buffer such as the sodium hydrogen carbonate/sodium carbonate buffer can be prepared by dissolving 0.025 mol (2.10 g) $NaHCO_3$ and 0.025 mol (2.65 g) Na_2CO_3 in 1 litre of water. This buffer has a pH of 10.3. The buffer equilibrium is:

$$CO_3^{2-}(aq) + H_2O(l) \rightleftharpoons HCO_3^-(aq) + OH^-(aq)$$

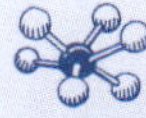

FIRSTHAND INVESTIGATION 3

Preparing and testing a buffer

» Students conduct a practical investigation to prepare a buffer and demonstrate its properties.

Aim

to prepare an acetic acid/acetate buffer and test its properties

Equipment/chemicals

pH electrode and meter
50 mL 0.20 mol/L acetic acid
50 mL 0.20 mol/L sodium acetate
0.20 mol/L HCl
0.20 mol/L NaOH
10 mL measuring cylinders
Beakers

Method

1. Mix 50 mL of 0.20 mol/L acetic acid with 50 mL of 0.20 mol/L sodium acetate to produce 100 mL of buffer solution in which the acetic acid and acetate ion have a concentration of 0.10 mol/L.
2. Use a calibrated pH electrode to measure the pH of this buffer solution.
3. Measure 50 mL of buffer into a new beaker and measure the pH of the buffer as small volumes (1 mL at a time) of 0.20 mol/L hydrochloric acid are added. Record the measurements.
4. Measure 50 mL of buffer into a new beaker and measure the pH of the buffer as small volumes (1 mL at a time) of 0.20 mol/L sodium hydroxide are added. Record the measurements.

Analysis of data

Examine the tabulated pH data and determine whether or not buffering has been demonstrated.

Sample results

Initial pH of buffer = 4.76

Addition of HCl

volume HCl (mL)	1	2	3	4
pH	4.74	4.73	4.71	4.69

Addition of NaOH

volume NaOH (mL)	1	2	3	4
pH	4.78	4.79	4.81	4.83

Conclusion

The tabulated results show that buffering has occurred as the pH changed very little as HCl or NaOH were added to the buffer.

EXAMPLE 14

Three experiments were conducted with a solution of sodium hydroxide.

Experiment 1: 150.0 mL of 0.100 mol/L NaOH was used as a control

Experiment 2: 50.0 mL of 0.100 mol/L NaOH was added to 100.0 mL of water

Experiment 3: 50.0 mL of 0.100 mol/L NaOH was added to 100.0 mL of 0.100 mol/L HF

a Calculate the pH of each solution (pKa(HF) = 3.17).

b **i** Explain why the pH in Experiment 3 was much lower than the pH in Experiments 1 and 2.

ii pH experiments conducted on the solution in Experiment 3 show that the addition of a few millilitres of NaOH solution produces only a very small rise in pH. Explain.

Answer:

Write the K_a expression for hydrofluoric acid

a **i** Experiment 1 (control):

$pOH = -\log_{10}[0.100] = 1.00$

$pH = 14 - 1.00 = 13.00$

ii Experiment 2

$n(NaOH) = cV = (0.100)((0.0500) = 0.005\,00$ mol

Total volume $V = 50.0 + 100.0 = 150.0$ mL

$= 0.1500$ L

$[NaOH]$ after dilution $= n/V = \dfrac{0.005\,00}{0.1500}$

$= 0.0333$ mol/L

$[OH^-] = [NaOH] = 0.0333$ mol/L

$pOH = -\log_{10}[OH^-] = -\log_{10}(0.0333) = 1.48$

$pH = 14 - pOH = 14 - 1.48 = 12.52$

iii Experiment 3

$$HF(aq) + NaOH(aq) \rightarrow NaF(aq) + H_2O(l)$$

Stoichiometry: HF : NaOH : NaF = 1 : 1 : 1

$n(NaOH) = cV = (0.100)((0.0500) = 0.005\,00$ mol

$n(HF) = cV = (0.100)(0.100) = 0.0100$ mol

The NaOH is limiting and the HF is in excess.

After neutralisation:

$n(HF) = 0.0100 - 0.005\,00 = 0.005\,00$ mol

$n(NaF) = n(F^-) = 0.005\,00$ mol

Total volume $V = 50.0 + 100.0 = 150.0$ mL

$= 0.1500$ L

$[HF] = n/V = \dfrac{0.005\,00}{0.1500} = 0.0333$ mol/L

$[F^-] = n/V = \dfrac{0.005\,00}{0.1500} = 0.0333$ mol/L

$HF(aq) + H_2O(l) \leftrightarrows F^-(aq) + H_3O^+(aq)$

$pK_a(HF) = 3.17$; $K_a = 10^{-3.17} = 6.76 \times 10^{-4}$

$K_a = [H_3O^+][F^-]/[HF] = 6.76 \times 10^{-4}$

$[H_3O^+][0.0333]/[0.0333] = 6.76 \times 10^{-4}$

$[H_3O^+] = 6.76 \times 10^{-4}$

$pH = -\log_{10}[H_3O^+] = -\log_{10}[6.76 \times 10^{-4}) = 3.17$

b **i** Experiment 1: pH = 13.00

Experiment 2: pH = 12.52

Experiment 3: pH = 3.17

In Experiment 3 all the NaOH has been neutralised and half the original quantity of hydrofluoric acid remains. Thus the pH of the solution is less than 7 as HF molecules remain to produce an acidic solution.

ii The solution in Experiment 3 is a buffer solution as equal amounts of HF and its conjugate base (F^-) are present. Thus the addition of a small amount of NaOH causes only a small pH rise.

→ KEY QUESTIONS

13 Explain how an equimolar solution of nitrous acid (HNO_2) and potassium nitrite can act as a buffer.

14 Potassium dihydrogen phosphate and potassium hydrogen phosphate are dissolved in water to form a buffer solution. Write an equation for this equilibrium and explain why the addition of a small amount of hydrochloric acid only produces a small decrease in pH.

15 An equimolar solution of sodium hydrogen carbonate and sodium carbonate has a pH of 10.0. How could the buffering ability of this solution be demonstrated using a laboratory experiment?

Answers ➲ p. 109

CHAPTER SYLLABUS CHECKLIST

Are you able to answer every question from the syllabus for this chapter? Tick each question as you go through the checklist if you are able to answer it. If you cannot answer a question, turn to the relevant page in the study guide to find the answer. For NESA key word meanings, go to www.educationstandards.nsw.edu.au and search 'key words'.

FOR A COMPLETE UNDERSTANDING OF THIS TOPIC:		PAGE NO.	✓
1	Can I describe the procedure for preparing a primary standard solution?	90	
2	Can I describe the steps to prepare glassware for use in a titration?	90	
3	Can I describe the steps in a titration?	91	
4	Can I calculate the concentration of the unknown based on data collected in a titration?	92	
5	Can I distinguish between the end point and equivalence point in a titration?	93	
6	Can I use pH graphs for acid–base titrations to select an appropriate indicator?	94	
7	Can I explain how the concentration of an unknown can be determined using a conductivity titration?	95	
8	Can I model neutralisation using a variety of media?	96	
9	Can I write an expression for the acid dissociation constant (K_a)?	97	
10	Can I perform calculations involving K_a?	98	
11	Can I describe applications of acid–base analysis?	100	
12	Can I define the term buffer and recall examples of natural buffers?	104	
13	Can I describe the preparation and testing of a laboratory buffer?	105	

HSC EXAM-TYPE QUESTIONS

Objective-response questions (1 mark each)

1 A student wishes to determine the concentration of a weak acid by titration with a base of accurately known concentration. The base should be:

A a weak base so the pH at the equivalence point is 7.

B a strong base so that there is a rapid change in pH at the equivalence point.

C a primary standard such as sodium hydroxide.

D diprotic.

2 Propanoic acid is a weak acid. Its dissociation equilibrium is:

$$CH_3CH_2COOH(aq) + H_2O(l) \leftrightharpoons CH_3CH_2COO^-(aq) + H_3O^+(aq)$$

A 0.10 mol/L propanoic acid solution is prepared. Given that the acid dissociation constant (K_a) for propanoic acid is 1.35×10^{-5}, calculate the hydronium ion concentration of the solution.

A 1.35×10^{-5} mol/L

B 1.35×10^{-6} mol/L

C 1.00×10^{-7} mol/L

D 1.16×10^{-3} mol/L

3 Identify which of the following pairs would form a buffer solution.

A $HNO_3(aq)/OH^-(aq)$

B $HF(aq)/F^-(aq)$

C $H_2SO_3(aq)/SO_3^{2-}(aq)$

D $H_3PO_4(aq)/PO_4^{3-}(aq)$

4 Identify the best indicator for a titration between nitric acid and sodium hydrogen carbonate solution.

A phenolphthalein

B methyl orange

C litmus

D universal indicator

5 Identify the glassware that should be used to prepare a standard solution of 1.00 mol/L anhydrous sodium carbonate and then dilute it accurately to a final concentration of 0.100 mol/L.

A measuring cylinder and beaker

B volumetric flask and measuring cylinder

C pipette and burette

D volumetric flask and pipette

Extended-response questions

6 Acetic acid is a weak acid with an acid dissociation constant (K_a) of 1.74×10^{-5} at 25 °C.

a Write an equation for the dissociation equilibrium of acetic acid. (1 mark)

b Write an expression for K_a. (1 mark)

c Calculate the pH of a 0.20 mol/L solution of acetic acid. (3 marks)

7 Sulfurous acid (H_2SO_3) is a weak acid that ionises in two steps in water.

a Write the dissociation equilibria for the two steps of the sulfurous acid ionisation. (2 marks)

b Each dissociation step has a different K_a value:

$K_{a1} = 1.58 \times 10^{-2}$; $K_{a2} = 6.46 \times 10^{-8}$

Explain why the first dissociation constant is larger than the second. (1 mark)

c Write a balanced equation for the reaction of sulfurous acid with potassium hydroxide solution. (1 mark)

8 5.00 mL of a 0.10 mol/L solution of nitric acid is placed in a flask and a 0.10 mol/L potassium hydroxide solution added slowly (in 10 × 1.00 mL amounts) to the flask from a burette.

a How could a student quantitatively detect changes in acidity during the addition of the potassium hydroxide? (1 mark)

b How will the concentration of hydrogen ions change throughout the experiment? (1 mark)

c In a separate experiment, 2.0 mL of the 0.10 mol/L nitric acid solution was diluted to a total volume of 400.0 mL. The diluted solution was tested with several drops of methyl orange indicator. Use your knowledge of the pH range of methyl orange to determine the colour of the indicator in the diluted solution. (3 marks)

9 Oxalic acid crystals ($C_2O_4H_2{\cdot}2H_2O$) can be used as a primary standard in acid–base titrations. Oxalic acid is a weak diprotic acid.

a Write an equation for the neutralisation reaction between an oxalic acid solution and a potassium hydroxide solution. (1 mark)

b A 0.10 mol/L oxalic acid solution was prepared. What mass of crystals will need to be weighed out to prepare 250 mL of this solution? (2 marks)

c The oxalic acid solution was placed in a burette and 25.0 mL of a potassium hydroxide solution placed in a conical flask using a pipette.

i Name a suitable indicator for this titration. (1 mark)

ii The mean titre was 18.40 mL. Calculate the concentration of the potassium hydroxide solution. (3 marks)

10 Describe the procedure used to prepare each of the following pieces of volumetric glassware for a titration.

a conical flask (1 mark)

b 25 mL pipette, to deliver the standard solution of sodium hydroxide to the conical flask (3 marks)

c 50 mL burette, to deliver the hydrochloric acid solution during the titration (3 marks)

ANSWERS

KEY QUESTIONS

Key questions ➲ p. 93

1 KOH crystals absorb moisture and carbon dioxide from the air and their mass and composition changes. Consequently they are not suitable as a primary standard.

2 The pipette is rinsed with water several times and then rinsed several times with the sodium carbonate prior to filling.

3 The end point is not initially known so a rough titration is performed to find the approximate position of the end point. The remaining titrations are then performed accurately.

Key questions ➲ p. 97

4 The end point in a titration occurs when the indicator just changes colour on the addition of the last drop of titrant. The equivalence point refers to the volume of titrant that is required for complete neutralisation, based on the reaction stoichiometry.

5 There is no steep inflexion in such titrations and so indicators only slowly change colour. pH titrations can identify the centre of the pH inflexion graph.

6 As neutralisation occurs the conductivity drops and reaches a minimum as the stoichiometric ratio is achieved.

Key questions ➲ p. 100

7 $HClO_2(aq) + H_2O(l) \leftrightarrows H_3O^+(aq) + ClO_2^-(aq)$;
$K_a = [H_3O^+][ClO_2^-]/[HClO_2]$

8 $H_2SeO_3(aq) + H_2O(l) \leftrightarrows HSeO_3^-(aq) + H_3O^+(aq)$
$HSeO_3^-(aq) + H_2O(l) \leftrightarrows SeO_3^{2-}(aq) + H_3O^+(aq)$

9 $HX(aq) + H_2O(l) \leftrightarrows X^-(aq) + H_3O^+(aq)$
At equilibrium: $[X^-] = [H_3O^+] = \frac{0.025}{100 \times 0.200} = 5.0 \times 10^{-5}$ mol/L
$[HX] = 0.200 - 5.0 \times 10^{-5} = 0.200$ mol/L (to 3 significant figures)
$K_a = [X^-][H_3O^+]/[HX] = \frac{(5.0 \times 10^{-5})^2}{(0.200)} = 1.3 \times 10^{-8}$ (2 significant figures)

Key questions ➲ p. 103

10 Standardised NaOH; phenolphthalein

11 The indicator's colour change would not be seen in a red wine.

12 Phenolphthalein; $CH_3COOH(aq) + KOH(aq) \rightarrow KCH_3COO(aq) + H_2O(l)$

Key questions ➲ p. 106

13 Nitrite ions are the conjugate bases of nitrous acid. This mixture will act as a buffer because the nitrous acid component will neutralise small amounts of strong bases and the nitrite ions will neutralise small amounts of strong acids.

14 $H_2PO_4^-(aq) + H_2O(l) \leftrightarrows HPO_4^{2-}(aq) + H_3O^+(aq)$
The conjugate base (HPO_4^-) neutralises the added acid and the equilibrium shifts to the left and consequently there is only a small decrease in pH.

15 Use a pH electrode and meter to monitor the pH of the buffer as small amounts of a strong acid or a strong base are added. The pH should only change slightly.

HSC EXAM-TYPE QUESTIONS

Objective-response questions

1 **B.** A strong base is required to drive the reaction to completion. **A** is incorrect as this reaction will result in an equilibrium. **C** is incorrect as NaOH is not a primary standard. **D** is incorrect as the base does not have to be diprotic.

2 **D.**
$K_a = [H_3O^+][CH_3CH_2COO^-]/[CH_3CH_2COOH]$
At equilibrium $[H_3O^+] = [CH_3CH_2COO^-] = x$
$1.35 \times 10^{-5} = x^2/0.10$
$x = 1.16 \times 10^{-3} = [H_3O^+]$
Thus **A**, **B** and **C** are mathematically incorrect.

3 **B.** The conjugate base has one less proton than the B/L acid and the acid should be a weak acid. **A** is incorrect as the hydroxide ion is not the conjugate base of nitric acid. **C** is incorrect as HSO_3^- is the conjugate base. **D** is incorrect as $H_2PO_4^{2-}$ is the conjugate base.

4 **B.** The solution is acidic at the equivalence point and methyl orange changes colour in the acidic region. **A** is incorrect as phenolphthalein does not change colour at pH < 7. **C** is incorrect as litmus is not used in titrations as the colour change is difficult to observe. **D** is incorrect as universal indicator has too many changes in colour.

5 **D.** The volumetric flask is used to prepare the solution and then a pipette can be used to transfer a fixed volume to a new volumetric flask for dilution. **A** is incorrect as measuring cylinders and beakers are too inaccurate. **B** is incorrect as the measuring cylinder is not sufficiently accurate. **C** is incorrect as these pieces of glassware cannot be used to prepare the standard solution.

Extended-response questions

6 EM Students need to be able to write a balanced equation for the dissociation of a monoprotic acid and demonstrate a deep understanding of the acid dissociation constant and the steps involved in relating pH to pK_a.

a $CH_3COOH(aq) + H_2O(l) \leftrightarrows CH_3COO^-(aq) + H_3O^+(aq)$ ✓

b $K_a = [CH_3COO^-][H_3O^+]/[CH_3COOH]$ ✓

c At equilibrium $[CH_3COO^-] = [H_3O^+] = x$; $[CH_3COOH] = 0.20 - x$ ✓
$1.74 \times 10^{-5} = x^2/(0.20 - x)$
$x = [H^+] = 1.86 \times 10^{-3}$ mol/L ✓
$pH = -\log(1.86 \times 10^{-3}) = 2.73$ ✓

7 EM Students are required to demonstrate how diprotic acids dissociate in water in two steps and explain how electrostatic attractions between oppositely charged ions influence the dissociation process.

a $H_2SO_3(aq) + H_2O(l) \leftrightarrows H_3O^+(aq) + HSO_3^-(aq)$ ✓
$HSO_3^-(aq) + H_2O(l) \leftrightarrows H_3O^+(aq) + SO_3^{2-}(aq)$ ✓

b The second dissociation requires a positive hydrogen ion to be removed from a negative HSO_3^- ion rather than a neutral H_2SO_3 molecule as in the first step. The attraction between the positive and negative ions makes this process more difficult. ✓

c $H_2SO_3(aq) + 2KOH(aq) \rightarrow K_2SO_3(aq) + 2H_2O(l)$ ✓

8 EM Students need to state how the measure of pH is a quantitative measurement involving neutralisation during a titration. They also need to apply the dilution formula to calculate the hydrogen ion concentration when nitric acid is diluted. They need to apply the pH formula to calculate the pH and use their knowledge of indicator pH ranges to predict the colour of an indicator added to the diluted solution.

a Use a pH electrode and meter to monitor acidity. ✓

b The hydrogen ion concentration would decrease as the alkali neutralises the acid. ✓

c Nitric acid is a strong acid and fully dissociated. To calculate the concentration following dilution:
$c_1V_1 = c_2V_2$
$(0.10)(0.002) = c_2(0.400)$
$c_2 = 5.0 \times 10^{-4}$ mol/L ✓
$[HNO_3] = [H^+] = 5.0 \times 10^{-5}$ mol/L
$pH = -\log_{10}[H^+] = -\log_{10}(5.0 \times 10^{-5}) = 3.30$ ✓
The pH range of methyl orange = 3.1–4.4. Between 3.1 and 4.4 the indicator is orange.
At a pH of 3.30 the indicator will be orange. ✓

9 EM Students must demonstrate a strong understanding of the calculation steps to determine the concentration of an unknown using mole theory. They also need to recall how equipment is prepared for a titration.

a $C_2O_4H_2(aq) + 2KOH(aq) \rightarrow K_2C_2O_4(aq) + 2H_2O(l)$ ✓

b $M(C_2O_4H_2{\cdot}2H_2O) = 126.068$ g/mol
$n = cV = 0.10 \times 0.250 = 0.0250$ mol $= m/M$ ✓
$m = nM = 0.0250 \times 126.068 = 3.152$ g ✓

c i Phenolphthalein (as this is a suitable indicator for a weak acid/strong base titration) ✓

ii $n(\text{oxalic acid}) = cV = 0.10 \times 18.40 \times 10^{-3} = 1.84 \times 10^{-3}$ mol ✓
$n(KOH) = 2n(\text{oxalic acid}) = 2 \times 1.84 \times 10^{-3}$
$= 3.68 \times 10^{-3}$ mol ✓
$c(KOH) = n/V = \dfrac{3.68 \times 10^{-3}}{0.0250} = 0.150$ mol/L ✓

10 EM Students must describe a logical sequence of actions in order to clean titration glassware and the steps of filling the pipette and burette so that the experiment will be as accurate as possible.

a The flask is thoroughly cleaned and rinsed several times with water. Discard the rinses. The flask does not need to be dried as water will be added during the titration. ✓

b Use a pipette filler to draw up distilled water to rinse the pipette. Repeat several times. Discard the rinses. ✓ Rinse the pipette with small volumes of the standard base and discard the washings.

Fill the pipette with the standard base. Lower the meniscus to the engraved line. ✓ The bottom of the meniscus should be on the engraved line and no air bubbles should be present. Wipe the outside of the pipette with a piece of paper towel. ✓

c Rinse the burette several times with distilled water and discard the rinses. Rinse with small volumes of the unknown acid to remove any water drops and discard the rinses. ✓ Fill the burette with the unknown acid to above the zero mark. The tap is then opened and the acid run into a waste beaker until the base of the meniscus is on the zero line. ✓ Ensure that the acid has filled the space below the tap and that there are no bubbles present. ✓

CHAPTER 8 NOMENCLATURE

MODULE 7 ORGANIC CHEMISTRY

INQUIRY QUESTION:

How do we systematically name organic chemical compounds?

Organic compounds are the basis of the living world. These molecules are based on the tetravalent carbon atom. Carbon atoms have the unique ability to form long, stable, straight and branched chains and ring compounds with other elements. Carbohydrates, fats, proteins and DNA are all organic molecules. Petrol, kerosene and diesel are hydrocarbon fractions (mixtures) extracted from petroleum by fractional distillation. Hydrocarbons are used as fuels or solvents and in the production of plastics. This chapter investigates the structure of hydrocarbons and their derivatives and shows how to name them using IUPAC nomenclature rules.

1 Hydrocarbons

» Students investigate the nomenclature of organic chemicals, up to C8, using IUPAC conventions, including simple methyl and ethyl branched chains, including alkanes, alkenes and alkynes.

Homologous series of hydrocarbons

→ Hydrocarbons are binary molecules composed of carbon and hydrogen. These hydrocarbon molecules belong to particular families or **homologous series**. Three of these families are the *alkanes*, *alkenes* and *alkynes*.

homologous series: a family of compounds that have similar physical and chemical properties and which can be represented by a general molecular formula and which contain the same functional group

Alkanes

→ Alkanes are **saturated hydrocarbons**.

saturated hydrocarbons: hydrocarbons containing only single C–C bonds and the maximum number of hydrogen atoms

- The general formula of alkanes is:

$$C_nH_{2n+2} \qquad (n = 1,2,3 \ldots)$$

- Saturated hydrocarbons contain only carbon–carbon single bonds and the maximum number of hydrogen atoms per hydrocarbon chain.
- Alkanes are the simplest of all the hydrocarbons. Straight chain (unbranched) alkanes do not contain a **functional group**.

functional group: a group of atoms which are part of or attached to a hydrocarbon chain that change the physical and chemical properties of the chain

- Alkanes are named using a stem and a suffix. The stem's name depends on the number of carbon atoms (n) in their chains. Table 8.1 lists the stems used for the first eight alkanes.

Table 8.1 Stems used in naming hydrocarbons

n	Stem	n	Stem
1	*meth-*	5	*pent-*
2	*eth-*	6	*hex-*
3	*prop-*	7	*hept-*
4	*but-*	8	*oct-*

- All alkanes are named using the suffix *-ane*.

→ Table 8.2 lists the IUPAC names of the first eight straight-chain members of the alkane homologous series. The acronym IUPAC refers to the International Union of Pure and Applied Chemists that makes decisions about chemical nomenclature.

Table 8.2 Alkane homologous series

n	Molecular formula (C_nH_{2n+2})	IUPAC name	n	Molecular formula (C_nH_{2n+2})	IUPAC name
1	CH_4	methane	5	C_5H_{12}	pentane
2	C_2H_6	ethane	6	C_6H_{14}	hexane
3	C_3H_8	propane	7	C_7H_{16}	heptane
4	C_4H_{10}	butane	8	C_8H_{18}	octane

→ Figure 8.1 shows the structural formulas and ball-and-stick models for ethane and butane. Although the alkane chains are referred to as being 'straight' they in fact form a 'zig-zag' pattern due to the tetrahedral nature of the bonding around each carbon atom.

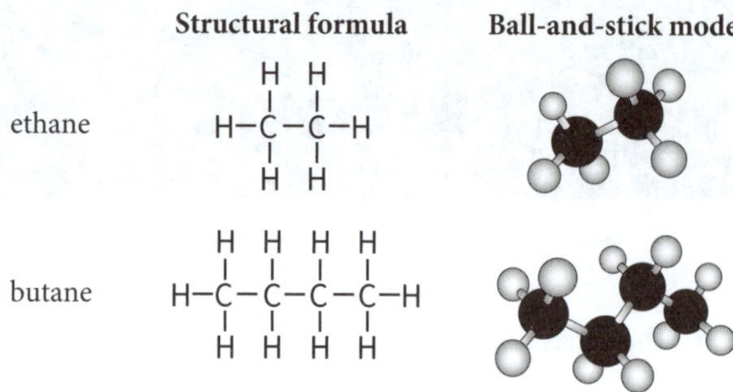

Figure 8.1 Alkanes' structural formulas and ball-and-stick models

➔ Branched chains are quite common structures for hydrocarbons. Branches are short chain functional groups attached to carbon atoms on the main chain. They are called *alkyl functional groups*. Methyl groups (CH_3) and ethyl groups (CH_2CH_3) are common **alkyl groups**.

> **alkyl groups:** functional groups composed of only carbon and hydrogen atoms and which are attached as branches to the hydrocarbon molecule

Figure 8.2 shows ball-and-stick models of hydrocarbons with methyl and ethyl groups.

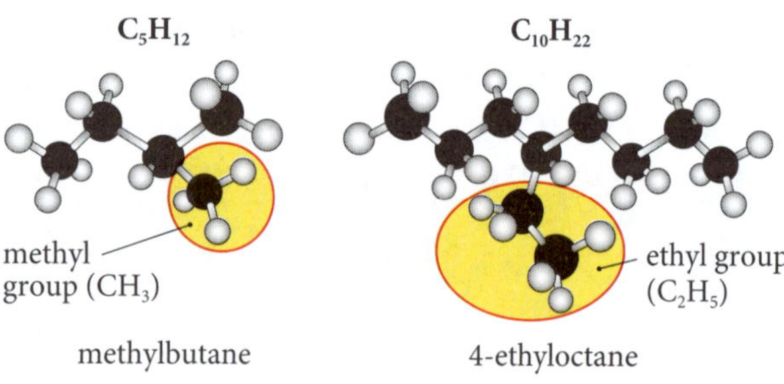

Figure 8.2 Methyl and ethyl groups attached to hydrocarbon chains

➔ The position of the branches along the carbon chain also needs to be specified. To do this the chain has to be numbered according to the following rules.

Rules of nomenclature for branched chain hydrocarbons:

1. Use the stem and suffix *-ane* to name the straight chain alkane.
2. Identify the location of the alkyl functional group using a number (**locant**) to identify the carbon atom where the functional group is located.

> **locant:** a number used to identify the position of a functional group along the carbon chain

3. Allocate locants to the chain from one end so that the functional groups have the lowest possible set of locants.
4. Name the alkyl groups and their locants as prefixes.
5. The alkyl groups are named alphabetically when naming the hydrocarbon.
6. Prefixes,such as *di-*, *tri-*, *tetra-*, etc., are used to indicate the number of each type of alkyl group.

EXAMPLE 1

Name the branched chain alkanes a and b shown in Figure 8.3.

a

```
    H  H  H  H  H   H
    |  |  |  |  |   |
  H-C--C--C--C--C---C-H
    |  |  |  |  |   |
    H  H  H  H  CH3 H
```

b

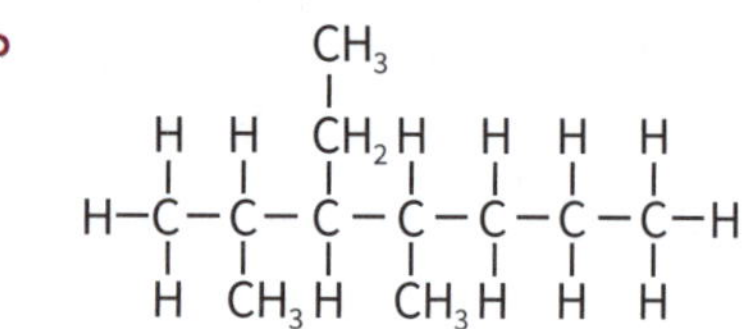

Figure 8.3 Branched chain alkanes

Determine which end of the alkane chain to start numbering

Answer:

a Stem = *hex-*; suffix = *-ane*; alkyl group = methyl; locant = 2 (numbering right to left to give the methyl group the lower locant)
Name = 2-methylhexane

b Stem = *hept-*; suffix = *-ane*; alkyl groups = methyl and ethyl; locants = 2,3 and 4 (numbering from left to right to give the alkyl groups the lower locant set (2,3,4 rather than 4,5,6)
Name = 3-ethyl-2,4-dimethylheptane

Note that locant numbers are separated from functional groups by dashes, commas are used to separate multiple locants, and there are no spaces in the name of the hydrocarbon.

Alkenes

➔ Alkenes are a family (or homologous series) of **unsaturated hydrocarbons**.

> **unsaturated hydrocarbon:** hydrocarbons containing double or triple carbon–carbon bonds and less than the maximum number of hydrogen atoms

- Alkenes have the general formula:

$$C_nH_{2n} \qquad (n = 2,3,4\ \dots)$$

- Unsaturated hydrocarbons contain at least one carbon–carbon double or triple bond (see alkynes). Simple alkenes contain one double bond. This double bond is the functional group that alters the properties of the hydrocarbon molecule.
- Alkenes are named using the same stems as those used to name alkanes, but they use the suffix *-ene*.
- When $n = 2$ the alkene has the *preferred* IUPAC name of 'ethylene' rather than ethene, which is the systematic name. Ethylene is a historical name for C_2H_4.
- The position of the double bond along the carbon chain also needs to be specified. To do this the chain has to be numbered according to the following rules.

  Rules of nomenclature:

  1. Use the stem and suffix *-ene* to name the straight chain alkene.
  2. Identify the location of the double bond using a number (locant) to identify the carbon atom where the double bond starts. Place this number after the stem name and before the suffix *-ene* (e.g. oct-1-ene).
  3. Allocate locants to the chain from one end so that the double bond has the lowest possible locant.
  4. If alkyl groups are present the double bond must be allocated the lowest locant.

EXAMPLE 2

Locants have been used to identify the alkene functional group. Name the following molecule represented in Figure 8.4.

```
    H     H   H   H   H   H
    |1  2 |3  |4  |5  |6  |7
H - C - C = C - C - C - C - C - H
    |   |       |   |   |   |
    H   H       H   H   H   H
```

Figure 8.4 Alkene nomenclature

The double bond functional group has priority in numbering the chain

Answer:

The locants have been allocated so the first carbon atom of the double bond has the lowest possible position (i.e. 2) along the hydrocarbon chain. If the chain had been numbered from right to left then the first carbon of the double bond would have a locant of 5.

Stem = *hept-*; suffix = *-ene*

Name = hept-2-ene

This name identifies on which carbon the double bond begins.

Alkynes

➔ Alkynes are a homologous series of unsaturated hydrocarbons.

- Alkynes have the general formula:

$$C_nH_{2n-2} \qquad (n = 2,3,4 \ldots)$$

- Simple alkynes contain one triple bond.
- The suffix *-yne* is used in naming alkynes.
- When $n = 2$ the alkyne has the *preferred* IUPAC name of acetylene rather than ethyne, which is its systematic name. Acetylene is a historical name for C_2H_2.
- The position of the triple bond along the carbon chain also needs to be specified. To do this the chain has to be numbered according to the previous rules discussed for alkenes.
- If alkyl groups are present the triple bond has priority and must be allocated the lowest locant.

EXAMPLE 3

Name the alkyne molecule represented in Figure 8.5.

```
    H   H   H   H           H   H
    |   |   |   |           |   |
H - C - C - C - C - C ≡ C - C - C - H
    |   |   |   |           |   |
    H   H   H   H           H   H
```

Figure 8.5 Alkyne molecule

Identify which end of the molecule to start numbering

Answer:

Stem = *oct-*; suffix = *-yne*

Numbering from right to left will give the triple bond's locant the lower number (i.e. 3).

Name = oct-3-yne

➔ KEY QUESTIONS

1. **Identify the homologous group of hydrocarbons to which each of the following molecules belong:**
   a. C_3H_4
   b. C_7H_{16}
   c. C_2H_4
2. **Draw the structural formula for pent-2-yne.**
3. **Draw the structural formula for 3,3-dimethyloctane.**

Answers ➲ p. 122

2 Alcohols and carboxylic acids

» Students investigate the nomenclature of organic chemicals, up to C8, using IUPAC conventions, including simple methyl and ethyl branched chains, including alcohols (primary, secondary and tertiary) and carboxylic acids.

Alcohols

- Alcohols are organic compounds that contain the hydroxyl functional group (OH). The hydroxyl functional group may be attached to different hydrocarbon chains such as alkanes, alkenes and alkynes.
- Alkanols are derivatives of alkanes and their general formula is:

$$C_nH_{2n+1}OH \qquad (n = 1,2,3 \ldots)$$

Examples:

1 (n = 1) CH_3OH methanol

2 (n = 2) C_2H_5OH ethanol

Nomenclature rules for alkanols:

- IUPAC rules
  Use these rules to name alkanols:
  1 Count the number of carbons in the straight chain. Select the correct stem to name the parent alkane. Remove the *-e* and replace it with the suffix *-ol*.
  2 For chains with three or more carbon atoms, number the chain from the end so as to give the hydroxyl group the lowest locant possible.
  3 Insert the locant of the OH group in front of the *-ol* suffix.

Example:
alkanol, $CH_3CH_2CH_2CH(OH)CH_3$

- There are five carbon atoms in the molecule: stem = *pent-*; alkane = pentane; suffix = *-ol*
- Number from right to left to give 2 as the locant for the OH group. (If numbering had been from left to right the locant for the OH group would be 4.)
- Name = pentan-2-ol
- Figure 8.6 shows structural formula for various alkanols. The hydroxyl group has priority over alkyl functional groups in determining the numbering of the hydrocarbon chain.

- In drawing structural formulas of alkanols it is important to show that the O atom (and not the H atom) of the hydroxyl group bonds to the carbon atom.
- Alkanols can be classified as *primary*, *secondary* or *tertiary*. This classification is based on the position of the hydroxyl functional group in the alkanol molecule.

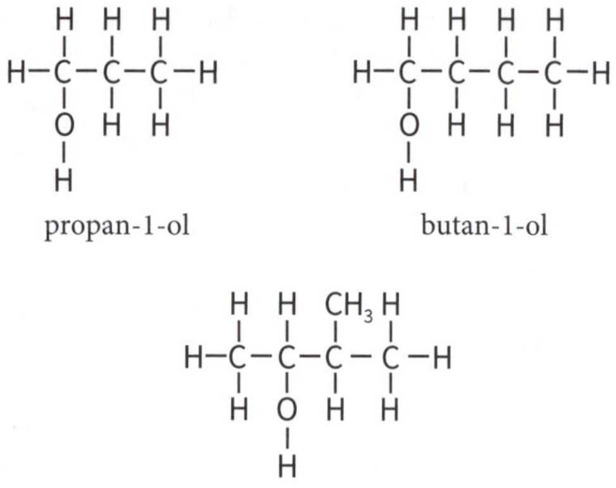

propan-1-ol

butan-1-ol

3-methylbutan-2-ol

Figure 8.6 Alkanols

- In a primary alkanol the hydroxyl functional group is attached to a carbon atom that is joined to only 1 other carbon atom in the chain. Thus the hydroxyl functional group is located at the end of a carbon chain.
- In a secondary alkanol the hydroxyl function group is attached to a carbon atom that is bonded to two other carbon atoms. This results in chain branching.
- In a tertiary alkanol the hydroxyl function group is attached to a carbon atom that is bonded to three other carbon atoms.
- Figure 8.7 shows these structural differences. The 'R' group represents a hydrocarbon chain or alkyl group of variable length. Methanol is a primary alkanol. In this case the R group is a hydrogen atom (CH_3OH).

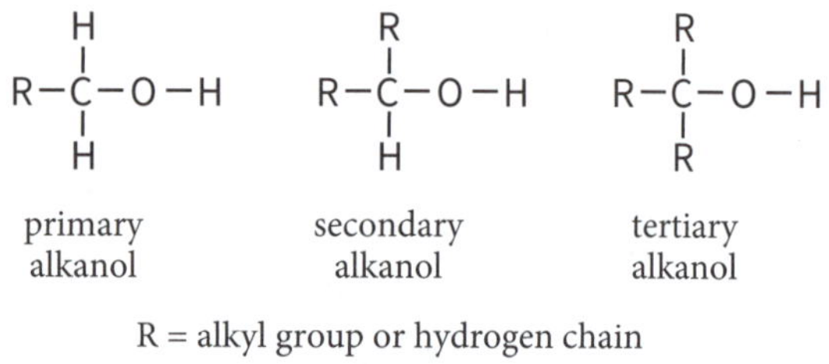

primary alkanol

secondary alkanol

tertiary alkanol

R = alkyl group or hydrogen chain

Figure 8.7 Alkanol classification

EXAMPLE 4

Name and classify each alkanol shown in Figure 8.8.

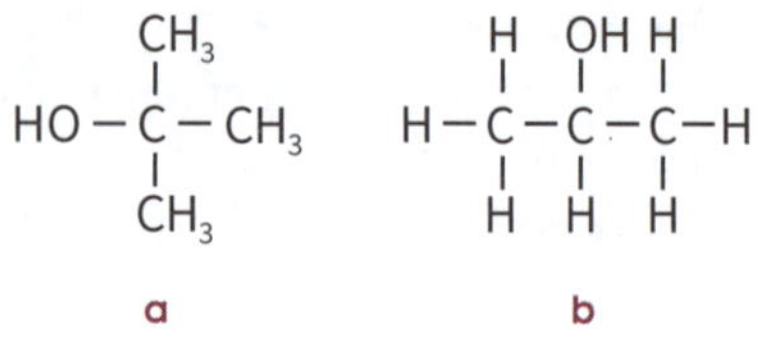

a

b

Figure 8.8 Name and classify these alkanols

Determine the number of carbon atoms attached to the carbon atom that has the alcohol functional group

Answer:

a Stem = *prop-*; alkane = propane;
functional groups = methyl and hydroxyl
Name = 2-methylpropan-2-ol
Classification = tertiary alkanol

b Stem = *prop-*; alkane = propane;
functional group = hydroxyl
Name = propan-2-ol
Classification = secondary alkanol

Carboxylic acids

- Carboxylic acids are organic molecules containing the carboxyl (COOH) functional group. This functional group may be present in a variety of organic compounds. When the carboxyl functional group is attached to an alkane chain then the acids are called *alkanoic acids*. Their general formula is:

$$C_nH_{2n+1}COOH \quad (n = 0,1,2,3\ \ldots)$$

Nomenclature rules for alkanoic acids:

- IUPAC nomenclature rules
  Use the following rule to systematically name alkanoic acids:
  1. Count the number of carbons in the straight chain.
  2. Select the correct stem to name the parent alkane.
  3. Remove the *-e* from the parent alkane's name and replace it with the suffix *-oic acid*.

Example:
alkanoic acid, $CH_3CH_2CH_2COOH$
There are four carbon atoms in the molecule:
stem = *but-*; alkane = butane; suffix = *-oic acid*
Name = butanoic acid

- The first two members of the alkanoic acid family are named according to the IUPAC *preferred* nomenclature (or PIN rules), where their historical common name is used. Their systematic names can also be used.

Examples:
1. HCOOH—IUPAC preferred name = formic acid (systematic name = methanoic acid)
2. CH_3COOH—IUPAC preferred name = acetic acid (systematic name = ethanoic acid)

Figure 8.9 shows the structural formula of propanoic acid.

```
    H  H
    |  |   //O
 H—C—C—C
    |  |   \
    H  H    O—H
```

Figure 8.9 Propanoic acid

- Short chain carboxylic acids are water soluble. They form acidic solutions and turn universal indicator red.

3 Aldehydes and ketones

» Students investigate the nomenclature of organic chemicals, up to C8, using IUPAC conventions, including simple methyl and ethyl branched chains, including aldehydes and ketones.

Aldehydes

- Aldehydes are organic compounds that contain the aldehyde functional group (CHO). This group contains a carbonyl centre (C=O).
- Alkanals are aldehyde derivatives of alkanes and their general formula is:

RCHO

(R = alkyl group or longer hydrocarbon chain)

The simplest alkanal has R=H. It is methanal (HCHO), which is a derivative of methane.

Examples:
CH_3CHO = ethanal
$CH_3CH_2CH_2CH_2CH_2CHO$ = hexanal

Nomenclature rules for alkanals:

- IUPAC Rules
  Use these rules to name alkanals:
  1. Count the number of carbons in the straight chain. Select the correct stem to name the parent alkane. Remove the *-e* from the alkane's name and replace it with the suffix *-al*.
  2. As the aldehyde functional group is always at the end of the alkane chain, numbering the chain is not required.

Example:
alkanal, $CH_3CH_2CH_2CH_2CHO$
There are five carbon atoms in the molecule:
stem = *pent-*; alkane = pentane; suffix = *-al*
Name: pentanal

Ketones

- Ketones are organic compounds that contain the carbonyl functional group (CO). This group contains a carbon atom joined to an oxygen atom with a double covalent bond.
- Alkanones are ketone derivatives of alkanes and their general formula is:

$$R_1COR_2$$

(R_1 and R_2 are alkyl groups or longer hydrocarbon chains)

Examples:

1 $(CH_3)_2CO$ = propanone

2 $CH_3COCH_2CH_3$ = butanone

Nomenclature rules for alkanones:

- IUPAC Rules

  Use these rules to name alkanones:

  1 Count the number of carbons in the straight chain. Select the correct stem to name the parent alkane. Remove the *-e* and replace it with the suffix *-one*.

  2 The carbonyl functional group is never at the end of the alkane chain and in most cases (except for alkanones with three or four carbon atoms) numbering the chain is required.

Example:

alkanone, $CH_3CH_2CH_2COCH_2CH_2CH_2CH_3$

There are eight carbon atoms in the molecule:

stem = *oct-*; alkane = octane; suffix = *-one*

The locant of the carbonyl group is 4 (fourth carbon from the left).

Name: octan-4-one

➔ Figure 8.10 shows the structural formulas of an alkanal and an alkanone.

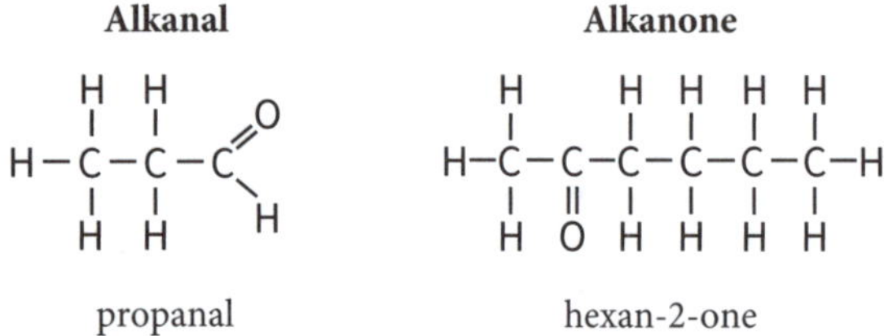

Figure 8.10 Alkanal and alkanone

EXAMPLE 5

Name molecules a and b in Figure 8.11.

a

```
    H  H  H  H  H     H
    |  |  |  |  |     |
 H—C—C—C—C—C—C—C—H
    |  |  |  |  |  ‖  |
    H  H  H  H  H  O  H
```

b

```
            H  H  H  H
  O         |  |  |  |
   ⩵C—C—C—C—C—H
  /         |  |  |  |
H—O         H  H  H  H
```

Figure 8.11 Molecules **a** and **b**

Determine the stem name of each molecule

Answer:

a The molecule is an alkanone with seven carbon atoms. Stem = *hept-*; alkane = heptane; suffix = *-one*. The carbonyl function group is at locant position 2. Name = heptan-2-one

b The molecule is an alkanoic acid with five carbon atoms. Stem = *pent-*; alkane = pentane. No numbering of the chain is needed as the carboxylic acid group is always on the first carbon. Suffix = *-oic acid* Name = pentanoic acid

➔ KEY QUESTIONS

4 **Name and classify the following organic compounds:**

a $CH_3CH_2CH_2CHO$

b $CH_3CH_2CH_2CH_2CH_2COOH$

5 **Draw a structural formula for 2-methylpentan-2-ol.**

6 **Draw a structural formula of 4-methyloctan-3-one.**

7 **Classify and name the following alcohol as a primary, secondary or tertiary alcohol:**

$CH_3CH_2CH_2C(CH_3)_2OH$

Answers ➲ p. 122

4 Amines and amides

» Students investigate the nomenclature of organic chemicals, up to C8, using IUPAC conventions, including simple methyl and ethyl branched chains, including amines and amides.

Amines

➔ Amines are organic molecules containing the amino (NH_2) functional group. This functional group may be present in a variety of organic compounds. When the amine functional group is attached to an alkane chain the general formula is:

$$C_nH_{2n+1}NH_2 \qquad (n = 1,2,3\ldots)$$

Nomenclature rules for amines:

1 Count the number of carbons in the straight chain. Select the correct name of the parent alkane. Remove the *-e* and replace it with the suffix *-amine*.

2 For chains with three or more carbon atoms, number the chain from the end so as to give the amino group the lowest locant possible.

3 Insert the locant of the NH_2 group in front of the *-amine* suffix.

Example:

amine, $CH_3CH_2CH_2CH_2CH(NH_2)CH_3$

There are six carbon atoms in the molecule:

Stem = *hex-*; alkane = hexane

Delete the *-e* from the alkane's name and add the suffix *-amine*.

Amine name = hexanamine

Number from right to left to give 2 as the locant for the NH_2 group.

Name = hexan-2-amine

→ Figure 8.12 shows the structural formulas and ball-and-stick models for two amines.

Structural formula **Ball-and-stick model**

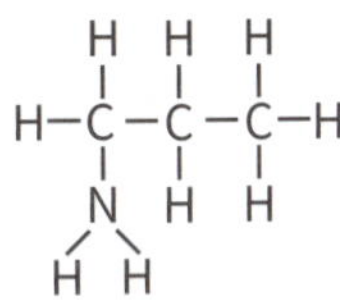

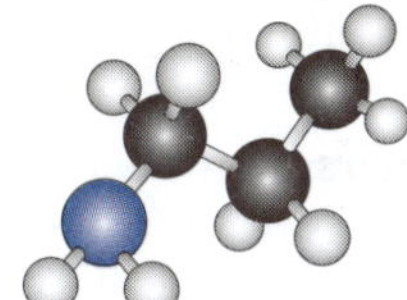

propan-1-amine

H H H H
H–C–C–C–C–H
H N H H
H H

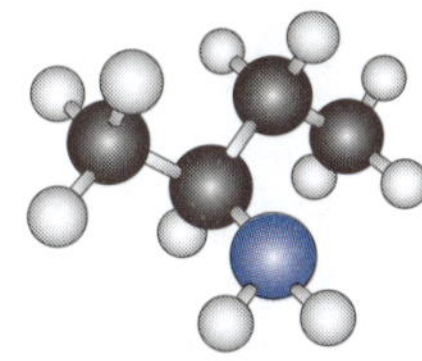

butan-2-amine

Figure 8.12 Amine examples

→ Amines, like alkanols, can be classified as primary, secondary or tertiary. This classification is based on the number of alkyl groups bonded to the nitrogen atom in the amine molecule. Figure 8.13 shows the structure of primary, secondary and tertiary amines.

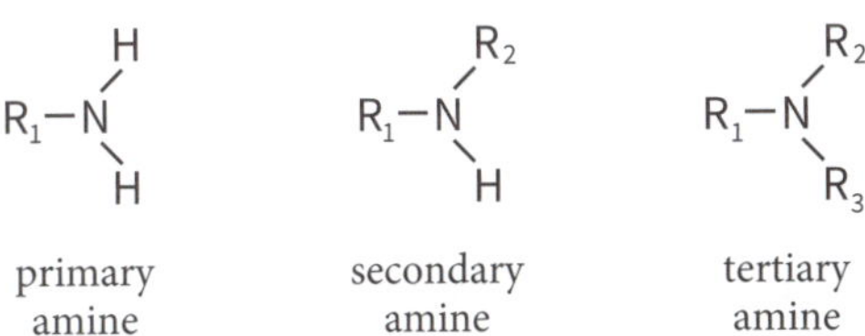

R = alkyl group or hydrocarbon chain

Figure 8.13 Amine classification

→ To name secondary and tertiary amines, the IUPAC nomenclature is modified as follows:

- Each alkyl group bonded to the N atom is named as an N-alkyl group using a prefix.

Examples:

1 $CH_3CH_2CH_2NHCH_3$
Name = N-methylpropan-1-amine

2 $(CH_3CH_2)_2NCH_2CH_2CH_3$
Name = N,N-diethylpropan-1-amine

Amides

→ Amides are derivatives of carboxylic acids. The OH group of the carboxylic acid is replaced by an amino (NH_2) group.

→ Figure 8.14 shows the structure of a primary amide. Two hydrogen atoms are bonded to the N atom of the amine group. In secondary amides one of the H atoms of the amine group is replaced by an alkyl group (R) and in a tertiary amide both H atoms are replaced by alkyl groups.

O
R – C – N (H)(H)

Figure 8.14 Primary amide structure

→ The following rules can be used to name primary amides. Nomenclature rules:

- Count the number of carbons in the straight chain.
- Select the correct stem to name the parent alkane.
- Remove the *-e* from the parent alkane's name and replace it with the suffix *-amide*.

Examples:

1 $HCONH_2$
Name = methanamide (or formamide)

2 CH_3CONH_2
Name = ethanamide (or acetamide)

3 $CH_3CH_2CONH_2$
Name = propanamide

5 Halogenated organic compounds

» Students investigate the nomenclature of organic chemicals, up to C8, using IUPAC conventions, including simple methyl and ethyl branched chains, including halogenated organic compounds.

→ Halogens are the elements of Group 17 of the periodic table. They can bond to hydrocarbon molecules to form halohydrocarbons.

→ Halogens are functional groups. The names of the halogen functional groups are:

Br = bromo; Cl = chloro; F = fluoro; I = iodo

→ The following rules can be used to name halogenated alkanes.

Nomenclature rules:

1 Count the number of carbons in the straight alkane chain.
2 Select the correct stem to name the parent alkane.
3 Number the chain from the end so as to give the halogen functional groups the lowest set of locants.*
4 The halogen functional groups are alphabetically named using prefixes. Locants are placed in front of these prefixes to specify the positions of the functional groups along the chain.
5 If the previous rules lead to more than one possible name then the correct name is the one in which the lowest locant is assigned to the functional group cited first alphabetically.

* The lowest set of locants is the set that, when compared term by term with other locant sets, each cited in order of increasing value, has the lowest term at the first point of difference.

Examples:

1 $CH_3CH_2CHBrCH_2CH_2CH_3$
Numbering from left to right gives the lower locant for bromine.
Name = 3-bromohexane
2 $CH_3CH_2CHBrCCl_2CH_2CH_3$
Numbering from right to left gives the lower locant set (3,3,4), whereas the set when numbering from left to right is (3,4,4).
The halogens are named alphabetically.
Name = 4-bromo-3,3-dichlorohexane
3 $CH_3CHFCH_2CHClCH_2$
Numbering from right to left or left to right gives the same locant set (2,4). Therefore the numbering will be from right to left based on the alphabetical order of the halogen functional groups.
Name = 2-chloro-4-fluoropentane

→ Figure 8.15 shows the structural formula of a halogenated alkane in which the lower locant set (2,2,3,3,4) is obtained by numbering the carbon chain from right to left. The functional groups are named in alphabetical order.

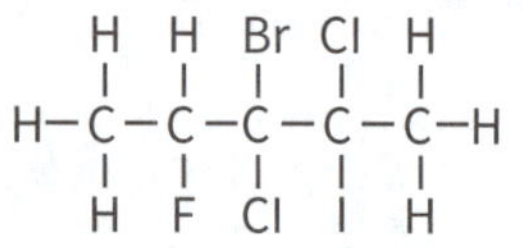

```
    H   H  Br  Cl  H
    |   |   |   |   |
H — C — C — C — C — C — H
    |   |   |   |   |
    H   F  Cl   I   H
```

3-bromo-2,3-dichloro-4-fluoro-2-iodopentane

Figure 8.15 Halogenated alkane

→ If halogen functional groups are present in alkenes then the double bond must have the lowest possible locant.

Examples:

1 $CH_3CH{=}CHCBr_2CH_3$
Name = 4,4-dibromopent-2-ene
2 $CH_3\,CH(CH_3)CHFCF{=}CHCH_3$
Name = 3,4-difluoro-5-methylhex-2-ene

→ KEY QUESTIONS

8 Name the following organic molecule:
$CH_3CH_2CH_2CONH_2$

9 Draw the structural formula for 2-bromo-4-fluoroheptane.

10 Draw the structural formula for 3,4-dibromohex-1-ene.

Answers ➲ p. 122

6 Isomers

» Students explore and distinguish the different types of structural isomers, including saturated and unsaturated hydrocarbons, including chain isomers, position isomers and functional group isomers.

→ **Isomers** are molecules that have the same molecular formulas but different structural formulas. Isomerism is therefore due to the difference in the arrangement of atoms in molecules. This results in variation in properties between isomers.

isomer: molecule with the same molecular formula but different structural formula

→ Isomers can be classified as *chain* isomers, *position* isomers and *functional group* isomers. Another classification of isomerism is stereoisomerism but this is not examined in this course.

Chain isomers

→ Chain isomer molecules form due to the branching of hydrocarbon chains. The carbon atoms have been rearranged in such isomers.

→ Figure 8.16 shows the three chain isomers of C_5H_{12}. These isomers have different physical properties and similar chemical properties.

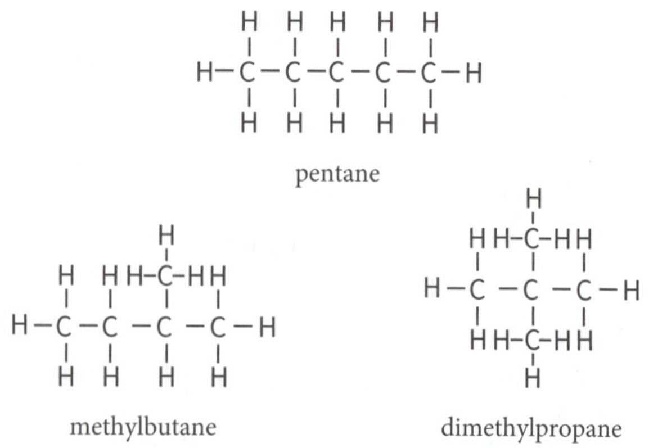

```
    H   H   H   H   H
    |   |   |   |   |
H — C — C — C — C — C — H
    |   |   |   |   |
    H   H   H   H   H
```

pentane

```
              H
              |
    H   H  H— C —H  H
    |   |     |     |
H — C — C  —  C  —  C — H
    |   |     |     |
    H   H     H     H
```

methylbutane

```
          H
          |
    H  H— C —H  H
    |     |     |
H — C  —  C  —  C — H
    |     |     |
    H  H— C —H  H
          |
          H
```

dimethylpropane

Figure 8.16 Chain isomers of C_5H_{12}

Position isomers

- Position isomers are formed when functional groups are located at different positions on the hydrocarbon chain.
- Position isomers can form in unsaturated hydrocarbons such as alkenes and alkynes. In these cases the double bond functional group or triple bond functional group is in a different position. Figure 8.17 shows examples of such isomers.

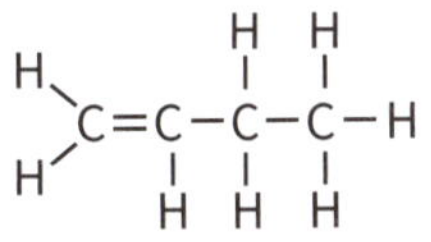

but-1-ene 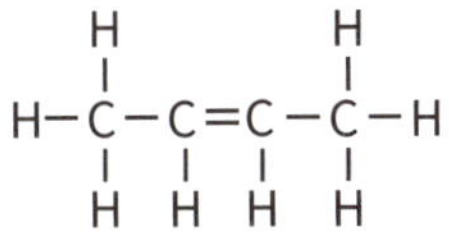 but-2-ene

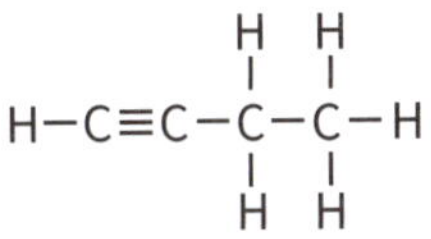

but-1-yne 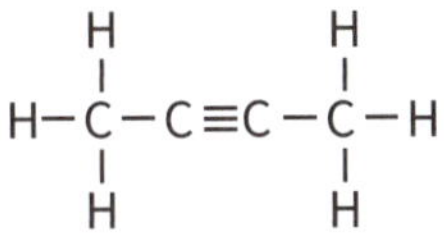 but-2-yne

Figure 8.17 Unsaturated isomers

- Figure 8.18 shows examples of straight chain position isomers for $C_4H_8Br_2$ in which the bromine functional groups occupy different positions. Each isomer has different physical properties and similar chemical properties.

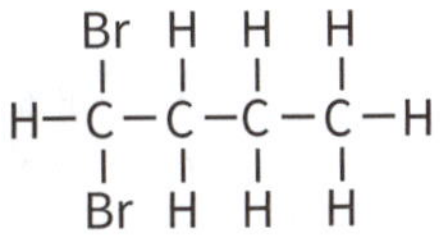

1,1-dibromobutane 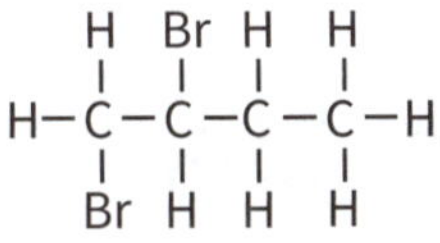 1,2-dibromobutane

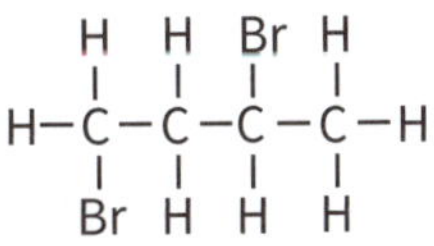

1,3-dibromobutane 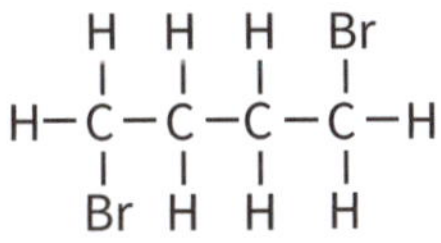 1,4-dibromobutane

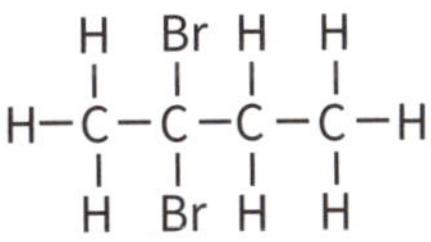

2,2-dibromobutane 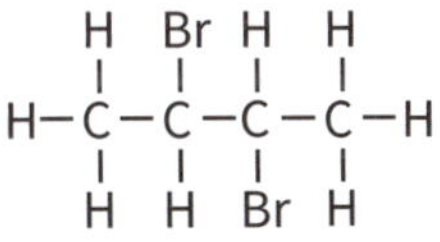 2,3-dibromobutane

Figure 8.18 Position isomers of $C_4H_8Br_2$

- Figure 8.19 shows examples of position isomers for an unsaturated straight-chain chlorinated hydrocarbon (C_4H_7Cl).

Functional group isomers

- Functional group isomers are isomers in which different functional groups are present.

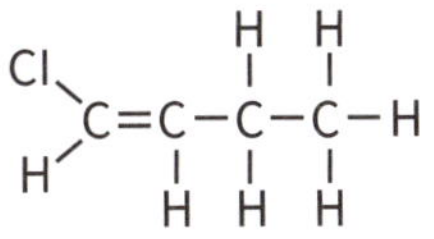

1-chlorobut-1-ene 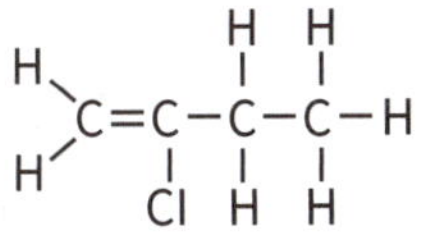 2-chlorobut-1-ene

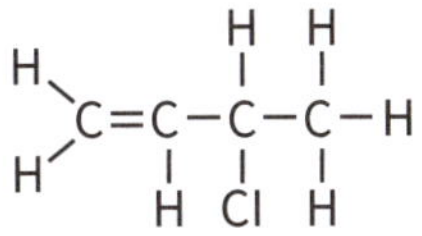

3-chlorobut-1-ene 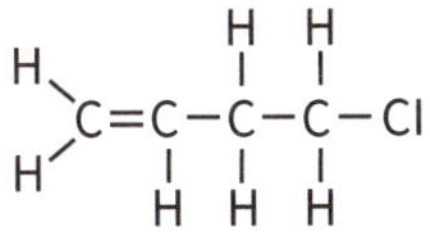 4-chlorobut-1-ene

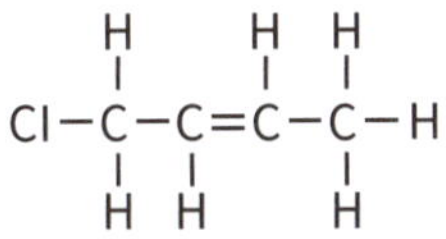

1-chlorobut-2-ene 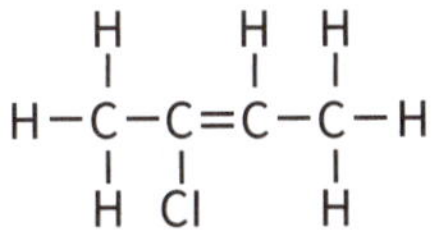 2-chlorobut-2-ene

Figure 8.19 Unsaturated position isomers of C_4H_7Cl

- Figure 8.20 shows examples of functional group isomers. Alkanals and alkanones with the same number of carbon atoms are functional group isomers. These molecules have different physical and chemical properties.

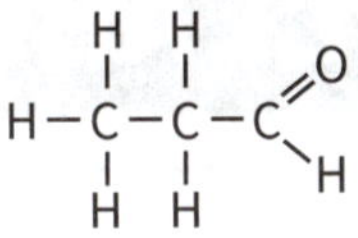

propanal 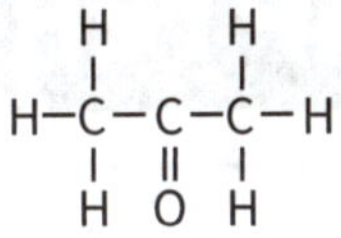 propanone

Figure 8.20 Functional group isomers of C_3H_6O

EXAMPLE 6

The following data was collected for an organic compound, X:
Molar mass: 88.104 g/mol
Composition (by weight): 54.53% C; 9.15% H; 36.32% O
Melting point: –5.4 °C; boiling point: 163.3 °C
X dissolves in water and the solution turns universal indicator red.

a Identify the functional group that is present in this compound.
b Calculate the empirical formula of the compound.
c Calculate the molecular formula of the compound.
d Draw and name two possible chain isomers of this compound.

Start the calculation with 100 g of X

Answer:

a Carboxyl functional group (COOH) is present as short chain carboxylic acids are acidic.
b In 100 g of X: $m(C) = 54.53$ g; $m(H) = 9.15$ g; $m(O) = 36.32$ g
Calculate the number of moles of each element:
$n(C) = \frac{54.53}{12.01} = 4.540$ mol

$n(H) = \frac{9.15}{1.008} = 9.077$ mol

$n(O) = \frac{36.32}{16.00} = 2.270$ mol

Mole ratio: C:H:O = 4.540:9.077:2.270 = 2:4:1

Empirical formula (EF) = C_2H_4O

c Empirical molar mass = 2(12.01) + 4(1.008) + 16.00 = 44.052 g/mol

∴ Molar mass = 2 × empirical molar mass

∴ MF = 2 × EF = $C_4H_8O_2$

d The compound's isomers are illustrated in Figure 8.21.

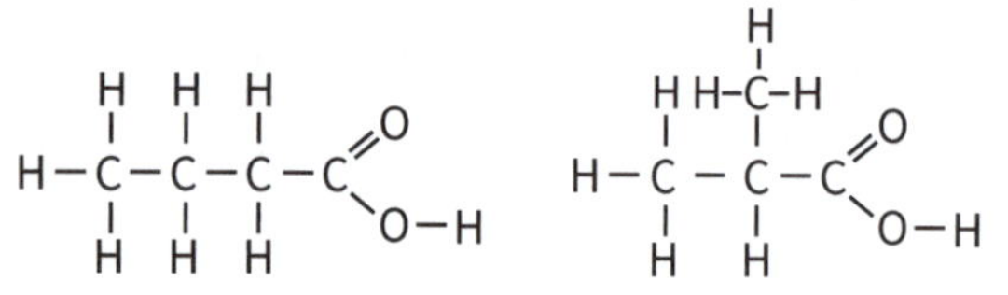

butanoic acid

2-methylpropanoic acid

Figure 8.21 Chain isomers

→ KEY QUESTIONS

11 Draw the structural formulae of the chain isomers of molecules with the molecular formula C_4H_{10}. Name each isomer.

12 Draw the structural formulae of the position isomers of molecules with the molecular formula $C_2H_4F_2$. Name each isomer.

13 Draw the structural formulae of the chain and position isomers of molecules with the molecular formula C_5H_8. Name each isomer.

Answers ➲ p. 122

CHAPTER SYLLABUS CHECKLIST

Are you able to answer every question from the syllabus for this chapter? Tick each question as you go through the checklist if you are able to answer it. If you cannot answer a question, turn to the relevant page in the study guide to find the answer. For NESA key word meanings, go to www.educationstandards.nsw.edu.au and search 'key words'.

	FOR A COMPLETE UNDERSTANDING OF THIS TOPIC:	PAGE NO.	✓
1	Can I use IUPAC nomenclature to name alkanes, alkenes and alkynes?	111	
2	Can I draw structural formulas of alkanes, alkenes and alkynes?	112	
3	Can I use IUPAC nomenclature to name alkanols?	114	
4	Can I draw structural formulas of alkanols?	114	
5	Can I use IUPAC nomenclature to name alkanoic acids?	115	
6	Can I draw structural formulas of alkanoic acids?	115	
7	Can I use IUPAC nomenclature to name alkanals and alkanones?	115	
8	Can I draw structural formulas of alkanals and alkanones?	116	
9	Can I use IUPAC nomenclature to name amines and amides?	116	
10	Can I draw structural formulas of amines and amides?	117	
11	Can I use IUPAC nomenclature to name halogenated hydrocarbons?	117	
12	Can I draw structural formulas of halogenated hydrocarbons?	118	
13	Can I distinguish the different types of structural isomers, including saturated and unsaturated hydrocarbons, including chain isomers, position isomers and functional group isomers?	118	
14	Can I draw structural formulas of isomers, including chain isomers, position isomers and functional group isomers?	119	

HSC EXAM-TYPE QUESTIONS

Objective-response questions

(1 mark each)

1 Name the hydrocarbon illustrated in Figure 8.22.

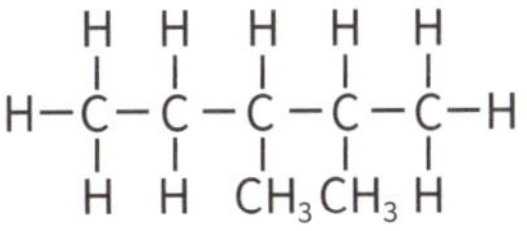

```
    H  H  H   H  H
    |  |  |   |  |
  H-C--C--C---C--C-H
    |  |  |   |  |
    H  H  CH3 CH3 H
```

Figure 8.22 Hydrocarbon

- **A** heptane
- **B** dimethylpentane
- **C** 3,4-dimethylpentane
- **D** 2,3-dimethylpentane

2 Name the unsaturated hydrocarbon illustrated in Figure 8.23.

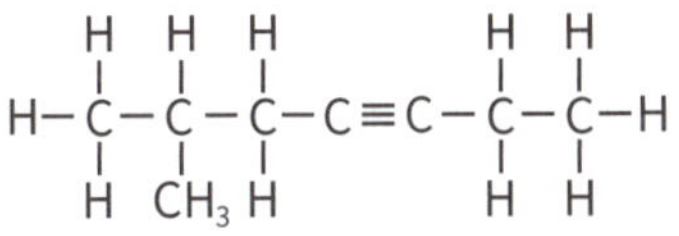

```
    H  H   H          H  H
    |  |   |          |  |
  H-C--C---C--C≡C--C--C-H
    |  |   |          |  |
    H  CH3 H          H  H
```

Figure 8.23 Unsaturated hydrocarbon

- **A** octyne
- **B** methylhept-4-yne
- **C** 6-methylhept-3-yne
- **D** 2-methylhept-4-yne

3 Name the following molecule:

$CH_3CH_2CH_2CH_2CH(OH)CH_2CH_2CH_3$

- **A** octan-5-ol
- **B** octan-4-ol
- **C** octanol
- **D** heptan-4-ol

4 Name the following molecule:

$CH_3CH_2CH_2CH_2CH_2COOH$

- **A** pentanoic acid
- **B** hexanal
- **C** hexan-1-oic acid
- **D** hexanoic acid

5 Name the following molecule:

$CH_3CH{=}CHCHClCHClCHClCH_3$

- **A** 2,3,4-chlorohept-5-ene
- **B** 4,5,6-chlorohept-2-ene
- **C** 4,5,6-chloroheptene
- **D** 4,5,6-trichlorohept-2-ene

Extended-response questions

6 Draw structural formulas and name the following unsaturated hydrocarbons:

- **a** CH_3CCCH_3 (2 marks)
- **b** $CH_3CH_2CH_2CH_2CH_2CH_2CHCH_2$ (2 marks)

7 A hydrocarbon has a molar mass of 72.146 g/mol. Draw structural formulas for isomeric molecules that have this molar mass. Name these molecules systematically. (4 marks)

8 Name the following molecules:

- **a** $CH_3CH_2CH(CH_3)CH_2CH_2OH$ (1 mark)
- **b** $CH_3CH_2CH(OH)CH_2CH_2CH_3$ (1 mark)
- **c** $CH_3CH_2CH_2CH_2CH_2CH_2COOH$ (1 mark)

9 Figure 8.24 shows a space-filling model of a chlorinated hydrocarbon. A space-filling model does not show the covalent bonds but shows the three-dimensional arrangement of the atoms in the molecule.

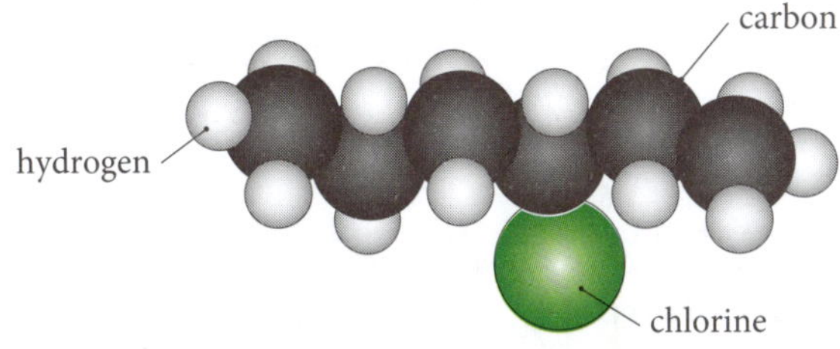

Figure 8.24 Chlorinated hydrocarbon

- **a** Write the molecular formula of the chlorinated hydrocarbon. (1 mark)
- **b** Draw the structural formula of the chlorinated hydrocarbon and use IUPAC nomenclature to name the molecule. (2 marks)
- **c** Draw the structural formulas and name the straight chain isomers of this molecule. (4 marks)

10 Figure 8.25 shows a ball-and-stick model of an organic molecule.

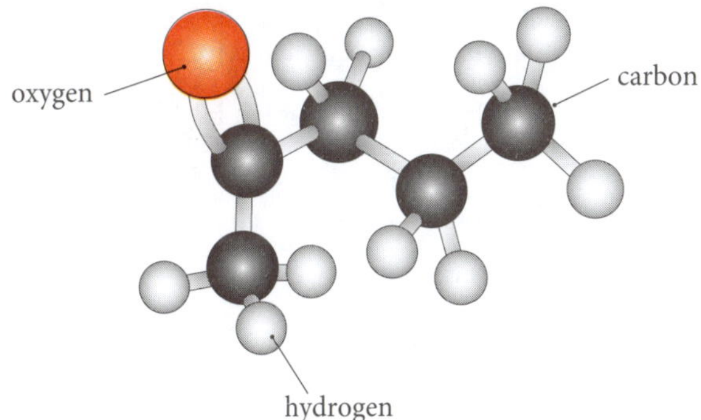

Figure 8.25 Organic molecule

- **a** Write the molecular formula of the molecule. (1 mark)
- **b** Write the condensed structural formula of the molecule and use IUPAC nomenclature to name the molecule. (2 marks)
- **c** Draw the structural formulas and name the five isomers of this molecule. (5 marks)

ANSWERS

KEY QUESTIONS

Key questions ⊃ p. 113

1 **a** alkyne **b** alkane **c** alkene

2 See Figure A8.1.

```
    H         H   H
    |         |   |
H - C - C≡C - C - C - H
    |         |   |
    H         H   H
```

Figure A8.1 Pent-2-yne

3 See Figure A8.2.

```
             H
             |
    H    H H-C-H H   H   H   H   H
    |    |   |   |   |   |   |   |
H - C -  C - C - C - C - C - C - C - H
    |    |   |   |   |   |   |   |
    H    H H-C-H H   H   H   H   H
             |
             H
```

Figure A8.2 3,3-dimethyloctane

Key questions ⊃ p. 116

4 **a** butanal (alkanal)
b hexanoic acid (alkanoic acid)

5 See Figure A8.3.

```
          H
          |
    H     O   H   H   H
    |     |   |   |   |
H - C  -  C - C - C - C - H
    |     |   |   |   |
    H   H-C-H H   H   H
          |
          H
```

Figure A8.3 2-methylpentan-2-ol

6 See Figure A8.4.

```
    H   H   O   H   H   H   H
    |   |   ||  |   |   |   |
H - C - C - C - C - C - C - C - H
    |   |       |   |   |   |
    H   H     H-C-H H   H   H
                |
                H
```

Figure A8.4 4-methyloctan-3-one

7 1,1-dimethylbutan-1-ol (tertiary alcohol)

Key questions ⊃ p. 118

8 butanamide

9 See Figure A8.5.

```
    H   Br  H   H   H   H   H
    |   |   |   |   |   |   |
H - C - C - C - C - C - C - C - H
    |   |   |   |   |   |   |
    H   H   H   F   H   H   H
```

Figure A8.5 2-bromo-4-fluoroheptane

10 See Figure A8.6.

```
             Br  Br  H   H
H            |   |   |   |
  \
   C = C - C - C - C - C - H
  /    |   |   |   |   |
H      H   H   H   H   H
```

Figure A8.6 3,4-dibromohex-1-ene

Key questions ⊃ p. 120

11 See Figure A8.7.

```
    H   H   H   H
    |   |   |   |
H - C - C - C - C - H
    |   |   |   |
    H   H   H   H
```

butane

```
            H
            |
      H   H-C-H  H
      |     |    |
H  -  C  -  C  - C - H
      |     |    |
      H     H    H
```

methylpropane

Figure A8.7 Isomers of C_4H_{10}

12 See Figure A8.8.

```
    F   H
    |   |
H - C - C - H
    |   |
    F   H
```

1,1-difluoroethane

```
    F   F
    |   |
H - C - C - H
    |   |
    H   H
```

1,2-difluoroethane

Figure A8.8 Isomers of $C_2H_4F_2$

13 See Figure A8.9.

```
          H   H   H
          |   |   |
H - C≡C - C - C - C - H
          |   |   |
          H   H   H
```

pent-1-yne

```
    H         H   H
    |         |   |
H - C - C≡C - C - C - H
    |         |   |
    H         H   H
```

pent-2-yne

```
            H
            |
          H-C-H H
            |   |
H - C≡C  -  C - C - H
            |   |
            H   H
```

3-methylbut-1-yne

Figure A8.9 Isomers of C_5H_8

HSC EXAM-TYPE QUESTIONS

Objective-response questions

1 **D.** The two methyl groups must be given the lower locant set. **A** is incorrect as the stem name is *pent-*. **B** is incorrect as the position of the methyl groups must be included in the name. **C** is incorrect as this is not the lower locant set.

2 **C.** The triple bond must get priority in numbering. **A** is incorrect as this is a branched chain based on heptyne. **B** is incorrect as the locant of the methyl group is not stated and the chain numbering must start from the right side to give the triple bond the lower locant. **D** is incorrect as numbering must favour the triple bond.

3 **B**. The hydroxyl functional group is located on the fourth carbon from the right. **A** is incorrect as the chain numbering starts at the right in this case to give the OH group the lower number. **C** is incorrect as the hydroxyl group has no locant. **D** is incorrect as the chain has eight carbons not seven.

4 **D**. This is a six-carbon alkanoic acid. **A** is incorrect as *pent-* is the incorrect stem. **B** is incorrect as this molecule is not an alkanal. **C** is incorrect as a locant number is unnecessary as the carboxylic acid group is always on the first carbon.

5 **D**. The numbering starts from the left to prioritise the double bond and the chloro groups are then on the fourth, fifth and sixth carbons. **A** is incorrect as the double bond has not been prioritised in numbering. **B** is incorrect as the prefix *tri-* has not been used. **C** is incorrect as the prefix *tri-* has not been used and the position of the double bond is not shown.

Extended-response questions

6 EM Students gain full marks when they demonstrate skills in drawing structural formulas and use IUPAC nomenclature correctly to name each molecule.
See Figure A8.10.

a

```
    H       H
    |       |
H - C - C≡C - C - H ✓    but-2-yne ✓
    |       |
    H       H
```

b

```
    H   H   H   H   H   H
    |   |   |   |   |   |        H
H - C - C - C - C - C - C - C = C/
    |   |   |   |   |   |   |     \   ✓    oct-1-ene ✓
    H   H   H   H   H   H   H      H
```

Figure A8.10 Unsaturated hydrocarbons

7 EM Students are required to determine by trial and error the molecular formula of the hydrocarbon. They then should understand that this molecule is an alkane and should demonstrate an understanding of the possible isomers.
MF = C_5H_{12} (M = 72.146 g/mol) ✓
See Figure A8.11.

```
    H   H   H   H   H
    |   |   |   |   |
H - C - C - C - C - C - H
    |   |   |   |   |
    H   H   H   H   H    ✓
         pentane
```

```
    H  CH3  H   H
    |   |   |   |
H - C - C - C - C - H
    |   |   |   |
    H   H   H   H    ✓
      methylbutane
```

```
    H  CH3  H
    |   |   |
H - C - C - C - H
    |   |   |
    H  CH3  H    ✓
   dimethylpropane
```

Figure A8.11 Isomers

8 EM Students use IUPAC nomenclature rules to name each molecule correctly.

a 3–methylpentan-1-ol ✓

b hexan-3-ol ✓

c heptanoic acid ✓

9 EM Students use IUPAC nomenclature rules to name each molecule correctly and to draw the other two straight chain isomers.

a $C_6H_{13}Cl$ ✓

b 3-chlorohexane ✓
See Figure A8.12.

```
    H   H   H   H   H   H
    |   |   |   |   |   |
H - C - C - C - C - C - C - H
    |   |   |   |   |   |
    H   H   H   Cl  H   H    ✓
```

Figure A8.12 Chlorinated alkane

c See Figure A8.13.

```
    H   H   H   H   H   H
    |   |   |   |   |   |
H - C - C - C - C - C - C - H
    |   |   |   |   |   |
    Cl  H   H   H   H   H    ✓
        1-chlorohexane ✓
```

```
    H   H   H   H   H   H
    |   |   |   |   |   |
H - C - C - C - C - C - C - H
    |   |   |   |   |   |
    H   Cl  H   H   H   H    ✓
        2-chlorohexane ✓
```

Figure A8.13 Chlorinated hydrocarbon straight chain isomers

10 EM Students use IUPAC nomenclature rules to name each molecule correctly and to draw the different structural formulas.

a $C_5H_{10}O$ ✓

b $CH_3COCH_2CH_2CH_3$; ✓ pentan-2-one ✓

c See Figure A8.14.

```
    H   H   O   H   H
    |   |   ||  |   |
H - C - C - C - C - C - H
    |   |       |   |
    H   H       H   H
     pentane-3-one ✓
```

```
    H   H   O   H
    |   |   ||  |
H - C - C - C - C - H
    |   |       |
    H  CH3      H
  3-methybutan-2-one ✓
```

```
    H   H   H   H   O
    |   |   |   |   ||
H - C - C - C - C - C - H
    |   |   |   |
    H   H   H   H
       pentanal ✓
```

```
    H   H   H   O
    |   |   |   ||
H - C - C - C - C - H
    |   |   |
    H   H  CH3
   2-methybutanal ✓
```

```
    H   H   H   O
    |   |   |   ||
H - C - C - C - C - H
    |   |   |
    H  CH3  H
   3-methybutanal ✓
```

Figure A8.14 Isomers

CHAPTER 9 HYDROCARBONS

MODULE 7 ORGANIC CHEMISTRY

INQUIRY QUESTION:

How can hydrocarbons be classified based on their structure and reactivity?

Hydrocarbons are used throughout society. Fractional distillation of petroleum is the main source of hydrocarbons. They are used as fuels to power machines and vehicles as well as in the production of a wide variety of organic chemicals, solvents and plastics. However, their use can lead to widespread environmental pollution. This chapter examines the structure of hydrocarbons and their properties.

1 Homologous series of hydrocarbons

» Students construct models, identify the functional group, and write structural and molecular formulae for homologous series of organic chemical compounds, up to C8—alkanes, alkenes and alkynes.

Models of hydrocarbons

- The homologous series of hydrocarbons were introduced in Chapter 8. These series are:
  - **Alkanes**—saturated hydrocarbons: general formula C_nH_{2n+2} (e.g. butane (n = 4); MF = C_4H_{10})
  - **Alkenes**—unsaturated hydrocarbons: general formula C_nH_{2n} (e.g. propene (n = 3); MF = C_3H_6)
  - **Alkynes**—unsaturated hydrocarbons: general formula C_nH_{2n-2} (e.g. pent-1-yne (n=5); MF=C_5H_8)
- Saturated hydrocarbons have single carbon–carbon bonds, whereas unsaturated hydrocarbons have either double or triple carbon–carbon bonds.
- Structural formulas are used to show the differences in bonding in these homologous series. Figure 9.1 shows examples of alkane, alkene and alkyne structural formulas.
- These structural formulas do not show the true bond angles of the three-dimensional molecules. In alkanes the bond angles are tetrahedral (109° 28′).

```
    H   H   H   H   H
    |   |   |   |   |
H - C - C - C - C - C - H
    |   |   |   |   |
    H   H   H   H   H
         pentane

        H   H   H                        H   H   H
H       |   |   |                        |   |   |
 \C = C - C - C - C - H      H - C≡C - C - C - C - H
 /    |   |   |   |                      |   |   |
H     H   H   H   H                      H   H   H
      pent-1-ene                         pent-1-yne
```

Figure 9.1 Structural formulas of alkanes, alkenes and alkynes

In alkenes the atoms attached to the carbon atoms of the double covalent bond are planar with bond angles close to 120°. In alkynes the atoms attached to the triple bond are linear with bond angles of 180°. Figure 9.2 shows the bond angles around single, double and triple bonds in hydrocarbons.

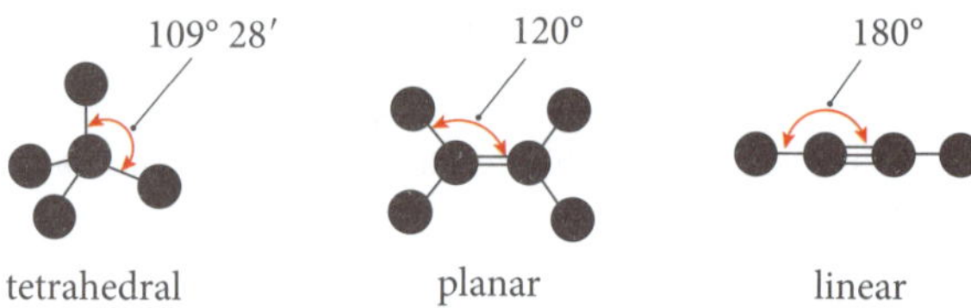

Figure 9.2 Bond angles in hydrocarbons

- Condensed structural formulas can also be used to show the arrangement of atoms in hydrocarbons:
  - $CH_3CH_2CH_2CH_2CH_3$ — pentane
  - $CH_2CHCH_2CH_2CH_3$ — pent-1-ene
  - $CHCCH_2CH_2CH_3$ — pent-1-yne

EXAMPLE 1

Write the condensed structural formulas for:

a **pent-2-ene**

b **hept-3-yne**

c **octane**

Ensure the carbon atoms maintain their valency of 4

Answer:

a $CH_3CHCHCH_2CH_3$

b $CH_3CH_2CCCH_2CH_2CH_3$

c $CH_3CH_2CH_2CH_2CH_2CH_2CH_2CH_3$

FIRSTHAND INVESTIGATION

Modelling hydrocarbons

» Students construct models, identify the functional group, and write structural and molecular formulas for homologous series of organic chemical compounds, up to C8—alkanes, alkenes, alkynes..

Aim

to use molecular model kits to investigate the three-dimensional structure of hydrocarbons

Equipment

Ball-and-stick model kit
Space-filling model kit

Method

1 Use both types of model kits to construct the following molecules:
   a propane (MF = C_3H_8)
   b propene (MF = C_3H_6)
   c propyne (MF = C_3H_4)
2 Take a photo of each model for your report and describe the structures of the models using terms such as tetrahedral, planar and linear.
3 Construct other molecules and include photos of these models in your report. Name these hydrocarbons using IUPAC nomenclature rules.
4 Use the following table headings in your report:

Name	MF	Structural formula/photo

Sample models

Figure 9.3 illustrates ball-and-stick models and space-filling models of hydrocarbons.

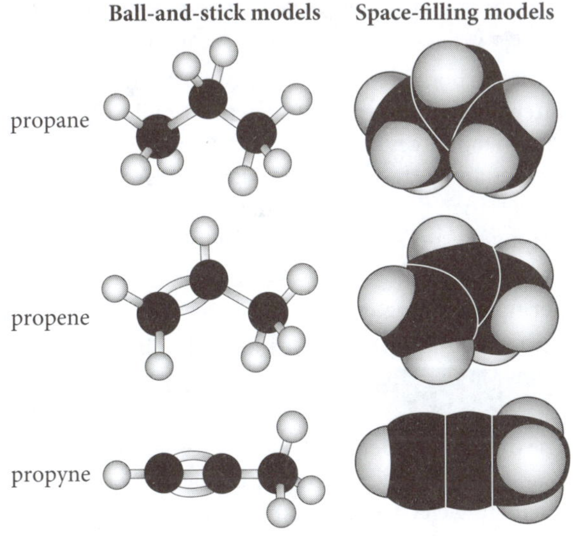

Figure 9.3 Ball-and-stick and space-filling models of propane, propene and propyne

2 Molecular shapes

» Students analyse the shape of molecules formed between carbon atoms when a single, double or triple bond is formed between them.

Hybridisation of orbitals in hydrocarbons

→ The modern atomic theory describes electrons in each electron shell (energy level) as occupying specific subshells and orbitals. Orbitals have specific shapes and orientations, and are designated by the letters s, p, d and f. This concept was introduced in the Year 11 Chemistry course.

→ Orbitals have variable shapes in three dimensions. A maximum of 2 electrons can occupy an orbital. Figure 9.4 shows that the s orbitals have a simple spherical shape.

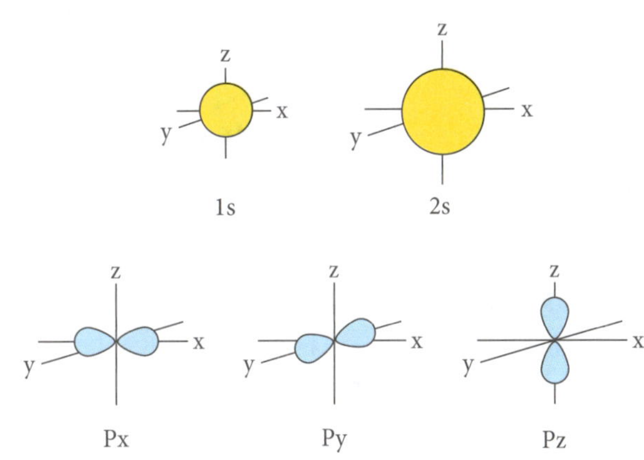

Figure 9.4 Shapes and orientations of atomic orbitals

→ To explain the three-dimensional shapes of hydrocarbon molecules, s and p atomic orbitals combine via a process called **hybridisation** to form a variety of molecular orbitals.

hybridisation: a procedure used in the valence bond theory in which atomic orbitals are combined to produce molecular orbitals

→ Figure 9.5 shows the shapes of the hybrid orbitals. The sp^3 orbitals are arranged tetrahedrally. The sp^2 orbitals are trigonally planar in orientation. The sp orbitals are linearly arranged.

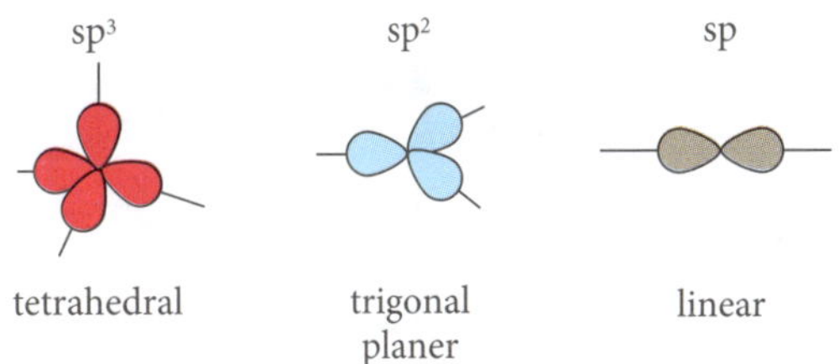

Figure 9.5 Hybrid molecular orbitals

→ Figure 9.6 shows bonding interactions between orbitals. Single C–C bonds form due to the interaction between sp^3 orbitals from each carbon atom. Double C=C bonds form due to interactions between sp^2 orbitals from each carbon atom as well as p orbitals from each. Triple C≡C bonds form due to interactions between sp orbitals from each carbon atom as well as two p orbitals from each.

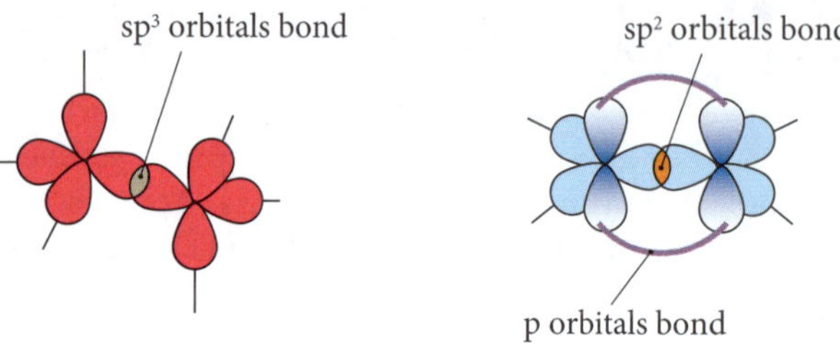

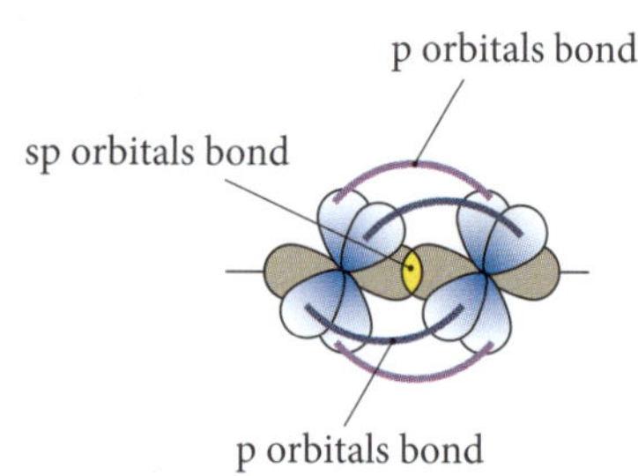

Figure 9.6 Bonding orbitals

→ The hybridisation process in the valence bond theory explains the variation in bond angles and molecular shapes in organic molecules.

- In *alkanes* the four carbon orbitals (one s orbital and three p orbitals) hybridise to form four sp^3 molecular orbitals arranged tetrahedrally around the nucleus of the carbon atom. Single C–H covalent bonds form due to the interaction of an s orbital of the hydrogen atom and an sp^3 hybrid orbital of the carbon atom. Figure 9.7 shows the orbital diagram of methane, in which four C–H bonds are present. In longer chain alkanes, single C–C covalent bonds form due to the interaction or overlap of sp^3 hybrid orbitals of each carbon atom.

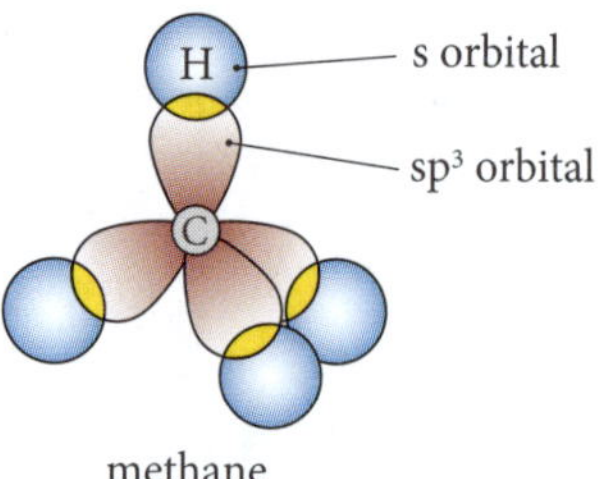

Figure 9.7 Orbital diagram of methane

- In *alkenes* the covalent bonds around the carbon atoms forming the double bond are arranged in a trigonal–planar orientation. One s orbital and two p orbitals hybridise to form three sp^2 molecular orbitals. The C=C double bond forms when one sp^2 orbital and the remaining p orbital from each carbon atom interact. The remaining sp^2 orbitals of the carbon atoms interact with s orbitals of the hydrogen atoms to form C–H bonds.
- In *alkynes* the covalent bonds around the carbon atoms forming the triple bond are arranged in a linear orientation. One s orbital and one p orbital hybridise to form two sp molecular orbitals. The C≡C triple bond forms when one sp orbital and the remaining two p orbitals from each carbon atoms interact. The remaining sp orbital of the carbon atoms interact with s orbitals of the hydrogen atoms to form C–H bonds.

→ KEY QUESTIONS

1 **Write the condensed structural formula for:**
 a **octane** b **oct-4-ene**

2 **Write the molecular formula and draw the structural formula for 3,4-dimethylhexane.**

3 **Explain how orbital hybridisation explains the shapes of hydrocarbon molecules.**

Answers ➲ p. 134

3 Intermolecular and intramolecular bonding

» Students explain the properties within and between the homologous series of alkanes with reference to the intermolecular and intramolecular bonding present.

Physical properties of hydrocarbons

→ Hydrocarbons are non-polar molecules. The **intermolecular** forces between these molecules are dispersion forces caused by the interaction of temporary dipoles and induced dipoles. These forces are much weaker than the **intramolecular** chemical covalent bonds joining the atoms together within the molecule.

intermolecular forces: attractive forces between molecules
intramolecular forces: covalent bonds joining atoms together in the molecule

→ The origin of dispersion forces was discussed in *Excel Year 11 Chemistry*. Temporary dipoles are caused by fluctuating electron distributions in atoms. These temporary dipoles induce dipoles in neighbouring atoms. The dispersion attractive force is the result. Figure 9.8 shows the steps in the formation of dispersion forces.

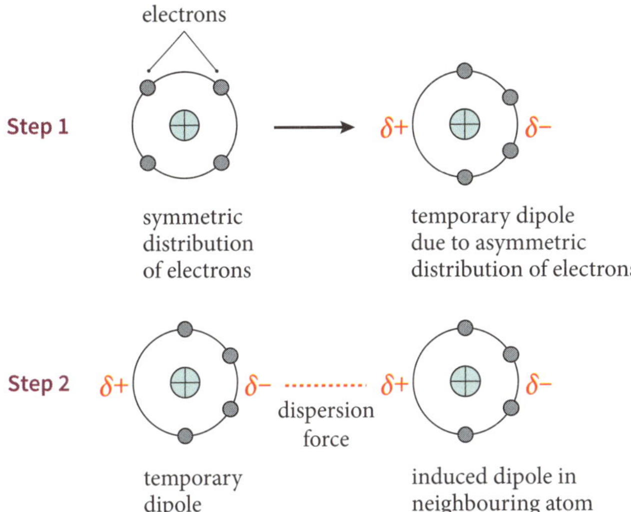

Figure 9.8 Formation of dispersion attractive forces

- The melting and boiling points of hydrocarbons increase as the molecular weight of the molecules increases. The longer the hydrocarbon chains, the stronger the dispersion forces that attract neighbouring molecules. The shape of the hydrocarbon molecules also affects the strength of the dispersion forces.
- Table 9.1 shows the increasing melting and boiling points of selected alkanes, alkenes and alkynes. The first four members of the alkane homologous series are gases and those with 5 to 16 carbon atoms per molecule are liquids. The rest are waxy solids. The presence of double or triple bonds changes the shape and dispersion forces between the molecules.
- Hydrocarbon isomers also vary in the melting and boiling points. This is mainly due to differences in the shapes of the molecules. The following examples show that straight chain hydrocarbons have higher boiling points than their branched chain isomers.

  Examples:
  - methylbutane: b.p = 28 °C; pentane: b.p = 36 °C
  - 2-methylpentane: b.p = 60 °C; hexane: b.p = 68 °C
  - 3-methylhexane: b.p = 92 °C; heptane: b.p = 98 °C
- Because hydrocarbons are non-polar molecules they do not dissolve in water but they do dissolve readily in organic solvents such as benzene (C_6H_6). Solid and gaseous alkanes, alkenes and alkynes readily dissolve in liquid hydrocarbons. The dissolution process involves dispersion interactions between solute and solvent molecules and is driven by an entropy increase.

SECONDARY-SOURCED INVESTIGATION

» Students conduct an investigation to compare the properties of organic chemical compounds within a homologous series, and explain these differences in terms of bonding.

Boiling points of alkanes

Aim

to graph the boiling points of straight chain alkanes as a function of molar mass and to explain the trend in the graphed data in terms of intermolecular forces

Method

1 Table 9.2 lists the boiling points of alkanes and their molar masses.

Table 9.2 Boiling points of alkanes

Alkane	C_5H_{12}	C_6H_{14}	C_8H_{18}	$C_{10}H_{22}$	$C_{12}H_{26}$
M (g/mol)	72.2	86.2	114.2	142.3	170.3
b.p (°C)	36.1	68.7	125.7	174.1	216.3

Alkane	$C_{14}H_{30}$	$C_{16}H_{34}$	$C_{18}H_{38}$	$C_{20}H_{42}$	$C_{22}H_{46}$
M (g/mol)	198.4	226.3	254.5	282.5	310.6
b.p (°C)	253.5	286.8	317.4	343.8	368.3

2 Use a spreadsheet program to plot the data as a line graph. Alternatively use a grid paper and pencil to plot this data. Points on the graph should be plotted as small crosses (x). Ensure the graph has a title and correct labels and units are used on each axis.

3 Describe and analyse the trend in the boiling point graph.

Sample results

Figure 9.9 shows a sample boiling point graph versus molar mass for alkanes.

Table 9.1 Melting and boiling points of hydrocarbons

Alkane	m.p (°C)	b.p (°C)	Alkene	m.p (°C)	b.p (°C)	Alkyne	m.p (°C)	b.p (°C)
ethane	−183	−89	ethylene	−169	−104	acetylene	−81	−84*
propane	−188	−42	propene	−185	−48	propyne	−102	−23
pentane	−130	36	pent-1-ene	−165	30	pent-1-yne	−109	40
octane	−54	151	oct-1-ene	−102	121	oct-1-yne	−74	125

(* sublimes)

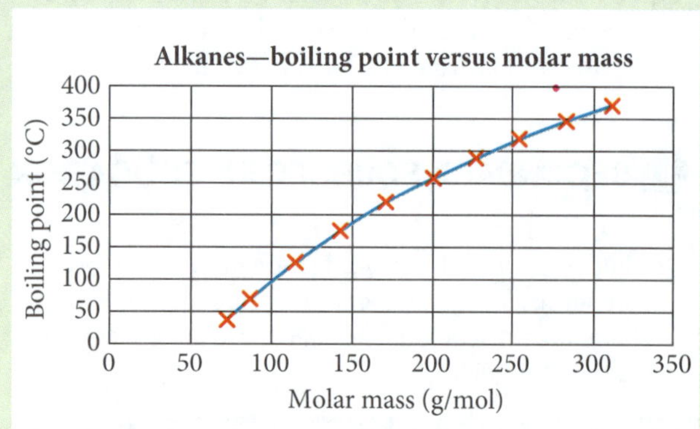

Figure 9.9 Boiling point of alkanes versus molar mass

Analysis

The boiling points of alkanes increase with increasing molar mass due to the increase in dispersion forces between molecules.

4 Safety in handling and disposal of organic substances

» Students describe the procedures required to safely handle and dispose of organic substances.

Volatility and flammability

- The weak intermolecular forces between hydrocarbon molecules of low molecular weight make them highly **volatile** substances. For example, a closed container of a liquid hydrocarbon such as hexane rapidly becomes filled with its vapour.
- The higher the ambient temperature the higher the volatility. This property makes many alkanes useful fuels for internal combustion engines. It also makes these substances dangerous in terms of their high flammability. The **equilibrium vapour pressure** is a measure of volatility. Figure 9.10 compares the equilibrium vapour pressures of hexane and heptane.

volatile: substances that readily vapourise and which have a high vapour pressure

equilibrium vapour pressure: the vapour pressure in a closed system due to the vapourisation of a substance into the gaseous state

- Generally the lower the molecular weight of a hydrocarbon, the greater its volatility. For example, short chain alkanes have weaker dispersion forces between its molecules. Table 9.3 shows that the equilibrium vapour pressure of liquid octane increases with increasing temperature.

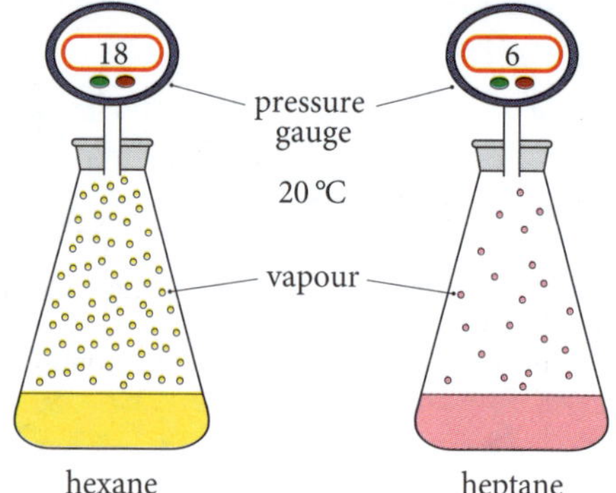

Figure 9.10 Equilibrium vapour pressures of hexane and heptane

Table 9.3 Octane vapour pressure as a function of temperature

Temperature (°C)	0	20	40	60
Vapour pressure (kPa)	0.4	1.3	4.1	10.4

Safety issues

- Safety issues related to the storage of alkanes can also be understood with reference to the low **flash points** of these fuels. As the composition of hydrocarbon fuels varies due to different proportions of hydrocarbon molecules in each fuel then the volatility of the fuel and its flash point also varies.

flash point: the lowest temperature at which the vapour pressure of the substance is just sufficient to form a combustible mixture of the vapour with air; flames or sparks can then ignite this combustible mixture

- Petrol has a flash point of −43 °C. This very low flash point indicates that combustible air/petrol vapour mixtures will form even in very cold environments. Kerosene has a higher flash point (48 °C) than petrol and therefore combustible air/kerosene vapour mixtures do not form under standard temperatures. In general the flash point increases with increasing molecular weight.
- Thus the storage and handling of petrol is more dangerous than the storage and handling of kerosene as sparks or glowing cigarettes are more likely to ignite the air/petrol vapour mixture. This is the reason that smoking is not permitted at petrol stations.
- High molecular weight alkanes (e.g. candle waxes) have a much lower volatility and flash point and are safer to store. Table 9.4 lists the typical flash points of some common vehicle fuels. Different hydrocarbon compositions in the fuel **fractions** will alter the flash point.

fractions: mixtures of hydrocarbons with a specific boiling point range that are produced by fractional distillation of petroleum

Table 9.4 Typical flash points of some common fuels

Fuel	Petrol	Kerosene	Diesel
Typical composition	C5–C8 hydrocarbons	C11–C16 hydrocarbons	C15–C18 hydrocarbons
Flash point (°C)	−43	+48	+65

- Some other important issues associated with the safe storage of flammable hydrocarbons are:
  - Store the liquids in strong metal containers (rather than plastic) with close-fitting lids and narrow necks to avoid excessive evaporation.
  - Ensure that gas cylinders and their fittings are regularly checked for leaks.
  - Keep the storage containers cool to avoid the build-up of vapours.
  - Store containers of liquid fuels in well-ventilated areas and do not transfer fuel from one container to another in closed spaces where vapours can build up.
- Hydrocarbons must be disposed of in a safe manner. Local authorised agents should be contacted to collect waste hydrocarbons so they are not dumped in the urban or rural environments.
- HAZCHEM signs and labels can be viewed online. These signs, which are found on storage containers, display safety information related to the use of dangerous chemicals. They are also found on vehicles that transport dangerous chemicals such as fuels and corrosive liquids.

KEY QUESTIONS

4 **Explain the difference between intermolecular and intramolecular forces.**

5 **Explain why the boiling points of alkanes increase with increasing molecular weight.**

6 **Define volatility and explain the safety issues in the handling of alkanes.**

Answers ➲ p. 134

5 Obtaining and using hydrocarbons

» Students examine the environmental, economic and sociocultural implications of obtaining and using hydrocarbons from the Earth.

Sources, production and uses of hydrocarbons

- The majority of hydrocarbons are extracted from organic sources such as **petroleum** and natural gas located in geological structures.

petroleum: oily fossil fuel composed of crude oil and natural gas which has formed over millions of years by geological processes acting on marine organisms

- Petroleum is a fossil fuel formed by the decay of the remains of plankton and algae in marine ecosystems over long periods of geological time (~500 million years).
- Coal is also a fossil fuel formed from the remains of ancient swamp plants. Coal is mainly used in Australia to produce electricity via combustion reactions. Coal is also used to manufacture coking coal, which is used in the smelting of iron ore to produce steel. Refer to Chapter 11 for information about different ranks of coal as fuels.
- Petroleum is a mixture of crude oil with dissolved hydrocarbon gases ('natural gas'). Petroleum deposits vary in the amounts of natural hydrocarbon gases.
- Australian natural gas is a mixture of gases. Methane (CH_4) is typically about 89% of the total number of molecules in natural gas. Other gases that are commonly present include ethane (~5%), carbon dioxide (~2%) and propane (~1%). Most of Australia's natural gas reserves are located in the Carnarvon and Bonaparte/Amadeus Basins of Western Australia and Northern Territory (see Figure 9.11).

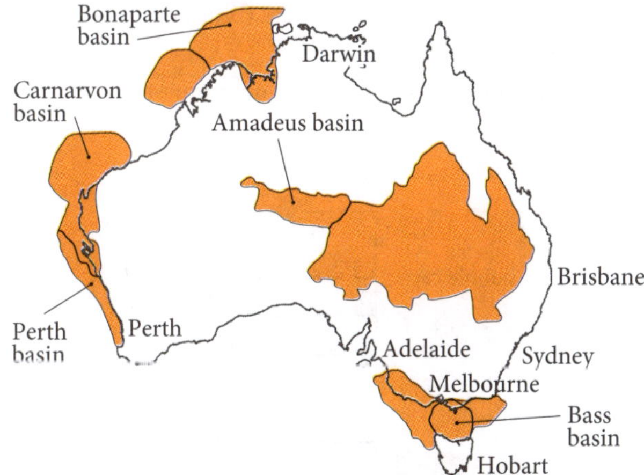

Figure 9.11 Australian oil and gas basins

Source: compilation based on information viewed online February 2018 at http://peakenergy.blogspot.com.au/2008/06/no-post-tonight.html and www.chemlink.com.au/gas.htm

- Natural gas can be compressed and liquefied to produce LNG (liquefied natural gas). In this liquefied form the gas can be exported overseas (particularly to Japan and China) in large tankers. Natural gas is also used in some power stations to generate electricity. Natural gas is also used in household gas cooktops, ovens and water heaters. When methane undergoes complete combustion, heat energy is produced:

$$CH_4(g) + 2O_2(g) \rightarrow CO_2(g) + 2H_2O(l);$$
$$\Delta H = -892 \text{ kJ/mol}$$

- Crude oil fractional distillation separates the hydrocarbon mixture according to differences in boiling points.

- The components of crude oil are too numerous to extract and separate them into separate pure compounds. They are separated into fractions containing mixtures of various hydrocarbons. This physical separation is performed in a steel fractionating column and is based on the boiling point ranges for each fraction. The higher the boiling point range, the greater the proportion of high molecular-weight hydrocarbons in the fraction.
- The fractionating process is carried out as follows:
  - The crude oil is heated and the vapours pass into the fractionating column.
  - The column contains trays at various levels.
  - The hot vapours together with steam move up the column and the pressure of these vapours pushes up 'bubble caps' as they rise higher in the column. As they cool they condense into liquids and collect in trays. The compounds with the lowest boiling points rise highest in the column before they condense.
  - The light, low molecular-weight components that do not condense (refinery gas) pass out the top of the column and are collected.
  - At each level the liquids that form have a narrow boiling point range. They collect in the overflow trays and are removed from the column.
  - High molecular weight components do not vapourise and collect at the base of the column. A residue containing a mixture of tars (bitumen), greases and waxes collect here.

Figure 9.12 shows a simplified diagram of a fractionating column.

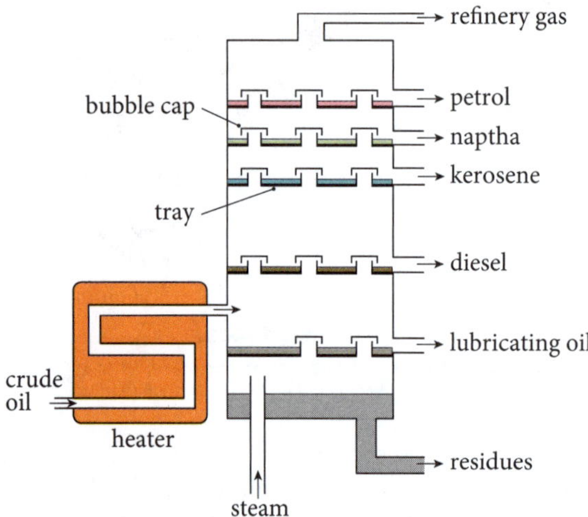

Figure 9.12 Fractionating column

- Table 9.5 shows the typical composition of fractions derived from crude oil distillation. The boiling point ranges for various fractions vary slightly from one refinery to another. The petrol fraction typically consists of hydrocarbons with 5 to 8 carbon atoms per molecule whereas the naphtha fraction typically consists of hydrocarbons with 7 to 13 carbon atoms per molecule.

The naphtha fraction is used as a petrochemical feedstock to manufacture ethylene by a process called *thermal (or steam) cracking* at 800 °C. In the thermal cracking process of decane ($C_{10}H_{22}$), both ethylene and hexane are produced. The hexane can be added to the petrol component of the gasoline fraction:

$$C_{10}H_{22}(l) \rightarrow 2C_2H_4(g) + C_6H_{14}(l)$$

Table 9.5 Crude oil fractions

Fraction	Boiling range (°C)	Composition (carbon atoms per chain)	Uses
Refinery gas	< 30	1–4	fuel; feedstock for plastic production
Petrol	30–125	5–8	car fuel
Naphtha	90–220	7–13	industrial solvents; feedstock for petrochemical and plastic production
Kerosene	175–275	11–16	aviation fuel; home heating; solvent; cracked to form petrol and petrochemical feedstocks
Diesel oil	260–340	15–18	diesel fuel; cracked to form petrol and petrochemical feedstocks; furnace fuel
Lubricating oils	> 350	16–20	lubrication
Residues	–	> 20	road surfaces; waxes

- The proportion of petrol derived from fractional distillation is too low to meet demand. Heavier fractions undergo further processing to break up the longer chains into shorter chains suitable for petrol. For example, $C_{16}H_{34}$ from the diesel oil fraction can be catalytically cracked at 500 °C to form heptane, ethylene and propene. The catalyst used is a porous aluminium silicate called zeolite. The heptane is suitable for use in petrol and the ethylene and propene can be used to make various petrochemicals. Polymer plastics such as polyethylene and polypropene are examples of petrochemicals:

$$C_{16}H_{34}(l) \rightarrow C_7H_{16}(l) + 3C_2H_4(g) + C_3H_6(g)$$

EXAMPLE 2

Name the fraction containing each of the following molecules and predict which of the listed molecules has the highest boiling point:

a $C_{12}H_{26}$

b C_7H_{16}

c $C_{16}H_{34}$

Use Table 9.5 to answer this question

Answer:

a kerosene fraction (C11–C16)

b petrol fraction (C5–C8)

c diesel oil fraction (C15–C18)

The molecule with the highest molecular weight will have the highest boiling point. This is $C_{16}H_{34}$.

Economic, environmental and sociocultural issues

- Petroleum products provide Australia and other nations with compounds for:
  - transport
  - heating
  - plastic production
  - lubrication greases
  - road surfacing.
- The economies of petroleum producing countries are significantly supported by the sale of petroleum products to other countries.
- Petroleum mining has led to environmental pollution when accidents have occurred.
- One of the major disadvantages of fossil fuels is that, on combustion, large amounts of carbon dioxide are produced. The following equation shows the production of carbon dioxide when octane (a component of petrol) undergoes complete combustion. Hydrocarbons release considerable amounts of energy on combustion:

  $2C_8H_{18}(l) + 25O_2(g) \rightarrow 16CO_2(g) + 18H_2O(l)$;
  $\Delta H = -5470$ kJ/mol
- During the combustion of hydrocarbon fuels, polluting gases such as carbon monoxide (CO), nitric oxide (NO) and nitrogen dioxide (NO_2) are also produced. Catalytic converters in the exhaust system of vehicles reduce the levels of these gases escaping to the atmosphere. For example, NO and CO can be catalytically converted to carbon dioxide and nitrogen. In this case the catalyst used is the metal rhodium:

  $2CO(g) + 2NO(g) \rightarrow 2CO_2(g) + N_2(g)$
- Carbon dioxide is classified as a greenhouse gas and as a result it contributes to climate change. Since the industrial revolution at the start of the 19th century there has been a significant rise in the levels of atmospheric carbon dioxide as well as a rise in average global temperatures. In addition, large amounts of carbon dioxide have dissolved in the oceans to form carbonic acid. Acidification of the oceans is the result:

  $CO_2(g) + H_2O(l) \leftrightarrows H_2CO_3(aq)$

  This has had a detrimental effect on organisms such as coral whose calcium carbonate exoskeletons are attacked by acidified water.
- The most effective way to reduce greenhouse emissions is to use non-carbon energy sources such as solar energy, wind energy and hydroelectricity. Solar energy can be transformed using photovoltaic cells into electrical energy. Wind energy can be converted to electrical energy using turbine generators. Hydroelectricity is generated when flowing water drives turbine generators. However, solar and wind energy are intermittent energy sources. The electrical energy generated by these energy sources can be transformed and stored as chemical potential energy in batteries and used to power our vehicles and appliances. Although large battery storage facilities are starting to be built, at the time of writing there is currently insufficient battery storage to meet the total energy requirements of Australian industries and society. Australia does not have nuclear power plants that could provide the back-up power that coal combustion currently provides.
- The transition from fossil fuels to renewable energy sources cannot occur quickly as national economies need time to adapt. The challenge for the future is to deliver reliable electricity supplies to all consumers using renewable energy sources. Continued scientific and engineering research and development is required.
- Plastic products that are manufactured from petrochemicals are causing environmental pollution due to their non-biodegradability when placed in waste landfill sites. An advanced technology called CAT-HTR has been developed to break down plastics. This catalytic process uses high-pressure and high-temperature water to break down the plastics to small hydrocarbon molecules that can be used to make new plastic products and hydrocarbon oils and fuels. This recycling process will reduce the need for mining petroleum.

KEY QUESTIONS

7 **Identify the composition and use of the kerosene fraction that is produced by fractional distillation of petroleum.**

8 **Explain the economic importance of petroleum mining.**

9 **Explain the environmental issues related to using petroleum fractions.**

Answers p. 134

CHAPTER SYLLABUS CHECKLIST

Are you able to answer these questions from the syllabus for this chapter? Tick each question as you go through the checklist if you are able to answer it. If you cannot answer a question, turn to the relevant page in the study guide to find the answer. For NESA key word meanings, go to www.educationstandards.nsw.edu.au and search 'key words'.

FOR A COMPLETE UNDERSTANDING OF THIS TOPIC:		PAGE NO.	✓
1	Can I classify hydrocarbons based on their structure and reactivity?	124	
2	Can I model the structures of hydrocarbons?	125	
3	Can I explain the shapes of hydrocarbons in terms of hybridised atomic orbitals?	125	
4	Can I explain why dispersion forces exist between hydrocarbon molecules?	126	
5	Can I explain why the melting and boiling points of hydrocarbons increase with increasing molecular weight?	127	
6	Can I explain the volatility of low molecular weight hydrocarbons?	128	
7	Can I recall the safety issued related to the use and storage of hydrocarbons?	128	
8	Can I name the fractions produced by fractional distillation of petroleum and describe their uses?	130	
9	Can I recall the economic and environmental issues related to the use of hydrocarbons?	131	

HSC EXAM-TYPE QUESTIONS

Objective-response questions (1 mark each)

1 Select the true statement concerning bonding in hydrocarbon molecules.

- A Carbon–carbon single bonds are formed when 4 electrons are shared.
- B Double covalent bonds between carbon atoms consist of one electron pair.
- C Triple covalent bonds between carbon atoms in alkynes consist of three electron pairs.
- D C–H bonds are formed when sp^3 hybrid orbitals from each atom interact.

2 The molecule C_8H_{18} can be extracted from which of the following crude oil fractions?

- A refinery gas fraction
- B petrol fraction
- C diesel fraction
- D kerosene fraction

3 Name the following molecule:

$CH_3CH_2CH_2CHCHCH_2CH_3$

- A heptane
- B oct-3-ene
- C hept-4-ene
- D hept-3-ene

4 Select the correct statement about hydrocarbons.

- A Oct-1-yne has a higher boiling point than pent-1-yne.
- B Heptane is a gas at 25 °C and 100 kPa.
- C C_8H_{18} is less volatile than $C_{12}H_{26}$.
- D Petrol has a higher flash point than kerosene.

5 The percentage by weight of carbon and hydrogen in a hydrocarbon is:

85.63% carbon; 14.37% hydrogen

The molar mass of the hydrocarbon is 42.078 g/mol. What is the molecular formula of the hydrocarbon?

- A CH
- B CH_2
- C C_3H_6
- D C_3H_8

Extended-response questions

6 Simple hydrocarbons can be obtained from crude oil.

- a Explain why crude oil is called a fossil fuel. (1 mark)
- b What property of the hydrocarbons is used to separate them into fractions? (1 mark)
- c Explain why molecules such as $C_{18}H_{38}$ are collected lower in the fractionating tower than such molecules as C_6H_{14}. (2 marks)

7 Figure 9.13 is a model of a hydrocarbon.

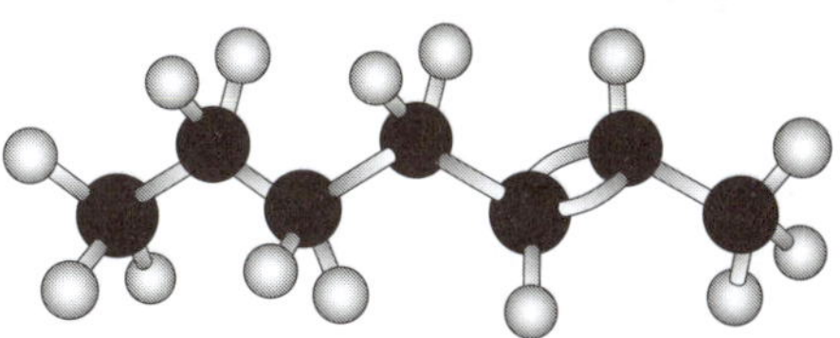

Figure 9.13 Model of a hydrocarbon

- a Name the homologous series to which this molecule belongs. (1 mark)
- b Write the molecular formula (MF) of the hydrocarbon. (1 mark)
- c Write the empirical formula (EF) of the hydrocarbon. (1 mark)
- d Use IUPAC nomenclature to name the hydrocarbon. (1 mark)

8 a Liquid fuel transfers from one container to another (e.g. pouring lawnmower fuel into the tank of the mower) should be performed outside in well-ventilated areas. Explain why this method is recommended. (1 mark)

- b It is not advisable to store large quantities of petrol in containers at home. Explain. (1 mark)

9 Figure 9.14 is a student's model of a hydrocarbon.

Figure 9.14 Hydrocarbon model

- a A student named this model as eth-1-yne. Explain why this name is incorrect. (1 mark)
- b The preferred IUPAC name for this hydrocarbon is acetylene. Explain why this name was selected. (1 mark)
- c Name the next member of the homologous series to which acetylene belongs and write the molecular formula of that hydrocarbon. (2 marks)

10 a Write the molecular formula of heptane and the formula of the next member of that homologous series. (2 marks)

- b i Identify the type of bond or force that exists between the atoms of the heptane molecule. (1 mark)
  - ii Identify the type of bond or force that exists between the neighbouring heptane molecules. (1 mark)
- c A sample of heptane is heated and is converted into a vapour. Identify the bonds or forces being broken during this process. (1 mark)
- d Hydrocarbons can be classified as saturated or unsaturated. State the classification for heptane. Justify your classification. (2 marks)

KEY QUESTIONS

Key questions ➲ p. 126

1 a $CH_3CH_2CH_2CH_2CH_2CH_2CH_2CH_3$
b $CH_3CH_2CH_2CHCHCH_2CH_2CH_3$

2 C_8H_{18}
Figure A9.1 Shows the structural formula of 3,4-dimethylhexane.

```
    H  H  CH3 H  H  H
    |  |  |   |  |  |
H — C— C— C — C— C— C— H
    |  |  |   |  |  |
    H  H  H  CH3 H  H
```

Figure A9.1 3,4-dimethylhexane

3 Atomic orbitals can be combined to produce molecular orbitals that have particular orientations. Thus four sp^3 orbitals are arranged tetrahedrally around carbon atoms. This explains the bond angles in alkanes. In alkenes the three sp^2 orbitals are trigonal planar in orientation and this explains the planar arrangements of atoms around carbon–carbon double bonds. The sp hybridisation explains the linear arrangement of atoms attached to a carbon–carbon triple bond.

Key questions ➲ p. 129

4 Intermolecular forces exist between molecules. They are dispersion forces, dipole–dipole forces or hydrogen bonding. These intermolecular forces are much weaker than covalent bonds, which are intramolecular forces.

5 Alkanes are non-polar and dispersion forces exist between their molecules. As the chain length increases there are more electrons, which results in a larger number of temporary and induced dipoles leading to greater attraction between molecules. Thus more energy is needed to boil the alkanes.

6 Volatile substances readily vapourise due to weak intermolecular forces. When the volatile substance is a hydrocarbon then vapour/air mixtures can undergo combustion in the presence of a spark or flame in poorly ventilated spaces. Thus storage of hydrocarbons requires containers that do not leak vapours and which are located in cool, well-ventilated environments.

Key questions ➲ p. 131

7 Kerosene is a mixture of hydrocarbons with molecules that have between 11 and 16 carbon atoms per molecule. Kerosene is used for home heating and as an aviation fuel.

8 Refining petroleum produces products such as petrol, kerosene and diesel, which are used to power vehicles. Modern economies rely on the jobs generated as well as the profits obtained by the production and sale of these fuels.

9 The combustion of hydrocarbon fractions generates vast quantities of carbon dioxide, which has led to ocean acidification and increased levels of global warming.

HSC EXAM-TYPE QUESTIONS

Objective-response questions

1 **C**. Triple bonds consist of three pairs of electrons. **A** is incorrect as single bonds have two shared electrons. **B** is incorrect as double bonds consist of two electron pairs. **D** is incorrect as C–H bonds are formed by the interaction of an sp^3 orbital from the carbon atom and an s atomic orbital from the hydrogen.

2 **B**. Octane is an important component of petrol. **A** is incorrect as octane is not a gas; it is a liquid. **C** is incorrect as the diesel fraction contains hydrocarbons with 15–18 carbon atoms per molecule. **D** is incorrect as kerosene is a mixture of C11–C16 hydrocarbons.

3 **D**. Number the chain from right to left to give the double bond the lowest locant. The stem name is *hept-* as there are seven carbon atoms. Therefore name = hept-3-ene. Thus **A**, **B** and **C** are incorrect.

4 **A**. The boiling points of alkynes increase with increasing molar mass as the size of the dispersion forces increase. **B** is incorrect as heptane is a liquid. **C** is incorrect as volatility decreases with increasing molar mass. **D** is incorrect as the flash point increases with increasing sizes of hydrocarbons in heavier fractions.

5 **C**. $m(C) = 85.63$ g; $m(H) = 14.37$ g

$n(C) = \frac{85.63}{12.01} = 7.13$ mol; $n(H) = \frac{14.37}{1.008} = 14.26$ mol

C : H = 7.1 : 14.26 = 1 : 2
EF = CH_2
MF = C_3H_6 (M = 42.072 g/mol)
A is incorrect as the ratio is not 1 : 1. **B** is incorrect as this is the empirical formula. **D** is incorrect as the molar mass is not 42.078 g/mol.

Extended-response questions

6 EM Students must demonstrate an understanding of the origins of petroleum and the physical principles behind the separation of its components.

a Crude oil is formed from the anaerobic decay of dead plankton and algae (under conditions of high pressure and heat) in subterranean geological structures over millions of years. ✓
b Boiling point differences ✓
c Higher molar mass hydrocarbons have higher boiling points due to greater dispersion forces between their molecules. ✓ Therefore the lighter hydrocarbons boil first and their vapours rise up the fractionating tower. ✓

7 EM Students should count the number of carbon and hydrogen atoms to determine the molecular formula. The empirical formula is the simplest ratio of atoms in the molecule.

a alkene ✓
b MF = C_7H_{14} ✓
c EF = CH_2 ✓
d hept-2-ene ✓

8 EM Students must show a good understanding of the dangers in using and storing hydrocarbons based on their volatility and flammability.

a This method will prevent the build-up of combustible vapour/air mixtures, which could ignite if sparked. ✓

b Petrol is highly volatile and dangerous. If leaks occur from the container then materials inside the home or garage become more combustible if flames or sparks from faulty electrical appliances occur. ✓

9 EM Students should remember that locants are used only when there could be a confusion over the position of a functional group in a molecule. Preferred IUPAC nomenclature is used for some molecules.

a As there are only 2 carbon atoms the locant number 1 is not required. Ethyne would be the systematic name. ✓

c Acetylene is the historical name and the IUPAC committee has decided that this is its preferred name. Ethyne is an acceptable systematic name. ✓

d propyne; ✓ C_3H_4 ✓

10 EM Students must show that they can distinguish between intermolecular and intramolecular forces as well as know the definition of a saturated hydrocarbon molecule.

a C_7H_{16}; ✓ C_8H_{18} ✓

b i intramolecular force or covalent bond ✓

ii intermolecular force or dispersion force ✓

c dispersion forces ✓

d saturated; ✓ only C–C single bonds are present ✓

CHAPTER 10

PRODUCTS OF REACTIONS INVOLVING HYDROCARBONS

MODULE 7 ORGANIC CHEMISTRY

INQUIRY QUESTION:

What are the products of reactions of hydrocarbons and how do they react?

Hydrocarbons are important compounds in our daily lives. Chemists have investigated the chemical properties of hydrocarbons over many centuries and new compounds have been produced based on this knowledge. Compounds formed from petroleum are called *petrochemicals*. This chapter examines the chemical reactions of hydrocarbons and the properties of products that form from them.

1 Hydrocarbon reactions

» Students:

- investigate, write equations and construct models to represent the reactions of unsaturated hydrocarbons when added to a range of chemicals, including but not limited to hydrogen (H_2), halogens (X_2), hydrogen halides (HX) and water (H_2O).
- investigate, write equations and construct models to represent the reactions of saturated hydrocarbons when substituted with halogens.

Addition reactions

- Unsaturated hydrocarbons have reactive double or triple carbon–carbon bonds. These bonds have high electron density and this property promotes reactions with other molecules such as halogens that have high electronegativity.
- Unsaturated hydrocarbons readily react by addition reactions in which an alkene or alkyne combines with another molecule to form one product.

Halogen addition reactions

- Alkenes and alkynes readily react with halogens such as chlorine via **addition reactions** where the double or triple bond breaks. Chlorinated alkanes are the products:

  $C_2H_4(g) + Cl_2(g) \rightarrow C_2H_4Cl_2(l)$

  Product = 1,2-dichloroethane

  $C_2H_2(g) + 2Cl_2(g) \rightarrow C_2H_2Cl_4(l)$

  Product = 1,1,2,2-tetrachloroethane

addition reaction: a reaction between an unsaturated hydrocarbon and another molecule to form a new product

EXAMPLE 1

Write a structural equation for the reaction of bromine vapour with hept-2-ene and name the product of the reaction.

Bromine vapour consists of diatomic molecules

Answer:

Figure 10.1 shows the structural equation. The product is 2,3-dibromoheptane.

```
    H   H   H   H   H   H   H
    |   |   |   |   |   |   |
H - C - C = C - C - C - C - C - H   +   Br-Br
    |           |   |   |   |
    H           H   H   H   H
        hept-2-ene                  bromine

          H   H   H   H   H   H   H
          |   |   |   |   |   |   |
 ---->  H-C - C - C - C - C - C - C - H
          |   |   |   |   |   |   |
          H   Br  Br  H   H   H   H
              2,3-dibromoheptane
```

Figure 10.1 Structural equation for the addition of bromine to hept-2-ene

- Various qualitative tests can be performed to identify unknown organic compounds using reactions that are specific to various functional groups. Unsaturated hydrocarbons, such as alkenes and alkynes, can be distinguished from alkanes using the bromine water test. Pure bromine is a toxic, red-brown fuming liquid. Bromine water is a solution of bromine in water. The colour of bromine water depends on the degree of dilution. It can vary from orange-brown to orange-yellow or yellow. The more dilute the solution, the safer it is to use. The following test uses a dilute bromine solution and describes how it is performed. It is essential that an excess of each hydrocarbon is used.

Bromine water test

- Use a dropper to add 10 drops of orange-yellow bromine water to two test tubes.

- 1 mL (20 drops) of hexane is added to the first tube, which is stoppered and shaken.
- 1 mL (20 drops) of hex-1-ene is added to the second tube, which is stoppered and shaken.
- The tubes are placed back into a test tube rack and the layers allowed to separate.

Sample observations are illustrated in Figure 10.2:

- Hexane: The orange-yellow colour of the bromine water diffuses into the upper organic layer. No decolourisation occurs.
- Hex-1-ene: The bromine water is decolourised (i.e. from orange-yellow to colourless).

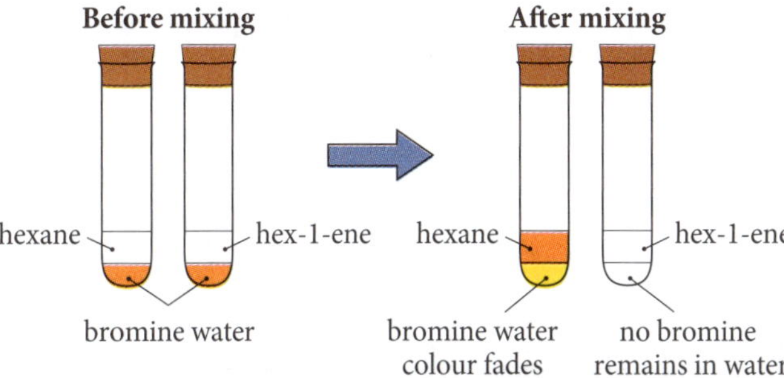

Figure 10.2 Bromine water test

➔ Various online videos can be used to reinforce your understanding of this qualitative investigation. Type the following headings into your browser:
- testing for alkenes using bromine water
- alkenes with bromine water

➔ Bromine water contains bromine (Br_2) molecules and hypobromous acid (HOBr) molecules. The double bond in the hex-1-ene reacts with these molecules via addition reactions, which breaks the double bond. Figure 10.3 shows the products that form. The colourless products are not formed in equal amounts. The decolourisation is the result of the bromine molecules being consumed in the reaction. HOBr is colourless so the addition reaction in that case does not result in a colour change.

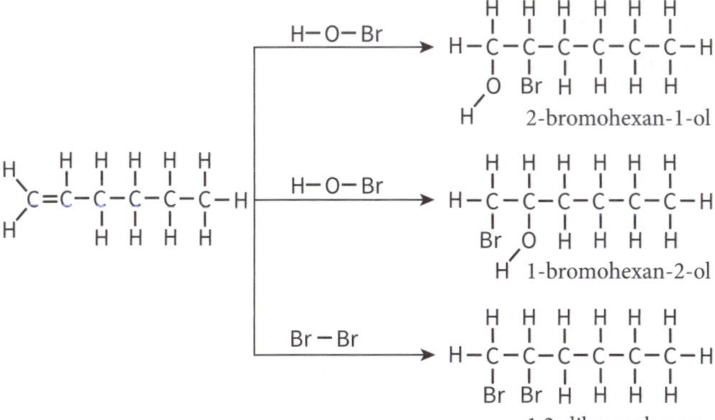

Figure 10.3 Addition reactions with hex-1-ene

Hydrogen halide addition reactions

➔ Hydrogen halides undergo addition reactions with unsaturated hydrocarbons.

➔ Figure 10.4 shows that the addition reaction of pent-2-ene with hydrogen chloride produces 2 different chloroalkane molecules, which are position isomers.

H–Cl
2-chloropentane
H–Cl
3-chloropentane

Figure 10.4 Addition of HCl

Addition of water (hydration)

➔ The addition of water to unsaturated hydrocarbons is called *hydration*. An acid catalyst (e.g. H_3PO_4) is required. Isomeric alkanol molecules are the products in the example illustrated in Figure 10.5. Note that the O atom of the hydroxyl group bonds to the carbon atom in the alkanol.

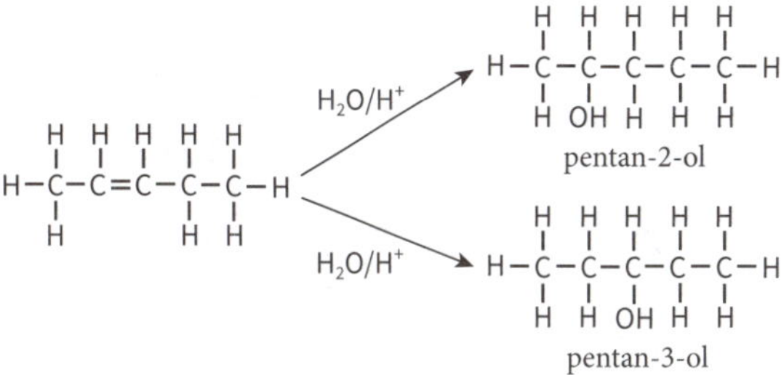

Figure 10.5 Addition of H_2O to pent-2-ene

Addition of hydrogen (hydrogenation)

➔ Alkenes or alkynes can be converted via an addition reaction with hydrogen to alkanes. This reaction is called *hydrogenation*. A nickel catalyst is used in hydrogenation reactions:
- but-2-ene → butane

  $CH_3CHCHCH_3(g) + H_2(g) \rightarrow CH_3CH_2CH_2CH_3(g)$
- but-2-yne → butane

  $CH_3CCCH_3(g) + 2H_2(g) \rightarrow CH_3CH_2CH_2CH_3(g)$

➔ KEY QUESTIONS

1 Write an equation using condensed structural formulas for the reaction of pure bromine with hex-1-ene. Name the product of the reaction.

2 Write equations using condensed structural formulas for the hydration reaction of pent-1-ene to demonstrate the production of isomeric products. Name the products.

3 Write an equation using condensed structural formulas for the hydrogenation reaction of propyne with excess hydrogen over a nickel catalyst. Name the product.

Answers ➲ p. 142

Substitution reactions

- Alkanes are commonly called *paraffins* (Latin for 'little affinity') because of their relative non-reactiveness towards many common chemical reagents.
- In order to make alkanes react with reactive molecules such as chlorine, UV light is used to produce reactive free radicals. In these reactions the halogen atoms replace hydrogen atoms. Such reactions are classified as **substitution reactions**. In the following reaction the methane and chlorine are mixed in a 1 : 1 mole ratio:

  methane → chloromethane

  $CH_4(g) + Cl_2(g) + UV \rightarrow CH_3Cl(l) + HCl(g)$

substitution reactions: reactions in which functional groups replace hydrogen atoms or other functional groups in an organic molecule

- In the chlorination of propane two isomeric products are formed, as shown in Figure 10.6.

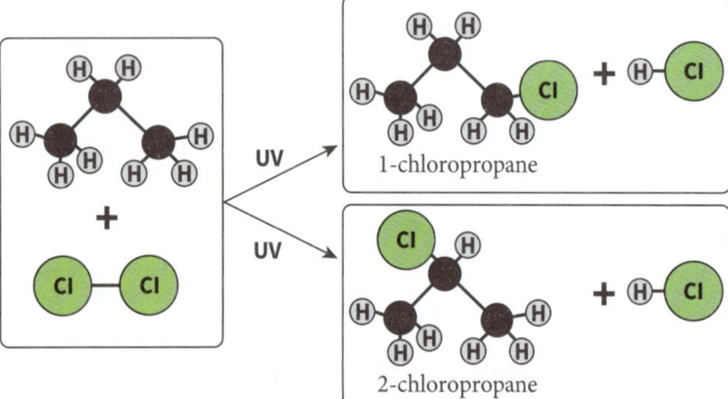

Figure 10.6 Chlorination of propane

- If an excess of chlorine is used to chlorinate ethane then all the hydrogen atoms are removed and a fully chlorinated product is formed:

  ethane → hexachloroethane

  $C_2H_6(g) + 6Cl_2(g) + UV \rightarrow C_2Cl_6(g) + 6HCl(g)$
- The halogenation of alkanes produces compounds that are more reactive. These chlorinated products can be converted to other useful chemicals via substitution reactions. For example, chloroethane reacts with sodium hydroxide solution to form ethanol (C_2H_5OH):

  $C_2H_5Cl(l) + NaOH(aq) \rightarrow C_2H_5OH\ (aq) + NaCl(aq)$
- When 2-bromobutane reacts with sodium hydroxide solution, butan-2-ol is formed. Figure 10.7 shows a model of this reaction.

→ KEY QUESTIONS

4 **Methane is reacted with excess bromine vapour in the presence of UV light. Write a balanced equation for the reaction and name the organic product.**

5 **2-chloropentane is reacted with sodium hydroxide solution. Write a balanced equation for the reaction and name the organic product.**

Answers ⮊ p. 142

2 Modelling reactions

- Molecular model kits can be used to help visualise molecular shape and to examine the structures of products of chemical reactions.

FIRSTHAND INVESTIGATION

Modelling reactions

» Students:

- investigate, write equations and construct models to represent the reactions of unsaturated hydrocarbons when added to a range of chemicals, including but not limited to hydrogen (H_2), halogens (X_2), hydrogen halides (HX) and water (H_2O).
- investigate, write equations and construct models to represent the reactions of saturated hydrocarbons when substituted with halogens.

Aim

to construct models and write equations to represent the addition reactions of unsaturated hydrocarbons and substitution reactions of saturated hydrocarbons

Equipment

Ball-and-stick model kit

Method

1 Use your model kit to construct the following molecules:
but-1-ene
but-2-ene

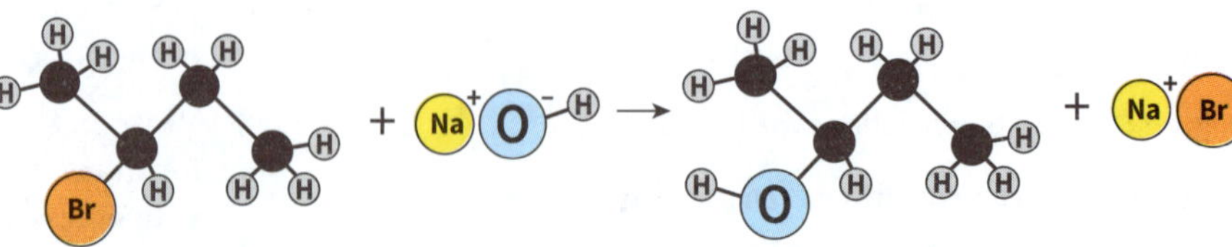

Figure 10.7 Production of butan-2-ol via a substitution reaction

2 Construct chlorine molecules as required.
3 Demonstrate the addition of chlorine to each hydrocarbon model. Write a structural equation for each reaction in your report.
4 Take photos of each model before and after the reaction. Include these photos in your report.
5 Repeat the experiment using hydrogen chloride molecules instead of chlorine molecules. Investigate the existence of possible isomers in the products.
6 Model substitution reactions using the chlorine molecules as well as models of methane and ethane.

Sample results

Figure 10.8 shows models of the addition reactions of chlorine with but-1-ene and but-2-ene in which isomeric products form.

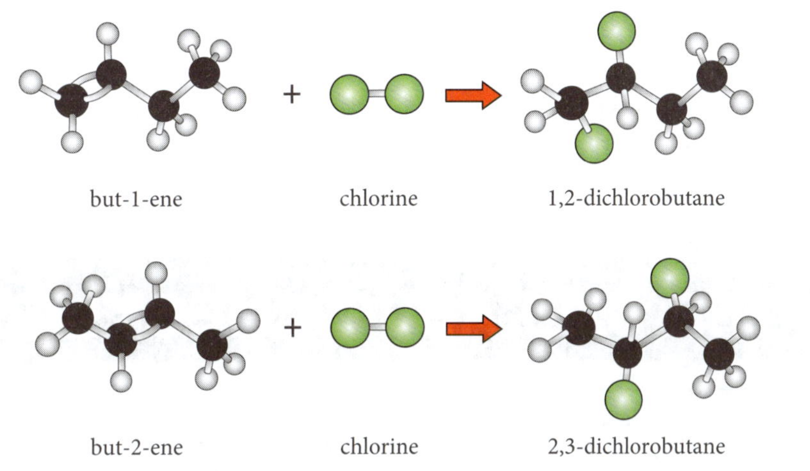

Figure 10.8 Addition reactions of chlorine

Figure 10.9 shows models of the addition reactions of hydrogen chloride and water with but-1-ene.

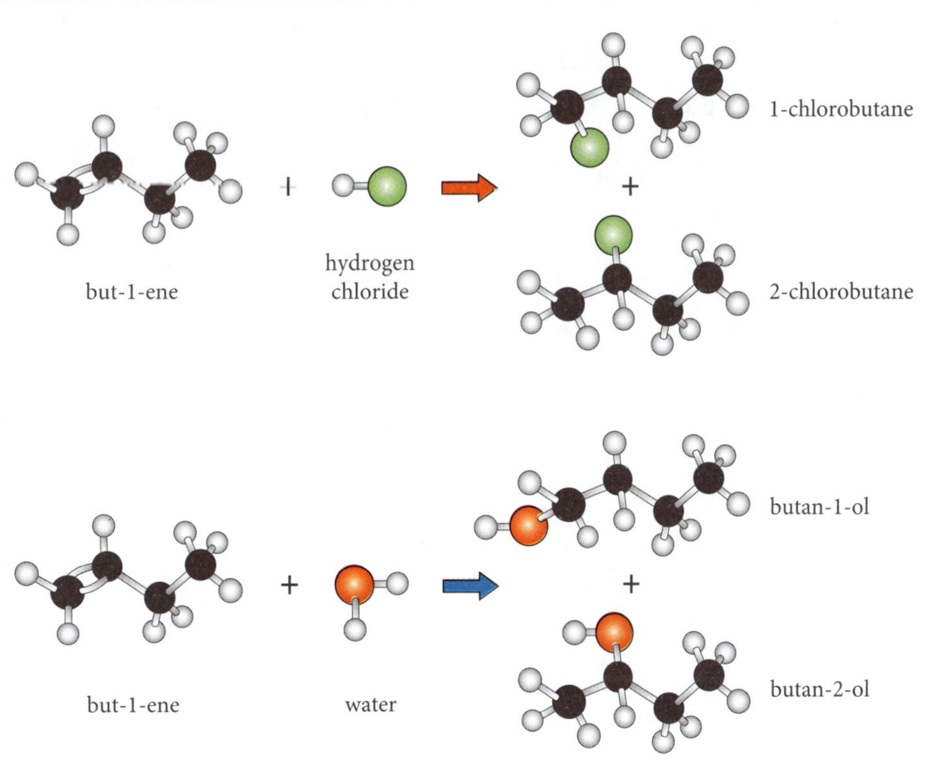

Figure 10.9 Addition of HCl and H_2O to but-1-ene

Figure 10.10 shows models of chlorine substituted products when chlorine reacts with methane and ethane under UV light.

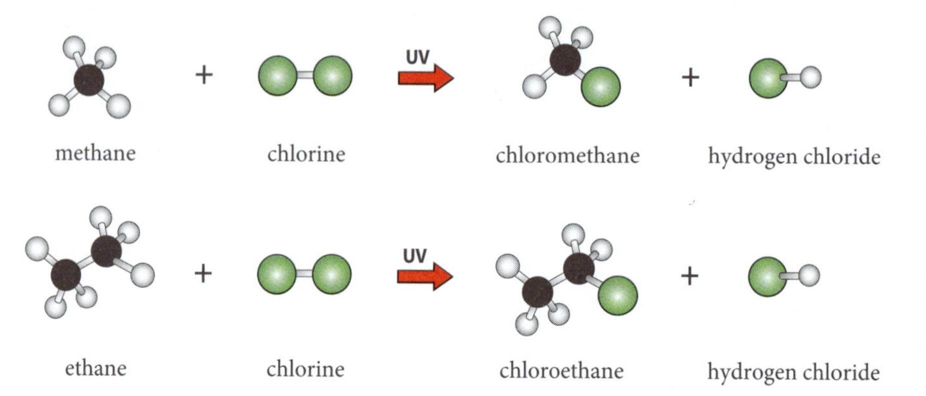

Figure 10.10 Substitution reactions

CHAPTER SYLLABUS CHECKLIST

Are you able to answer these questions from the syllabus for this chapter? Tick each question as you go through the checklist if you are able to answer it. If you cannot answer a question, turn to the relevant page in the study guide to find the answer. For NESA key word meanings, go to www.educationstandards.nsw.edu.au and search 'key words'.

FOR A COMPLETE UNDERSTANDING OF THIS TOPIC:		PAGE NO.	✓
1	Can I define the term *addition reaction*?	136	
2	Can I write equations and name the products for addition reactions of unsaturated hydrocarbons with halogens?	137	
3	Can I write equations and name the products for addition reactions of unsaturated hydrocarbons with hydrogen halides?	137	
4	Can I write equations and name the products for addition reactions of unsaturated hydrocarbons with hydrogen?	137	
5	Can I write equations and name the products for addition reactions of unsaturated hydrocarbons with water?	137	
6	Can I write equations for substitution reactions of hydrocarbons?	138	
7	Can I use model kits to model addition and substitution reactions?	139	

HSC EXAM-TYPE QUESTIONS

Objective-response questions (1 mark each)

1 A hydrocarbon has the following condensed structural formula:

$CH_3CH_2CH_2CH_2CH_2CHCHCH_3$

It is mixed with pure bromine liquid. Name the product of the addition reaction.

A dibromooctane

B 2,3-dibromooctane

C 6,7-dibromooctane

D 3-bromooctan-2-ol

2 A hydrocarbon has the following condensed structural formula:

$CH_3CHCHCH_2CH_2CH_3$

It is mixed with bromine water and decolourisation of the bromine water occurs. How many different addition products can form?

A 2

B 3

C 4

D 5

3 Name the product of the reaction between one mole of pent-2-yne and one mole of hydrogen.

A pentane

B pent-2-ene

C pent-1-ene

D pent-3-ene

4 Name the product(s) of the reaction between hex-1-ene and water in the presence of phosphoric acid catalyst.

A hexan-1-ol and hexan-2-ol

B hexan-2-ol and hexan-3-ol

C hexan-1-ol

D hexan-2-ol

5 2-chlorohexane reacts with potassium hydroxide solution. What is the condensed structural formula of the organic product?

A $CH_3CH(OH)CH_2CH_2CH_2CH_3$

B $CH_2(OH)CH_2CH_2CH_2CH_2CH_3$

C $K^+(CH_3CHOCH_2CH_2CH_2CH_3)^-$

D $C_6H_{13}OH$

Extended-response questions

6 Figure 10.11 shows the structure of cyclohexane and cyclohexene. They are both colourless liquids that are insoluble and less dense than water.

a Describe the observations in which an excess of both liquids is shaken with bromine water in separate stoppered test tubes. (2 marks)

b Write the formula of possible products that will form when cyclohexene reacts with bromine water. (2 marks)

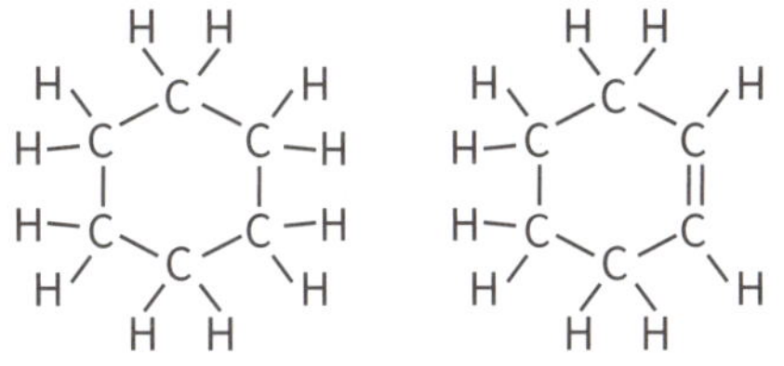

cyclohexane cyclohexene

Figure 10.11 Cyclohexane and cyclohexene

7 a Distinguish between the terms *saturated* and *unsaturated* hydrocarbons. (1 mark)

b Classify a molecule with the molecular formula $C_{14}H_{26}$ as saturated or unsaturated. (1 mark)

c Gaseous but-1-ene is brominated with bromine vapour (Br_2) via an addition reaction. The mixture is cooled back to room temperature after the reaction. A colourless liquid forms. Write a balanced equation for this reaction and name the liquid product formed. (2 marks)

d One mole of ethane reacts with one mole of chlorine in the presence of UV light. Name the organic product and identify the type of reaction that has occurred. (2 marks)

8 a Hept-2-ene is converted by a chemical reaction into heptane. Name the process that has occurred and name the other reactant. (2 marks)

b One mole of heptane vapour is mixed with one mole of chlorine and the mixture exposed to UV light. How many position isomers could form as organic products? Name these isomers. (4 marks)

9 Write equations using condensed structural formulae for the following reactions and name the organic reaction product(s):

a Oct-3-ene is reacted with hydrogen bromide. (4 marks)

b Hept-1-ene is reacted with bromine water. (6 marks)

10 a Explain how but-1-ene gas can be converted to butane gas. Write an equation for the reaction. (2 marks)

b One mole of butane gas is mixed with excess chlorine gas and the mixture exposed to UV light. Draw a structural formula of the product and name the product. (2 marks)

ANSWERS

KEY QUESTIONS

Key questions ➲ p. 137

1 $CH_2CHCH_2CH_2CH_2CH_3 + Br_2 \rightarrow CH_2BrCHBrCH_2CH_2CH_2CH_3$
Product = 1,2-dibromohexane

2 $CH_2CHCH_2CH_2CH_3 + H_2O \rightarrow CH_3CH(OH)CH_2CH_2CH_3$
Product = pentan-2-ol
$CH_2CHCH_2CH_2CH_3 + H_2O \rightarrow CH_2(OH)CH_2CH_2CH_2CH_3$
Product = pentan-1-ol

3 $CHCCH_3 + 2H_2 \rightarrow CH_3CH_2CH_3$
Product = propane

Key questions ➲ p. 138

4 $CH_4 + 4Br_2 \rightarrow CBr_4 + 4HBr$
Organic product = tetrabromomethane

5 $CH_3CHClCH_2CH_2CH_3 + NaOH$
$\rightarrow CH_3CH(OH)CH_2CH_2CH_3 + NaCl$
Organic product = pentan-2-ol

HSC EXAM-TYPE QUESTIONS

Objective-response questions

1 **B**. Bromine is a diatomic molecule (Br_2). The molecule adds across the double bond located between carbon atoms with locants 2 and 3. Thus the product is 2,3-dibromooctane. Number the chain from right to left to give the lower locant set. Thus A, C and D are incorrect: **A** is incorrect as the locants are not shown; **C** is incorrect as 6 and 7 are not the lowest locants; and **D** is incorrect as no oxygen atoms are present in the reactants.

2 **B**. Possible products are:
$CH_3CHBrCHBrCH_2CH_2CH_3$ (2,3-dibromohexane)
$CH_3CH(OH)CHBrCH_2CH_2CH_3$ (3-bromohexan-2-ol)
$CH_3CHBrCH(OH)CH_2CH_2CH_3$ (2-bromohexan-3-ol)
Therefore **A**, **C** and **D** are incorrect.

3 **B**. One mole of hydrogen will convert the alkyne to an alkene with the double bond at locant 2. **A** is incorrect as two moles of hydrogen would be needed. **C** and **D** are incorrect as the double bond formed must be in the position of the original triple bond.

4 **A**. The OH group will join onto hex-1-ene in the locant position 1 or 2. Thus two isomers form. **B** is wrong as the OH group cannot move to the locant position 3. **C** and **D** are incorrect as more than one product can form.

5 **A**. The Cl functional group on carbon-2 is substituted by the OH functional group. **B** is incorrect as the OH group will not be located on carbon-1. **C** is incorrect as the potassium ion forms KCl as the inorganic reaction product. **D** is incorrect as this formula is not a condensed structural formula.

Extended-response questions

6 EM Students need to show an understanding that bromine water contains both Br_2 and HOBr molecules that can add across the double bond in cyclohexene. The cyclohexane is unreactive under the conditions used.

a Decolourisation of the bromine water will occur with cyclohexene but not with cyclohexane. ✓
The orange colour of the bromine water persists in the latter case and the bromine diffuses into the organic upper layer to form an orange solution with the cyclohexane. ✓

b $C_6H_{10}Br_2$ (1,2-dibromocyclohexane) ✓
$C_6H_{10}BrOH$ (2-bromocyclohexan-1-ol) ✓

7 EM Students must demonstrate an understanding of the difference between saturated and unsaturated in terms of structure as well as being able to predict the product of an addition reaction. Students need to distinguish substitution reactions from addition reactions.

a Saturated hydrocarbons have molecules with only C–C single bonds. Unsaturated hydrocarbons can have double or triple bonds between carbon atoms. ✓

b Unsaturated. It is an alkyne (C_nH_{2n-2}, where $n = 14$). ✓

c $C_4H_8(g) + Br_2(g) \rightarrow C_4H_8Br_2(l)$ ✓
1,2-dibromobutane forms (as the double bond was between C1 and C2) ✓

d chloroethane; ✓ substitution reaction ✓

8 EM Double bonds react via addition reactions. Alkanes are unreactive and UV light creates reactive radicals that result in substitution reactions. Students need to review the concept of position isomers.

a addition reaction; ✓ hydrogen gas ✓

b Chlorine atoms substitute for one of the H atoms:
($C_7H_{16} + Cl_2 \rightarrow C_7H_{15}Cl + HCl$)
There are four position isomers. The organic products are:
$CH_2ClCH_2CH_2CH_2CH_2CH_2CH_3$; 1-chloroheptane ✓
$CH_3CHClCH_2CH_2CH_2CH_2CH_3$; 2-chloroheptane ✓
$CH_3CH_2CHClCH_2CH_2CH_2CH_3$; 3-chloroheptane ✓
$CH_3CH_2CH_2CHClCH_2CH_2CH_3$; 4-chloroheptane ✓

9 EM Students must show an understanding that some addition reactions can produce more than one product. They must also show a good understanding of the IUPAC nomenclature rules.

a $CH_3CH_2CHCHCH_2CH_2CH_2CH_3 + HBr$
$\rightarrow CH_3CH_2CH_2CHBrCH_2CH_2CH_2CH_3$ ✓
Product = 4-bromooctane ✓
$CH_3CH_2CHCHCH_2CH_2CH_2CH_3 + HBr$
$\rightarrow CH_3CH_2CHBrCH_2CH_2CH_2CH_2CH_3$ ✓
Product = 3-bromooctane ✓

b Bromine water contains HOBr molecules and Br_2 molecules.
$CH_2CHCH_2CH_2CH_2CH_2CH_3 + HOBr$
$\rightarrow CH_2(OH)CHBrCH_2CH_2CH_2CH_2CH_3$ ✓
Product = 2-bromoheptan-1-ol ✓
$CH_2CHCH_2CH_2CH_2CH_2CH_3 + HOBr$
$\rightarrow CH_2BrCH(OH)CH_2CH_2CH_2CH_2CH_3$ ✓
Product = 1-bromoheptan-2-ol ✓
$CH_2CHCH_2CH_2CH_2CH_2CH_3 + Br_2$
$\rightarrow CH_2BrCHBrCH_2CH_2CH_2CH_2CH_3$ ✓
Product = 1,2-dibromoheptane ✓

10 EM Students are to show their understanding that hydrogenation is a slow addition reaction and a catalyst is required. With excess chlorine the substitution reaction with butane will lead to the complete chlorination of the butane.

a Mix but-1-ene with hydrogen gas and heat. A nickel catalyst is required to increase the rate of the addition reaction. ✓
$C_4H_8(g) + H_2(g) \rightarrow C_4H_{10}(g)$ ✓

b decachlorobutane ✓
Figure A10.1 shows the structural formula of decachlorobutane.

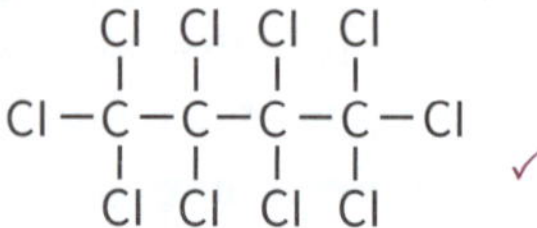

Figure A10.1 Chlorinated hydrocarbon

CHAPTER 11

ALCOHOLS

MODULE 7 ORGANIC CHEMISTRY

INQUIRY QUESTION:

How can alcohols be produced and what are their properties?

Alcohols are organic compounds in which the hydroxyl functional group is present. Ethanol is an alcohol present in beer and wine. Methylated spirits is a mixture of methanol and ethanol. Vinyl alcohol is an unstable alcohol that can be polymerised to form a polymer called PVA (polyvinyl alcohol), which is used to manufacture paper, textiles and coatings. The structure, properties and reactions of alcohols are examined in this chapter.

1 Structure and properties of alcohols

» Students:
- investigate the structural formulae, properties and functional group including primary, secondary and tertiary alcohols.
- explain the properties within and between the homologous series of alcohols with reference to the intermolecular and intramolecular bonding present.

Classification of alcohols

➔ Alcohols are organic compounds that contain the hydroxyl functional group (OH). The structure and classification of alcohols was discussed briefly in Chapter 8.

➔ The hydroxyl functional group is present in many **homologous series** of organic compounds. Two common hydrocarbons that form alcohols are alkanes and alkenes.
- *Alkanols* are alkanes in which hydroxyl groups have replaced one or more hydrogen atoms.
- *Alkenols* are alkenes in which hydroxyl groups have replaced one or more hydrogen atoms.

homologous series: a series of organic compounds with the same general formula and similar structures

➔ Figure 11.1 shows some representative examples of alkanols and alkenols. Priority is given to the double bond when numbering the locant positions of functional groups in alkenols.

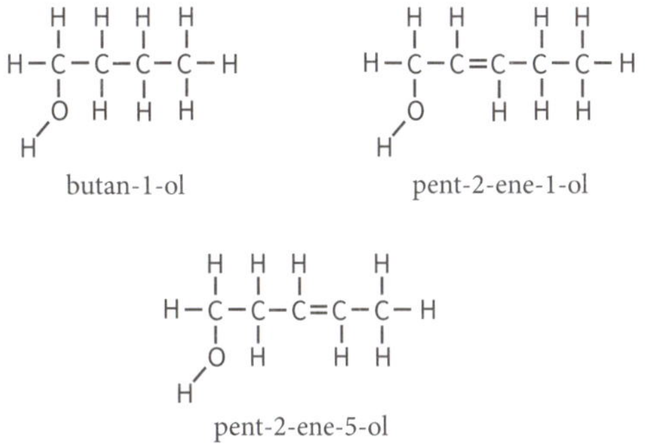

butan-1-ol

pent-2-ene-1-ol

pent-2-ene-5-ol

Figure 11.1 Alkanols and alkenols

➔ In Chapter 8 alkanols were classified as primary, secondary or tertiary. These classifications are reviewed here.
- Primary alcohols have the general structure:

  R_1CH_2OH (R_1 = hydrocarbon chain)
- Secondary alcohols have the general structure:

  R_1R_2CHOH (R_1,R_2 = hydrocarbon chains)
- Tertiary alcohols have the general structure:

  $R_1R_2R_3COH$ (R_1,R_2,R_3= hydrocarbon chains)

➔ Figure 11.2 shows examples of these three classifications.

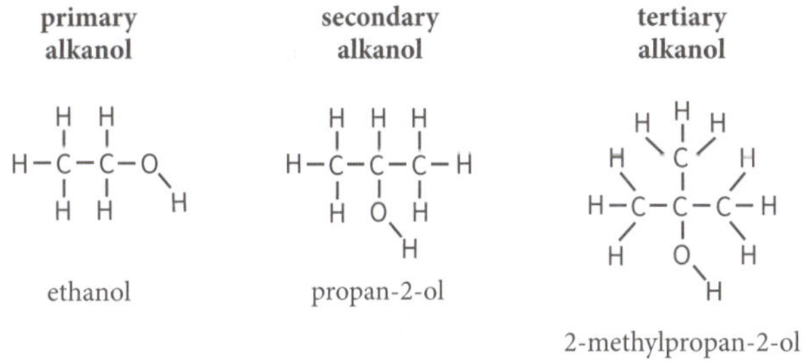

primary alkanol	secondary alkanol	tertiary alkanol
ethanol	propan-2-ol	2-methylpropan-2-ol

Figure 11.2 Classification of alkanols

Properties of alcohols

➔ Alkanols are polar molecules. The presence of the hydroxyl group creates asymmetry in an alkane molecule. The O–H and C–O bonds are polar and the hydrocarbon chain is non-polar. Dipole–dipole attractive forces exist between OH groups in neighbouring alkanol molecules but as the chain length increases, the dipole–dipole attraction decreases and the dispersion forces between the hydrocarbon chains increase. Alkanols are less volatile and much more water soluble than the corresponding alkane. Methanol, ethanol and propan-1-ol are completely miscible in water.

- These properties are explained by the high polarity of the hydroxyl functional group. Apart from dipole–dipole attractions between alkanol molecules, hydrogen bonding is also a significant intermolecular force for short chain alkanols. Table 11.1 compares the boiling points of alkanols with their parent alkane. Methanol and ethanol show significant hydrogen bonding and so their boiling points are much higher than their parent alkanes.

Table 11.1 Boiling points

Alkane	Boiling point (°C)	Alkanol	Boiling point (°C)
methane	−162.0	methanol	65
ethane	−89.0	ethanol	78
propane	−42.0	propan-1-ol	97
butane	−0.5	butan-1-ol	117
pentane	36.0	pentan-1-ol	138

- Figure 11.3 shows the hydrogen bonding interactions between ethanol molecules as well as when ethanol is dissolved in water. The electropositive hydrogen atoms bonded to the oxygen atoms hydrogen-bond with the oxygen atom's non-bonding electrons in neighbouring molecules.

Hydrogen bonding between ethanol molecules

hydrogen bonding

δ− δ+ δ+ δ−

Hydrogen bonding between ethanol and water

hydrogen bonding

δ− δ+ δ+ δ− δ+

Figure 11.3 Hydrogen bonding between ethanol and water molecules

- Ethanol is currently used as a motor fuel supplement (E10) and as a solvent in cosmetics, antiseptics and perfumes. Ethanol is an important industrial solvent for lacquers and paints and this use is related to the polar nature of the ethanol molecule.
- The non-polar ethyl (C_2H_5) group in ethanol assists some low molecular-weight non-polar solutes to dissolve in ethanol. Ethanol–water mixtures are often used to dissolve various natural oils and organic substances that have very low solubility in water.

SECONDARY-SOURCED INVESTIGATION

Physical properties of alkanols

» Students investigate the structural formulae, properties and functional group, including primary, secondary and tertiary alcohols.

Aim

to compare the physical properties of primary, secondary and tertiary alkanols

Method

1 The physical data was collected and tabulated for a range of alkanols.

Alkanol	Molar mass (g/mol)	Solubility in water (g/100 mL)	m.p (°C)	b.p (°C)	Vapour pressure (kPa) (at 20 °C)
propan-2-ol	60.1	miscible	-89	+83	4.4
pentan-2-ol	88.2	4.5	-73	+119	0.8
2-methylbutan-2-ol	88.2	12.0	-9	+102	1.6
hexan-1-ol	102.2	0.6	-45	+157	0.1
octan-1-ol	130.2	0.05	-16	+195	0.009
octan-2-ol	130.2	0.11	-38	+179	0.03

2 Draw structural formulas of each listed alkanol and classify each alkanol as primary, secondary or tertiary.

3 Discuss the trend in solubility in water as molar mass increases.

4 Identify the physical states of these alkanols at 25 °C.

5 Explain why 2-methybutan-2-ol has a higher vapour pressure than pentan-2-ol.

6 Explain why octan-1-ol has a higher boiling point but lower vapour pressure at 20 °C than octan-2-ol.

Sample answers: Physical properties of alkanols

2 The structural formulas and classification are illustrated in Figure 11.4.

3 As molar mass increases the solubility of alkanols in water decreases. The alkanols become less polar as the length of the hydrocarbon chain increases and dipole–dipole attractions with the polar solvent decrease. The differences in shape of isomeric alkanols also causes changes in solubility.

4 All the listed alkanols are liquids at 25 °C as their melting points are less than 25 °C but their boiling points are above 25 °C.

5 2-methybutan-2-ol and pentan-2-ol are isomers. The intermolecular forces are different due to their different structures. The higher vapour pressure of 2-methylbutan-2-ol indicates that the intermolecular forces are weaker than the intermolecular forces between pentan-2-ol molecules. The methyl

functional group leads to a change in molecular shape such that 2-methylbutan-2-ol molecules cannot make close contact with other molecules in the liquid state. Thus the molecules vapourise more readily.

6 Octan-1-ol and octan-2-ol are isomers with different structures and molecular shapes. Octan-1-ol has slightly higher intermolecular forces than octan-2-ol and therefore a higher boiling point. At 20 °C the stronger intermolecular forces will reduce the vapour pressure above the liquid.

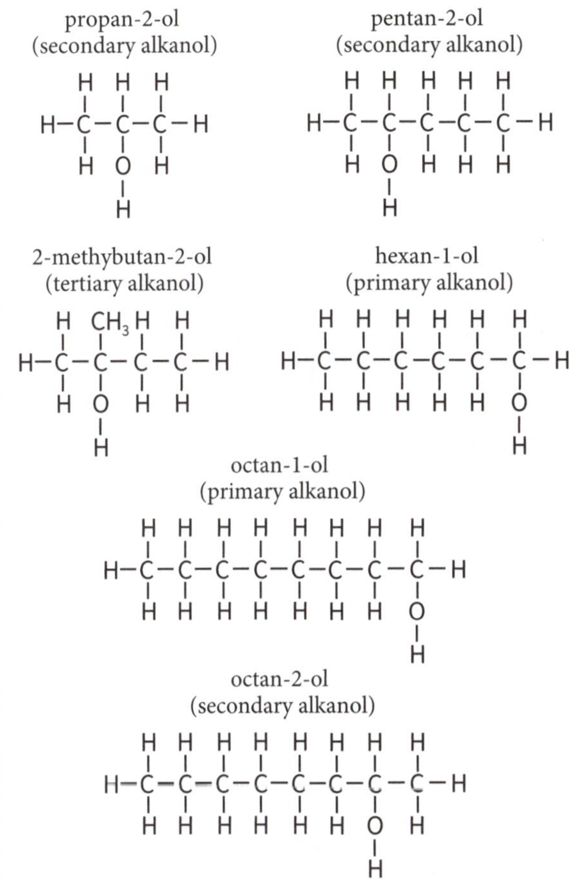

Figure 11.4 Alkanol structural formulas and classification

➔ KEY QUESTIONS

1 Classify the following alkanols as primary, secondary or tertiary:
a hexan-2-ol
b 3-methyloctan-3-ol
c $CH_3CH_2CH_2CH_2OH$

2 Explain why pentan-1-ol has a higher boiling point than pentane.

3 Explain why ethanol is completely miscible in water.

Answers ➲ p. 156

2 Combustion of alcohols

» Students conduct a practical investigation to measure and reliably compare the enthalpy of combustion for a range of alcohols.

- Alkanols undergo combustion reactions in air and oxygen.
- Complete combustion to form carbon dioxide and water is achieved as long as a plentiful supply of oxygen is available. In a limited oxygen supply, carbon monoxide and carbon (soot) may also form.
- As the chain length of the alkanol increases there is a noticeable tendency towards incomplete combustion.

FIRSTHAND INVESTIGATION 1

Enthalpy of combustion of alkanols

Aim

to measure the enthalpies of combustion of various alkanols

Apparatus

Figure 11.5 illustrates the apparatus for the investigation.

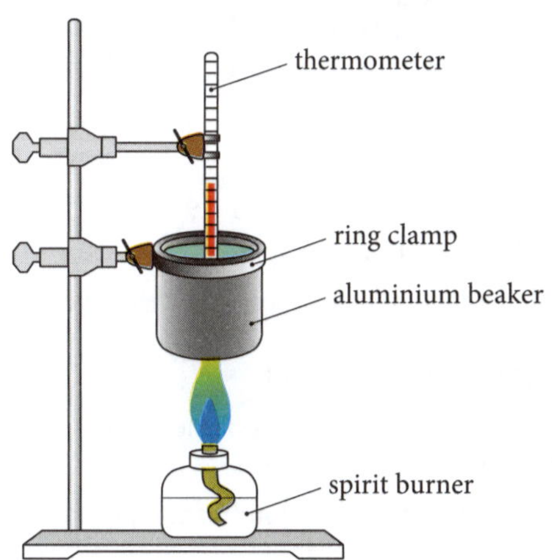

Figure 11.5 Enthalpy of combustion apparatus

Method

1 Use an aluminium (or copper) beaker as a calorimeter. Alternatively use an aluminium soft-drink can as a calorimeter. Support the calorimeter using a ring clamp attached to a retort stand or place the calorimeter on a tripod (Figure 11.5).

2 Add 200 g (200 mL) of water to the calorimeter and measure the initial temperature. (Note that the volume of water used will depend on the size of the calorimeter.)

3 Fill the spirit burner with methanol so the wick is immersed and weigh the burner.

4 Place the burner just below the calorimeter and light the wick. The flame should touch the bottom of the calorimeter. Allow the heat from the flame to heat the calorimeter until the temperature of the water has increased by about 10 °C.
5 Extinguish the burner, stir the water and measure the final temperature.
6 Reweigh the burner.
7 Repeat the experiment with ethanol, butan-1-ol and pentan-1-ol.

Sample results for methanol

Mass of water in calorimeter = 200 g = 0.200 kg
Mass of aluminium calorimeter = 25 g = 0.025 kg
Initial temperature of water = 23.6 °C
Final temperature of water = 33.1 °C
Initial mass of spirit burner = 222.75 g
Final mass of spirit burner =222.11 g
Specific heat capacity of water = 4.18×10^3 J/kg/K
Specific heat capacity of aluminium = 9.04×10^2 J/kg/K

Sample calculations for methanol

The water and the aluminium vessel *both* absorb heat from the combustion of methanol.

1 Calculate the heat (q_{water}) absorbed by the water.

$$q_{water} = mc\Delta T_{water} = (0.200)(4.18 \times 10^3)(33.1 - 23.6) = 7942 \text{ J} = 7.942 \text{ kJ}$$

2 Calculate the heat (q_{Al}) absorbed by the aluminium.

$$q_{Al} = mc\Delta T_{aluminium} = (0.0250)(9.04 \times 10^2)(33.1 - 23.6) = 215 \text{ J} = 0.215 \text{ kJ}$$

3 Calculate the total heat absorbed by the calorimeter and contents.

$$q = 7.942 + 0.215 = 8.157 \text{ kJ}$$

4 Calculate the mass of methanol fuel burnt.

$$m(\text{methanol}) = 222.75 - 222.11 = 0.64 \text{ g}$$

5 Calculate the enthalpy of combustion per mole of methanol.

$$n(CH_3OH) = m/M = \frac{0.64}{(12.01 + (4 \times 1.008) + 16.00)} = \frac{0.64}{32.042} = 0.0200 \text{ mol}$$

$$\Delta H = -q/n = \frac{-8.157}{0.0200} = -408 \text{ kJ/mol}$$

(Note: The heat absorbed by the aluminium is quite small compared with the heat absorbed by the water.)

Sample results for alkanols

The accompanying table contains some typical results for the experimental enthalpy data as well as the standard enthalpies of combustion for the four alkanols.

Alkanol	methanol	ethanol	butan-1-ol	pentan-1-ol
Experimental enthalpy of combustion (kJ/mol)	−408	−771	−1482	−1795
Standard enthalpy of combustion (kJ/mol)	−726	−1367	−2676	−3331

Discussion

The experimental results are much lower than the standard enthalpy data. This is due to heat loss to the surroundings as well as incomplete combustion due to the restricted supply of oxygen from the air. Loss of heat to the surroundings can be reduced (but not eliminated) by using non-combustible insulation on the sides and top of the calorimeter. The apparatus can also be surrounded by shields that prevent draughts.

Combustible oxygen/alcohol vapour mixtures will burn incompletely if the mole ratio of oxygen to alcohol vapour is too low. As the molar mass of the alkanol increases, the degree of incomplete combustion increases and considerable soot is produced in the flame of the spirit burner.

The apparatus used is inadequate to achieve the aim of the experiment to measure the enthalpy of combustion. Considerable heat is lost to the surroundings and the alkanols do no burn completely. The procedure is therefore invalid. The experimental reliability can be improved by five or more repetitions of each experiment. It is important in each repetition to control the variables as far as the method and equipment allows.

EXAMPLE 1

a Complete the table by calculating the molar mass of each alkanol and use the data to plot a line graph of the standard enthalpy of combustion (vertical axis) versus alkanol molar mass.

b Use the line of best fit to estimate the standard enthalpy of combustion of propan-1-ol.

Alkanol	methanol	ethanol	butan-1-ol	pentan-1-ol	propan-1-ol
Molar mass (g/mol)					
Standard enthalpy of combustion (kJ/mol)	−726	−1367	−2676	−3331	?

Answer:

Draw the line of best fit

a The following table shows the values of the molar masses of the five alkanols. Figure 11.6 shows the graph of enthalpy of combustion versus molar mass. Note that data points are shown by small crosses.

Alkanol	methanol	ethanol	butan-1-ol	pentan-1-ol	propan-1-ol
Molar mass (g/mol)	32.04	46.07	74.13	88.15	60.10
Standard enthalpy of combustion (kJ/mol)	-726	-1367	-2676	-3331	?

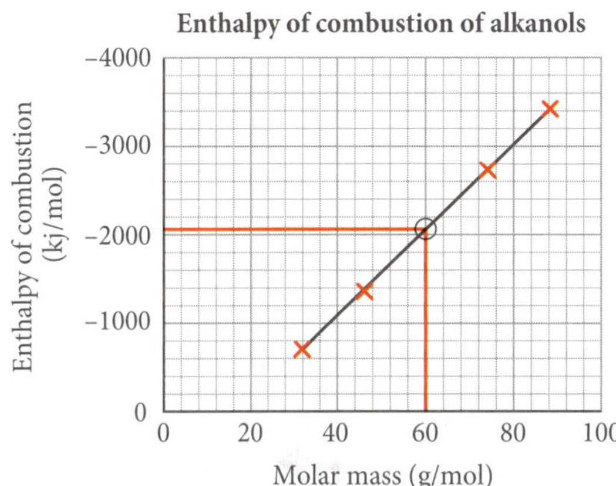

Figure 11.6 Enthalpy of combustion of alkanols versus molar mass

b From Figure 11.6, the enthalpy of combustion of propan-1-ol is –2050 kJ/mol. The literature value for the enthalpy of combustion of propan-1-ol is -2021 kJ/mol.

→ KEY QUESTIONS

4 **Compare the heat released by the combustion of alkanols as the length of the hydrocarbon chain increases.**

5 **Explain why the enthalpy of combustion for butan-1-ol as measured using a spirit burner and a metal flask of water is much less than the accepted literature value.**

Answers ➲ p. 156

3 Reactions of alcohols

» Students:
- write equations, state conditions and predict products to represent the reactions of alcohols, including but not limited to combustion, dehydration, substitution with HX and oxidation.
- investigate the products of the oxidation of primary and secondary alcohols.

Combustion

→ In a plentiful supply of oxygen, alkanols undergo complete combustion to form carbon dioxide and water. (Ensure physical states are shown in the equation.)

Complete combustion:

$$2CH_3OH(l) + 3O_2(g) \rightarrow 2CO_2(g) + 4H_2O(l)$$

→ In a restricted-oxygen supply, alkanols burn incompletely to form a mixture containing variable amounts of carbon dioxide, carbon monoxide, carbon (soot) and water. The longer the hydrocarbon chain, the more likely incomplete combustion will occur.

Incomplete combustion:

$$C_4H_9OH(l) + 4O_2(g) \rightarrow CO_2(g) + 2CO(g) + C(s) + 5H_2O(l)$$

Dehydration

→ When alkanols are dehydrated, alkenes are formed. Concentrated phosphoric acid or concentrated sulfuric acid are common dehydrating agents.

→ To dehydrate ethanol it is mixed with an excess of concentrated sulfuric acid and heated to 170 °C. The ethylene gas is collected by bubbling it though sodium hydroxide solution. This method is used to remove gases such as carbon dioxide and sulfur dioxide that form due to side reactions. The following equation shows the net **dehydration reaction** of ethanol:

$$C_2H_5OH(l) \rightarrow C_2H_4(g) + H_2O(l)$$

dehydration reaction: a chemical reaction in which water is removed from a reactant

Substitution

→ Alkanols undergo **substitution reactions** when they react with hydrogen halides. An acid catalyst is required.

substitution reaction: a chemical reaction in which one functional group in a molecule is replaced by another functional group

→ In the following examples a hydrogen halide is used. This leads to the removal of the hydroxyl functional group forming water and the halogen atom replacing the –OH group.
- The following substitution reaction involves a tertiary alkanol that reacts with concentrated hydrochloric acid. A model of a substitution reaction is shown in Figure 11.7:

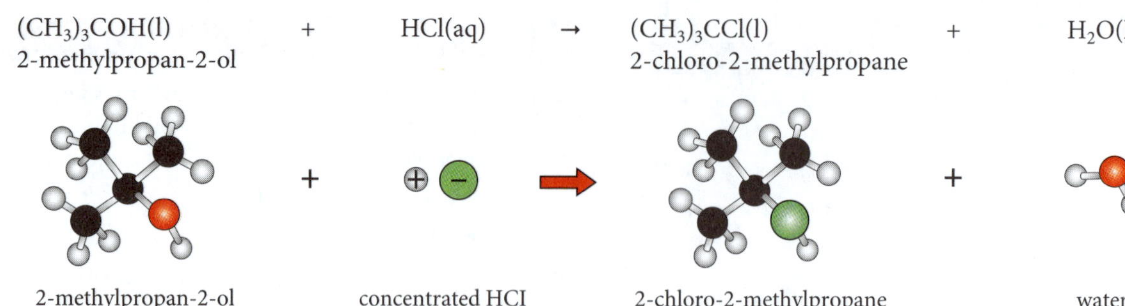

Figure 11.7 Substitution reaction

- Ethanol can be converted to iodoethane in a substitution reaction with hydrogen iodide. The HI is formed by reaction concentrated phosphoric acid with potassium iodide. The iodoethane is recovered by distillation of the reaction mixture.

$$CH_3CH_2OH(l) + HI(g) \rightarrow CH_3CH_2I\ (l) + H_2O(l)$$
ethanol — iodoethane

Oxidation

➔ Primary and secondary alkanols can be oxidised by a variety of oxidants. Tertiary alkanols cannot be oxidised under normal laboratory conditions. Alkanes, tertiary alkanols and alkanoic acids are not oxidised by these oxidants and consequently these compounds can be distinguished from primary and secondary alkanols.

➔ Common oxidants include acidified CrO_3 solution, acidified potassium permanganate ($H^+/KMnO_4$) and acidified potassium dichromate ($H^+/K_2Cr_2O_7$) solutions. CrO_3 is classified as a mild oxidant. Acidified $KMnO_4$ and $K_2Cr_2O_7$ solutions are classified as strong oxidants.

- The acidified solution of potassium permanganate is violet in colour. During the **oxidation** reaction it is converted into colourless manganese (II) ions. The reduction half-equation is:

$$MnO_4^-(aq) + 8H^+(aq) + 5e^- \rightarrow Mn^{2+}(aq) + 4H_2O(l)$$

- The acidified solution of potassium dichromate is orange in colour. During the oxidation reaction it is converted into green chromium (III) ions. The reduction half-equation is:

$$Cr_2O_7^{2-}(aq) + 14H^+(aq) + 6e^- \rightarrow 2Cr^{3+}(aq) + 7H_2O(l)$$

oxidation: a chemical reaction in which a chemical substance loses electrons

Oxidation of primary alkanols

➔ Using a mild oxidant such as an acidified solution of CrO_3, primary alkanols are oxidised to form alkanals. Alkanal molecules have the aldehyde (CHO) functional group, as shown in Figure 11.8. The carbonyl (C=O) group is part of the aldehyde group. If stronger oxidants such as an acidified solution of potassium permanganate are used then the primary alkanols are oxidised to alkanoic acids. Figure 11.8 compares the differences in products formed.

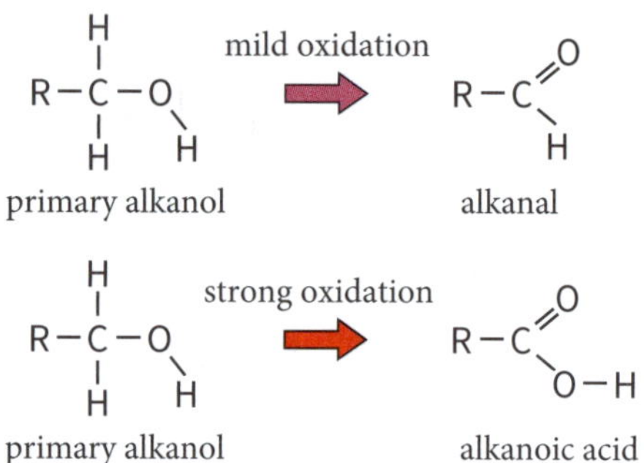

Figure 11.8 Oxidation of primary alkanols

Examples:

- Ethanol can be oxidised to form ethanal using an acidified solution of CrO_3:

$$C_2H_5OH(l) \rightarrow CH_3CHO(l)$$

- Propan-1-ol can be oxidised to form propanoic acid using an acidified solution of potassium dichromate:

$$C_3H_7OH(l) \rightarrow C_2H_5COOH(l)$$

EXAMPLE 2

Figure 11.9 shows ball-and-stick models of two organic compounds, A and B. A chemical reaction converts A to B.

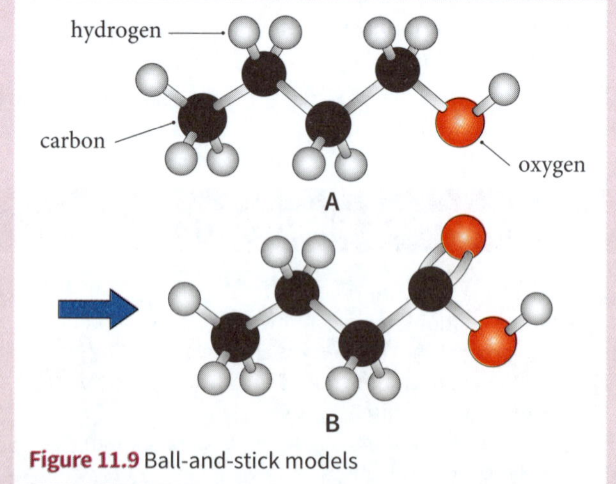

Figure 11.9 Ball-and-stick models

a Write the condensed structural formulas of A and B.
b Name the molecules A and B.
c Identify the type of reaction in which A is converted to B.

Identify the functional groups present

Answer:

a A = $CH_3CH_2CH_2CH_2OH$; B = $CH_3CH_2CH_2COOH$
b A = butan-1-ol; B = butanoic acid
c oxidation reaction

EXAMPLE 3

White wine contains ethanol. Ethanol can be oxidised by acidified potassium dichromate solution. This reaction is the basis of the determination of the ethanol content of wine. This example can be used as the start of a depth study in which students could determine the ethanol content of different alcoholic drinks.

The ethanol content of white wine was determined by titration. A dilute solution of the white wine was prepared by adding 20 mL of wine to a 1 L volumetric flask and water added to the mark. 5.00 mL of this diluted wine was placed in a conical flask and acidified with sulfuric acid. The ethanol in the sample was oxidised using 10.0 mL of 0.020 mol/L potassium dichromate solution. The mixture was heated for 1 hour to ensure complete oxidation of the ethanol (Figure 11.10).

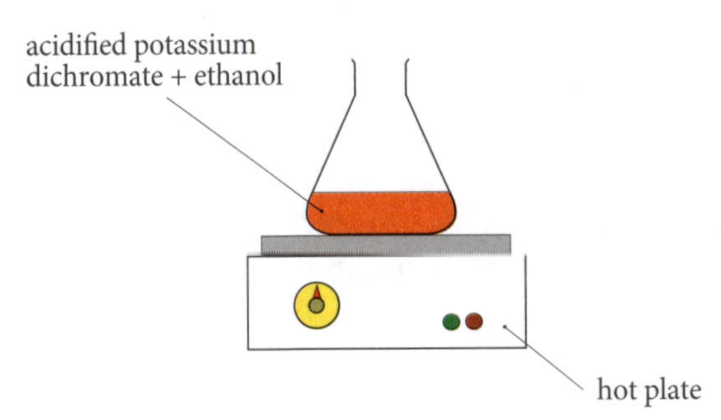

Figure 11.10 Oxidation of ethanol

After cooling, the excess dichromate was reduced by the addition of 20.0 mL of a 0.0500 mol/L iron (II) ion solution. This amount is an excess of Fe^{2+}. The solution formed will be pale yellow-green due to the iron (III) ions produced and the excess of the pale-green iron (II) ions. A back titration using 0.0100 mol/L potassium permanganate was performed to determine the amount of iron (II) ions remaining. This titration is an oxidation–reduction titration. The end point occurs when the solution just turns a pale pink. The mean permanganate titre was 11.10 mL. Figure 11.11 shows the titration apparatus.

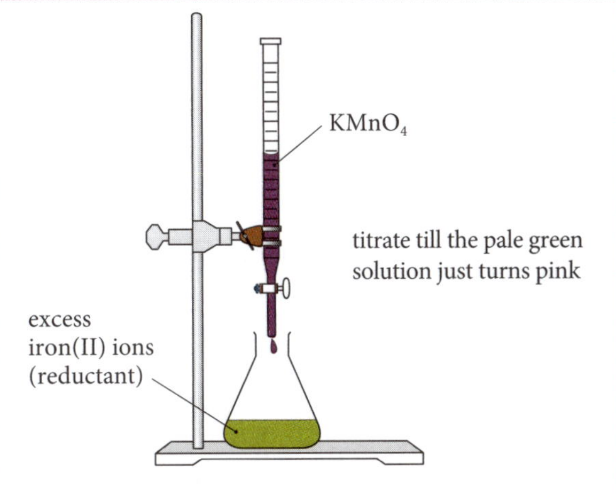

Figure 11.11 Back titration

a Explain how the 2 %v/v solution of the white wine was prepared.
b Use the following equations for the reactions involved to determine the percentage by volume (%v/v) of ethanol in the white wine (density of ethanol = 0.780 g/mL):
- Oxidation of ethanol
  1: $3C_2H_5OH(aq) + 2Cr_2O_7^{2-}(aq) + 16H^+(aq) \rightarrow 3CH_3COOH(aq) + 4Cr^{3+}(aq) + 11H_2O(l)$
- Reduction of excess dichromate
  2: $Cr_2O_7^{2-}(aq) + 6Fe^{2+}(aq) + 14H^+(aq) \rightarrow 2Cr^{3+}(aq) + 6Fe^{3+}(aq) + 7H_2O(l)$
- Oxidation of excess Fe^{2+}
  3: $5Fe^{2+}(aq) + MnO_4^-(aq) + 8H^+(aq) \rightarrow 5Fe^{3+}(aq) + Mn^{2+}(aq) + 4H_2O(l)$

For each equation determine the stoichiometry of the reactants

Answer:

a Use a 20 mL pipette to deliver 20.00 mL of the white wine to a 1 L volumetric flask. Fill with deionised water to the mark and mix thoroughly ($c = \frac{20}{1000} \times 100 = 2$ %v/v).

b Calculate the amount of permanganate used in the back titration:
$n(MnO_4^-) = cV = (0.0100)(11.10 \times 10^{-3})$
$= 1.11 \times 10^{-4}$ mol

From equation 3:
Stoichiometry $Fe^{2+} : MnO_4^- = 5 : 1$
$n(Fe^{2+})$ excess $= 5(1.11 \times 10^{-4}) = 5.55 \times 10^{-4}$ mol
$n(Fe^{2+})$ initially added $= cV = (0.0500)(20.0 \times 10^{-3})$
$= 1.00 \times 10^{-3}$ mol

Calculate the amount of Fe^{2+} that reacted with dichromate:
$n(Fe^{2+})$ reacted with dichromate
$= 1.00 \times 10^{-3} - 5.55 \times 10^{-4}$
$= 4.45 \times 10^{-4}$ mol

From equation 2:

Stoichiometry $Fe^{2+} : Cr_2O_7^{2-} = 6:1$

$n(\text{dichromate}) \text{ excess} = \left(\frac{1}{6}\right)(4.45 \times 10^{-4})$

$= 7.42 \times 10^{-5}$ mol

$n(\text{dichromate}) \text{ initially} = cV = (0.0200)(10.0 \times 10^{-3})$

$= 2.00 \times 10^{-4}$ mol

Calculate the amount of dichromate that reacted with the ethanol:

$n(\text{dichromate})$ reacted with ethanol =

$(2.00 \times 10^{-4} - 7.42 \times 10^{-5})$ mol $= 1.258 \times 10^{-4}$ mol

From equation 1:

Stoichiometry $Cr_2O_7^{2-}$: ethanol = 2 : 3

$n(\text{ethanol}) \text{ reacted} = \left(\frac{3}{2}\right)(1.258 \times 10^{-4})$

$= 1.887 \times 10^{-4}$ mol

$m(\text{ethanol}) = nM = (1.887 \times 10^{-4})(46.07)$

$= 8.69 \times 10^{-3}$ g in 5.0 mL of diluted wine

$V(\text{ethanol}) = m/D = \frac{(8.69 \times 10^{-3})}{(0.780)}$

$= 1.11 \times 10^{-2}$ mL per 5.0 mL of diluted wine

$V(\text{ethanol}) = \left(\frac{100}{5}\right)(1.11 \times 10^{-2})$ mL

$= 0.222$ mL in 100 mL of diluted wine

$= 0.222\%$v/v

The diluted wine is 50 times less concentrated than the original wine as 20 mL was diluted to 1000 mL in the volumetric flask.

%(ethanol) in undiluted wine = $50 \times 0.222 = 11.1\%$v/v

Oxidation of secondary alkanols

➔ Secondary alkanols can be oxidised to form alkanones using mild or strong oxidising agents. The alkanones cannot be further oxidised. Alkanones contain the carbonyl functional group (C=O), as shown in Figure 11.12.

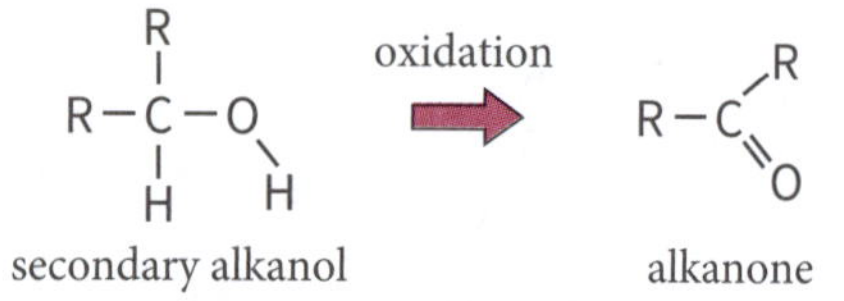

secondary alkanol —oxidation→ alkanone

Figure 11.12 Oxidation of secondary alkanols

➔ If butan-2-ol is heated with an acidified solution of potassium dichromate, the product formed is butanone, as shown in Figure 11.13.

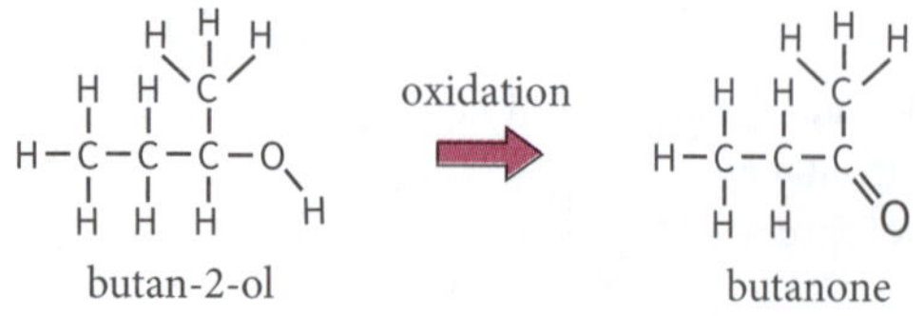

butan-2-ol —oxidation→ butanone

Figure 11.13 Oxidation of butan-2-ol

➔ KEY QUESTIONS

6 Write an equation for the complete combustion of propan-2-ol.

7 Write an equation for the reaction between propan-1-ol and concentrated sulfuric acid.

8 Write an equation for the reaction of hydrogen fluoride gas with ethanol.

9 Identify the product of the reaction between butan-1-ol and an acidified solution of potassium dichromate.

Answers ➲ p. 156

4 Production of alcohols

» Students investigate the production of alcohols, including substitution reactions of halogenated organic compounds and fermentation.

Fermentation and ethanol production

➔ In Queensland and northern New South Wales, large amounts of sugarcane are grown for sugar (sucrose) production. A dark-brown syrup called *molasses* is produced during table-sugar production from the juice of the sugarcane. Some of this molasses syrup is also used to produce ethanol by **fermentation**. The sucrose sugar ($C_{12}H_{22}O_{11}$) solution is first hydrolysed to produce glucose ($C_6H_{12}O_6$) and fructose ($C_6H_{12}O_6$):

$$C_{12}H_{22}O_{11}(aq) + H_2O(l) \rightarrow 2C_6H_{12}O_6(aq)$$

fermentation: a reaction in which sugars are converted into ethanol and carbon dioxide by microbes such as yeast or bacteria

➔ The glucose/fructose solution can then be anaerobically fermented by yeast to produce ethanol (C_2H_5OH):

$$C_6H_{12}O_6\,(aq) \rightarrow 2C_2H_5OH(l) + 2CO_2(g)$$

➔ During anaerobic fermentation, air is excluded so that the yeast produces only ethanol and carbon dioxide. The yeast strains that are used are genetically modified to be alcohol tolerant up to levels of about 15–17% ethanol.

➔ The fermentation reaction rate is optimised at 37 °C. The ethanol has to be separated from the water solution by fractional distillation. The distillate is 95% ethanol and is called *hydrous ethanol*.

➔ Anhydrous (100%) ethanol can be obtained from hydrous ethanol by absorbing the water into porous solids (such as aluminium silicate compounds called zeolites) that are called molecular sieves. The flowchart in Figure 11.14 summarises the production of ethanol from sugarcane.

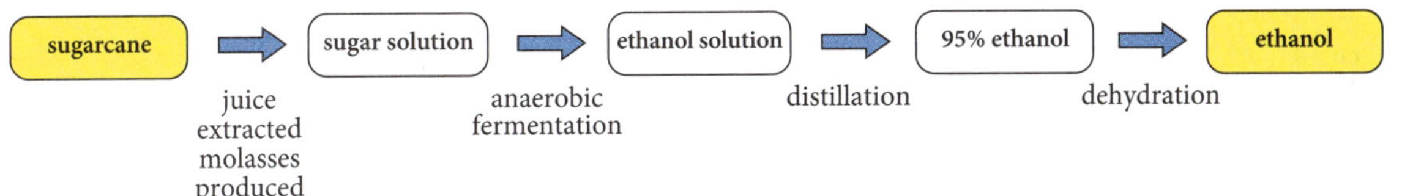

Figure 11.14 Fermentation flowchart

FIRSTHAND INVESTIGATION 2

Monitoring mass changes in the fermentation of glucose

» Students investigate the production of alcohols, including fermentation.

Aim

to monitor mass changes during fermentation of glucose

Equipment

Stoppered flask with outlet tube
Water bath (~35 °C) *or* incubator
Limewater test tube
Electronic balance

Materials

20 mL glucose solution (10 %w/v)
Limewater
Yeast (~2 g)

Apparatus

Figure 11.15 shows the apparatus used.

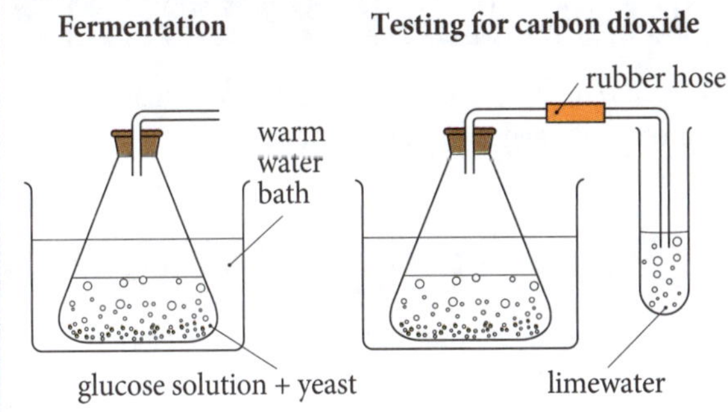

Figure 11.15 Fermentation apparatus

Method

1. Place the glucose solution and yeast in the flask. Insert the stopper and outlet tube. Record the initial mass.
2. Place the flask in the warm water bath or incubator and maintain the temperature at 30–35 °C. Allow the mixture to ferment.
3. Attach a tube of limewater to the outlet tube from the flask and allow the carbon dioxide to bubble through the limewater. Record the change observed. Remove the limewater tube after this test.
4. Record the final mass of the *dried* flask and its contents in the next lesson. Calculate the total decrease in mass of the flask and its contents.

Sample results

Limewater turns milky. This is a positive test for carbon dioxide.
Initial mass of apparatus = 95.65 g
Final mass of apparatus = 94.90 g
Loss in mass = 95.65 – 94.90 = 0.75 g

Sample analysis

Mass of glucose used = $\frac{10.0}{100} \times 20.0 = 2.0$ g

Fermentation equation: $C_6H_{12}O_6(aq) \rightarrow 2C_2H_5OH(aq) + 2CO_2(g)$

Stoichiometry: glucose : carbon dioxide = 1 : 2

$n(\text{glucose}) = m/M = \frac{2.0}{((6 \times 12.01) + 12(1.008) + (6 \times 16.00))}$

$= \frac{2.0}{180.156} = 0.0111$ mol

$n(\text{carbon dioxide}) = 2(0.0111) = 0.0222$ mol

$m(CO_2) = n.M = (0.0222)(12.01 + 2(16.00))$

$= (0.0222)(44.01) = 0.95$ g

The calculated mass of carbon dioxide evolved (0.95 g) is greater than the experimental mass loss (0.75 g). The possible reasons for this lower mass in the experiment is that some carbon dioxide has remained dissolved in the fermenting mixture and in the air inside the flask. As well, not all the glucose has fermented.

EXAMPLE 4

Barley seeds are rich in starch. Starch is a complex polymeric carbohydrate. Barley is used to make beer. The starch in the barley is converted to maltose, which is a disaccharide sugar formed from two molecules of glucose ($C_6H_{12}O_6$). Figure 11.16 shows the structures of starch, maltose, glucose and ethanol.

Beer-making involves the conversion of barley into ethanol:

a **Step A involves the conversion of starch to maltose. The formula for starch is $(C_6H_{10}O_5)_n$, where n is a large integer. Write an equation for the hydrolysis of starch to form maltose.**

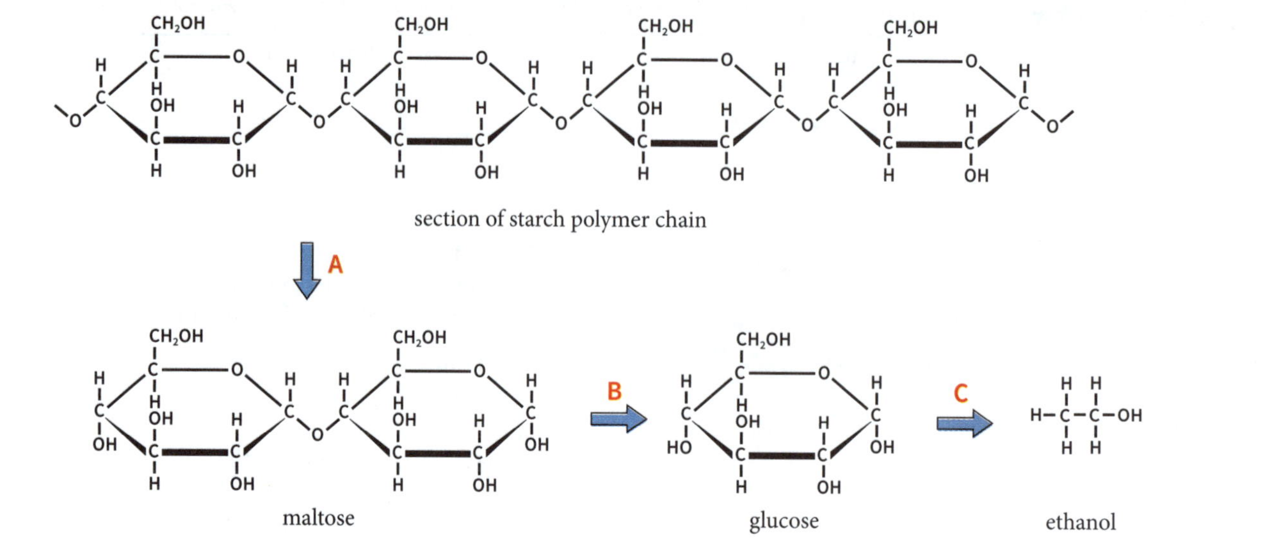

section of starch polymer chain

A

maltose

B

glucose

C

ethanol

Figure 11.16 Conversion of starch to ethanol

b Yeast is used to ferment the sugars. Step B is a hydrolysis reaction catalysed by the enzyme *maltase* produced by the yeast. Write an equation for the hydrolysis of maltose to glucose and explain the role of enzymes.

c Step C is the fermentation of glucose to form ethanol. This reaction is catalysed by *zymase* enzymes produced by the yeast. How many moles of ethanol are produced per mole of glucose?

Answer:

Check atom balance in terms of *n*

a $2(C_6H_{10}O_5)_n(s) + nH_2O(l) \rightarrow nC_{12}H_{22}O_{11}(aq)$

b $C_{12}H_{22}O_{11}(aq) + H_2O(l) \rightarrow 2C_6H_{12}O_6(aq)$
Enzymes are biological catalysts that increase the rate of the reaction by lowering the activation energy.

c $C_6H_{12}O_6\,(aq) \rightarrow 2C_2H_5OH(l) + 2CO_2(g)$
Therefore one mole of glucose produces two moles of ethanol on fermentation.

Substitution reactions and alcohol production

- Alkanes are quite unreactive whereas the halogenation of alkanes produces compounds that are more reactive. The halogenated products can be converted to other useful chemicals via substitution reactions. For example, chloroethane can be reacted with sodium hydroxide solution to form ethanol (C_2H_5OH):

$$C_2H_5Cl(l) + NaOH(aq) \rightarrow C_2H_5OH(aq) + NaCl(aq)$$

- When 2-bromobutane reacts with sodium hydroxide solution, butan-2-ol is formed. Figure 11.17 shows a model of this reaction.

5 Fuel production

» Students compare and contrast fuels from organic sources to biofuels, including ethanol.

Fossil fuels

- A fuel is a substance that burns in air or oxygen to release useful energy. Many common fuels such as coal, petroleum and natural gas are called fossil fuels. They are formed by the fossilisation of ancient life forms. Petroleum and natural gas were discussed in Chapter 9.

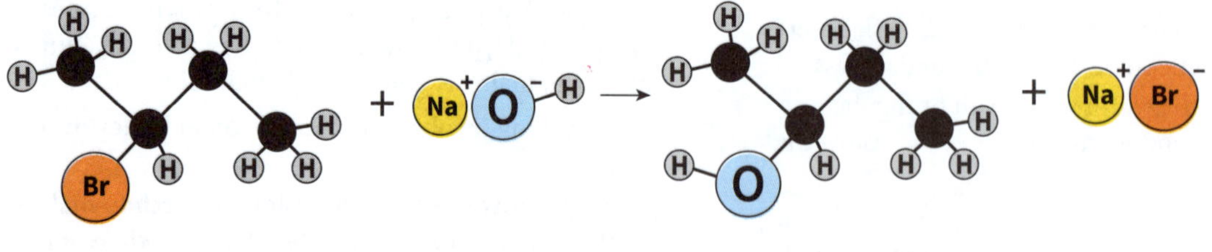

2-bromobutane

butan-2-ol

Figure 11.17 Production of butan-2-ol via a substitution reaction.

Summaries of the major ideas about fuels derived from petroleum are:

- Petroleum fractional distillation separates the mixture of hydrocarbons according to differences in boiling points. The mixture is separated into fractions containing mixtures of various hydrocarbons. The higher the boiling point range, the heavier the fraction.
- The fuel fractions include refinery gas (C1–C4 hydrocarbons), petrol (C5–C8 hydrocarbons), naphtha (C7–C13 hydrocarbons), kerosene (C11–C16 hydrocarbons) and diesel (C15–C18 hydrocarbons).
- The petrol component of gasoline is insufficient to meet the demand for car fuel and heavier fractions are cracked into smaller molecules to meet this demand.

➔ Coal is a rock that contains organic compounds that will burn in air. Coal is a major energy source in Australia. Coal is the fossilised remains of ancient swamp plants. These plants grew in the distant geological past. Most of the New South Wales and Queensland black coal began its formation in the Permian period about 250–300 million year ago, whereas Victorian brown coal is younger (~15–50 million years). The following map (Figure 11.18) shows the locations of coal basins in Australia. The amount of coal in each basin varies and mining occurs where it is economically viable.

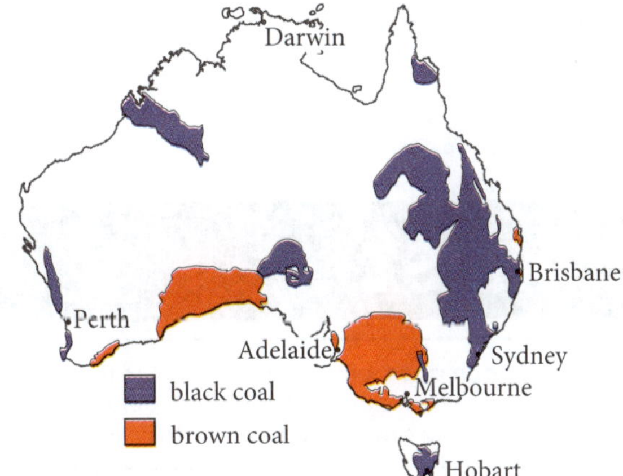

Figure 11.18 Coal basins in Australia

Source: compiled from information based on Geoscience Australia (2012), Coal, viewed online February 2018 at www.ga.gov.au/data-pubs/data-and-publications-search/publications/australian-minerals-resource-assessment/coal and blogspot https://woollydays.files.wordpress.com/2013/04/sbn160412partnerships.jpg.

➔ In order to form coal the remains of dead plants must be prevented from aerobically decomposing. If the plant debris is buried in the absence of oxygen (anaerobic conditions) it is converted gradually over millions of years to different forms or ranks of coal. These ranks of coal are peat, brown coal, black coal and anthracite coal.

- Peat
  Peat is the first stage of coal formation. Some plant leaves and stems are still observable after the relatively short periods of fossilisation after burial, usually in bogs or swamps. The excavated peat needs to be dried before it can be used as a fuel. Dried peat has a typical carbon content of 50–60% by dry weight. It is a low-grade fuel.
- Brown coal (lignite)
  As the fossilisation process continues for longer periods, peat is converted by increasing heat and pressure due to overlying sediments into brown coal. Commercial brown coal is mined mainly in Victoria and has been used for electricity production, but its use is being phased out and renewable energy sources are increasingly being used. Brown coal's typical dry weight carbon content is 70–75%. This higher carbon content and lower moisture content makes brown coal superior to peat.
- Black coal (bituminous coal)
  The continued fossilisation of brown coal over several hundred millions of years leads to the production of black coal. Over this long period, most of the moisture and volatile compounds (e.g. low molecular-weight hydrocarbons) are driven off and the black coal that forms has a high carbon content (80–90%). Commercial black coal is mined mainly in New South Wales and Queensland. Black coal is the typical fuel used in coal-fired electric power stations in these states.
- Anthracite coal
  Anthracite coal forms from black coal after further fossilisation under intense pressure and heat. It is the most metamorphosed form of coal, with the lowest percentage of volatile materials. Its typical carbon content is 90–95%. Anthracite coal can be found in Queensland.

➔ The combustion of coal as a source of electrical energy results in significant amounts of carbon dioxide into the atmosphere. Carbon dioxide is a greenhouse gas that contributes to global warming and climate change. Alternative technologies using renewable energy sources such as solar cells and wind turbines are being promoted to reduce dependence on fossil fuels.

Biofuels

➔ Supplies of fossil fuels such as coal and petroleum will become exhausted over time and so alternative fuel sources need to be developed. One alternative is to use biofuels.

➔ Biofuels include liquids such as ethanol. Ethanol can be manufactured from biomass. Biomass is mainly carbonaceous matter (e.g. polymeric carbohydrates such as cellulose and starch) derived from the photosynthetic activity of plants. The long chain carbon structure of cellulose (Figure 11.19) gives it the potential to replace petroleum as the feedstock for the production of biofuels. The cellulose is hydrolysed to produce glucose. The glucose is fermented anaerobically to produce ethanol. This preparation of ethanol from cellulose is further discussed in Chapter 12.

Cellulose—a glucose polymer

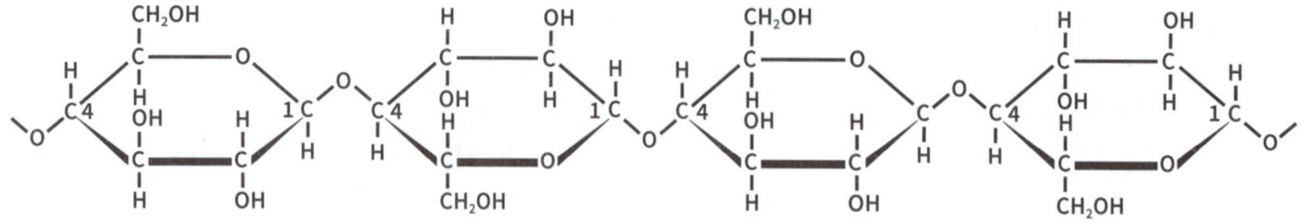

Figure 11.19 Cellulose

- Biodiesel is a biofuel manufactured from organic sources such as fats and oils. Biodiesel is a renewable alternative to petrodiesel derived from crude oil. Biodiesel production is discussed in Chapter 16.
- Ethanol is a flammable liquid that burns with a yellow-blue flame in air. Ethanol produces less energy (30 MJ/kg) on combustion than petrol (47 MJ/kg), but it burns more cleanly in air to produce less soot and carbon monoxide.
- Ethanol forms flammable air–vapour mixtures and consequently can be used as a petrol extender (e.g. E10 fuel = 10% ethanol in petrol) in car engines without engine modifications. Ethanol helps the petrol burn more efficiently and cleanly.
- A major drawback in using biofuels is the large areas of arable land that must be devoted to the growing of the biomass-crops. Energy must also be expended in fertilising and cultivating the crops and then distilling the ethanol. If this energy for cultivation and distillation is derived from fossil fuel combustion then there is virtually no reduction in carbon dioxide emissions by using biofuels. However, biofuels have much greater efficiency in energy production than fossil fuels on combustion.
- A benefit of using biofuels is that large amounts of the carbon dioxide produced on combustion is recycled via photosynthesis back into organic compounds in plants. Although this process is not perfectly carbon-neutral it is preferable to the combustion of carbon compounds from fossil sources.
- Various research projects are investigating weeds as a source of biofuels. Land that is not suitable for food crop production could be used to grow weeds that would yield considerable supplies of bioethanol on fermentation. Another area of research is the development of biofuels by farming marine algae in coastal oceans or in coastal chemical factories and ponds.

→ KEY QUESTIONS

10 Explain how sugarcane can be used to generate ethanol.

11 Identify how biofuels are produced and explain whether such fuels will contribute to global warming.

Answers ➲ p. 156

CHAPTER SYLLABUS CHECKLIST

Are you able to answer every question from the syllabus for this chapter? Tick each question as you go through the checklist if you are able to answer it. If you cannot answer a question, turn to the relevant page in the study guide to find the answer. For NESA key word meanings, go to www.educationstandards.nsw.edu.au and search 'key words'.

FOR A COMPLETE UNDERSTANDING OF THIS TOPIC:		PAGE NO.	✓
1	Can I classify alkanols as primary, secondary or tertiary?	143	
2	Can I explain the properties of alcohols in terms of their polarity?	143	
3	Can I use experimental data to calculate the enthalpy of combustion of alkanols?	145	
4	Can I write equations for combustion reactions of alkanols?	147	
5	Can I write equations for dehydration reactions of alkanols?	147	
6	Can I write equations for substitution reactions of alkanols?	147	
7	Can I write equations for oxidation reactions of alkanols?	148	
8	Can I write an equation for the anaerobic fermentation of glucose?	150	
9	Can I write equations for substitute reactions leading to the production of alcohols?	152	
10	Can I compare fossil fuels with biofuels such as ethanol?	152	

HSC EXAM-TYPE QUESTIONS

Objective-response questions (1 mark each)

1 Select the true statement concerning the production of ethanol.

A Ethanol is produced by the anaerobic fermentation of glucose derived from molasses.

B Genetically modified bacteria are used to produce ethanol by the aerobic fermentation of sugarcane solution.

C Anhydrous ethanol can be obtained by the fractional distillation of an alcohol solution.

D Cellulose can be converted to ethanol using yeast fermentation.

2 Select the true statement about fuels and their uses.

A A major advantage of fossil fuels over biofuels is that, on combustion, only carbon dioxide is produced.

B Ethanol is a non-polluting fuel used to generate electricity in power stations.

C Coal is a renewable fuel.

D Ethanol can be used as a petrol extender in car engines without engine modifications.

3 The heat of combustion of ethanol is 1367 kJ/mol. The heat of combustion in the units of MJ/kg is:

A 0.0297 MJ/kg

B 29.7 MJ/kg

C 2970 MJ/kg

D 2.97 MJ/kg

4 Pentan-2-ol was heated with an acidified solution of potassium dichromate. Name the product(s) of this reaction.

A pentanoic acid

B pentanal

C pentan-2-one

D propanoic acid and ethanoic acid

5 Ethanol is reacted with concentrated phosphoric acid. Identify the two products of this reaction.

A ethylene and water

B acetic acid and water

C ethylene and hydrogen

D ethanal and water

Extended-response questions

6 72.1 g of glucose is fermented anaerobically.

a Identify the organism used to ferment the glucose. (1 mark)

b Calculate the volume of carbon dioxide produced at 25 °C and 100 kPa, given that one mole of gas occupies 24.79 L. (3 marks)

7 The apparatus shown in Figure 11.20 was used to measure the mass of carbon dioxide produced during the anaerobic fermentation of glucose solution. Drying crystals in the first U-tube removes water vapour.

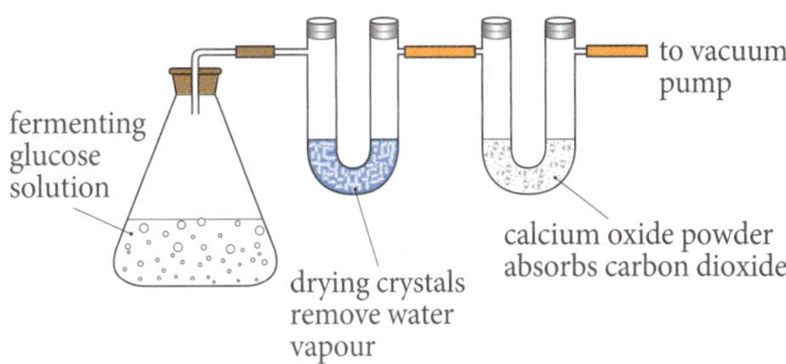

Figure 11.20 Anaerobic fermentation

Calcium oxide in the second U-tube absorbs and reacts with the carbon dioxide gas. The initial and final mass of the U-tube are recorded after fermentation is complete.

Initial mass = 88.560 g

Final mass = 89.537 g

a Write an equation for the reaction of carbon dioxide and calcium oxide. (1 mark)

b Calculate the mass of glucose that has fermented. (3 marks)

8 The table compares the experimentally measured enthalpies of combustion using spirit burners with standard literature values for ethanol and pentan-1-ol.

Alkanol	ethanol	pentan-1-ol
Experimental enthalpy of combustion (kJ/mol)	−771	−1795
Standard enthalpy of combustion (kJ/mol)	−1367	−3331

a Explain why the experimental values of the enthalpies of combustion are lower than the standard literature values. (2 marks)

b Calculate the heat released on the complete combustion of 10.0 g of ethanol and 10.0 g of pentan-1-ol. (4 marks)

c When heptan-1-ol was burnt in a spirit burner, it was found to be more difficult to ignite than other alkanols tested and the flame was quite yellow and sooty. Explain these observations. (2 marks)

9 An alkanol (X) has the molecular formula $C_4H_{10}O$. The alkanol was mixed with acidified potassium permanganate and the mixture turns from violet to colourless on heating. The liquid product (Y) of this reaction was isolated from the reaction mixture and tested with sodium carbonate solution. An effervescence of a colourless gas occurred.

a Identify the product Y. (2 marks)

b Identify the alkanol X. (2 marks)

10 Compounds W and X are isomers with a molecular formula of C_3H_8O. Compound W can be oxidised with acidified potassium dichromate to form compound Y, with a molecular formula C_3H_6O. Compound X on similar oxidation yields compound Z, with a molecular formula of $C_3H_6O_2$. Write structural formulas and name each compound. (4 marks)

ANSWERS

KEY QUESTIONS

Key questions ➲ p. 145

1 a secondary b tertiary c primary

2 Pentan-1-ol is a polar molecule whereas pentane is non-polar. The intermolecular forces between pentan-1-ol molecules are dipole–dipole forces and dispersion forces whereas dispersion forces only exist between pentane molecules. Thus the boiling point of pentan-1-ol is much higher.

3 Ethanol can hydrogen-bond with water and this interaction drives the dissolution process in water so that ethanol is soluble in water in all proportions.

Key questions ➲ p. 147

4 The heat released increases as the length of the hydrocarbon chain increases.

5 The heat is not completely absorbed by the water and the flask as much is lost to the environment. The combustion is not complete and so less heat is produced.

Key questions ➲ p. 150

6 $2C_3H_7OH(l) + 9O_2(g) \rightarrow 6CO_2(g) + 8H_2O(l)$

7 $C_3H_7OH(l) \rightarrow C_3H_6(g) + H_2O(l)$

8 $HF(g) + C_2H_5OH(l) \rightarrow C_2H_5F(g) + H_2O(l)$

9 butanoic acid

Key questions ➲ p. 154

10 The juice from the sugarcane contains sugars. Molasses is produced from this juice. The molasses solution can be anaerobically fermented using yeast. The solution of ethanol is distilled to produce hydrous ethanol, which can be dehydrated to form pure ethanol.

11 Biofuels such as ethanol are produced from biomass (e.g. plants) via anaerobic fermentation reactions. Biofuels produce carbon dioxide on combustion and so would contribute to global warming.

HSC EXAM-TYPE QUESTIONS

Objective-response questions

1 **A.** Molasses is a product of sugar production and yeast is used to anaerobically ferment the sugar to produce ethanol. **B** is incorrect as the fermentation is anaerobic. **C** is incorrect as hydrous ethanol is formed. **D** is incorrect as yeast does not ferment cellulose.

2 **D.** E10 petrol contains 10% ethanol and this can be used without engine modification. **A** is incorrect as CO and C are also produced. **B** is incorrect as liquid hydrogen is dangerous and too expensive to be used. **C** is incorrect as coal is a non-renewable fossil fuel.

3 **B.** $M(C_2H_5OH) = 46.068$ g/mol
1 mol = 46.068 g = 0.046 068 kg
1367 kJ = 1.367 MJ

Heat of combustion $= \frac{1.367}{0.046\,068} = 29.7$ MJ/kg

Thus **A**, **C** and **D** are incorrect.

4 **C.** The oxidation causes the secondary alcohol to form a ketone. **A** is incorrect as this is the product of pentan-1-ol oxidation. **B** is incorrect as pentanal cannot form from a secondary alkanol. **D** is wrong as the carbon chain is not broken in the reaction.

5 **A.** This is a dehydration reaction in which an alkene and water are formed. **B** is incorrect as acetic acid is formed by oxidation. **C** is incorrect as hydrogen is not produced on dehydration. **D** is incorrect as ethanal is an oxidation product.

Extended-response questions

6 EM Students must demonstrate a knowledge of the fermentation products and to use a balanced equation to determine the reaction stoichiometry and then relate the moles of gas to the volume of the gas.

a yeast ✓

b $C_6H_{12}O_6(aq) \rightarrow 2C_2H_5OH(l) + 2CO_2(g)$ ✓
M(glucose) = 180.156 g/mol
$n(\text{glucose}) = m/M = \frac{72.1}{180.156} = 0.400$ mol ✓
$n(CO_2) = 0.800$ mol
$V(CO_2) = (0.800)(24.79) = 19.8$ L ✓

7 EM Students need to demonstrate an understanding of the mass change in the U tube to the fermentation reaction. Students must show a mass–mole calculation to determine the mass of glucose fermented.

a $CO_2(g) + CaO(s) \rightarrow CaCO_3(s)$ ✓

b $m(CO_2) = 0.977$ g
$n(CO_2) = m/M = \frac{0.977}{44.01} = 0.0222$ mol ✓
$C_6H_{12}O_6(aq) \rightarrow 2C_2H_5OH(l) + 2CO_2(g)$
$n(\text{glucose}) = \frac{0.0222}{2} = 0.0111$ mol ✓
$M(C_6H_{12}O_6) = 180.156$ g/mol
$m(\text{glucose}) = (0.0111)(180.156) = 1.98$ g ✓

8 EM Students should review the experiment on enthalpies of combustion and consider how the flame colour of the spirit burner changed as the molar mass of the alkanol increased.

a In a spirit burner flame, the combustion is incomplete and less heat is produced compared with complete combustion. ✓ Considerable heat is lost to the surroundings. ✓

b M(ethanol) = 46.068 g/mol
$n(\text{ethanol}) = m/M = \frac{10.0}{46.068} = 0.217$ mol ✓
$q = -n.\Delta_c H = -(0.217)(-1367) = 29.5$ kJ ✓
M(pentan-1-ol) = 88.15 g/mol
$n(\text{pentan-1-ol}) = m/M = \frac{10.0}{88.15} = 0.113$ mol ✓
$q = -n.\Delta_c H = -(0.113)(-3331) = 376$ kJ ✓

c Heptan-1-ol has very low volatility as the dispersion forces are greater than in the alkanols of lower molar mass. Thus a combustible fuel vapour/air mixture is slow to form. ✓ When the mixture ignites it burns incompletely to generate a lot of soot. Soot particles colour the flame yellow. ✓

9 EM **Students need to recall the product of the reaction of primary alkanols and how the alkanoic acid formed reacts with a sodium carbonate solution.**

a Y is an alkanoic acid as the sodium carbonate solution effervesced. Y has 4 carbon atoms per molecule. ✓ It is butanoic acid ($CH_3CH_2CH_2COOH$). ✓

b X is a primary alkanol because primary alkanols can be oxidised to alkanoic acids. ✓ Thus X is butan-1-ol. ✓

10 EM **Students need to recall that primary and secondary alkanols are oxidised to form different products. The structural formulas of reactants and products should show the bonds in the alcohol and carbonyl functional groups.**

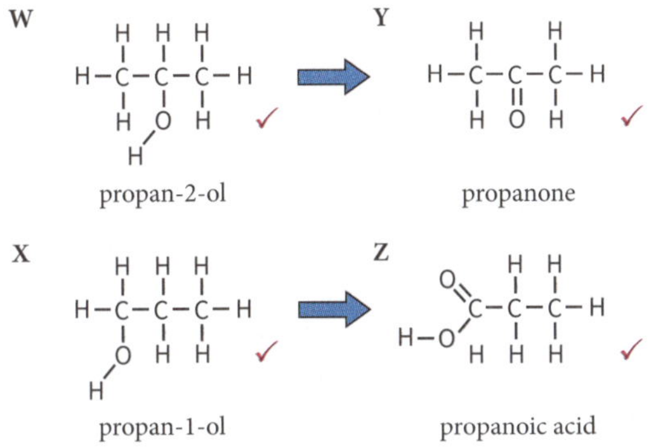

W: propan-2-ol ✓ → Y: propanone ✓

X: propan-1-ol ✓ → Z: propanoic acid ✓

Figure A11.1 Products of reactions

CHAPTER 12 REACTIONS OF ORGANIC ACIDS AND BASES

MODULE 7 ORGANIC CHEMISTRY

INQUIRY QUESTION:

What are the properties of organic acids and bases?

Organic acids such as alkanoic acids are weak acids. Organic bases such as amines are weak bases. Alcohols are neutral molecules in water solution. Alkanoic acids can react with alcohols to form molecules called *esters*. Fats are examples of complex esters. Fats react with strong bases to form soap. Alkanoic acids can react with amines to form amides. This chapter investigates the properties of these compounds.

1 Functional group properties

» Students:

- investigate the structural formulae, properties and functional group, including primary, secondary and tertiary alcohols, aldehydes and ketones, amines and amides, and carboxylic acids.
- explain the properties within and between the homologous series of carboxylic acids, amines and amides, with reference to the intermolecular and intramolecular bonding present.
- investigate the differences between an organic acid and organic base.

Hydroxyl functional group

➔ Chapters 8 and 11 investigated the structure and properties of alcohols. These concepts are summarised as follows:

- Alcohols are organic molecules containing the hydroxyl (OH) functional group.
- Alkanols are molecules in which the hydroxyl group is located on an alkanes chain.
- Alcohols can be classified as primary, secondary and tertiary.
- Primary alkanols have the general formula R_1CH_2OH (R_1 = alkyl group or alkane chain).
- Secondary alkanols have the general formula R_1R_2CHOH (R_1, R_2 = alkyl groups or alkane chains).
- Tertiary alkanols have the general formula $R_1R_2R_3COH$ (R_1, R_2, R_3 = alkyl groups or alkane chains).
- Alcohols are polar molecules as discussed in Chapter 11. This polarity increases their melting and boiling points compared with their parent alkanes. In short chain alkanols, hydrogen bonding between molecules is an important intermolecular force.
- The polarity of short chain alkanols and their ability to hydrogen bond with water makes them water soluble, but the solubility quickly decreases as the alkane chain length increases.
- Soluble alkanols form neutral solutions.

➔ Alkanols can be dehydrated by concentrated sulfuric acid to form alkenes.

Example:
butan-2-ol forms but-1-ene and but-2-ene isomers on dehydration

$$CH_3CHOHCH_2CH_3 \rightarrow CH_2CHCH_2CH_3 + H_2O$$
$$CH_3CHOHCH_2CH_3 \rightarrow CH_3CHCHCH_3 + H_2O$$

➔ Primary and secondary alkanols can be oxidised whereas tertiary alkanols cannot. Students are advised to review the information about oxidation in Chapter 11.

- Primary alkanols can be oxidised to form alkanals, which can be further oxidised by a strong oxidant to form alkanoic acids.
- Secondary alkanols can be oxidised to form alkanones.

EXAMPLE 1

Draw structural formulas and classify the following alkanols:

a **butan-2-ol**
b **2-methylbutan-1-ol**
c **2-methylpropan-2-ol**

Revise the structures of primary, secondary and tertiary alkanols

Answer:

The structures and classification are shown in Figure 12.1.

```
    H  H  H  H           H CH3 H  H            H CH3 H
    |  |  |  |           |  |  |  |            |  |  |
 H—C—C—C—C—H        H—C—C—C—C—H         H—C—C—C—H
    |  |  |  |           |  |  |  |            |  |  |
    H O-HH  H          H-O  H  H  H            H O-HH
```

a	b	c
butan-2-ol Secondary alkanol	2-methylbutan-1-ol Primary alkanol	2-methylpropan-2-ol Tertiary alkanol

Figure 12.1 Alkanols

EXAMPLE 2

Name and classify the alkanols in Figure 12.2.

```
   H  CH3 H  H  H  CH3  H             H  H   CH3 H
   |  |   |  |  |  |    |             |  |   |   |
H─ C─ C ─ C─ C─ C─ C ── C─H        H─ C─ C ─ C ─ C─H
   |  |   |  |  |  |    |             |  |   |   |
   H  H   H  H  H  O─H  H             H  O─H H   H
              a                             b
```

Figure 12.2 Name and classify the alkanols

The stem name of the alkanol must include the carbon atom bonded to the OH group

Answer:

a 2,6-dimethylheptan-2-ol (tertiary alkanol). The hydrocarbon chain is numbered right to left to give the hydroxyl group priority.

b 3-methylbutan-2-ol (secondary alkanol). The hydrocarbon chain is numbered left to right to give the hydroxyl group priority.

Carbonyl functional group

- The carbonyl functional group (CO) is present in aldehydes and ketones. This functional group was briefly examined in Chapter 8.
  - The aldehydes formed from alkanes are called *alkanals*. The general formula of alkanals is RCHO or $C_nH_{2n+1}CHO$.
    Examples:
    CH_3CH_2CHO propanal;
    $CH_3CH_2CH_2CH_2CH_2CHO$ hexanal
- The ketones formed from alkanes are called *alkanones*. The general formula of alkanones is R_1COR_2.
  Examples:
  CH_3COCH_3 = propanone;
  $CH_3CH_2COCH_2CH_2CH_3$ = hexan-3-one
- Aldehydes and ketones are polar molecules. Dispersion forces and dipole–dipole attractions are the intermolecular forces. Their melting and boiling points are higher than their parent alkanes.
  Boiling point examples:
  propane (–42.1 °C); propanal (47.9 °C);
  propanone (56.1 °C)
- Aldehydes and ketones become less soluble in water as their chain length increases. In aqueous solution the carbonyl group of low molecular weight (1–3 carbon atoms) aldehydes and ketones can hydrogen bond with the electropositive hydrogen atoms of water molecules and are quite miscible with water. Figure 12.3 shows this hydrogen-bonding interaction.

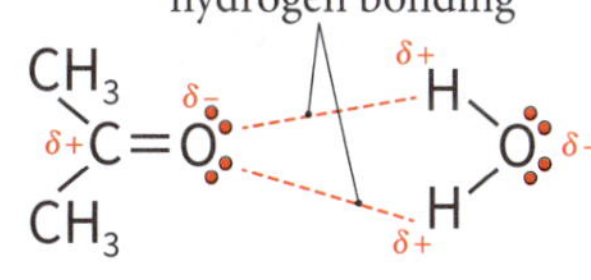

Figure 12.3 Hydrogen bonding between propanone and water

→ KEY QUESTIONS

1 **Distinguish between the hydroxyl and carbonyl functional groups.**

2 **Explain why alkanals have higher boiling points than their parent alkanes.**

3 **Name the following molecules:**
  a $CH_3CH_2CH_2CH_2COCH_3$
  b $CH_3CH_2CH(CH_3)CH_2CH_2OH$

Answers ➲ p. 171

Amine functional group

- Amines are weak organic bases.
- The amine (NH_2) functional group was briefly examined in Chapter 8. Revise that section here now.
- Amines can be classified as primary, secondary or tertiary. This classification is based on the number of hydrocarbon chains attached to the nitrogen atom of the amine functional group.
  - Primary amines have the general formula R_1NH_2 or $C_nH_{2n+1}NH_2$.
    Example:
    CH_3NH_2 methanamine
  - Secondary amines have the general formula R_1NHR_2.
    Example:
    $CH_3CH_2CH_2CH_2$–NH–CH_3; N-methylbutan-1-amine
  - Tertiary amines have the general formula $R_1NR_2R_3$.
    Example:
    $CH_3CH_2CH_2CH_2CH_2$–N–$(CH_3)_2$;
    N,N-dimethylpentan-1-amine
  - Amines are polar molecules.
  - This polarity increases their melting and boiling points compared with their parent alkanes.
  - Primary and secondary amines form hydrogen bonds between their molecules. Amines do not have as high a melting or boiling point as the corresponding alcohol because the hydrogen bonding is not as strong. Figure 12.4 shows the hydrogen-bonding interactions between methanamine molecules.
    Boiling point examples:
    ethanol (78.3 °C); ethanamine (16.6 °C)

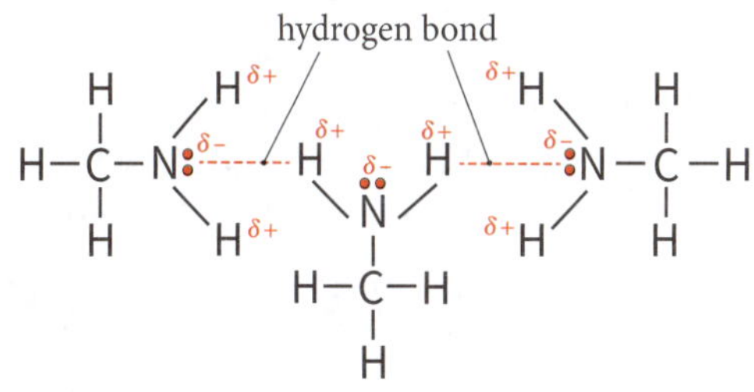

Figure 12.4 Hydrogen bonding between methanamine molecules

- In aqueous solution, low molecular weight amines can hydrogen bond with water molecules and are quite miscible with water.

➔ Amines are weak organic bases in water solution. The lone electron pair on the nitrogen atom interacts with water to form hydroxide ions. A proton is donated from the water to the amine. The amine is a Brønsted–Lowry base and the water is a Brønsted–Lowry acid.

Example:

$$CH_3NH_2(aq) + H_2O(l) \leftrightarrows CH_3NH_3^+(aq) + OH^-(aq)$$

➔ Amine solutions are neutralised by solutions of hydronium ions. Hydrochloric acid neutralises a solution of ethanamine to form a solution of ethylammonium chloride.

Example:

$CH_3CH_2NH_2(aq) + HCl(aq))$
$\rightarrow CH_3CH_2NH_3Cl(aq)$ ethylammonium chloride

Amide functional group

➔ Amides are weak organic bases and they are derivatives of carboxylic acids. The amide group is located at the end of the hydrocarbon chain. The general formula of a primary amide is R-$CONH_2$.

Example:

$CH_3CH_2CH_2CH_2CONH_2$ = pentanamide

- Amides are very polar molecules. Their boiling points are significantly higher than the equivalent alcohol or carboxylic acid.

  Boiling point examples:
  methanol (64.7 °C); formic acid (100.8 °C)
  methanamide (210 °C)

- Low molecular weight amides are soluble in water due to their ability to hydrogen bond with water, as shown in Figure 12.5.

Figure 12.5 Amide hydrogen bonding with water molecules

➔ Amides are very weak bases in water solution. They are much less basic than amines. Hydrochloric acid reacts with propanamide to produce propanoic acid and ammonium chloride:

$CH_3CH_2CONH_2(aq) + HCl(aq) + H_2O(l)$
$\rightarrow CH_3CH_2COOH(aq) + NH_4Cl(aq)$

Carboxyl functional group

➔ Carboxylic acids are organic molecules containing the carboxyl (COOH) functional group. As discussed in Chapter 8, alkanoic acids are carboxylic acids that have a carboxyl group at the end of an alkane chain.

Example:

$CH_3CH_2CH_2CH_2CH_2CH_2COOH$ = heptanoic acid

- Alkanoic acids are highly polar molecules. Their melting and boiling points are significantly higher than the equivalent alkanol.

  Boiling point examples:
  ethanol (78.3 °C); acetic acid (117.9 °C)

- Formic acid and acetic acid form strong hydrogen bonds between their molecules, resulting in the formation of **dimers**, as shown in Figure 12.6. The boiling points of formic acid and acetic acid are raised by this extensive hydrogen bonding.

dimer: two structurally similar molecules (called monomers) that are joined by covalent bonds or intermolecular forces such as hydrogen bonds

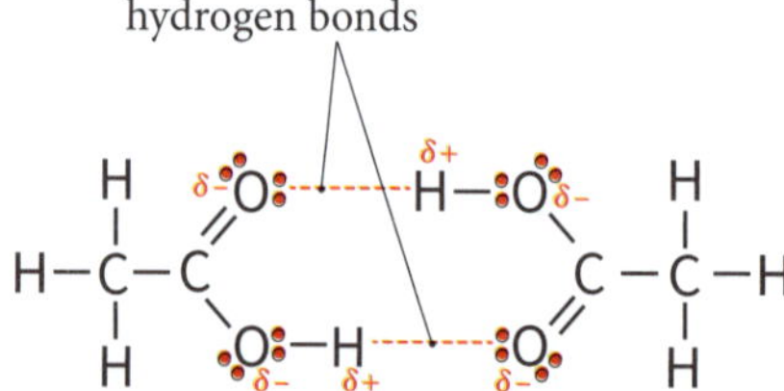

Figure 12.6 Acetic acid dimers

- Low-molecular weight alkanoic acids are soluble in water due to their ability to hydrogen bond with water. As the chain length increases they become decreasingly soluble.

➔ Some organic acids such as formic acid and acetic acid were investigated in Chapter 7. These acids are weak acids and in water solution they are incompletely dissociated:

Formic acid: $HCOOH(aq) + H_2O(l)$
$\leftrightarrows H_3O^+(aq) + HCOO^-(aq)$

Acetic acid: $CH_3COOH(aq) + H_2O(l)$
$\leftrightarrows H_3O^+(aq) + CH_3COO^-(aq)$

➔ The smaller the value of the acid dissociation constant (K_a), the weaker the acid:
Formic acid: $K_a = 1.8 \times 10^{-4}$
Acetic acid: $K_a = 1.7 \times 10^{-5}$

Therefore, acetic acid is weaker than formic acid. As the hydrocarbon chain length increases the alkanoic acids become weaker. Solubility in water also decreases with increasing length of the hydrocarbon chain. The first four alkanoic acids are completely miscible in water whereas pentanoic acid (C_4H_9COOH) has a solubility of 5.0 g/100 g water and hexanoic acid ($C_5H_{11}COOH$) has a solubility of 1.1 g/100 g water.

- Alkanoic acids are neutralised by alkaline solutions to form salts and water. Potassium hydroxide solution neutralises an aqueous solution of propanoic acid to form a solution of potassium propanoate.

$$CH_3CH_2COOH(aq) + KOH(aq) \rightarrow KCH_3CH_2COO(aq) + H_2O(l)$$

- Alkanoic acids react with amines to form amides. The reaction mixture is heated to drive off the water produced in the reaction. Figure 12.7 shows the structural formulas of an alkanoic acid reacting with a primary amine to form a secondary amide and water. This reaction is an example of a **condensation reaction**.

condensation reaction: a condensation reaction is one in which two molecules combine together with the elimination of a smaller molecule such as water

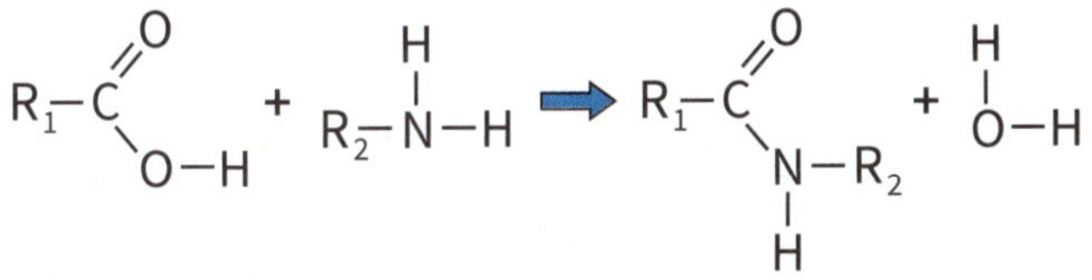

```
      O              H                  O         H
     //              |                 //         |
R1—C       +    R2—N—H    ➡    R1—C        +   O—H
     \                                 \
      O—H                               N—R2
                                        |
                                        H
```

Figure 12.7 Formation of an amide

EXAMPLE 3

Name the following molecules and classify them according to their functional group.

a $CH_3CH_2CH_2CH_2CH_2NH_2$
b $CH_3CH_2CH_2COOH$
c $CH_3CH_2CONH_2$

Identify the parent alkane

Answer:

a pentanamine (amine)
b butanoic acid (alkanoic acid)
c propanamide (amide)

EXAMPLE 4

Amino acids are the building blocks of proteins. Amino acid molecules contain the amine and carboxyl functional groups. Two common amino acids are alanine (2-aminopropanoic acid) and serine (2-amino-3-hydroxypropanoic acid). The amino group of serine reacts with carboxyl group of alanine to form a molecule called *alanylserine*, as well as water. This reaction is an example of a condensation reaction. Write a structural equation for this reaction.

Water is formed from the OH of the carboxyl group and an H atom from the amine group

Answer:

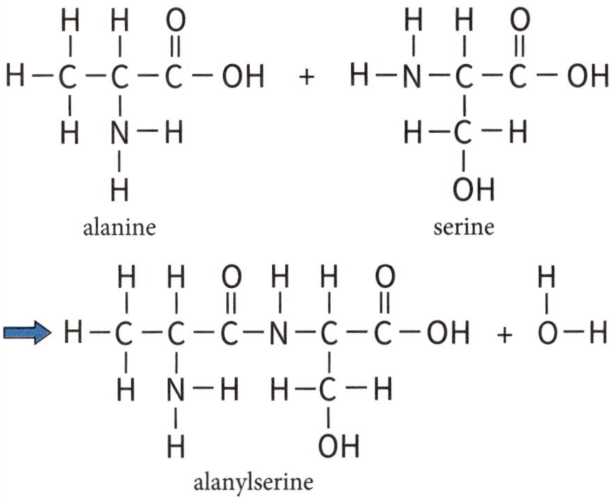

```
   H  H  O                 H  H  O
   |  |  ||                |  |  ||
H—C—C—C—OH    +    H—N—C—C—OH
   |  |                       |
   H  N—H                  H—C—H
      |                       |
      H                       OH
   alanine                  serine

      H  H  O  H  H  O               H
      |  |  ||  |  |  ||              |
➡ H—C—C—C—N—C—C—OH    +    O—H
      |  |        |
      H  N—H   H—C—H
         |        |
         H        OH
         alanylserine
```

Figure 12.8 Condensation reaction

→ KEY QUESTIONS

4 **Name the following molecules:**
a $CH_3CH_2CH_2CONH_2$
b $CH_3CH(CH_3)CH_2CH_2CH_2CH_2CH_2NH_2$

5 **Draw the structural formula for pent-2-yne.**

Answers ➲ p. 171

2 Esters

» Students investigate the production, in a school laboratory, of simple esters.

Preparation of esters

- Alcohols and carboxylic acids can react to form compounds called *esters*. These reactions are examples of condensation reactions in which the two reactants combine, with the elimination of a small molecule such as water.
- When alkanols and alkanoic acids condense the ester that forms is called an *alkyl alkanoate*. The reaction does not proceed to completion. An equilibrium is established:

alkanol + alkanoic acid ⇆ alkyl alkanoate + water

Example:

$$CH_3CH_2OH + CH_3CH_2CH_2COOH \rightleftharpoons CH_3CH_2CH_2COOCH_2CH_3 + H_2O$$

Figure 12.9 shows a structural equation for the esterification reaction.

```
    H   H              H   H   H   O
    |   |              |   |   |   ||
H — C — C — O — H  +  H — C — C — C — C — O — H
    |   |              |   |   |
    H   H              H   H   H
    ethanol              butanoic acid

        H   H   H   O       H   H            H
        |   |   |   ||      |   |            |
⇆   H — C — C — C — C — O — C — C — H   +   O — H
        |   |   |           |   |
        H   H   H           H   H
              ethyl butanoate                water
```

Figure 12.9 Esterification

➔ Radioisotope studies show that the atoms in the water that is formed come from the OH of the carboxyl group of the alkanoic acid and the H of the hydroxyl group in the alkanol.

➔ Figure 12.10 shows the structure of an ester. The R_1 group is part of the original alkanoic acid and the R_2 group is part of the original alkanol. The COO group is the **ester** functional group.

> **ester:** the condensation product of an alcohol and a carboxylic acid

```
        O
       //
R1 — C
       \
        O — R2
```

Figure 12.10 Ester structure

➔ The following rules can be used to name esters. Nomenclature rules for alkyl alkanoates (esters):

- Count the number of carbon atoms in the alkyl group that is derived from the original alkanol. Name this alkyl group by deleting the *-anol* suffix from the alkanol and replacing it with the suffix *-yl* (e.g. ethanol becomes ethyl).
- Identify the number of carbon atoms in the alkanoate chain that is derived from the alkanoic acid. Name this alkanoate chain by deleting the *-oic acid* suffix and replacing it with the suffix *-oate* (e.g. propanoic acid becomes propanoate).
- The name of the ester is two separate words. The first name comes from the alkanol and the second from the alkanoic acid.

  Example:
  $CH_3CH_2CH_2COOCH_3$
  There is one carbon atom in the alkyl group.
  Name of alkyl group = methyl
  There are four carbon atoms in the alkanoate group.
  Name of the alkanoate group = butanoate
  Ester name = methyl butanoate

➔ **Esterification** is the procedure used to make an ester. The reaction is kinetically slow and the yield of ester is quite variable for different esters.

> **esterification:** the preparation of an ester by the reaction of a carboxylic acid with an alcohol

- The rate of the reaction can be increased by moderate heating (using an electric heating mantle) as well as adding a catalyst such as concentrated sulfuric acid. The concentrated sulfuric acid also acts as a dehydrating agent to a limited extent (due to the small amounts used) to remove water as it forms. This shifts the equilibrium to increase the yield of the ester.
- The equilibrium yield of ester can be increased by adding an excess of one reactant (e.g. alkanoic acid).
- To avoid a loss of the volatile components the reaction is performed using a reflux condenser. During **refluxing** the rising hot vapours condense and droplets are returned to the reaction flask. The condenser is open to the atmosphere and so dangerous vapour pressure build-up is avoided. Cooling water flows through a hose from the water tap through the outer tube of the condenser and the heated water exits into the sink through a second hose.

> **refluxing:** the heating of a mixture of liquids in a flask with an attached condenser such that the rising vapours are cooled and condensed back to a liquid in order to prevent the loss of volatile reactants or products

- Small pieces of crushed ceramic are added to the reaction flask. These are called *boiling chips*. They are used to prevent dangerous vibrations during reflux. The ceramic has a high surface area, which promotes the formation of small vapour bubbles and therefore even boiling. Without these chips large vapour bubbles form that cause the apparatus to vibrate.

➔ Figure 12.11 shows an apparatus in which an ester can be prepared. Do not use a Bunsen burner as the risk of fire is increased if vapours escape from the condenser. An electric heating mantle or electric hotplate is safer.

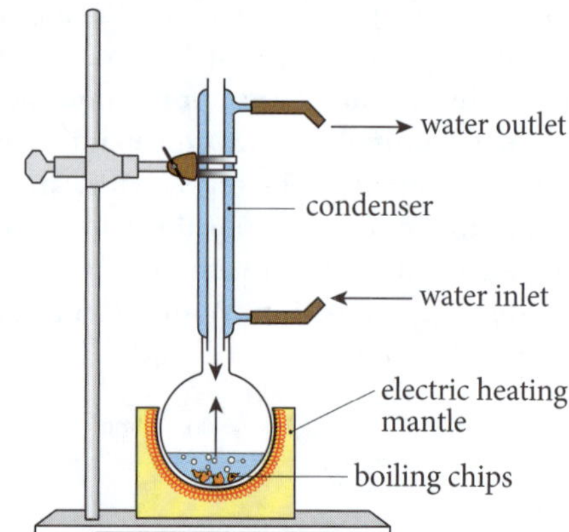

Figure 12.11 Reflux apparatus

- Once equilibrium is achieved the reaction mixture is poured into water to separate the reactants from the ester. The ester is less polar and less soluble in the water than the reactants. The mixture is placed in a separating funnel to remove the ester layer (Figure 12.12). The crude ester is further purified by washing with dilute sodium carbonate solution, which removes any acidic contaminants. Finally the ester can be distilled to produce a pure sample.

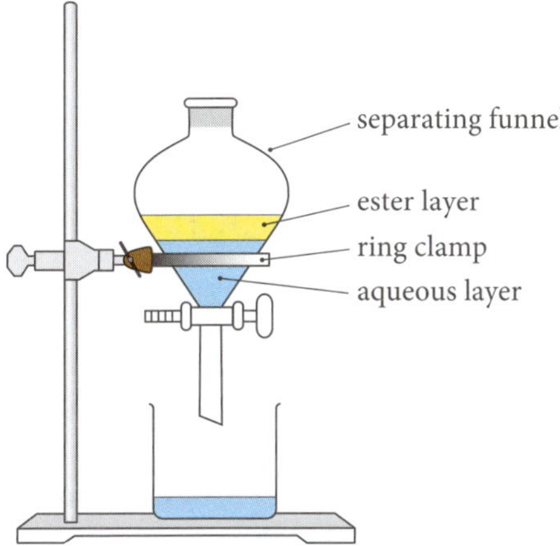

Figure 12.12 Separating ester layer

Properties and uses of esters

- Esters are less polar than the alkanoic acids or alkanols. Esters also do not hydrogen bond because they have no hydrogen atoms attached to an electronegative oxygen atom.
- Table 12.1 demonstrates that esters with the same molecular weight as alkanoic acids and alkanols have the lowest boiling points.

Table 12.1 Boiling point comparisons

Molecule	Molecular weight (u)	Boiling point (°C)
propanoic acid	74.1	141
butan-1-ol	74.1	118
methyl acetate	74.1	57

- Low molecular weight esters are water soluble. As the molecular weight increases the esters become immiscible in water. High molecular weight esters are solids rather than liquids.
- Esters have a wide variety of properties and uses. Esters occur widely in nature. They are also manufactured for a wide variety of uses, including flavourings in food and fragrances in perfumes. For example, 1-octyl acetate ($CH_3COOC_8H_{17}$) is used as a synthetic orange flavour in foods.

  Esters are also used as solvents in industry. Esters can be used as emulsifying agents in cosmetics and in food.

FIRSTHAND INVESTIGATION

Preparation of an ester

» Students investigate the production, in a school laboratory, of simple esters.

Aim

to prepare ethyl acetate

Equipment and chemicals

Glacial acetic acid (teacher use only)
Ethanol
Ethyl acetate
Concentrated sulfuric acid (teacher use only)
Round-bottom boiling flask
Reflux condenser/hoses
Electric heating mantle or hotplate
Retort stand, boss heads and clamps
Measuring cylinder
Boiling chips

Safety issues

Wear safety glasses throughout this experiment.
Sulfuric acid and glacial acetic acid are corrosive.
These can be used only by teachers.

Organic chemicals are flammable. Do not allow liquids or vapours to come into contact with sparks or flames.

Method

1. Measure out 10 mL of ethanol and add it to the flask. The teacher will add 30 mL of glacial acetic acid to the flask, followed by 2 mL of concentrated sulfuric acid.
2. Add two or three boiling chips to the flask.
3. Set up the flask and condenser for reflux (Figure 12.10). Support the condenser and flask with a retort stand, boss heads and clamps.
4. Ensure that cold water is flowing from the tap to the base of the condenser jacket and then out the top and back into the sink.
5. Heat the flask and ensure that vapour condensation occurs no higher than halfway up the condenser.
6. Continue reflux for at least 30 minutes.
7. Turn off the heat and allow the apparatus to cool until cold.
8. Remove the boiling flask and carefully pour the reaction mixture into a beaker containing 80 mL of water. Stir with a glass rod. The mixture contains the ester plus unreacted alkanol and alkanoic acid. Compare the odour of this solution with a sample of pure ethyl acetate. The ester will have quite a distinct odour similar to nail polish remover that contains ethyl acetate.

Conclusion

Write a suitable conclusion for this investigation.

EXAMPLE 5

Ethyl butanoate has a pineapple flavour. The ester is prepared from ethanol and butanoic acid. The table gives information about the boiling points of the reactants and the product.

Chemical	ethanol	butanoic acid	ethyl butanoate
Boiling point (°C)	78	164	121

a **Draw a structural equation for the reaction.**

b **Explain how the ester can be extracted from the equilibrium mixture.**

Consider the differences in boiling points

Answer:

a Figure 12.13 shows structural equation for the esterification equilibrium.

```
   H  H  H  O             H  H
   |  |  |  ||            |  |
H– C– C– C– C–OH   +   H– C– C–OH
   |  |  |                |  |
   H  H  H                H  H

     H  H  H  O     H  H         H
     |  |  |  ||    |  |         |
⇆ H– C– C– C– C–O– C– C–H   +   O–H
     |  |  |        |  |
     H  H  H        H  H
```

Figure 12.13 Esterification equation

b The ester has a different boiling point than the reactants. Fractional distillation of the reaction mixture can be used to extract the ester. The ethanol will distil off first and the ester will distil off second.

EXAMPLE 6

Figure 12.14 shows the structural formula of an organic compound (X) that was synthesised in a laboratory.

```
X
   H  H  O     H  H  H  H
   |  |  ||    |  |  |  |
H– C– C– C– O– C– C– C– C–H
   |  |        |  |  |  |
   H  H        H  H  H  H
```

Figure 12.14 Structural formula of compound X

Figure 12.15 shows the structural formulas of organic compounds (A, B, C, D, E) involved in the synthesis of X as well as the reagents (F, G, H) used.

```
(A)  H  H  O             (B)    H  H  H  H
     |  |  ||                   |  |  |  |
  H– C– C– C–OH             HO– C– C– C– C–H
     |  |                       |  |  |  |
     H  H                       H  H  H  H

(C)  H     H  H          (D)    H  H  H
     |     |  |                 |  |  |
  H– C= C– C– C–H            H– C– C– C–OH
        |  |  |                 |  |  |
        H  H  H                 H  H  H

(E)  H  H  H  H          (F) concentrated H2SO4
     |  |  |  |          (G) KMnO4/H+
  H– C– C– C– C–H        (H) H2O/H3PO4
     |  |  |  |
     H  OH H  H
```

Figure 12.15 Reactants and reagents

a **Use this information to complete the following flowchart (Figure 12.16) by inserting letters in the correct box.**

b **Name the compounds A, B, C, D, E and X.**

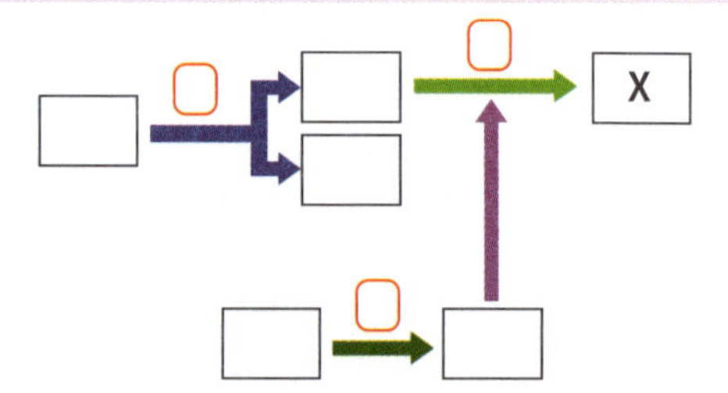

Figure 12.16 Flowchart

Review reactions discussed in Chapter 10

Answer:

a The alkene C undergoes an addition reaction with water (catalysed by phosphoric acid H) to form two alkanol position isomers (B and E), as summarised in Figure 12.17. The alkanol D is oxidised by acidified potassium permanganate solution G to form an alkanoic acid A. Then A and B undergo esterification (catalysed by concentrated sulfuric acid F) to form the ester X.

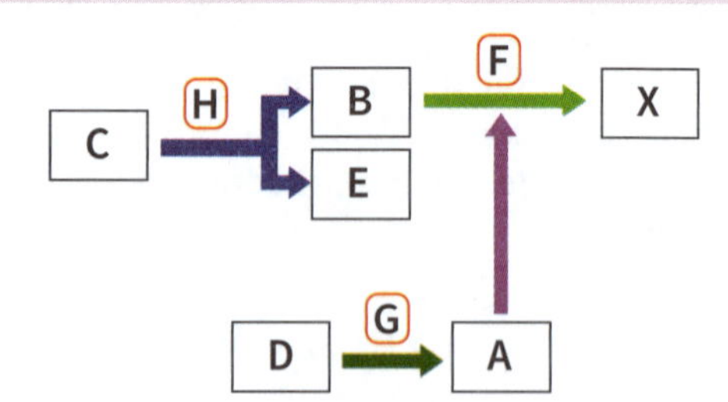

Figure 12.17 Completed flowchart

b A = propanoic acid; B = butan-1-ol; C = but-1-ene; D = propan-1-ol; E = butan-2-ol
X = 1-butyl propanoate

→ KEY QUESTIONS

6 **Name the ester formed in the reaction of hexanoic acid and butan-2-ol.**

7 **Explain why esters with the same molecular weight as alkanoic acids and alkanols have the lowest boiling points.**

8 **Explain the purpose of concentrated sulfuric acid and boiling chips in the preparation of an ester.**

Answers ➲ p. 171

3 Soaps and detergents

» Students investigate the structure and action of soaps and detergents.

Soap

→ Soap is an important cleaning product.

→ **Saponification** is the process used to manufacture soap.

→ Saponification involves the alkaline hydrolysis of fat. Fats are esters typically formed from glycerol (propan-1,2,3-triol) and long chain carboxylic acids called **fatty acids**.

saponification: the chemical reaction in which fatty esters are hydrolysed in alkaline solution to produce soap
fatty acid: a carboxylic acid molecule with a long hydrocarbon chain (~12–20 carbon atoms)

→ Figure 12.18 shows the structures of glycerol, stearic acid (a common fatty acid) and the fat called *tristearin* (or glyceryl tristearate).

$CH_3(CH_2)_{15}CH_2-COOH$

stearic acid

H_2C-OH
$HC-OH$
H_2C-OH

glycerol

$H_2C-OOC-CH_2(CH_2)_{15}CH_3$
$HC-OOC-CH_2(CH_2)_{15}CH_3$
$H_2C-OOC-CH_2(CH_2)_{15}CH_3$

glyceryl tristearate

Figure 12.18 Stearic acid, glycerol and glyceryl tristearate

→ Sodium stearate (or sodium octadecanoate) is a common soap. Soap is the salt of the fatty acid. Its structure is shown in Figure 12.19.

$H-\left(CH_2\right)-\left(CH_2\right)_{16}-C(=O)-O^-Na^+$

Figure 12.19 Soap structure

→ The soap is made by heating tristearin fat with concentrated sodium hydroxide solution. The equation for this reaction is shown in Figure 12.20.

$H_2C-OOC-CH_2(CH_2)_{15}CH_3$
$HC-OOC-CH_2(CH_2)_{15}CH_3 \quad + \quad 3NaOH$
$H_2C-OOC-CH_2(CH_2)_{15}CH_3$

glyceryl tristearate — sodium hydroxide

H_2C-OH
$\rightarrow \quad HC-OH \quad + \quad 3NaCH_3(CH_2)_{15}CH_2COO$
H_2C-OH

glycerol — sodium stearate soap

Figure 12.20 Production of soap

→ The soap anion consists of a long hydrocarbon chain called the *tail*, and a carboxylate end-group (COO^-) called the *head*. In a soap–water solution the long tails of the soap anion have little affinity for polar water molecules. They are said to be *hydrophobic* (water-hating).

→ The charged head group attracts the water dipoles and is said to be *hydrophilic* (water-loving).

→ In a dilute soap–water solution the soap anions move to the water surface and the soap tails form an oily layer on the surface while the head groups interact with water dipoles. At higher concentrations of soap, the soap forms stable spherical structures called **micelles** in the bulk of the water (Figure 12.21). In the centre of the micelle the tails of the soap anions are stabilised by dispersion forces. At the surface of the micelle the charged head groups attract water dipoles.

micelle: a spherical structure (between 1 and 1000 nm diameter) in which large numbers of soap anions cluster together to lower their potential energy

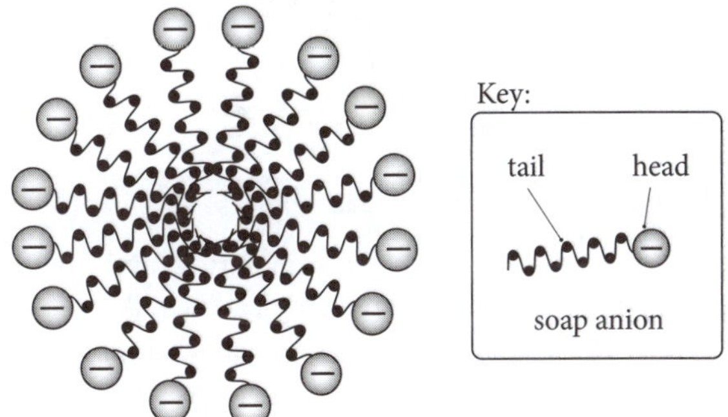

Figure 12.21 Soap micelle

→ Soap is an efficient cleaning agent. Soap powder is added to a washing machine to remove oily stains from clothing. The tails of the soap anions penetrate the oil stain and, following agitation, the oil is lifted from the fabric and oil–soap micelles form in the bulk of the washing water. Figure 12.22 shows the structure of the oil–soap micelle. The mixture of oily soap micelles and water is called an **emulsion**. The negative charges of the head groups of the soap prevent micelles from combining. The oily micelles remain in the water until the washing water is drained out of the machine and fresh water is added in the next cycle.

emulsion: a colloidal dispersion of one liquid in another

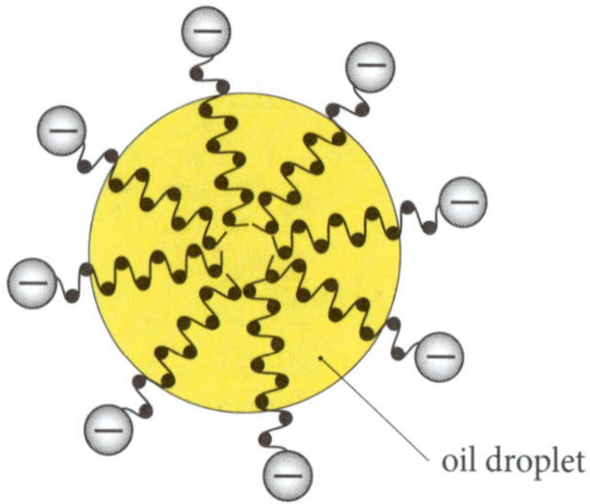

Figure 12.22 Oil–soap micelle

→ There are a number of procedures used in commercial soap making. One of these is the kettle-boiled batch process, in which stainless steel containers called *kettles* are used. Figure 12.23 shows the essential features of the process.

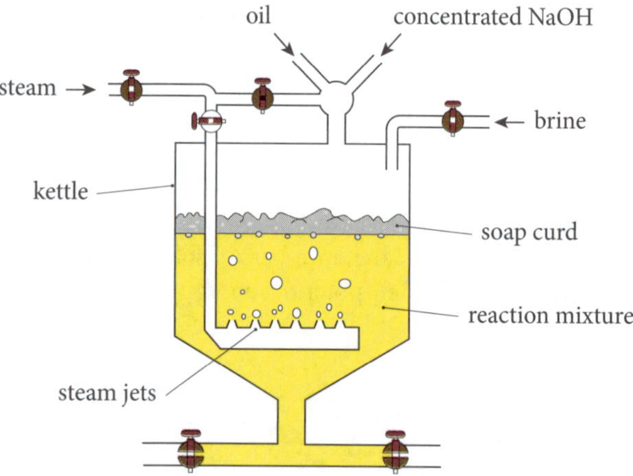

Figure 12.23 Kettle-boiled batch process

→ The stages of this process are summarised in the flowchart in Figure 12.24. Steam is used to heat the mixture of fats and alkali. Concentrated brine is used to precipitate out the soap. Glycerol is also extracted and used in many personal care products. The crude soap curd has to be washed to remove alkali and excess salt impurities.

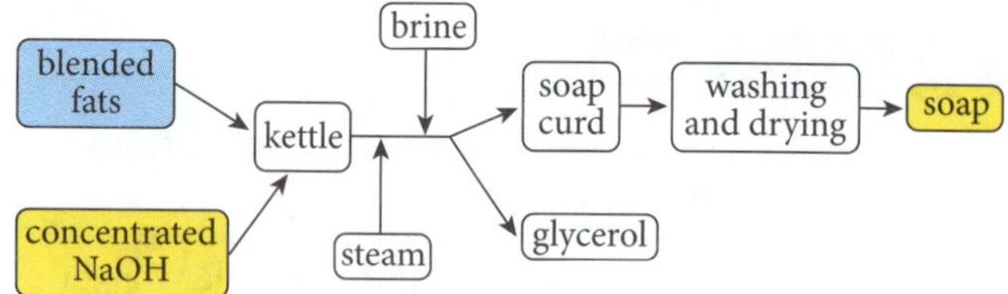

Figure 12.24 Production of soap flowchart

→ Soap can be made in the school laboratory. A solid fat such as coconut fat or lard (pig fat) makes good soap as the soap is hard and easier to extract. Oils, such as olive oil, make softer soap. Ethanol is added to help the fat and aqueous solution form an emulsion so that the alkali can attack small droplets of fat. The steps are:

1. Wear safety glasses as concentrated NaOH is used.
2. Weigh about 10 g of fat into a 500 mL beaker.
3. Prepare a concentrated alkaline solution of NaOH, to which ethanol is added.
4. Add the alkaline solution to the fat in the beaker.
5. Cover the beaker with a clock glass. Heat on a hotplate for 30 minutes, while stirring occasionally.
6. Remove the beaker from the heat and allow to cool.
7. 'Salt out' the soap with saturated salt water (brine). Thick curds of soap should form on the top of the mixture.
8. Filter using a Buchner funnel and vacuum pump. Wash the soap curds with a little water. Allow the soap to dry.

EXAMPLE 7

Glyceryl trilaurate is a fat formed when glycerol ($C_3H_5(OH)_3$) and 3 molecules of lauric acid ($CH_3(CH_2)_{10}COOH$) are esterified. 500 g of glyceryl trilaurate is saponified in excess sodium hydroxide until all the fat has been converted into glycerol and soap.

a **Write a balanced equation for this saponification reaction.**

b **Calculate the theoretical yield of soap.**

Count the number of carbon, hydrogen and oxygen atoms in the fat and the soap

Answer:

a Formula of fat = $C_3H_5(OOC(CH_2)_{10}CH_3)_3$
$= C_{39}H_{74}O_6$

Balanced equation:

$$C_3H_5(OOC(CH_2)_{10}CH_3)_3(s) + 3NaOH(aq) \rightarrow C_3H_5(OH)_3(aq) + 3NaCH_3(CH_2)_{10}COO(s)$$

b Calculate the molar mass of the glyceryl trilaurate fat.
$M(\text{fat}) = (39 \times 12.01) + (74 \times 1.008) + (6 \times 16.00)$
$= 638.982$ g/mol

Calculate the molar mass of the soap ($NaC_{12}H_{23}O_2$).
$M(\text{soap}) = (22.99) + (12 \times 12.01) + (23 \times 1.008) + (2 \times 16.00)$
$= 222.294$ g/mol

Use the stoichiometry to determine the mass relationship:
1 mole of fat produces 3 moles of soap
638.982 g of fat produces 3 × 222.294 = 666.882 g of soap
Calculate the yield from 500 g of fat.

$$M(\text{soap}) = \left(\frac{666.82}{638.982}\right) \times 500 = 522 \text{ g}$$

Thus the theoretical mass of soap produced = 522 g

Detergents

- Soap is classified as a non-petrochemical **detergent**.

detergent: a cleaning agent

- The petrochemical industry has developed a range of synthetic detergents to provide alternatives to soap. Soap has the disadvantage that it will not lather if the water contains high levels of calcium ions. Such water is called *hard water*. Magnesium ions and aluminium ions can also cause hardness in water. If the concentration of calcium ions is less than 75 ppm then the water is called *soft water* and soap will lather. In hard water the soap anions precipitate as a calcium salt:

$$2NaCH_3(CH_2)_{10}COO(aq) + Ca^{2+}(aq) \rightarrow Ca(CH_3(CH_2)_{10}COO)_2(s) + 2Na^+(aq)$$

- Petrochemical detergents, however, are generally able to lather in hard water. Synthetic detergents are manufactured from organic molecules derived from petroleum cracking. They can be classified into a number of classes:
  - anionic detergents—the head group is anionic
  - cationic detergents—the head group is cationic
  - non-ionic detergents—the head group is polar
- Figure 12.25 shows some typical synthetic detergents in each class.

Anionic detergent: $H-(CH_2)_{17}-C_6H_4-SO_3^-Na^+$
sodium stearyl benzene sulfonate

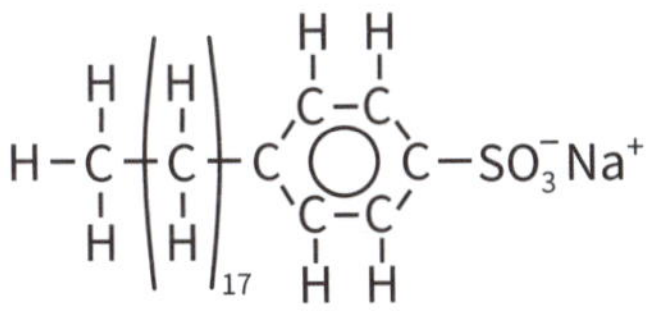

Cationic detergent: $H-(CH_2)_{11}-N^+(CH_3)_2-CH_3Cl^-$
lauryltrimethylammonium chloride

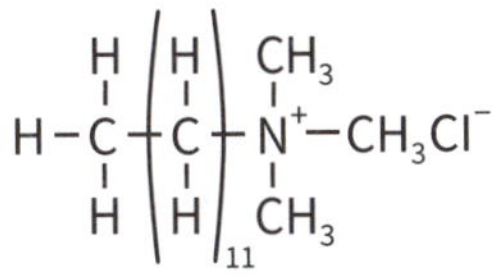

Non-ionic detergent: $H-(CH_2)_{15}-O-(CH_2-CH_2-O)_n-H$
palmityl alcohol ethoxylate

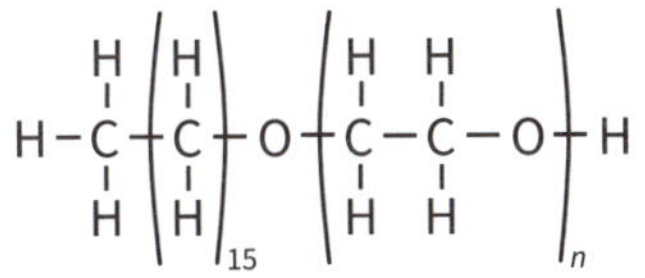

Figure 12.25 Detergent classes

- Anionic detergents are strongly foaming, like soap. Historically these detergents were made from whale oil, coconut oil and palm oil, but in modern times they are produced from petrochemicals. The anionic detergents produced before the 1960s caused considerable problems as their foams created problems in rivers and sewage works. They did not biodegrade readily due to their highly branched hydrocarbon tails.
- Modern anionic detergents have non-branched alkyl chains so that they can be readily biodegraded by microbes when waste water is discharged. These detergents are widely used in laundry detergents, dishwashing liquids and oven cleaners.
- Cationic detergents are used as fabric softeners and conditioners, and hair conditioners. Cationic detergents have biocidal properties as they bind to the negatively charged membrane surfaces of bacterial cells where they disrupt the cellular processes.
- Non-ionic detergents have hydrophilic head groups. They are widely used in automatic dishwashing detergents, paints, cosmetics and adhesives as they improve the contact between polar and non-polar substances or surfaces. Non-ionic detergents have low-lathering properties. They are often blended with anionic or cationic detergent to reduce the foaming level of the anionic component.
- Detergents contain chemicals called *builders*. The first compounds that acted as builders were phosphate and polyphosphate compounds that softened the water and made it slightly alkaline. These builders complexed with mineral cations in hard water and prevented them from flocculating any clay colloids that may have been present in the water. Flocculated clay particles can soil the garments being washed.
- The discharge of phosphate-containing detergents into waterways can lead to **eutrophication**. The phosphates and polyphosphates in detergent waste water can lead to algal blooms and the depletion of oxygen levels in waterways when the algae die and decompose. Detergents are now marketed as 'phosphate free' as they contain alternative compounds called *zeolites* that do not cause eutrophication. Sodium zeolite is a sodium aluminium silicate that replaces any calcium ions with sodium ions in the washing water.

eutrophication: the presence of nutrients such as nitrate, ammonium and phosphate ions in waterways causes aquatic plants and algae to grow abundantly

→ KEY QUESTIONS

9 **Explain how soap removes an oily stain from a fabric during the washing process.**

10 **Identify the chemicals required for the preparation of soap.**

11 **Classify the following compounds as anionic, cationic or non-ionic detergents:**

a $CH_3(CH_2)_{16}O(CH_2CH_2O)_3H$

b $CH_3(CH_2)_{14}(C_6H_4)SO_3K$

c $CH_3(CH_2)_{12}N(CH_3)_3Br$

Answers ⊃ p. 171

4 Chemical synthesis

» Students draft and construct flowcharts to show reaction pathways for chemical synthesis, including those that involve more than one step.

→ This section investigates some industrial processes involving chemical synthesis.

Bioethanol production from cellulosic biomass

→ Biomass is biological material derived from living things. The great bulk of terrestrial biomass is found in plants in the form of polymers such as cellulose, hemicellulose, lignin and starch.

→ Bioethanol can be obtained from cellulose via series of chemical reactions. The steps can be summarised as follows. Figure 12.26 shows a flowchart for this process:

- Cellulosic biomass is obtained from a number of sources such as bagasse (sugarcane waste) and corn fibre.
- The biomass is heated with water and acid to prepare the cellulose for the next enzymic step.
- The cellulose $(C_6H_{10}O_5)_n$ is enzymatically hydrolysed using cellulolytic enzymes to break down the cellulose into simple sugars:

  $$(C_6H_{10}O_5)_n(s) + nH_2O(l) \rightarrow nC_6H_{12}O_6(aq)$$

- The simple sugars (e.g. glucose) are anaerobically fermented using genetically modified yeasts or bacteria to produce a solution of ethanol in water.

  $$C_6H_{12}O_6(aq) \rightarrow 2C_2H_5OH(aq) + 2CO_2(g)$$

- The ethanol solution is distilled to produce hydrous ethanol (~95% ethanol)
- Anhydrous ethanol is produced by removal of the water using water-absorbent compounds (molecular sieves such as zeolites).
- The anhydrous ethanol is a biofuel.

→ Further information about addition polymerisation can be found in various online videos. For example, enter the following titles in a search bar:

- bioethanol production from rice straw
- biofuel and ethanol.

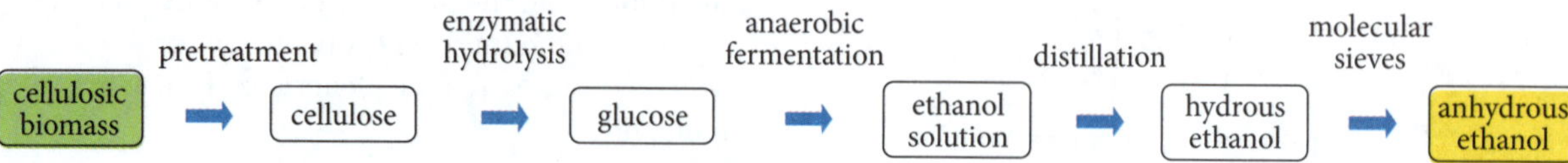

Figure 12.26 Production of ethanol

Polyvinyl acetate production

→ Polyvinyl acetate is a synthetic polymer used as a wood glue or other adhesive. It is also used in the preparation of latex paints and in the lamination of metal foils.

→ The steps used in its industrial production from crude oil are summarised below.

- Crude oil undergoes fractional distillation and the naphtha hydrocarbon fraction (7–13 carbon atoms) are collected.
- The naphtha fraction undergoes thermal (steam) cracking (~800 °C) to produce smaller molecules, including ethylene gas.
- The ethylene gas is reacted with acetic acid and oxygen (using a palladium catalyst) to form the vinyl acetate ($CH_3CO_2CHCH_2$) **monomer**.
- The vinyl acetate monomer undergoes free radical addition polymerisation to form polyvinyl acetate.

monomer: small molecules that bond to other similar molecules to form long chain molecules called *polymers*

→ Figure 12.27 shows the structures of the vinyl acetate monomer and the repeating structure of the polyvinyl acetate polymer composed of n-monomers. Polymers and the stages of polymerisation will be discussed further in Chapter 13.

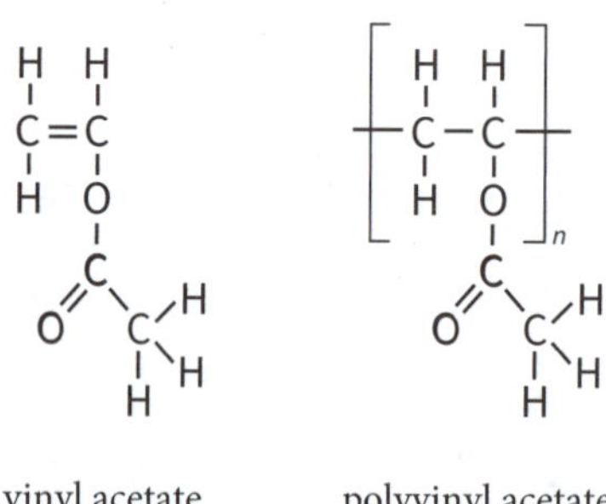

Figure 12.27 Vinyl acetate and polyvinyl acetate

➔ The flowchart for the preparation of polyvinyl acetate is shown in Figure 12.28.

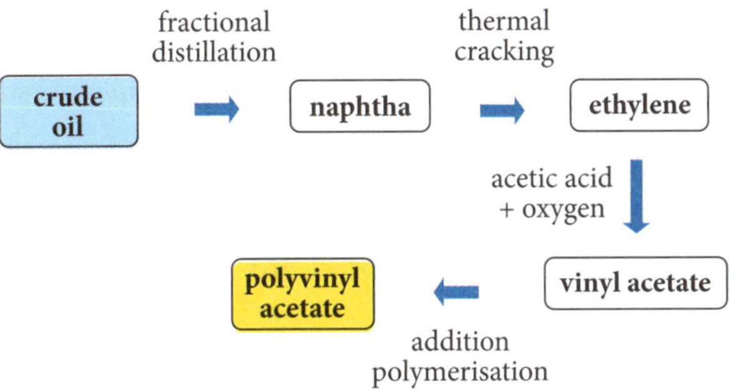

Figure 12.28 Flowchart for the production of polyvinyl acetate

➔ KEY QUESTIONS

12 Identify the purpose of a cellulolytic enzyme in the production of ethanol from cellulosic biomass.

13 Explain the role of thermal cracking in the production of polyvinyl acetate from crude oil.

Answers ➲ p. 171

CHAPTER SYLLABUS CHECKLIST

Are you able to answer these questions from the syllabus for this chapter? Tick each question as you go through the checklist if you are able to answer it. If you cannot answer a question, turn to the relevant page in the study guide to find the answer. For NESA key word meanings, go to www.educationstandards.nsw.edu.au and search 'key words'.

	FOR A COMPLETE UNDERSTANDING OF THIS TOPIC:	PAGE NO.	✓
1	Can I classify alkanols as primary, secondary or tertiary alcohols?	158	
2	Can I recall the properties of alkanols?	158	
3	Can I classify carbonyl compounds as aldehydes or ketones?	159	
4	Can I recall the properties of alkanals and alkanones?	160	
5	Can I distinguish the structural differences between amines and amides?	160	
6	Can I recall the properties of amines and amides?	160	
7	Can I identify the structure of the carboxyl group?	160	
8	Can I recall the properties of carboxylic acids?	161	
9	Can I write an equation for the preparation of a named ester?	162	
10	Can I describe the experimental procedure used to prepare an ester in the school laboratory?	163	
11	Can I recall the structure of soap and the procedure to manufacture soap?	165	
12	Can I classify detergents as anionic, cationic or non-ionic?	167	
13	Can I explain the cleaning actions of soap and detergents?	167	
14	Can I construct flowcharts to show reaction pathways for chemical synthesis?	168	

HSC EXAM-TYPE QUESTIONS

Objective-response questions (1 mark each)

1 1-propyl acetate is prepared by reacting propan-1-ol with acetic acid. The rate of the reaction is quite slow without a suitable catalyst. The catalyst that is commonly used is:

A dilute hydrochloric acid
B platinum
C concentrated sulfuric acid
D ceramic

2 Figure 12.29 shows the structure of an organic molecule.

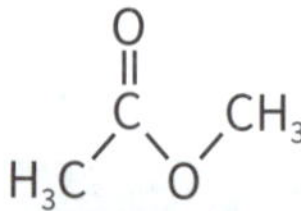

Figure 12.29 Name this molecule

The IUPAC name for this molecule is:

A propanone
B methyl acetate
C ethyl formate
D methyl propanoate

3 Identify the true statement about compound X, which has the condensed structural formula:

$X = NaCH_3(CH_2)_{14}COO$

A X is a soap called sodium heptadecanoate.
B X is formed when a fat called glyceryl trihexadecanoate is hydrolysed with sodium hydroxide.
C X is a detergent commonly found in cosmetics and adhesives.
D X can be dehydrated using dilute sulfuric acid.

4 Select the statement that is true concerning alkanols, alkanoic acids and esters.

A Alkanols, esters and alkanoic acids are all acidic in water solution.
B The boiling points of alkanoic acids are higher than the boiling points of alkanols with the same number of carbon atoms per molecule.
C Esters have significant hydrogen bonding between their molecules.
D Octanoic acid forms dimers due to the extensive hydrogen bonding between their molecules.

5 In the laboratory preparation of esters the reaction flask is heated with an electric hotplate or electric heating mantle rather than a Bunsen burner. The reason for this is:

A the heating rate is too difficult to control with a Bunsen burner.
B much higher temperatures can be reached using an electric heating mantle.
C the danger of fires is reduced in the absence of a flame if flammable organic vapours escape from the condenser.
D the yield of ester is greater when electrical heating is used.

Extended-response questions

6 An ester has the empirical formula C_3H_6O. The molar mass of the ester is 116.16 g/mol.

a Determine the molecular formula of the ester. (2 marks)
b Draw structural formulas and name five esters with this molecular formula. (5 marks)

7 Describe the steps of a laboratory procedure to convert:

a propan-1-ol to 1-propyl ethanoate. (2 marks)
b butanal to potassium butanoate. (4 marks)
c a vegetable oil into soap. (5 marks)

8 Dodecyl alcohol ethoxylate is a synthetic detergent. Its structure is:

$CH_3(CH_2)_{11}O(CH_2CH_2O)_nCH_2CH_2OH$ (n = 5–10)

a To what class of detergents does this substance belong? (1 mark)
b Describe the properties and uses of this class. (2 marks)
c Why is this surfactant unaffected by the presence of calcium ions in water? (1 mark)

9 The ester ethyl pentadecanoate was hydrolysed in hot sodium hydroxide solution. A soap was formed.

a Write a balanced equation for the reaction. (1 mark)
b Name the two reaction products that form. (2 marks)

10 1-butyl acetate is to be prepared in a laboratory. The following chemicals are available:
ethanol (b.p = 78 °C); acetic acid (b.p = 118 °C);
butan-1-ol (b.p = 117 °C); butanoic acid (b.p = 164 °C)

a Name the chemicals that will be used to prepare butyl acetate. (2 marks)
b The mixture is refluxed for three hours in the presence of 2 mL of concentrated sulfuric acid. What is the function of the concentrated sulfuric acid? (1 mark)
c Explain why a reflux condenser is used. (2 marks)
d 1-butyl acetate has a density of 0.88 g/mL and is insoluble in water. The reaction mixture is poured into 100 mL of water in a separating funnel. Two layers form. Identify the major chemicals present in the upper layer and lower layer in the separation funnel. (2 marks)
e The crude 1-butyl acetate is removed and washed with a solution of sodium carbonate. Some effervescence occurs. Explain why an effervescence occurred. (2 marks)
f The crude 1-butyl acetate is now purified. Its boiling point is 124 °C. What method is used to purify the 1-butyl acetate? (1 mark)

ANSWERS

KEY QUESTIONS

Key questions ⊃ p. 159

1 hydroxyl functional group = OH; carbonyl functional group = CO

2 Dispersion forces and hydrogen bonding exist between alkanal molecules, whereas only dispersion forces exist between alkanes. Thus alkanals have higher boiling points.

3 **a** hexan-2-one **b** 3-methylpentan-1-ol

Key questions ⊃ p. 161

4 **a** butanamide **b** 6-methylheptan-1-amine

5 There is significant hydrogen bonding between the carboxyl groups of these molecules and there is no hydrocarbon chain to disrupt this interaction.

Key questions ⊃ p. 165

6 2-butyl hexanoate

7 There is no hydrogen bonding between ester molecules and this reduces their boiling points compared with alkanols and alkanoic acids of the same molecular weights.

8 The concentrated sulfuric acid is a catalyst and the boiling chips prevent bumping or dangerous vibrations due to large vapour-bubble formation.

Key questions ⊃ p. 168

9 The hydrocarbon tails of the soap dissolve in the oil stain. During washing the charged heads remain at the surface of the oil so they can interact with the water molecules. The oil–soap mixture starts to lift off the fabric and these droplets become stabilised in the water by the formation of micelles.

10 animal or plant fats and alkaline solutions such as sodium hydroxide

11 **a** non-ionic **b** anionic **c** cationic

Key questions ⊃ p. 169

12 The enzymes break down the cellulose polymer chains to simple sugars.

13 The thermal cracking breaks longer hydrocarbon chains into smaller molecules, including ethylene, which is required for the production of vinyl acetate.

HSC EXAM-TYPE QUESTIONS

Objective-response questions

1 **C.** Concentrated sulfuric acid is the appropriate catalyst in esterification. Thus **A**, **B** and **D** are not the preferred catalysts.

2 **B.** The longest chain is the acetate (2-carbon chain) and the methyl group is attached to the oxygen atom. **A** is incorrect as the ester functional rather than the carbonyl group is present. **C** is incorrect as formates are derived from formic acid. **D** is incorrect as propanoates have three carbon atoms in the chain.

3 **B.** Sodium hexadecanoate ion is a soap formed when the fat undergoes alkaline hydrolysis. **A** is incorrect as heptadecanoate would have 17 carbon atoms. **C** is incorrect as cosmetics and adhesives use non-ionic detergents. **D** is incorrect as water is not removed from such a salt.

4 **B.** Alkanoic acids have greater hydrogen bonding than alkanols. **A** is incorrect as alkanols and esters are not acidic. **C** is incorrect as no hydrogen bonding exits between esters. **D** is incorrect as long chain alkanoic acids do not dimerise.

5 **C.** To avoid possible fires from the accidental escape of flammable vapours, electric heating is used. **A** is incorrect as heating rate is readily controlled using a Bunsen. **B** is incorrect as very high temperatures are not required. **D** is incorrect as the type of heating does not affect ester yield.

Extended-response questions

6 EM Students need to use their Year 11 course knowledge to determine the empirical and molecular formulas. Students need to recall the correct IUPAC nomenclature for esters.

a $M(C_3H_6O) = 58.08$ g/mol ✓
This empirical mass is half the molar mass of the ester.
Thus the molecular formula is twice the empirical formula:
MF = $C_6H_{12}O_2$ ✓

b Figure A12.1 shows the structural formulas of the esters.

A
```
     O     H   H   H   H   H
     ‖     |   |   |   |   |
 H — C — O — C — C — C — C — C — H
             |   |   |   |   |
             H   H   H   H   H
```
A = 1-pentyl formate ✓

B
```
     H   O       H   H   H   H
     |   ‖       |   |   |   |
 H — C — C — O — C — C — C — C — H
     |           |   |   |   |
     H           H   H   H   H
```
B = 1-butyl acetate ✓

C
```
     H   H   O       H   H   H
     |   |   ‖       |   |   |
 H — C — C — C — O — C — C — C — H
     |   |           |   |   |
     H   H           H   H   H
```
C = 1-propyl propanoate ✓

D
```
     H   H   H   O       H   H
     |   |   |   ‖       |   |
 H — C — C — C — C — O — C — C — H
     |   |   |           |   |
     H   H   H           H   H
```
D = ethyl butanoate ✓

E
```
     H   H   H   H   O       H
     |   |   |   |   ‖       |
 H — C — C — C — C — C — O — C — H
     |   |   |   |           |
     H   H   H   H           II
```
E = methyl pentanoate ✓

Figure A12.1 Esters

Extended-response questions

7 EM Students need to demonstrate a deep understanding of reaction pathways studied in this course. The intermediates formed need to be identified.

a The two steps are:
- i Reflux a mixture of propan-1-ol and ethanoic acid in a reaction flask, which is heated using an electric heating mantle. Use 2 mL of concentrated sulfuric acid as a catalyst. ✓
- ii Fractionally distill the reaction mixture and collect each component as separate fractions. ✓

b The four steps are:
- i Oxidise the butanal in an acidified solution of potassium dichromate to form butanoic acid. ✓
- ii Separate the butanoic acid from the reaction mixture by distillation. ✓
- iii Add potassium hydroxide solution to the butanoic acid until it is neutralised. ✓
- iv Isolate the potassium butanoate from the solution by crystallisation. ✓

- **c** The five steps are:
  - **i** Place 10 mL of the vegetable oil in a large beaker. ✓
  - **ii** Add 50 mL of a concentrated NaOH solution and a small volume of ethanol. ✓
  - **iii** Place a clock glass on the beaker and heat moderately on a hotplate, stirring occasionally. ✓
  - **iv** Leave to cool and add concentrated brine to salt out the soap. ✓
  - **v** Filter off the soap curds. Wash the crude soap with a little warm water and allow to dry. ✓

8 EM Students need to recall the properties and uses of a non-ionic detergent. The students need to state that calcium ions produce hard water, which precipitates soap but not synthetic detergents.

- **a** non-ionic detergent ✓
- **b** Non-ionic detergents have hydrophilic head groups. They are widely used in automatic dishwashing detergents, paints, cosmetics and adhesives because they improve the contact between polar and non-polar substances or surfaces. ✓ Non-ionic detergents and have low lathering properties. ✓
- **c** The non-ionic detergent is not precipitated by the calcium ions in hard water and so retains its properties. ✓

9 EM Students need to apply their knowledge of saponification to a simple ester and predict the formulas of the products formed.

- **a** $CH_3(CH_2)_{13}COOC_2H_5 + NaOH \rightarrow NaCH_3(CH_2)_{13}COO + C_2H_5OH$ ✓
- **b** sodium pentadecanoate; ✓ ethanol ✓

10 EM Students need to demonstrate a high-level understanding of a laboratory experiment, including appropriate chemical selection. Students then need to use known properties to isolate a pure sample of the ester.

- **a** butan-1-ol ✓ and acetic acid ✓
- **b** catalyst ✓
- **c** The reflux condenser ensures that there is no build-up of pressure as the heating occurs because it is open to the atmosphere. ✓ The condenser cools the vapours back to a liquid and they drip back into the reaction flask to continue to react. ✓
- **d** upper layer = 1-butyl acetate ✓
  lower layer = aqueous solution of butan-1-ol, acetic acid and sulfuric acid ✓
- **e** The upper ester layer contains some dissolved acetic and sulfuric acid. ✓ These acids are neutralised by the sodium carbonate solution and carbon dioxide gas is evolved. This procedure removes the acids from the ester before further purification occurs. ✓
- **f** distillation ✓

CHAPTER 13 POLYMERS

MODULE 7 ORGANIC CHEMISTRY

INQUIRY QUESTION:

What are the properties and uses of polymers?

Polymers are long chain organic compounds composed of repeating subunits called *monomers*. Some common natural polymers include starch, cellulose, proteins and DNA. Synthetic polymers have been manufactured and include common plastics such as polyethylene, polyvinyl chloride and polyvinyl acetate. Such polymers are derived from compounds extracted from crude oil. They are not biodegradable. Biodegradable synthetic polymers have been manufactured. This chapter investigates the structure, properties and uses of polymers.

1 Addition polymers

» Students model and compare the structure, properties and uses of addition polymers of ethylene and related monomers; for example, polyethylene (PE), polyvinyl chloride (PVC), polystyrene (PS) and polytetrafluoroethylene (PTFE).

Petrochemical polymers

➔ **Polymers** are large molecules made from smaller molecules joined in long chains. The smaller molecules are called **monomers**.

polymer: long chain molecules composed of repeating units called monomers
monomer: a molecule that is the basic unit of long chain molecules called polymers

➔ Chemists have synthesised many types of polymers. Many of these polymers are made from compounds extracted from petroleum.

➔ Ethylene (C_2H_4) is an important monomer produced by the thermal or catalytic cracking of the naphtha fraction produced by fractional distillation of petroleum.

➔ The naphtha fraction consists of hydrocarbon molecules with 7–13 carbon atoms per molecule. In the cracking process the longer hydrocarbon chains are broken into smaller chains, such as ethylene.

Polyethylene (PE)

➔ Ethylene monomers can join together in long chains to form polymers. This type of polymerisation is called *addition polymerisation*. The monomer's double bonds are broken during the addition polymerisation process. Addition polymerisation is achieved in two ways:

- using initiator molecules
- using catalytic surfaces.

➔ The overall reaction for the production of polyethylene (PE) by addition polymerisation is shown in Figure 13.1. In this structure *n* refers to the number of monomers that polymerise. PE chains typically vary in length from 2000 to 30 000 monomers, depending on the reaction conditions. Figure 13.1 shows the repeating unit of the polymer as well as a ball-and-stick model of a section of the chain.

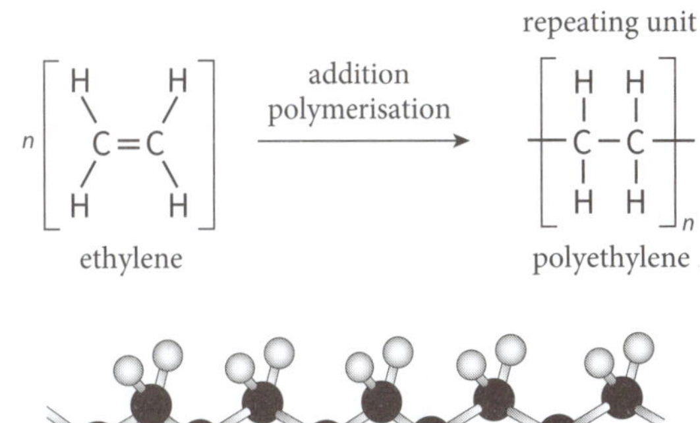

Figure 13.1 Polymerisation of ethylene

Initiator method

➔ Low density polyethylene (LDPE) is produced in this way. The process occurs at a high pressure (100–300 MPa) and a moderately high temperature (300 °C). A common **radical initiator** is benzoyl peroxide. On heating the initiator decomposes into **free radicals**. These radicals attack the double bond of the ethylene monomer and form a monomer radical. Figure 13.2 shows this initiation stage. Note the dot symbolism (•) of the radical species.

radical initiator: substances that can produce radical species under moderate conditions and promote further radical reactions

free radical: a highly reactive uncharged molecule or atom having an unpaired valence electron

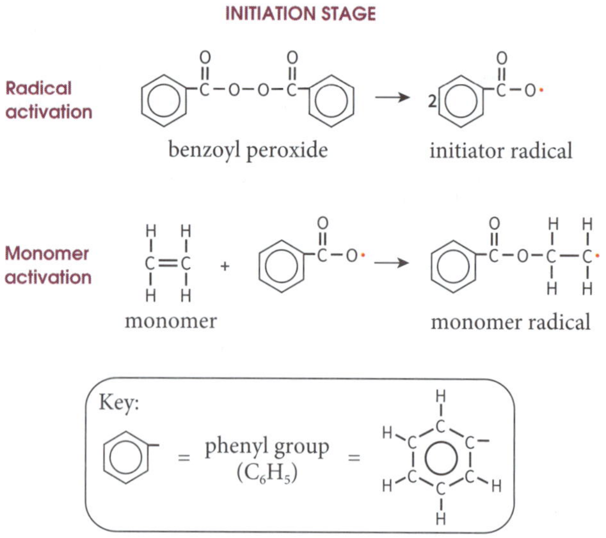

INITIATION STAGE

Radical activation — benzoyl peroxide → 2 initiator radical

Monomer activation — monomer + initiator radical → monomer radical

Key: phenyl group (C_6H_5)

Figure 13.2 Initiation stage of free radical polymerisation of ethylene

The monomer radical then attacks another monomer and a dimer radical forms. This propagation process continues and a polymer radical chain grows longer. Figure 13.3 shows the early stages of propagation.

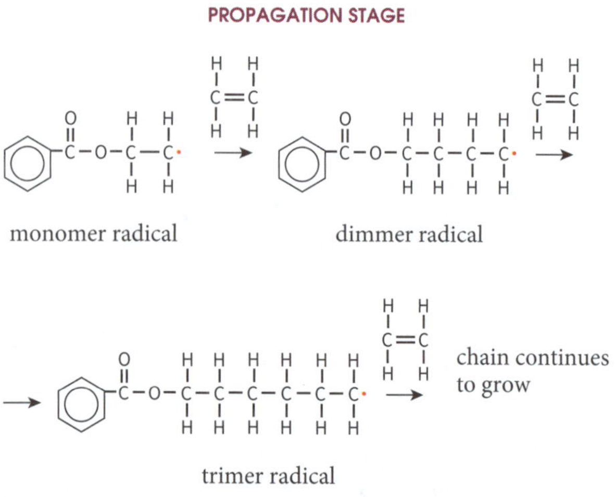

PROPAGATION STAGE

monomer radical → dimmer radical → trimer radical → chain continues to grow

Figure 13.3 Propagation stage of the polymerisation of polyethylene

If two polymer radicals combine then termination occurs and the polymer has reached its maximum size. This process is random and so polymer chains of different lengths form. Figure 13.4 illustrates the termination process.

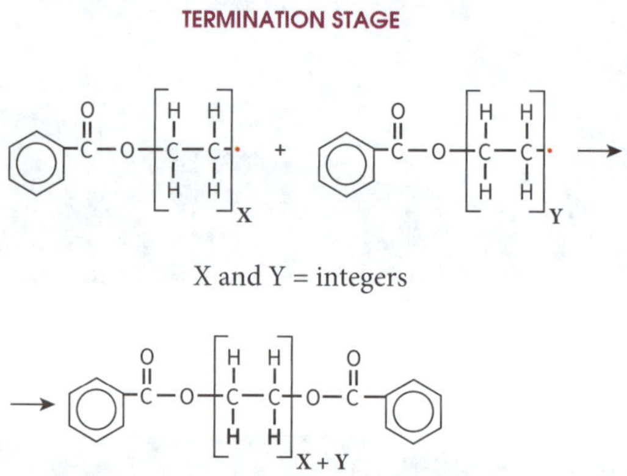

TERMINATION STAGE

X and Y = integers

Figure 13.4 Termination of polymer radical chains

- LDPE is characterised by branching polymer chains. During the propagation phase the polymer may form side branches if the radical site moves from the growing tip to a carbon atom back along the chain.

Catalytic method

- High density polyethylene (HDPE) is produced in this way. The common catalysts used are the Ziegler–Natta catalyst or the metallocene catalyst.
- The Ziegler–Natta catalyst is composed of $TiCl_4$ together with an aluminium-based co-catalyst.
  - The first ethylene monomer binds to the active site on the surface of the catalyst. Electron rearrangement occurs and a monomer radical forms with the unpaired electron at the end furthest from the active site.
  - The next monomer then reacts with this radical species and gradually a growing polymer radical chain forms, with the unpaired electron moving further away from the catalyst's active site. Figure 13.5 models this process.

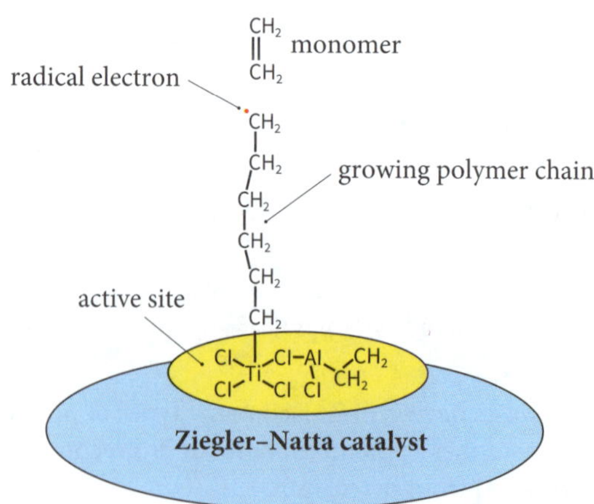

Figure 13.5 Model of a polyethylene chain growing at active site of catalyst

- Metallocene catalysts consist of a positively charged zirconium ion sandwiched between two negatively charged cyclopentadienyl anions.
  - The first monomer binds to the zirconium active site. The electron rearrangement leads to the breaking of the double bond.
  - The next monomer reacts at this active site and link onto the first monomer. This process continues with the polymer chain growing out from the active site. Figure 13.6 models this process.

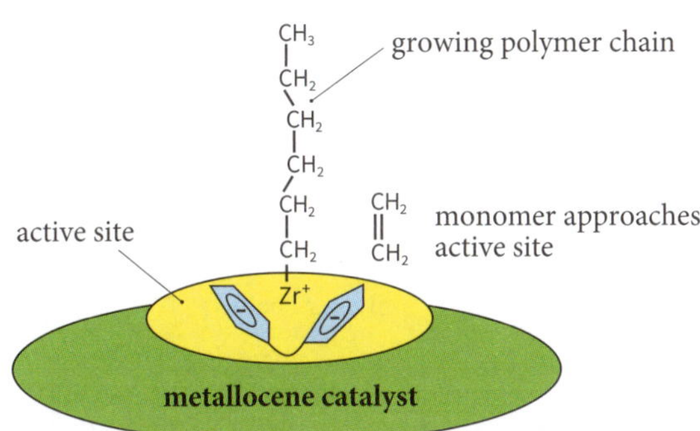

Figure 13.6 Model of a polyethylene chain growing from the catalyst's active site

- Unlike LDPE, HDPE has very few side branches along the polymer chain. Figure 13.7 models the different structures of LDPE and HDPE.

→ KEY QUESTIONS

1. **Identify the source of the ethylene used to manufacture polyethylene.**
2. **Name the three steps involved in the free radical polymerisation of ethylene.**
3. **Explain the main structural difference between low density polyethylene and high density polyethylene.**

Answers ➲ p. 184

Polyvinyl chloride (PVC)

- Polyvinyl chloride (PVC) is another important addition polymer. Ethylene is initially converted to chloroethylene (CH_2CHCl), which is commonly called *vinyl chloride*.
- PVC is produced using the same addition polymerisation process as polyethylene. Figure 13.8 shows the structure of the monomer and the PVC polymer.

```
 [ H     Cl ]          [   H  Cl   ]
 [  \   /   ]          [   |  |    ]
 [   C=C    ]          [ —C — C—   ]
 [  /   \   ]          [   |  |    ]
 [ H     H  ]          [   H  H    ]
   monomer          repeating polymer unit

     H  Cl  H  H  H  Cl  H  H
     |  |   |  |  |  |   |  |
   —C — C — C — C — C — C — C — C—
     |  |   |  |  |  |   |  |
     H  H   H  Cl H  H   H  Cl
         section of polymer chain
```

Figure 13.8 Polyvinyl chloride (PVC)

- Commercial PVC polymers typically vary in length from 2000 to 40 000 monomer units. Their chains are linear and the polymer is very strong and rigid.

Polystyrene (PS)

- Polystyrene (PS) is another important commercial polymer. It is manufactured from the styrene monomer ($CH_2CHC_6H_5$). The IUPAC name for this monomer is phenylethylene.
- Commercial PS polymers typically vary in length from 10 000 to 40 000 monomer units. The polymer is quite rigid due to the presence of the large phenyl group. The phenyl group (C_6H_5) consists of a hexagonal ring of carbon atoms in which hybridised carbon orbitals form bonds between C and H atoms.
- Figure 13.9 shows the styrene monomer and the addition polymer formed from styrene monomers. The phenyl group consists of a hexagon of carbon atoms with

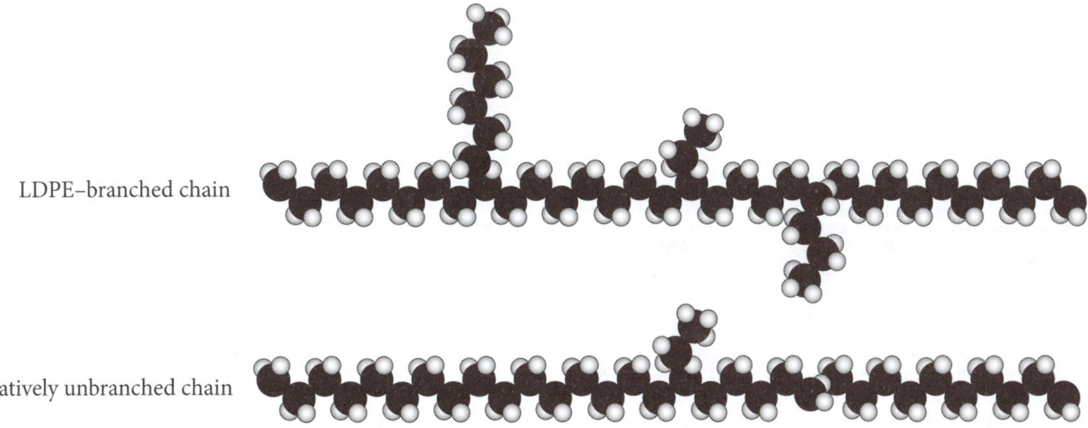

Figure 13.7 LDPE and HDPE polymer chains

p-electron orbitals forming a ring of delocalised electrons inside the hexagon. The phenyl groups are arranged in variable orientations along the hydrocarbon chain (see Example 1).

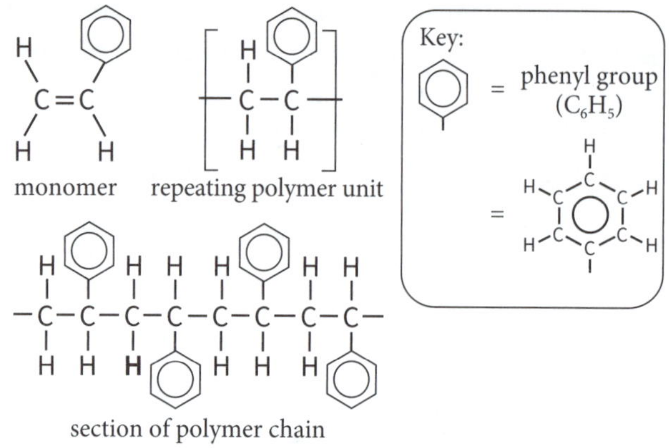

monomer   repeating polymer unit

Key: = phenyl group (C_6H_5)

section of polymer chain

Figure 13.9 Polystyrene polymer

- Polystyrene is a transparent, hard plastic. It can be converted into polystyrene foam by heating and using blowing agents such as propane gas to expand the polymer droplets and create holes in the plastic. The polystyrene foam is white and has a very low density.

EXAMPLE 1

Figure 13.10 shows a section of a syndiotactic polystyrene polymer chain and an atactic polymer chain. In the syndiotactic form of the polymer the phenyl groups alternate on opposite sides of the chain. This form is crystalline. In the atactic form the phenyl groups are randomly arranged. This form has very low crystallinity.

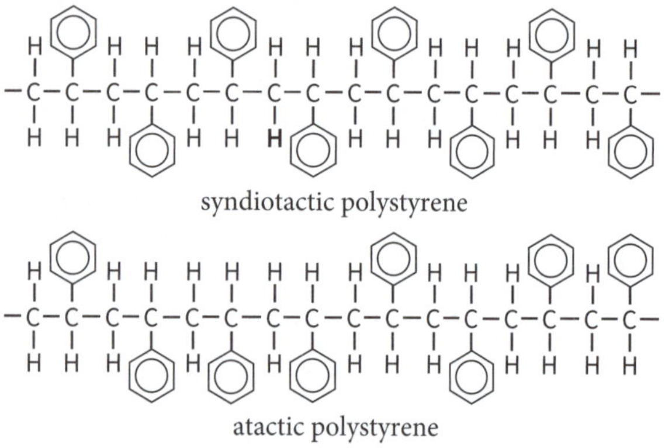

syndiotactic polystyrene

atactic polystyrene

Figure 13.10 Polystyrene chain

Explain why the syndiotactic structure leads to maximum crystallinity and increased stiffness in the polymer.

Consider the folding arrangement of polymer chains in crystals

Answer:

To maximise crystallinity the polymer chains will fold and pack in ordered layers in the polymer.

The dispersion forces between the layers will be maximised if the chains have ordered structures, as in the syndiotactic form. The chains will fold and stack closely together to produce crystalline regions. Such crystalline regions create a polymer that is strong and less flexible.

If the chains fail to pack in ordered arrangements the polymer is said to be amorphous. The random arrangement of phenyl groups leads to poor folding and stacking of polymer chains. The polymer chains are then arranged randomly and intertwined. The dispersion forces are weaker in such solids. The greater the proportion of amorphous regions, the more flexible the polymer. Figure 13.11 compares crystalline and amorphous regions.

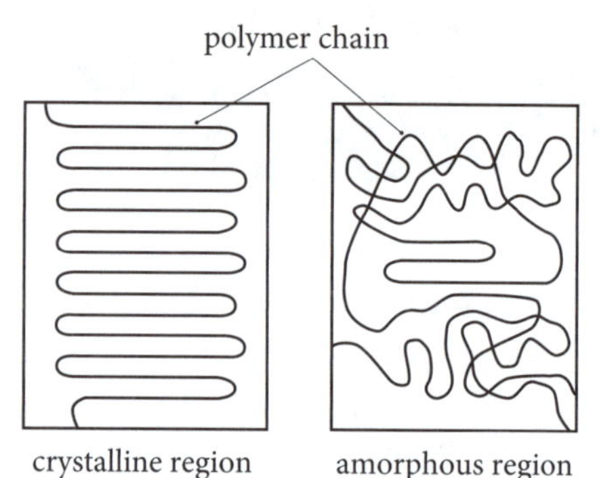

Figure 13.11 Crystalline and amorphous regions in a polymer

Polytetrafluoroethylene (PTFE)

- Polytetrafluoroethylene is an important commercial polymer that is used for non-stick, low friction coatings on cookware. It is manufactured from the tetrafluoroethylene monomer (CF_2CF_2), which is formed via a series of reactions beginning with methane.
- Figure 13.12 shows the monomer and the addition polymerisation process for tetrafluoroethylene monomers. The commercial polymer is called Teflon.

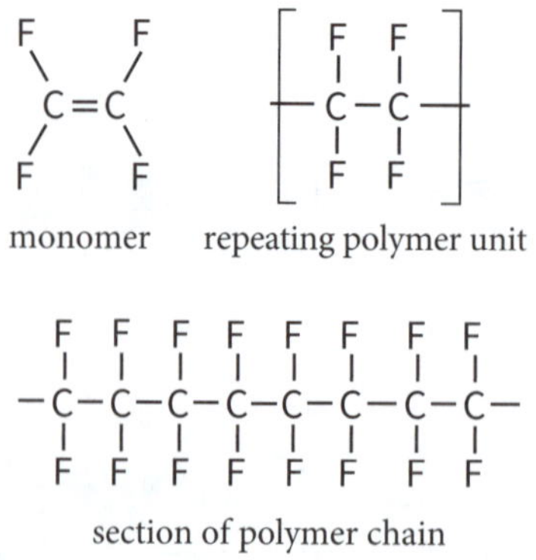

monomer   repeating polymer unit

section of polymer chain

Figure 13.12 Polytetrafluoroethylene

➔ KEY QUESTIONS

4 State the common name and the IUPAC systematic name of the monomer used to make PVC.

5 Explain how polystyrene can be converted to polystyrene foam.

6 Name the monomer used to produce PTFE polymers.

Answers ➲ p. 184

EXAMPLE 2

Polypropene (or polypropylene) is an addition polymer formed from propene monomers. Draw structural formulas of the monomer and a short section of the polymer.

The methyl group is a branch off the main chain

Answer:

Figure 13.13 shows the monomer and polymer structural formulas. Methyl groups can be arranged in different orientations along the polymer chain. The most common arrangement is the isotactic form, with all methyl groups on the same side of the polymer chain.

```
H       CH3         H  CH3 H  CH3 H  CH3
 \     /            |   |  |   |  |   |
  C = C           — C — C — C — C — C — C —
 /     \            |   |  |   |  |   |
H       H           H   H  H   H  H   H

  propene              polypropene
                     (isotactic form)
```

Figure 13.13 Propene and polypropene

Structure, properties and uses of addition polymers

➔ The properties of various commercial polymers are linked to their use. Table 13.1 summarises the properties and uses of five common commercial addition polymers.

➔ Further information about addition polymerisation can be found in various online videos. Enter the following titles in a search bar:

- free radical polymerisation animation
- addition polymerisation PVC
- polymer chemistry Fritz Klatte & Hermann Staudinger.

Table 13.1 Structure, properties and uses of addition polymers

Structure and properties	Uses
High density polyethylene (HDPE) • Linear polymer chains (little side branching) • Chains pack closely together and the dispersion forces between chains hold the chains tightly, resulting in polymers that are quite rigid.	• High density polyethylene is rigid and tough and is used to make bottle caps for milk and carbonated drinks, plastic toys, agricultural pipes and petrol tanks.
Low density polyethylene (LDPE) • LDPE has many branching side chains. • Side chains cause disruption in the packing of the polymer chains in the lattice. • LDPE molecules are softer and more flexible than HDPE molecules owing to the weaker dispersion forces between the polymer chains.	• Low density polyethylene is commonly used to make plastic bags, squeeze bottles, electrical insulation and food clingwrap.
Polyvinyl chloride (PVC) • The polymer chains are linear. • The chlorine atoms are large and create stiffness, and the polymer is inflexible • Additives called *plasticisers* may be added to produce flexibility by reducing the dispersion forces between the polymer chains. Plasticisers increase softness and elongation properties as well as low temperature flexibility.	• Rigid PVC is used to manufacture agricultural and drainage pipes, window frames and door frames. • Flexible PVC is used as coatings for electrical wires and soles of shoes. • The presence of chlorine in the polymer makes it resistant to atmospheric oxidation and it maintains its properties for a long time.
Polystyrene (PS) • Polystyrene is a clear polymer that exhibits high stiffness and good electrical insulation properties. The high stiffness results from the presence of the large phenyl groups that are attached along the polymer chains. • Polystyrene has a high refractive index. • Expanded polystyrene foam (EPS) is a good thermal insulator.	• The high stiffness of PS makes it suitable for car battery cases, plastic cutlery, handles for tools and heat-pressed food packaging. Its high refractive index makes it suitable for clear containers such as plastic drinking glasses and packaging of CDs and DVDs. • EPS is used for building insulation materials in house walls and roof spaces. The foam can be moulded to produce packing material for cushioning fragile items during transport.
Polytetrafluoroethylene (PTFE) • PTFE chains are unbranched, with polymer chains forming helical structures that pack tightly together. • PTFE is quite inert with a low friction surface. • PTFE is water insoluble and has a high melting point, high thermal stability and high electrical resistance.	• The high thermal stability and low friction make PTFE an excellent coating for non-stick frypans. • The low friction and inert properties are used in industrial coatings (e.g. gears and bearings) as well as to prevent corrosion of steel. • Its high electrical resistance makes it a useful insulator in electrical cables.

> **KEY QUESTIONS**
>
> 7 **Identify the property of low density polyethylene that makes it suitable for use in plastic bags and clingwrap.**
>
> 8 **Identify the property of polystyrene that makes it suitable for use as cases for DVDs.**
>
> Answers ⮕ p. 184

2 Condensation polymers

» Students model and compare the structure, properties and uses of condensation polymers; for example, nylon and polyesters.

Condensation reactions

- A condensation reaction is one in which two molecules combine together, with the elimination of a smaller molecule such as water.
- The formation of an ester from the reaction between an alkanoic acid and an alkanol is a condensation reaction. Figure 13.14 shows that water is removed in the condensation reaction.

$$R_1\text{–C(=O)–OH} + \text{H–O–CH}_2\text{–}R_2 \rightarrow R_1\text{–C(=O)–O–CH}_2\text{–}R_2 + H_2O$$

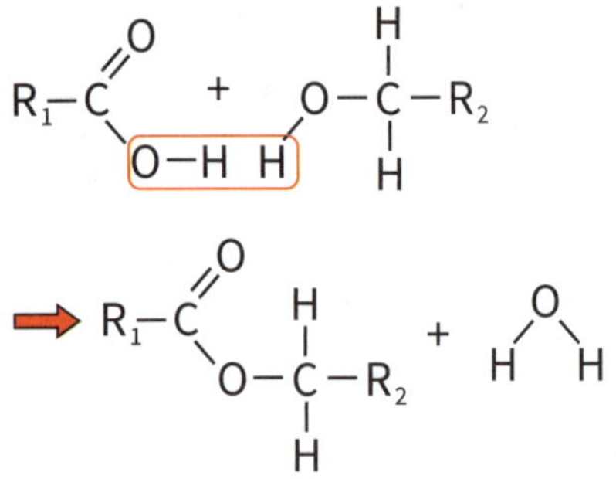

Figure 13.14 Condensation reaction

- In order to form a polymer using a condensation reaction, the monomers must have two reactive functional groups.

Polyesters

- Polyesters are an important group of commercial polymers formed by condensation polymerisation reactions.
- Polyesters form when an alkandiol monomer bonds (via an ester linkage COO) with a dialkanoic acid monomer. The monomers are linked by the ester functional group.
- The following reaction shows the formation of the dimer molecule (illustrated in Figure 13.15). The dimer's functional groups continue the process and the polymer chain grows larger. The OH group from the dialkanoic acid and the H from the hydroxyl functional group of the alkandiol combine to form water.

Example:

$HOCH_2CH_2OH$ + $HOOCCH_2COOH$
ethan-1,2-diol   propanedioic acid

$\rightarrow HOCH_2CH_2COOCH_2COOH + H_2O$
dimer

polymer = $(\text{–}OCH_2CH_2COOCH_2CO\text{–})_n$

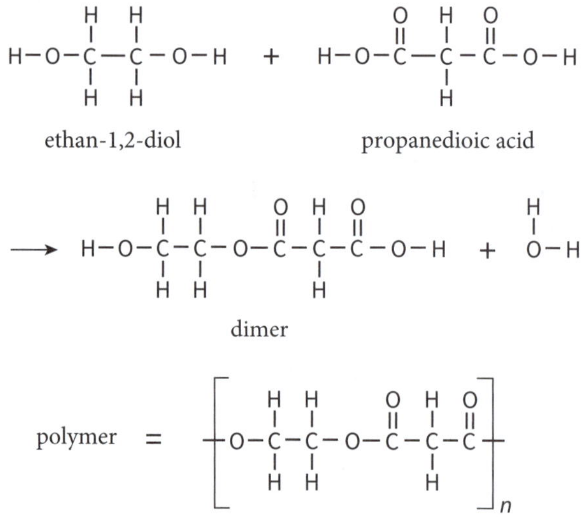

$$\text{H–O–CH}_2\text{–CH}_2\text{–O–H} + \text{H–O–C(=O)–CH}_2\text{–C(=O)–O–H}$$

ethan-1,2-diol   propanedioic acid

$$\rightarrow \text{H–O–CH}_2\text{–CH}_2\text{–O–C(=O)–CH}_2\text{–C(=O)–O–H} + \text{H–O–H}$$

dimer

$$\text{polymer} = \left[\text{–O–CH}_2\text{–CH}_2\text{–O–C(=O)–CH}_2\text{–C(=O)–}\right]_n$$

Figure 13.15 Condensation polymerisation

- Polyester fibres are strong and elastic and are extensively used in the production of textiles as they also have low water absorption and minimal shrinkage after washing. A common polyester used to make clothing is PET (polyethylene terephthalate). The structure of the polymer is shown in Figure 13.16.

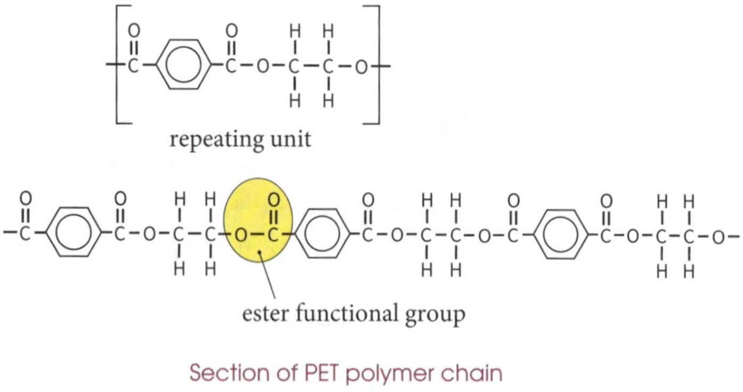

$$\left[\text{–C(=O)–C}_6\text{H}_4\text{–C(=O)–O–CH}_2\text{–CH}_2\text{–O–}\right]$$

repeating unit

ester functional group

Section of PET polymer chain

Figure 13.16 PET polyester

Nylon

- Nylon plastic is an example of a polyamide. It is a condensation polymer formed when dialkanoic acid and diamine monomers condense with the elimination of a water molecule. An amide linkage (CONH) binds the monomers as the polymer chain grows.

- The following equation shows the formation of the dimer. Nylon 6,6 is a common plastic that is formed by the condensation polymerisation of hexanedioic acid and hexan-1,6-diammine. The polymer has high tensile strength and is commonly used in the production of textiles. Its high rigidity also finds application in the production of machine parts and pipes:

$HOOC(CH_2)_4COOH + H_2N(CH_2)_6NH_2$

hexanedioic acid   hexan-1,6-diammine

$\rightarrow HOOC(CH_2)_4CONH(CH_2)_6NH_2 + H_2O$

dimer

- The structure of nylon 6,6 is shown in Figure 13.17.

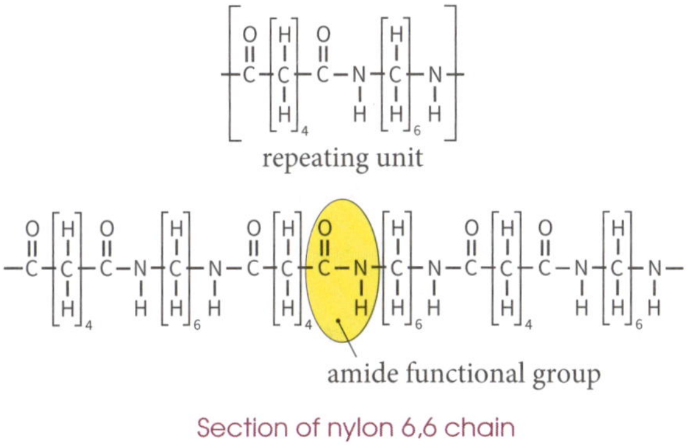

repeating unit

amide functional group

Section of nylon 6,6 chain

Figure 13.17 Nylon 6,6

EXAMPLE 3

Figure 13.18 shows how three chains of a synthetic polymer stack and line up in layers to form a crystalline structure.

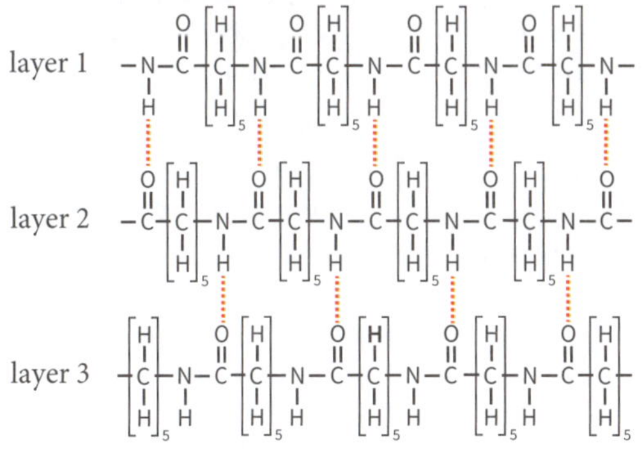

layer 1

layer 2

layer 3

Figure 13.18 Three layers of polymer chains

a **Draw the structural formula of the monomer.**

b **Classify the type of synthetic polymer illustrated.**

c **Explain the major type of intermolecular forces that allows the chains to line up in an orderly fashion, as shown in Figure 13.18.**

Examine the functional groups present in the polymer

Answer:

a Figure 13.19 shows the monomer structure.

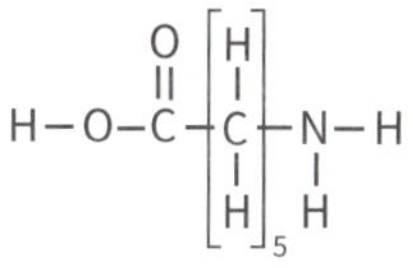

Figure 13.19 Monomer

b polyamide

c Hydrogen bonding between the oxygen atom of the C=O group and the hydrogen atom of the and N–H group on the next layer of the polymer chains allows the chains to stack and line up in a regular and orderly arrangement. This intermolecular force will increase the strength of the polymer.

KEY QUESTIONS

9 **Draw the condensed structural formulas of the two monomers that produce the following polyester: $(-OCH_2CH_2CH_2CH_2COOCH_2CH_2CH_2CO-)_n$**

10 **Identify the name of synthetic polymers containing the amide (CONH) functional group.**

Answers p. 184

Biodegradable condensation polymers

- Cellulose is an example of a natural condensation biopolymer. It consists of up to 10 000 glucose units that form the unbranched cellulose chain.
- The monomer is called *beta-glucose*. The structure of this monomer and a section of the cellulose polymer chain is shown in Figure 13.20. The bond that links each monomer is called the beta-1,4-glycosidic bond. This bond links carbon-1 in 1 glucose molecule to carbon-4 in another glucose molecule. The cellulose structure in Figure 13.20 also shows the inversion of each alternate monomer.

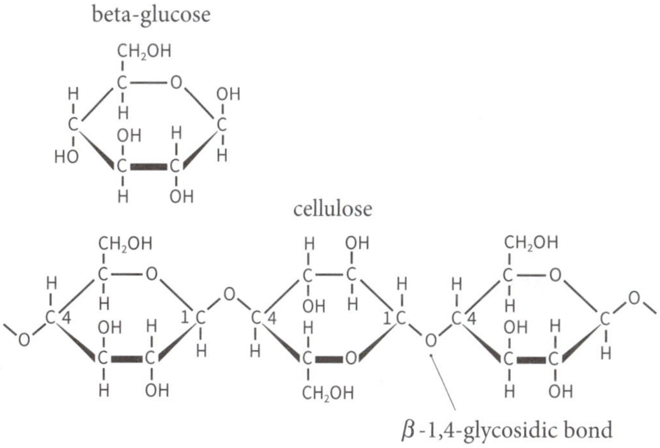

beta-glucose

cellulose

β-1,4-glycosidic bond

Figure 13.20 Beta-glucose and cellulose structures

- ➔ Cellulose biopolymers are readily degraded in the environment by decomposers such as bacteria and moulds.
- ➔ A major problem with polymers derived from petroleum is that they are not biodegradable. When they are buried in rubbish tips they are not decomposed by bacteria and moulds. To overcome this problem, chemists have developed a range of synthetic biodegradable polymers from renewable resources in the biosphere. They are examples of synthetic biopolymers.
- ➔ Polylactic acid (PLA) is an example of a synthetic biopolymer. PLA is a biodegradable condensation polymer that is strong and flexible. The monomer is lactic acid ($CH_3CHOHCOOH$), which is produced from starch waste (typically corn, potato or sorghum). The starch waste is initially converted to simple sugars (such as glucose). The sugars are fermented by bacteria (*Lactobacillus*) to produce lactic acid. The monomer then reacts to form a cyclic lactide dimer. In the bacterial cells the cyclic dimer undergoes further chain growth via condensation polymerisation to from the linear PLA polymer. PLA is a polyester and can be extracted from the bacterial cells.
- ➔ The flowchart in Figure 13.21 summarises the steps of the industrial process of PLA.
- ➔ Plastic film, plastic bags, composting bags, disposable plates and tableware can be made from PLA.

Comparisons of PLA with LDPE

- PLA is biodegradable and will be decomposed in rubbish tips when discarded. LDPE is not biodegradable.
- PLA is made from a renewable resource whereas LDPE is made from petroleum, which is non-renewable.
- PLA is **biocompatible** and can be used in medical applications inside the body. The PLA polymer slowly dissolves as wounds repair themselves. LDPE cannot be used in this way.
- PLA is similar to LDPE in that it is a **thermoplastic** and can be moulded and recycled.
- PLA cannot be used to hold hot water as the plastic deforms on heating. HDPE can be used for hot liquids as long as the temperature is below 70 °C.

biocompatible: a material that is compatible with living tissue and does not produce adverse reactions or rejection

thermoplastic: a substance that can become plastic (soften) on heating and harden on cooling

- ➔ The following equation shows the first condensation reaction to form the dimer composed of the two lactic acid monomers. Here the water is formed from the OH group of the carboxyl functional group and the H atom of the hydroxyl group:

$$CH_3CHOHCO\mathbf{OH} + CH_3CHO\mathbf{H}COOH \rightarrow CH_3CHOHCOOCH(CH_3)COOH + \mathbf{H_2O}$$

- ➔ The structure of the PLA polymer is shown in Figure 13.22.

```
        H
     [  |  ]
     [ H-C-H ]
     [   |   ]
    -[-C-C-O-]-
     [ || |  ]
     [ O  H  ]
```

repeating unit

```
   H                 H
   |                 |
 H-C-H      H      H-C-H      H
   |        |        |        |
-C-C-O-C-C-O-C-C-O-C-C-O-
 ||  |   ||  |   ||  |   ||  |
 O   H   O H-C-H O   H   O H-C-H
             |               |
             H               H
```

section of PLA polymer

Figure 13.22 Polylactic acid

EXAMPLE 4

3-hydroxybutanoic acid undergoes condensation polymerisation to form a polyester commonly called polyhydroxybutyrate (PHB). It is an example of a bacterial plastic in which a bacterium (*Ralstonia eutropha*) is used to manufacture the polyester, which is stored by the bacterium in its cell wall.

a **Draw the structural formula of the monomer and the repeating structure of the polymer.**

b **PHB is biodegradable and biocompatible. Explain the benefits of these properties.**

Water is lost when the monomers combine

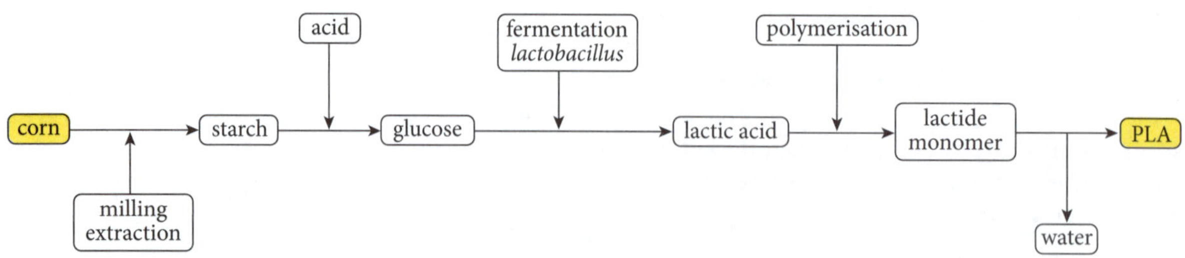

Figure 13.21 Production of PLA flowchart

Answer:

a Figure 13.23 shows the monomer and repeating unit structures.

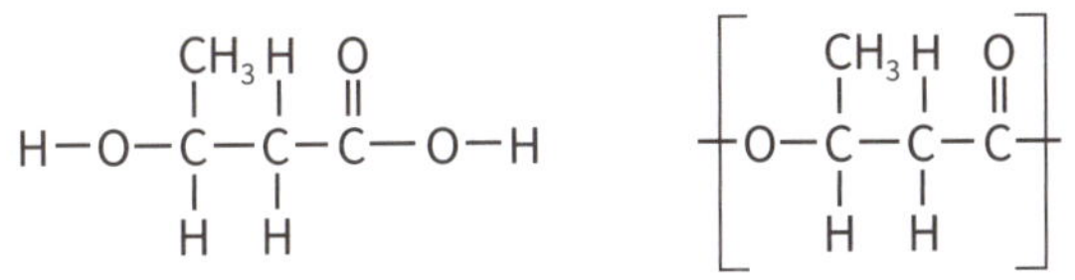

3-hydroxybutanoic acid

repeating unit

Figure 13.23 Monomer and polymer (PHB)

b Biodegradable plastics are readily broken down by decomposers and do not damage the environment like petrochemical plastics. As PHB is biocompatible it can be used in thread form as sutures during operations. It will not cause body rejection and inflammation problems.

→ Further information about addition polymerisation can be found in various online videos. Enter the following titles in a search bar:

- condensation polymers sample
- conversion of food waste into polylactic acid.

→ KEY QUESTIONS

11 Cellulose is a natural condensation polymer. Name the monomer and the bond that joins the monomers together.

12 Polylactic acid (PLA) is synthetic biopolymer. Name the bacterium used in its production and draw a condensed structural formula of the monomer.

Answers ➲ p. 184

CHAPTER SYLLABUS CHECKLIST

Are you able to answer every question from the syllabus for this chapter? Tick each question as you go through the checklist if you are able to answer it. If you cannot answer a question, turn to the relevant page in the study guide to find the answer. For NESA key word meanings, go to www.educationstandards.nsw.edu.au and search 'key words'.

FOR A COMPLETE UNDERSTANDING OF THIS TOPIC:		PAGE NO.	✓
1	Can I name examples of addition polymers?	173	
2	Can I draw a structure for polyethylene and compare the properties of LDPE and HDPE?	173	
3	Can I describe the differences between the free radical polymerisation and the catalytic method of making addition polymers?	174	
4	Can I draw a structure for polyvinyl chloride and describe its properties and uses?	175	
5	Can I draw a structure for polystyrene and describe its properties and uses?	176	
6	Can I draw a structure for polytetrafluoroethylene and describe its properties and uses?	178	
7	Can I use equations to demonstrate the formation of condensation polymers?	178	
8	Can I distinguish between polyesters and polyamides and name examples of each?	178	
9	Can I explain the importance of the development of biodegradable condensation polymers?	179	
10	Can I construct flowcharts to summarise the steps in the production of a synthetic biopolymer?	180	
11	Can I compare the properties of a synthetic biopolymer with petrochemical polymers?	180	

HSC EXAM-TYPE QUESTIONS

Objective-response questions

(1 mark each)

1 **Identify a common use of low density polyethylene (LDPE).**

A plastic bags
B building insulation
C agricultural pipes
D DVD cases

2 **Select the statement that is true of polylactic acid (PLA).**

A It is a non-biodegradable synthetic polymer.
B It is a polyamide with high tensile strength.
C It is used extensively in the textile industry.
D It is manufactured from biomass.

3 **Select the true statement about polymers.**

A Polyvinyl chloride is biodegradable.
B High density polyethylene is manufactured using free radical polymerisation.
C Nylon 6,6 is a polyester.
D Cellulose is formed from beta-glucose monomers.

4 **Figure 13.24 is a model of a stage in a polymerisation process.**

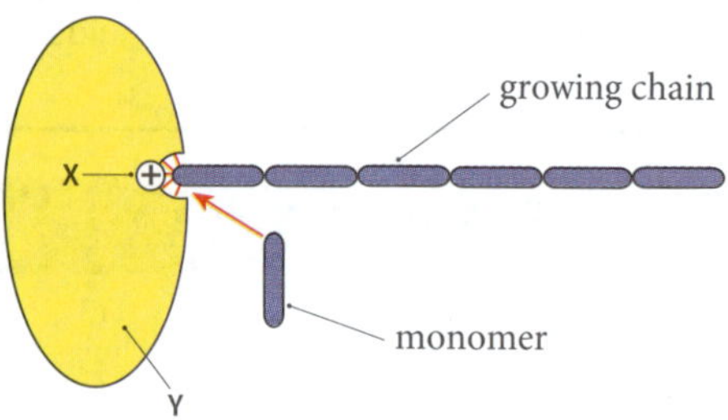

Figure 13.24 Polymerisation stage

Select the response that correctly identifies structures X and Y and the name of the polymerisation process.

	X	Y	Polymerisation process
A	active site	catalyst	metallocene catalyst polymerisation
B	free radical	initiator	free radical polymerisation
C	active site	catalyst	Ziegler–Natter catalyst polymerisation
D	initiator	free radical	free radical polymerisation

5 **The graph in Figure 13.25 shows the distribution of molecular weights of polymer chains in a sample of polyethylene produced by free radical polymerisation. Select the statement that is true.**

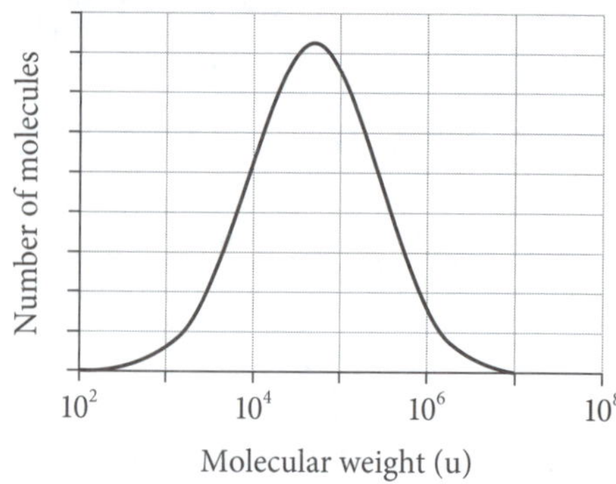

Figure 13.25 Distribution of molecular weights

A Most polyethylene molecules have molecular weights greater than 100 000 u.
B All polyethylene polymers have molecular weights greater than 1000 u.
C The polymer molecules in the polyethylene sample are identical in size.
D The polymer molecules are not the same size because the termination of chain growth occurs randomly.

Extended-response questions

6 **The process of addition polymerisation can be modelled using a molecular model kit. Figure 13.26 shows the atomic models available to use. The stepwise process of modelling the free radical polymerisation of vinyl chloride monomers to form PVC is to be illustrated using the atomic models.**

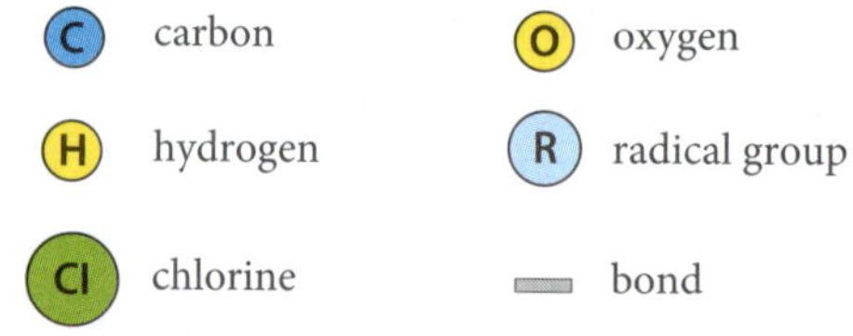

Figure 13.26 Atomic models

a **Figure 13.27 shows the initial modelling steps. Use similar diagrams to show the remaining stages of the polymerisation process.** (4 marks)

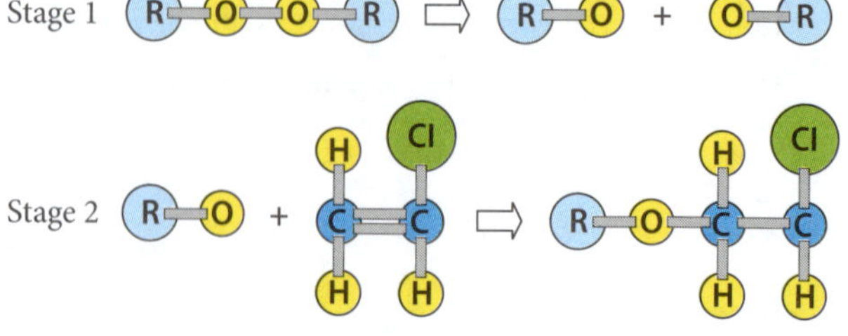

Figure 13.27 Initial stages of the polymerisation process

b Describe an alternative modelling procedure that will increase our understanding of the polymerisation process. (2 marks)

7 a Propan-1,3-diol and butanedioic acid monomers undergo condensation polymerisation.

i Write a balanced equation using condensed structural formulas to show the formation of the dimer. (1 mark)

ii To which class of polymers does the resulting polymer belong? (1 mark)

b i The vinyl acetate monomer has the following semi-structural formula:

$CH_2{=}CHCOOCH_3$

Draw the structural formula of this monomer. (1 mark)

ii Draw a section of the polyvinyl acetate polymer that would form on addition polymerisation. (1 mark)

8 Answer the following questions about addition polymers.

a Compare the flexibility and melting point of HDPE to LDPE and account for these differences. (2 marks)

b Explain how the presence of a chlorine atom alters the flexibility of PVC compared with LDPE. (2 marks)

c Compare the physical properties of polystyrene to expanded polystyrene (EPS). (2 marks)

d PVC is to be used to make a garden hose. Explain how the properties of PVC can be modified to make it suitable for this use. (2 marks)

9 Draw a 3-monomer section of the chain of an addition polymer formed from the following monomer:

CHF=CBrF (1 mark)

10 Glycine is an amino acid. It can undergo condensation polymerisation to form polyglycine. The structure of glycine is shown in Figure 13.28.

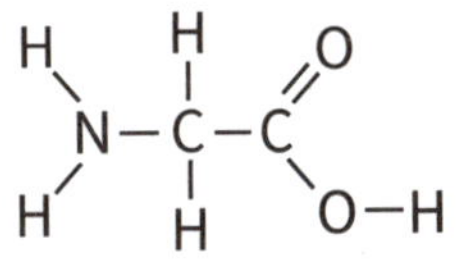

H H O
N–C–C
H H O–H

Figure 13.28 Glycine

a Draw a section of the polymer chain that forms. (1 mark)

b Classify the polymer as a polyester or polyamide. (1 mark)

ANSWERS

KEY QUESTIONS

Key questions ➲ p. 175

1 Ethylene is derived from the naphtha petroleum fraction of crude oil. The naphtha is thermally cracked to produce ethylene.

2 initiation, propagation, termination

3 LDPE has branched chains and HDPE has very few branches along its chain.

Key questions ➲ p. 177

4 vinyl chloride, chloroethylene

5 The melted polymer is injected with propane gas to form bubbles that on cooling leave holes in the polymer.

6 tetrafluoroethylene

Key questions ➲ p. 178

7 flexible and soft

8 transparent and rigid

Key questions ➲ p. 179

9 $HOCH_2CH_2CH_2CH_2COOH$ and $HOCH_2CH_2CH_2COOH$

10 nylon

Key questions ➲ p. 181

11 beta-glucose, beta-1,4-glycosidic bond

12 *Lactobacillus*, $CH_3CHOHCOOH$

HSC EXAM-TYPE QUESTIONS

Objective-response questions

1 **A**. LDPE is very flexible and therefore is suited for plastic bags. **B**, **C** and **D** are incorrect as all these products are rigid and need to be made from rigid polymers.

2 **D**. PLA is commonly made from waste biomass such as corn fibre. **A** is incorrect as PLA is biodegradable. **B** is incorrect as PLA is a polyester. **C** is incorrect as its uses include bags and tableware.

3 **D**. Cellulose forms from beta-glucose. **A** is incorrect as PVC is non-biodegradable. **B** is incorrect as HDPE is made using a catalyst. **C** is incorrect as nylon 6,6 is a polyamide.

4 **A**. The metallocene catalyst has an active site where new monomers join onto the active site. **B** and **D** are incorrect as no initiator is present. **C** is incorrect as the Ziegler–Natta catalyst creates a growing chain with new monomers growing from the radical tip and not back at the active site.

5 **D**. The termination step can occur at any time during the polymerisation reaction. Therefore a mixture of molecular sizes results. **A** is incorrect as the maximum in the graph occurs at molecular weights between 10 000 u and 100 000 u. **B** is incorrect as the graph shows that a small number of polymer molecules have molecular weights between 100 u and 1000 u. **C** is incorrect as the graph shows a wide range of molecular weights and therefore the size of the molecules is variable.

Extended-response questions

6 EM Students need to recall modelling experiments performed in the laboratory and recall the three stages of the free-radical polymerisation reaction.

a Figure A13.1 shows the propagation steps and termination steps.

Stage 3: Propagation steps

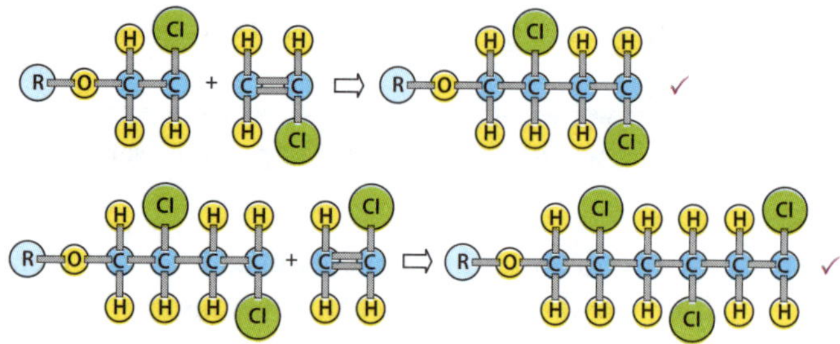

Stage 4: Termination steps

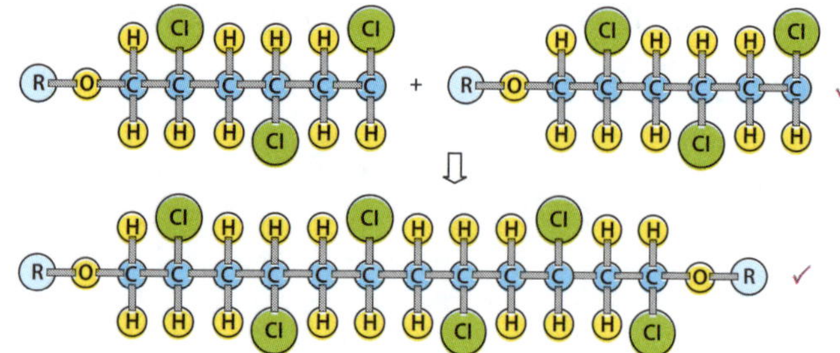

Figure A13.1 Propagation and termination

b Computer animation is an excellent modelling method to aid our understanding. ✓ The animation sequences can show collision of molecules, bond breaking and bond forming. ✓

7 EM Students need to demonstrate an understanding of the differences between condensation and addition polymerisation and being able to correctly use condensed structural formulas to write a balanced equation.

a i $HOCH_2CH_2CH_2OH + HOOCCH_2CH_2COOH \rightarrow HOCH_2CH_2CH_2OOCCH_2CH_2COOH + H_2O$ ✓

ii polyester ✓

b i Figure A13.2 shows the structure of vinyl acetate.

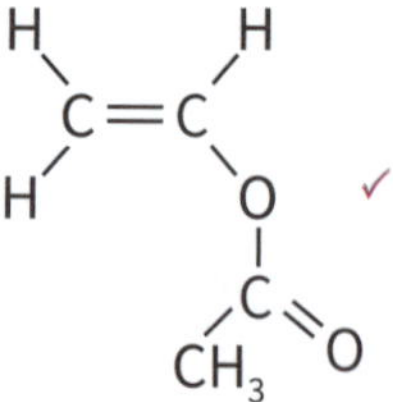

Figure A13.2 Vinyl acetate

ii Figure A13.3 shows a section of the polyvinyl acetate polymer chain.

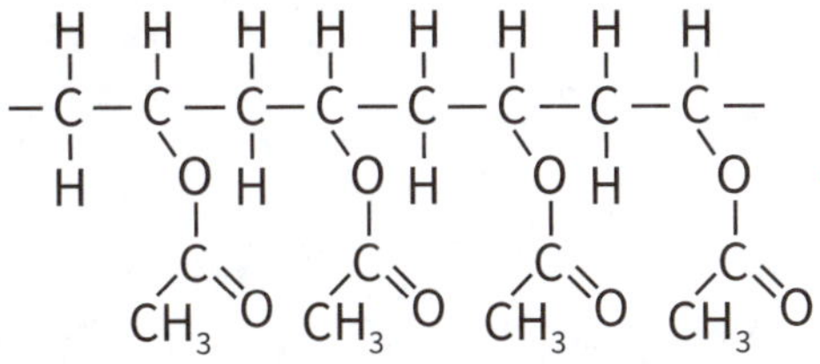

```
  H   H   H   H   H   H   H   H
  |   |   |   |   |   |   |   |
 -C - C - C - C - C - C - C - C -
  |    \  |    \  |    \  |    \
  H     O H     O H     O H     O     ✓
        |       |       |       |
        C       C       C       C
       / \\    / \\    / \\    / \\
    CH3   O CH3   O CH3   O CH3   O
```

Figure A13.3 Section of polyvinyl acetate polymer

8 EM Students need to demonstrate an understanding between structure, properties and uses of different polymers to score full marks.

a LDPE has a lower melting point and greater flexibility than HDPE. The LDPE chains are branched and this results in less uniform packing in the solid state. ✓ The dispersion forces between chains is therefore weaker and so flexibility increases and the melting point decreases compared with HDPE. ✓

b Chlorine has a large atomic radius and so the polymer chain has decreased flexibility and increased stiffness when PVC is bent. ✓ PVC is a linear polymer with a non-crystalline structure, whereas LDPE has a crystalline structure and is much more flexible than PVC as LDPE is a branched chain polymer. ✓

c Polystyrene is a clear, stiff and rigid crystalline polymer. ✓ Expanded polystyrene is white and opaque to light. It is a rigid polymer that is very light and a thermal insulator. ✓

d Plasticiser molecules are added to the melted polymer and the mixture is allowed to set. ✓ The resulting plastic is more flexible and would be suitable now for use in a garden hose as the hose has to be able to bend and be coiled. Common plasticisers are phthalate compounds. ✓

9 EM Students need to demonstrate the ability to examine the structure of a monomer and then predict the structure of the addition polymer that would form.

Figure A13.4 shows the structure of the polymer chain.

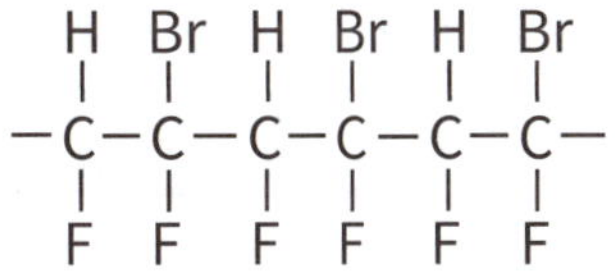

```
  H  Br  H  Br  H  Br
  |  |   |  |   |  |
 -C- C - C- C - C- C-   ✓
  |  |   |  |   |  |
  F  F   F  F   F  F
```

Figure A13.4 Polymer section

10 EM Students need to demonstrate an understanding of condensation polymerisation using a new monomer. They need to recognise the repeating functional group that results and then determine the correct classification of the polymer.

a Figure A13.5 shows a section of the condensation polymer chain.

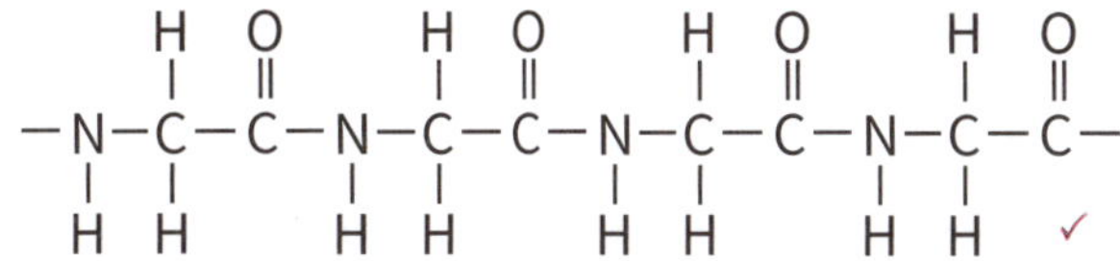

Figure A13.5 Condensation polymer

b polyamide (CONH is the amide linkage) ✓

MODULE 8 APPLYING CHEMICAL IDEAS

CHAPTER 14 ANALYSIS OF INORGANIC SUBSTANCES

INQUIRY QUESTION:

How are the ions present in the environment identified and measured?

Analytical chemists utilise a wide range of qualitative and quantitative techniques to analyse material. Environmental chemists use analytical techniques in monitoring water and air quality to ensure that the environment does not become contaminated with pollutants. This chapter investigates various examples of environmental monitoring.

1 Monitoring the environment

» Students analyse the need for monitoring the environment.

Investigating environmental monitoring

- Monitoring the environment and its wide range of ecosystems allows scientists to determine whether or not human activities cause damaging changes.
- Environmental monitoring is crucial to know if the quality of our environment is getting better or worse. Systematic sampling of air, water, soil and living organisms are conducted to observe and study the environment.
- The data obtained can be used to understand the state and composition of the environment. Environmental monitoring uses a variety of specialised equipment and techniques.

Testing and monitoring water quality

- Water that is fit to drink is called *potable water*. Large dams are constructed to collect water for human consumption. Water that is to be used for other purposes (e.g. irrigation of crops) does not have to be of the quality required for potable water. It is the job of specialist analytical chemists as well as microbiologists, physical chemists and environmental chemists employed by water authorities to monitor and manage our water supplies.
- Some examples of environmental measurements are investigated below.

Total dissolved solids (TDS)

- The measurement of total dissolved solids (TDS) provides information about the suitability of water for drinking or agricultural purposes.
- Dissolved solids are mainly ionic but small quantities of organic molecules can be present. Ocean water has very high concentration of ions, including sodium, chloride and sulfate ions. Potable water must have low ion concentrations otherwise the water will taste slightly salty. Gravimetric analysis can be used to determine TDS. TDS values are reported in parts per million (ppm), where 1 ppm = 1 mg/L. TDS measurements involve the evaporation of a known volume of filtered water to dryness and measuring the mass of dry residue left behind.
- Typical values for TDS are: potable water, 40–100 ppm; bushland streams, < 100 ppm; Sydney water supplies, 40–100 ppm; **artesian water**, 500–1200 ppm; and ocean water, 30 000–40 000 ppm.
- As most dissolved solids in waterways are ionic, conductivity measurements can also be used to monitor the TDS.

artesian water: water that is present in an underground permeable rock layer

EXAMPLE 1

Artesian water is groundwater present in underground porous rocks. Drilling into these rocks allows the water to flow under pressure to the surface and be collected.

a **A sample of artesian water was collected and filtered. 5.00 L of the water was slowly evaporated to dryness in a large beaker and the mass of the dry residue was 4.10 g. Calculate the TDS.**

b **Conductivity electrodes were also used to determine the TDS of the filtered sample of the artesian water. The conductivity was found to be 1060 microsiemens/cm. Use the calibration graph in Figure 14.1 to determine the TDS and compare the answer with the gravimetric method in a.**

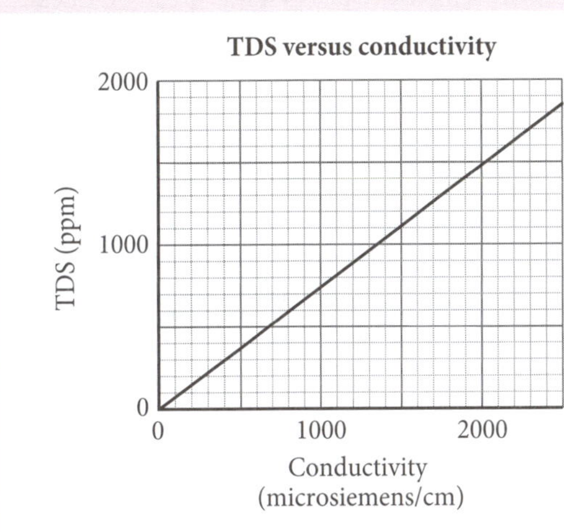

Figure 14.1 TDS calibration graph

Answer:

Convert the mass of residue in method **a** to milligrams

a TDS = mass (mg)/volume (L)
m = 4.10 g = 4100 mg
TDS = 410/5 = 820 mg/L = 820 ppm

b Using the calibration graph, TDS = 800 ppm.
The two measurements give results that are very similar. The gravimetric procedure is a more accurate method but is very time consuming and has to be conducted in a laboratory using an accurate electronic balance. The conductivity method is suitable for quick measurements that can be performed when testing waterways onsite.

Water acidity

- Acidity in water is caused by hydronium ions. Water acidity can be readily measured using calibrated pH probes. Potable water should have a pH range of 6.5 to 8.5. This pH range is compatible with our body chemistry.
- Release of acidic mining or factory wastes or acid rain produced by air pollution can raise the hydronium ion concentration and lower the pH of water. If town water is too acidic it will cause corrosion of water pipes and tanks. Acid rain can fall into lakes and prevent fish eggs from hatching, and acid rain can damage tree leaves. Environmental chemists can use portable electronic pH electrodes (Figure 14.2) during field work. They typically have an accuracy of ±0.2 pH units.

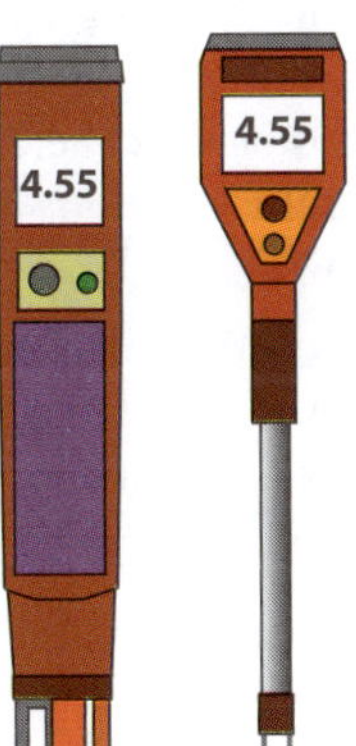

Figure 14.2 Illustrations of typical portable electronic pH electrodes

Water hardness

- *Water hardness* refers to the difficulty of soap to lather in this water. The presence of high levels of calcium ions and magnesium ions makes water hard. Instead of the soap lathering the soap anions are precipitated by the calcium or magnesium ions.
- Sydney water is classified as soft (the hardness in various dams varies from 10–40 ppm). Adelaide water is moderately hard (up to 140 ppm).
  Soft water: < 75 ppm
  Hard water: 150–500 ppm
- *Total hardness* is a measure of the total concentration of calcium and magnesium ions in water expressed as the equivalent amount of dissolved $CaCO_3$ (1 mg $CaCO_3$/L = 1 ppm).
- Total hardness can be determined by titration with the complexing agent called EDTA (ethylene diamine tetraacetic acid), which binds to cations:
  $$Ca^{2+} + EDTA^{4-} \rightarrow CaEDTA^{2-}$$
- Monitoring total dissolved solids, water acidity and water hardness are important ongoing measurements conducted by environmental and analytical chemists. In this way we can take quick action to remedy any emerging water quality issues.

EXAMPLE 2

20.00 mL of hard water was pipetted into a conical flask. The solution was titrated with a standard solution of 0.0100 mol/L EDTA using Eriochrome Black T indicator, as shown in Figure 14.3. At the red-to-blue end point the titre was 8.90 mL. Calculate the total hardness of the water. Assume the hardness is totally due to dissolved calcium carbonate.

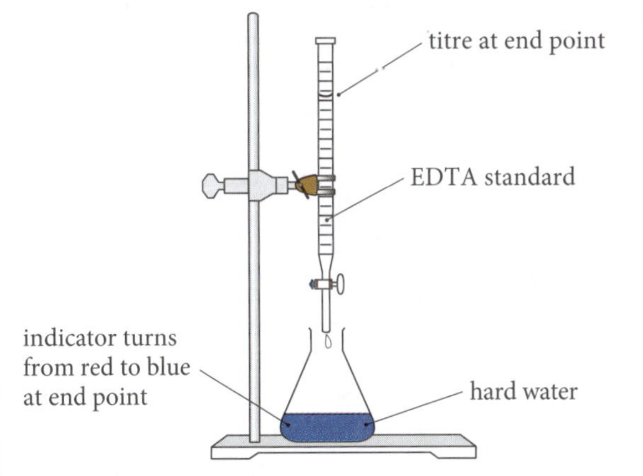

Figure 14.3 Titration of hard water

Answer:

Determine the concentration of dissolved calcium carbonate

Calculate the number of moles of EDTA.

$n(EDTA) = cV = (0.0100)(0.00890) = 8.90 \times 10^{-5}$ mol
Reaction stoichiometry = 1 : 1
$n(Ca^{2+}) = 8.90 \times 10^{-5}$ mol

The calcium ions are present due to calcium carbonate undergoing dissolution in the water.
$n(CaCO_3) = 8.90 \times 10^{-5}$ mol

Calculate the mass of $CaCO_3$.
$M(CaCO_3) = 40.08 + 12.01 + 3(16.00) = 100.09$ g/mol
$m(CaCO_3) = n.M = (8.90 \times 10^{-5})(100.09) = 8.91 \times 10^{-3}$ g
$= 8.91$ mg

$c(CaCO_3) = n/V = \frac{8.91}{0.0200} = 446$ mg/L $= 446$ ppm

This water sample is very hard.

Testing and monitoring air quality

- Increasing pollution of the atmosphere in the 1960s led to governments introducing pollution controls. The increasing combustion of coal to make electricity and the increased use of motor vehicles after World War II and the resulting increase in oil consumption led to large increases in the levels of SO_2 and NO_2 in the atmosphere. Gas analysis technologies such as gas chromatography were introduced in the 1970s and this has allowed chemists to monitor the global increase in acidic oxide emissions in the atmosphere.
- The gaseous emissions from the combustion of coal can be 'scrubbed' to remove acidic oxides such as sulfur dioxide. Scrubbing involves the passing of emissions through a calcium hydroxide slurry. The slurry is a mixture of solid and dissolved calcium hydroxide. The alkaline slurry neutralises acidic gases and converts them into soluble or insoluble salts:

$$SO_2(g) + Ca(OH)_2(aq) \rightarrow CaSO_3(s) + H_2O(l)$$

- Carbon monoxide is a lower-atmospheric pollutant. Carbon monoxide is produced by the process of incomplete combustion. Road traffic emissions account for up to 90% of carbon monoxide emissions. Carbon monoxide is toxic because it is readily absorbed by haemoglobin molecules and transported by the blood. This causes tiredness and headaches as less oxygen reaches the brain.
- Exhaust gas emissions from road traffic are also the principal sources of nitrogen oxides such as NO and NO_2. In the combustion chamber of a car, nitrogen and oxygen combine in the presence of a spark to form nitric oxide (NO). When NO is emitted from the car's exhaust it reacts with oxygen to form NO_2. Ozone (O_3) is a toxic gas above 20 ppm and it is produced by the interaction of nitrogen dioxide and UV light from the Sun. Oxygen free radicals (O•) are intermediates:

$$NO_2(g) + \text{UV radiation} \rightarrow NO(g) + O\bullet(g)$$
$$O\bullet(g) + O_2(g) \rightarrow O_3(g)$$

- The Environment Protection Agency (EPA) monitors the levels of pollutant gases in the atmosphere in many regions across Australia. The introduction of catalytic converters in vehicle exhaust systems greatly reduced carbon monoxide emissions from vehicle exhaust. Catalytic converters are made from alloys of rhodium and platinum that operate at temperatures as low as 150 °C. They speed up reactions that convert pollutant gases to materials that are present in the air naturally.

Table 14.1 compares pollutant gas levels during a weekday in November 2017 and January 2018 in various areas of Sydney. These recordings are classified as very good (0–33 ppm), good (34–66 ppm) or fair (67–99 ppm) air quality.

Table 14.1 Comparison of average levels of O_3, NO_2 and CO in various monitoring sites in Sydney on two weekdays in November 2017 and January 2018

Monitoring site	1 hour average concentration of O_3 (ppm)	1 hour average concentration of NO_2 (ppm)	8 hour average concentration of CO (ppm)
Rozelle (east)	11 (Nov); 45 (Jan)	10 (Nov); 37 (Jan)	1 (Nov); 1 (Jan)
Chullora (central west)	24 (Nov); 80 (Jan)	1 (Nov); 23 (Jan)	2 (Nov); 4 (Jan)
Campbelltown (south west)	35 (Nov); 79 (Jan)	3 (Nov); 23 (Jan)	3 (Nov); 5 (Jan)

Source: New South Wales Government Office of Environment and Heritage (2018), Air quality data; viewed online March 2018 at www.environment.nsw.gov.au/AQMS/hourlydata.htm.

- Constant monitoring of air quality is essential in order to ensure that warnings can be given to the public when air quality is very poor. In particular, people who suffer from lung disease need to take immediate action.

Testing and monitoring soil salinity

- Soil salinity is a major problem in some parts of Australia. Australian soils naturally contain various levels of salts which have been deposited by winds carrying salt water from the oceans. The native Australian vegetation evolved to be salt-tolerant but this soil is unsuitable for agriculture as the excess salt prevents pasture and grass growth. Deforestation that causes salty groundwater to rise to the surface also increases soil salinity and prevents growth of grasses. The unauthorised removal of trees on properties is therefore illegal.
- Chemists can determine salinity levels in soil by measuring the total dissolved solids (TDS) using conductivity electrodes. The salts in a known mass of dry soil are extracted into a known volume of water and the salty solution's conductivity is measured and compared with standards.

Salinity calculation:
A 20 g soil sample was added to 100 mL of water and the salts extracted. The TDS of the solution was found to be 760 ppm (760 mg/L). Thus the mass of salts in 100 mL was 76 mg. Thus the salinity of the soil $= \frac{76 \text{ mg}}{20 \text{ g}} = 3800$ mg/kg
$= 3800$ ppm.

Below 3000 ppm the soil is classified as non-saline. The salinity of the soil in this experiment is classified as slightly saline.

- Soil quality monitoring is vital to determine whether the quality of agricultural soil is declining and whether industrial waste dumping has affected soil quality.

KEY QUESTIONS

1 **Explain the importance of monitoring the quality of potable water.**

2 **A water sample from a stream was found to have a pH of 5.20. Describe possible causes of this acidity.**

3 **Identify the cause of water hardness and describe how the hardness of the water can be determined.**

4 **Explain how an analytical chemist can measure the total dissolved solids in a river during field work.**

Answers ➲ p. 207

2 Qualitative analysis of ions

» Students conduct qualitative investigations—using flame tests, precipitation and complexation reactions as appropriate—to test for the presence in aqueous solution of the following ions:
- Cations—barium (Ba^{2+}), calcium (Ca^{2+}), magnesium (Mg^{2+}), lead (II) (Pb^{2+}), silver ion (Ag^{+}), copper (II) (Cu^{2+}), iron (II) (Fe^{2+}), iron (III) (Fe^{3+}).
- Anions—chloride (Cl^{-}), bromide (Br^{-}), iodide (I^{-}), hydroxide (OH^{-}), acetate (CH_3COO^{-}), carbonate (CO_3^{2-}), sulfate (SO_4^{2-}), phosphate (PO_4^{3-}).

Cation analysis

- One qualitative test alone is not sufficient to identify a cation. Confirmatory tests must be conducted. These tests are normally conducted in a specific order so that false conclusions are not made. This type of testing is called *elimination testing*.
- Solutions of some cations are coloured:
  - Copper (II) ions in aqueous solution are blue or bluish-green.
  - Iron (III) ions in aqueous solution are yellow or yellow-orange.
- Table 14.2 lists some precipitation tests for the identification of cations. Drops of dilute hydrochloric acid, sulfuric acid and sodium hydroxide are added to solutions containing the nitrate salts of various cations.

Table 14.2 Precipitation tests for cations

Cation	1 Dilute hydrochloric acid	2 Dilute sulfuric acid	3 Dilute sodium hydroxide
Ag^{+}	white precipitate	faint white precipitate*	brown precipitate#
Pb^{2+}	white precipitate* (dissolves on heating)	white precipitate	white precipitate
Cu^{2+}	no reaction	no reaction	blue precipitate (slowly darkens)
Fe^{2+}	no reaction	no reaction	green precipitate (quickly turns brown at surface)
Fe^{3+}	no reaction	no reaction	orange-brown precipitate
Ca^{2+}	no reaction	white precipitate	faint white precipitate*
Ba^{2+}	no reaction	white precipitate	no reaction
Mg^{2+}	no reaction	no reaction	white precipitate

(* If the cation concentration is too low, no precipitate may form.)
(# Hydrated silver oxide forms.)

- Confirmatory tests can also be performed to identify specific cations.
- Table 14.3 describes tests to confirm the presence of lead (II) and silver ions, as well as distinguishing between calcium and barium ions and iron (II) and iron (III) ions. Silver ions and copper (II) ions form a complex ion with ammonia solution.
- Flame tests can also be used as confirmatory tests. They were investigated in the Year 11 course. Flame tests can be conducted by using a spray bottle to spray a solution of the cation into a blue Bunsen flame, or by dipping a platinum wire in a solution of the compound dissolved in hydrochloric acid and then placing the wire in the flame. Figure 14.4 demonstrates these methods. The following cations produce characteristic colours:
  - calcium ions—orange-red flame
  - barium ions—pale yellow-green flame
  - copper (II) ions—green (or blue-green) flame
- Lead (II) ion solutions should **not** be flame tested because lead vapour is toxic.

Table 14.3 Confirmatory tests for cations

Cation	Test
Pb^{2+}	1 Add KI solution. A bright-yellow precipitate of lead (II) iodide forms. 2 Add NaOH to form a white precipitate of $Pb(OH)_2$. Add excess NaOH and the white precipitate dissolves to form a colourless solution of PbO_2^{2-}. 3 Add ammonia solution and a white precipitate of $Pb(OH)_2$ forms. Add excess ammonia but the precipitate does not dissolve.
Ag^+	1 Collect the white precipitate of AgCl formed when HCl is added. Place the AgCl in a test tube and add ammonia solution. The precipitate dissolves to form the colourless $Ag(NH_3)_2^+$ ion. 2 Collect the brown precipitate of silver oxide when NaOH is added to the silver ion solution. Place the brown solid in a test tube and add excess ammonia solution. The precipitate dissolves to form the colourless $Ag(NH_3)_2^+$ ion.
Ca^{2+}/Ba^{2+}	Test separate solutions of calcium ions and barium ions with sodium fluoride solution. Only the calcium solution will form a white precipitate (CaF_2).
Fe^{2+}	1 Add $K_3Fe(CN)_6$ solution. A deep-blue solution containing $Fe_3(Fe(CN)_6)_2$ forms. 2 Add ammonia solution. A green precipitate ($Fe(OH)_2$) forms.
Fe^{3+}	1 Add KSCN solution. A blood-red solution containing $FeSCN^{2+}$ forms. 2 Add ammonia solution. A brown precipitate ($Fe(OH)_3$) forms.
Cu^{2+}	Add NH_3 solution. A blue precipitate of $Cu(OH)_2$ initially forms, which dissolves in excess ammonia to form a deep-blue solution of $Cu(NH_3)_4^{2+}$.

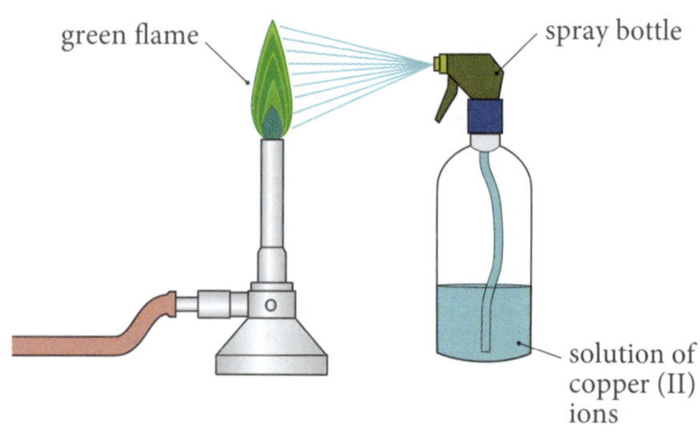

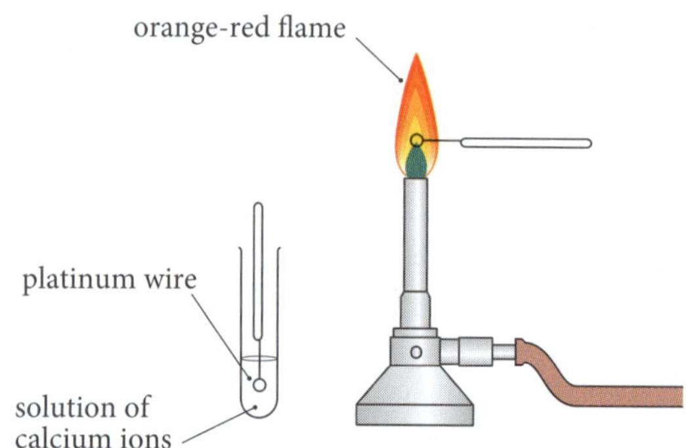

Figure 14.4 Flame test for copper (II) ions and calcium ions

➔ The flame test has various limitations, which can lead to inconclusive results. These include the following:

- The cation concentration is too low.
- Some cations such as sodium ions often contaminate samples and the bright yellow of its flame may mask the presence of other ions.
- Not all cations produce a visible flame colour.

EXAMPLE 3

A colourless solution was tested to determine the two cations present. The results of the investigation are:

- **Test 1: Excess hydrochloric acid produces a white precipitate.**
- **Test 2: The precipitate dissolves on heating and reforms on cooling.**
- **Test 3: The white precipitate from test 1 was filtered off and the filtrate divided in two. The first half was tested with sulfuric acid and a white precipitate formed.**
- **Test 4: The second sample of filtrate was tested with sodium hydroxide and a white precipitate formed.**
  - **a Determine the cations that are present.**
  - **b Write ionic equations for the precipitation reactions.**
  - **c Describe additional tests that will confirm the presence of the two cations.**

Review Tables 14.1 and 14.2 to answer this question

Answer:

a HCl test: Ag^+ and Pb^{2+} may be present as both form white chloride precipitates.

The heating and cooling test suggests $PbCl_2$ is the precipitate and that Pb^{2+} is present.

H_2SO_4 test: Ba^{2+} and Ca^{2+} may be present as both form white sulfate precipitates.

NaOH test: Ca^{2+} and not Ba^{2+} is present as barium ions do not form insoluble hydroxides.

b $Pb^{2+}(aq) + 2Cl^-(aq) \rightarrow PbCl_2(s)$

$Ca^{2+}(aq) + 2OH^-(aq) \rightarrow Ca(OH)_2(s)$

c Confirmatory test for Pb^{2+}: Add KI solution to the unknown and a bright-yellow precipitate of PbI_2 will form.

$Pb^{2+}(aq) + 2I^-(aq) \rightarrow PbI_2(s)$

Confirmatory test for Ca^{2+}: Flame test will produce an orange-red flame.

EXAMPLE 4

A yellow solution contains the following salts: barium nitrate, iron (III) nitrate and silver nitrate.

Describe tests and confirmatory tests that will show the presence of barium ions, iron (III) ions and silver ions. Include relevant equations.

Answer:

Some ions may need to be removed before testing for remaining ions

- Test 1

  Add excess HCl to a sample of the original solution. A white precipitate will show the presence of silver ions because magnesium and iron (III) ions have soluble chlorides:

  $Ag^+(aq) + Cl^-(aq) \rightarrow AgCl(s)$

- Test 2

  Filter off the AgCl precipitate and collect the filtrate. Add excess sulfuric acid to the filtrate. A white precipitate shows the presence of barium ions. Iron (III) ions have a soluble sulfate:

  $Ba^{2+}(aq) + SO_4^{2-}(aq) \rightarrow BaSO_4(s)$

- Test 3

  Filter off the $BaSO_4$ precipitate and test the yellow filtrate with excess NaOH solution. An orange-brown precipitates shows the presence of iron (III) ions:

  $Fe^{3+}(aq) + 3OH^-(aq) \rightarrow Fe(OH)_3(s)$

- Confirmatory tests

  Add ammonia solution to the white AgCl formed in test 1. The precipitate dissolves to form a colourless solution of diamminesilver (I) ions:

  $AgCl(s) + 2NH_3(aq) \rightarrow Ag(NH_3)_2^+(aq) + Cl^-(aq)$

  A pale yellow-green flame test will confirm barium ions are present.

  The yellow colour of the original solution suggests iron (III) ions are present. If KSCN solution is added a blood-red solution forms, which confirms iron (III) ions are present:

  $Fe^{3+}(aq) + SCN^-(aq) \rightarrow FeSCN^{2+}(aq)$

FIRSTHAND INVESTIGATION 1

Cation identification

» Students conduct qualitative investigations—using flame tests, precipitation and complexation reactions as appropriate—to test for the presence in aqueous solution of cations: barium (Ba^{2+}), calcium (Ca^{2+}), magnesium (Mg^{2+}), lead (II) (Pb^{2+}), silver ion (Ag^+), copper (II) (Cu^{2+}), iron (II) (Fe^{2+}) and iron (III) (Fe^{3+}).

Aim

to identify common cations in solution using qualitative tests

Method

A Three solutions are formed containing the following cations as their nitrate salts for solutions 1 and 2. For solution 3 the sulfate salts are used.

Solution 1: Ag^+, Pb^{2+}

Solution 2: Ca^{2+}, Ba^{2+}, Mg^{2+}

Solution 3: Cu^{2+}, Fe^{2+}, Fe^{3+}

B To identify each cation, elimination tests are performed on samples of each solution.

- Solution 1: Ag^+/Pb^{2+} testing
  1. Prepare a mixed solution of silver nitrate and lead (II) nitrate.
  2. Add HCl to the solution of silver nitrate and lead (II) nitrate. A mixed white precipitate (containing AgCl and $PbCl_2$) is observed. Add excess HCl until all precipitation ceases.
  3. Heat the mixture and the lead (II) chloride will dissolve. Filter off the white AgCl and collect the filtrate. Test the filtrate with KI solution. A bright-yellow precipitate of PbI_2 forms.
  4. Add the white AgCl precipitate from step 2 to a test tube and add excess ammonia solution. The precipitate dissolves to form a colourless solution of silver diammine ions ($Ag(NH_3)_2^+$).
- Solution 2: $Ca^{2+}/Ba^{2+}/Mg^{2+}$ testing
  1. Prepare a mixed solution of calcium nitrate, barium nitrate and magnesium nitrate.
  2. Divide the solution of calcium nitrate, barium nitrate and magnesium nitrate into two test tubes for testing.
  3. Add H_2SO_4 to the first tube. A white mixed precipitate of $BaSO_4$ and $CaSO_4$ forms. Filter off the white precipitate and collect the filtrate containing the Mg^{2+} ion. Add excess NaOH solution to neutralise excess acid and make the solution basic. Once the solution is basic a white precipitate of $Mg(OH)_2$ forms.

4. Add NaF solution to the second tube. A white precipitate of calcium fluoride forms. The fluorides of barium and magnesium ions are sparingly soluble. Filter off the calcium fluoride precipitate and perform a flame test on the filtrate. A pale yellow-green flame indicates barium ions.

- Solution 3: $Cu^{2+}/Fe^{2+}/Fe^{3+}$ testing
  1. Prepare separate solutions of copper (II) sulfate, iron (II) sulfate and iron (III) sulfate. Use these solutions for tests in step 6. Create a mixed solution of these salts for tests in steps 3, 4 and 5.
  2. Divide the solution of copper (II) sulfate, iron (II) sulfate and iron (III) sulfate into three test tubes.
  3. To the first tube, add potassium hexacyanoferrate (III) (also called potassium ferricyanide ($K_3Fe(CN)_6$)) solution. A deep-blue solution containing $Fe_3(Fe(CN)_6)_2$ forms. This colour is a positive test for Fe^{2+}.
  4. To the second tube, add potassium thiocyanate (KSCN) solution. A blood-red solution forms containing the $FeSCN^{2+}$ ion. This colour is a positive test for Fe^{3+}.
  5. To the third tube add *excess* ammonia (NH_3) solution. A deep-blue solution and an orange solid forms. The solid is a mixture of iron (II) hydroxide and iron (III) hydroxide. The deep-blue supernatant solution is a positive test for Cu^{2+}. The deep-blue $Cu(NH_3)_4{}^{2+}$ ion has formed.
  6. Repeat the experiment with separate solutions of each cation. Report your observations. See Figure 14.5 to compare your results.

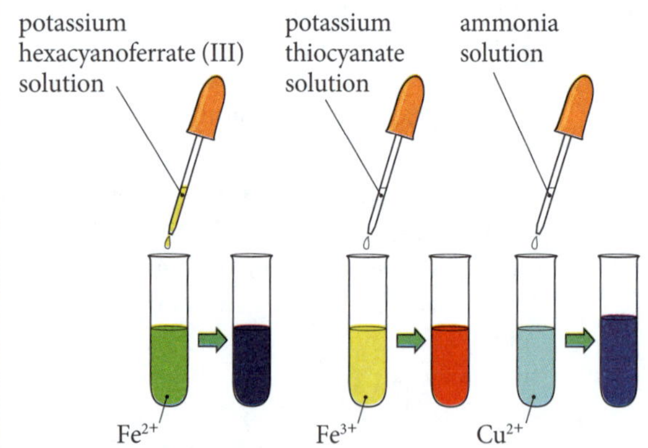

Figure 14.5 Testing for iron (II) ions, iron (III) ions and copper (II) ions

Analysis

Tabulate all your observations and write equations as appropriate.

➔ Various online videos can be used to reinforce your understanding of this qualitative investigation. Type the following headings into a search bar:
- ammonia solution used to test for solutions
- cation test—lead (II) ions
- cation test—calcium ions
- testing for ions and determining ions in unknown samples.

EXAMPLE 5

A solution contains four cations: Ca^{2+}, Fe^{2+}, Pb^{2+}, Cu^{2+}. The flowchart in Figure 14.6 shows an elimination procedure to identify these ions. Use this information to identify the three precipitates and the ion that produced the deep-blue solution C. Write relevant equations for each reaction.

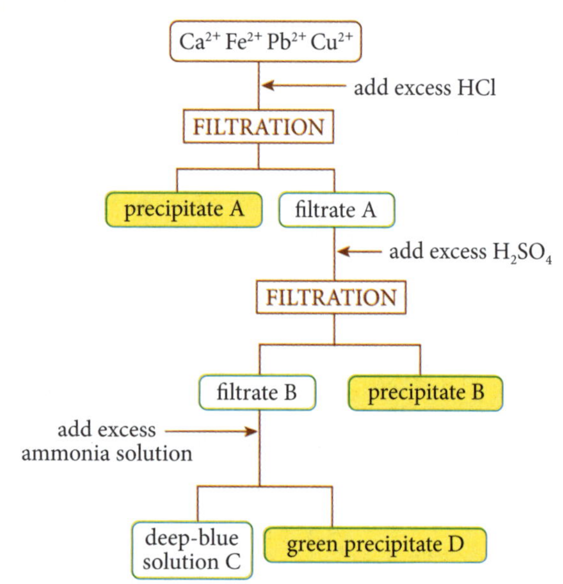

Figure 14.6 Flowchart to identify ions

Identify ions that form insoluble chlorides and sulfates

Answer:

Precipitate A is lead (II) chloride ($PbCl_2$) because the other cations have soluble chlorides:

$$Pb^{2+}(aq) + 2Cl^-(aq) \rightarrow PbCl_2(s)$$

Filtrate A contains Ca^{2+}, Fe^2 and Cu^{2+}.

Precipitate B is calcium sulfate ($CaSO_4$) because the other cations have soluble sulfates:

$$Ca^{2+}(aq) + SO_4{}^{2-}(aq) \rightarrow CaSO_4(s)$$

Filtrate B contains Fe^{2+} and Cu^{2+}.

The deep-blue solution C contains $Cu(NH_3)_4{}^{2+}$ and the green precipitate D is $Fe(OH)_2$:

$$Cu^{2+}(aq) + 4NH_3(aq) \rightarrow Cu(NH_3)_4{}^{2+}(aq)$$

$$Fe^{2+}(aq) + 2OH^-(aq) \rightarrow Fe(OH)_2(s)$$

➔ KEY QUESTIONS

5 **Identify qualitative tests that can be used to identify silver and lead (II) ions in solution.**

6 **A solution contains both iron (II) and iron (III) ions. Describe tests that will show the presence of both ions.**

7 **A solution contains magnesium, calcium and barium ions. Describe tests that will show the presence of these ions.**

Answers ➲ p. 207

Anion analysis

➔ The anions to be identified are OH^-, CO_3^{2-}, CH_3COO^-, SO_4^{2-}, PO_4^{3-}, Cl^-, Br^- and I^-.

➔ Various acid–base and precipitation reactions will be used to identify these ions, which are present in aqueous solution as sodium or potassium salts.

- Test 1: pH test

  Use a pH meter or universal indicator to determine whether the separate anion solutions are acidic, basic or neutral. The results are:

  Neutral: Cl^-, Br^-, I^-, SO_4^{2-}

  Basic: OH^-, CH_3COO^-, CO_3^{2-}, PO_4^{3-}

- Test 2: Reaction with nitric acid

  Add dilute nitric acid to each anion solution. Only the carbonate ion solution produces a visible reaction in which effervescence occurs with the production of carbon dioxide gas. The limewater test confirms the identity of the carbon dioxide:

  $$CO_3^{2-}(aq) + 2H_3O^+(aq) \longrightarrow 3H_2O(l) + CO_2(g)$$

  Hydroxide ions are neutralised by hydronium ions. No colour change is seen:

  $$OH^-(aq) + H_3O^+(aq) \longrightarrow 2H_2O(l)$$

- Test 3: Precipitation reactions

  Table 14.4 shows the results of solubility tests in which drops of various cation solutions are added to the separate solutions of the sodium salts of various anions. Figure 14.7 shows a test for lead (II) ions using KI solution.

Table 14.4 Anion precipitation tests

Anion	1 Dilute $Ba(NO_3)_2$	2 Dilute $AgNO_3$	3 Dilute $Pb(NO_3)_2$
CO_3^{2-}	white precipitate	white precipitate	white precipitate
OH^-	no reaction	brown precipitate	white precipitate
CH_3COO^-	no reaction	no reaction	no reaction
SO_4^{2-}	white precipitate	no reaction	white precipitate
PO_4^{3-}	white precipitate	yellow precipitate	white precipitate
Cl^-	no reaction	white precipitate	white precipitate
Br^-	no reaction	cream precipitate	white precipitate
I^-	no reaction	pale yellow precipitate	bright-yellow precipitate

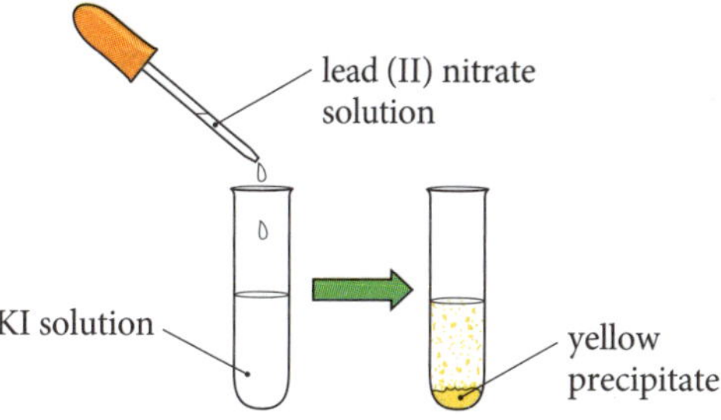

Figure 14.7 Identification of iodide ions

➔ Confirmatory tests can also be performed to identify specific anions. Table 14.5 describes tests to confirm the presence of phosphate and acetate ions.

Table 14.5 Confirmatory tests for selected anions

Anion	Test
PO_4^{3-}	1 Phosphates vary in solubility in acidic and alkaline solution. Place sodium phosphate solution into two test tubes. Acidify the first tube with nitric acid and make the second tube alkaline with ammonia solution. Add barium nitrate solution to each tube. No precipitate forms in the acidic solution but white barium phosphate precipitates in the alkaline solution. 2 Phosphate ions can be detected by the molybdate test. Add ammonium molybdate solution to an acidified phosphate solution. Warm the mixture gently. A yellow precipitate of ammonium phosphomolybdate ($(NH_4)_3PO_4{\cdot}12MoO_3{\cdot}3H_2O$) forms.
CH_3COO^-	1 Add hydrochloric acid to a concentrated sodium acetate solution and heat the mixture. The vapours emitted smell of vinegar (acetic acid). Hold a blue litmus paper in the vapours and the acetic acid turns the paper red. 2 $FeCl_3$ solution will produce a red-coloured precipitate on reaction with acetate ions. (The $FeCl_3$ reagent is prepared by mixing 1 mL of $FeCl_3$ solution in a test tube with 1 mL of ammonia solution to give a red precipitate. Then dilute HCl is then added to the test tube till the precipitate has just dissolved. It is this solution that is then added to the acetate solution and the mixture warmed to produce the red precipitate.)

FIRSTHAND INVESTIGATION 2

Anion identification

» Students conduct qualitative investigations—using flame tests, precipitation and complexation reactions as appropriate—to test for the presence in aqueous solution of anions: chloride (Cl^-), bromide (Br^-), iodide (I^-), hydroxide (OH^-), acetate (CH_3COO^-), carbonate (CO_3^{2-}), sulfate (SO_4^{2-}) and phosphate (PO_4^{3-}).

Aim

to identify common anions in solution using qualitative tests

Method

A All the following anions are present in an aqueous solution as sodium salts:

$$OH^-, CO_3^{2-}, CH_3COO^-, SO_4^{2-}, PO_4^{3-}, Cl^-, Br^-, I^-$$

B To identify each anion, elimination tests are performed on samples of this solution.

- Test 1: Testing pH

  Use universal indicator paper to determine whether the anion solution is alkaline (pH > 7). Apart from the strong base NaOH, various salts can make a solution alkaline (e.g. Na_2CO_3, $NaCH_3COO$, Na_3PO_4) due to hydrolysis of the anion in the water.

- Test 2: Testing with strong acids

  a Add dilute nitric acid to the anion solution. An effervescence indicates the presence of carbonate ions, which react with hydronium ions to produce gaseous carbon dioxide.

  b Add an excess of HNO_3 to remove all carbonate ions and hydroxide ions. Heat the mixture and test the vapours for acetic acid using a piece of damp blue litmus paper, which should turn red. Thus the presence of acetate ions has been demonstrated.

  c Allow the solution to cool and use this solution for test 3.

  d The $FeCl_3$ confirmation test for acetate ions (see Table 14.5) can also be conducted on a separate sample.

- Test 3: Testing with barium nitrate solution

  Use the acidified solution from test 2.

  a Add excess barium nitrate solution. A white precipitate of barium sulfate will form. Other ions will not precipitate. The sulfate ion has been detected.

  b Filter off the precipitate and collect the filtrate for test 4.

- Test 4: Testing with ammonia solution

  Use the filtrate from test 3.

  a Neutralise the acidic filtrate with ammonia solution to produce an alkaline solution (~pH 11). The excess barium ions will now precipitate white barium phosphate. Ensure all the phosphate ions are precipitated by adding additional barium nitrate.

  b Filter the mixture and collect the filtrate for test 5.

  c The phosphomolybdate test for phosphate ions (see Table 14.5) can also be performed on the filtrate from test 3.

- Test 5: Testing for halide ions

  Re-acidify the filtrate with nitric acid.

  a Add excess silver nitrate solution to a sample of the filtrate. A mixed white-yellow precipitate of AgCl, AgBr and AgI forms.

  b Filter off the mixed precipitate and place it in a new test tube. Add dilute ammonia solution. Stir to mix and heat. The AgCl will dissolve. Filter this mixture and retain the solid residue and filtrate.

  Reacidify the filtrate with nitric acid and white silver chloride should precipitate, thus confirming the presence of chloride ions. The white AgCl is light sensitive and turns grey in sunlight as silver forms.

  c Place the solid residue from step **b** in a tube and add concentrated ammonia solution. The AgBr will dissolve, leaving the pale yellow AgI solid. Thus iodide ions are detected. The AgI is not light sensitive.

  d The ammoniated solution from step **c** is acidified with excess hydrobromic acid (HBr), and creamy AgBr precipitate will form. Thus bromide ions are detected. The cream AgBr is light sensitive and turns grey in sunlight as silver forms.

Analysis

Tabulate all your observations and write equations as appropriate.

➔ Various online videos can be used to reinforce your understanding of this qualitative investigation. Type the following headings into a search bar:

- chemical tests for acetate
- halides in solution—test using acidified silver nitrate
- halides in solution—test using acidified silver nitrate and ammonia.

EXAMPLE 6

A solution contains three anions, PO_4^{3-}, CO_3^{2-} and Br^-. The flowchart in Figure 14.8 shows a procedure to identify these ions. Use this information to identify the ion that caused the effervescence and the ions that produced precipitates B and C. Write relevant equations.

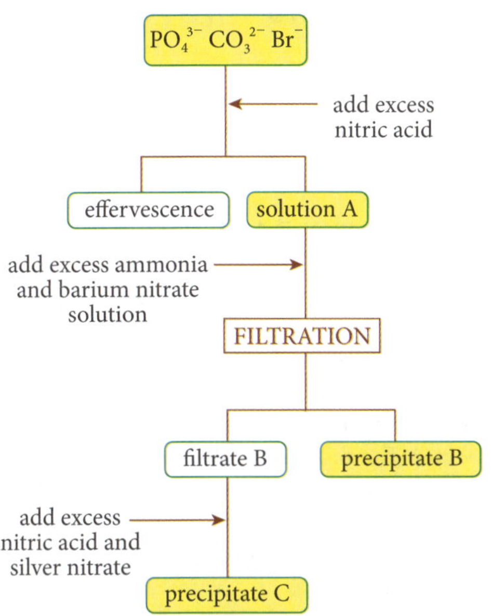

Figure 14.8 Flow chart for anion identification

The gas evolved gives a positive limewater test

Answer:

The carbonate ion reacts with hydronium ions in the nitric acid and carbon dioxide gas is released as a gaseous **effervescence**:

$$CO_3^{2-}(aq) + 2H_3O^+(aq) \rightarrow 3H_2O(l) + CO_2(g)$$

effervescence: foaming or fizzing in a liquid due to the production of gas bubbles

Solution A contains phosphate and bromide ions. The addition of ammonia to solution A raises the pH and barium ions precipitate the phosphate ions. Precipitate B is barium phosphate:

$$3Ba^{2+}(aq) + 2PO_4^{3-}(aq) \rightarrow Ba_3(PO_4)_2(s)$$

The nitric acid neutralises excess ammonia and lowers the pH. The silver nitrate causes the bromide ion to precipitate as cream silver bromide. Precipitate C is silver bromide:

$$Ag^+(aq) + Br^-(aq) \rightarrow AgBr(s)$$

➔ KEY QUESTIONS

8 **Identify a simple test to show the presence of carbonate ions in an aqueous solution.**

9 **A solution contains both sulfate and phosphate ions. Describe tests that will show the presence of both anions.**

10 **A solution contains both chloride ions and iodide ions. Describe tests that will show the presence of both anions.**

Answers ➲ p. 207

3 Quantitative analysis of ions

» Students conduct investigations and/or process data involving gravimetric analysis and precipitation titrations.

Gravimetric analysis

➔ The composition of ionic compounds or mixtures or the concentration of ions in solution can often be determined by gravimetric analysis involving precipitation reactions.

➔ The sulfate ion content of a soluble fertiliser can be determined gravimetrically.

FIRSTHAND INVESTIGATION 3

Sulfate ion content of a fertiliser

» Students conduct investigations and/or process data involving gravimetric analysis.

Aim

to determine the weight percentage of sulfate ions in a soluble fertiliser

Method

1 Accurately weigh out 1.25 g of the soluble fertiliser into a clean 250 mL beaker. Add 25 mL of warm water to dissolve the crystals. Acidify the solution with dilute nitric acid. Heat the mixture until it is near boiling.

2 Add an excess of barium nitrate solution slowly to the hot sulfate solution until no further white precipitate forms. Allow the precipitate to coagulate during heating. Allow the mixture to cool after heating.

3 Weigh a circle of filter paper. Fold the filter paper into a cone and set up the apparatus for filtration.
4 Filter the suspension of barium sulfate. Wash the precipitate with warm water.
5 Transfer the opened filter paper to a pre-weighed clock glass and allow to dry in a low temperature oven.
6 Weigh to constant dryness. Calculate the mass of barium sulfate collected.

Validity

Quantitative precipitation will occur if the solution is acidified before excess barium chloride solution is added. If not then other anions such as phosphate and carbonate ions may precipitate when barium nitrate is added.

Accuracy issues

The accuracy of a measurement is measured by how close the measurement is to the true or accepted value. The accuracy of a gravimetric analysis can be improved in several ways:

1 Make sure all the precipitate is transferred to the filter paper otherwise the final weight will be too low and the calculated sulfate content will be too low.
2 Make sure the precipitate is washed and dried to constant weight. (Any water remaining will lead to an overestimate of the sulfate content.)
3 Use quantitative, very fine-pore filter papers that have been dried to constant mass prior to use. (If fine particles pass through the filter paper then the calculated sulfate content will be too low.)
4 Add excess barium nitrate solution to make sure all sulfate is precipitated. (If not all the sulfate is precipitated then the calculated sulfate content will be too low.)
5 Use a precision, pre-calibrated electronic balance that weighs to two or three decimal places.

Reliability issues

It is necessary to show that repeated measurements achieve the same result within the limits of experimental design if the results are to be reliable. Statistically it is important to repeat measurements a minimum of five times to ensure reliability. Control all the variables and perform each repeat experiment with the same level of accuracy and precision. When repeated experiments are within a narrow range (i.e. the standard deviation is small) then reliability has been achieved.

Sample results

mass of fertiliser = 1.25 g
mass of dry $BaSO_4$ precipitate = 0.585 g

Sample calculation

$Ba^{2+}(aq) + SO_4^{2-}(aq) \rightarrow BaSO_4(s)$

Calculate the number of moles of barium sulfate.

$$n(BaSO_4) = m/M = \frac{0.585}{(137.3 + 32.07 + 4 \times 16.00)} = \frac{0.585}{233.37}$$
$$= 2.51 \times 10^{-3} \text{ mol}$$

$n(BaSO_4) = n(SO_4^{2-})$
$n(SO_4^{2-})$ in fertiliser $= 2.51 \times 10^{-3}$ mol

Calculate the mass of sulfate.

$$m(SO_4^{2-}) = nM = (2.51 \times 10^{-3})(32.07 + 4 \times 16.00)$$
$$= 0.241 \text{ g}$$

$$\%SO_4^{2-} \text{ by weight in fertiliser} = \frac{0.241}{1.25} \times \frac{100}{1} = 19.3\%\text{w/w}$$

EXAMPLE 7

Sulfur dioxide is used as a preservative in some dried fruits such as dried apricots and peaches. It has strong anti-microbial and anti-oxidant properties.
The amount of sulfur dioxide in a sample of dried apricots was determined gravimetrically. 40.0 g of powdered dried apricots and 100 mL of acidified water were mixed in a blender. The sulfur dioxide dissolved in the water. The mixture was filtered and the dissolved sulfur dioxide in the filtrate was first oxidised to sulfate ions using hydrogen peroxide solution and the the sulfate ions were then precipitated using barium chloride solution.

a The half-equations for the oxidation of sulfur dioxide with hydrogen peroxide are:

$$SO_2(aq) + 2H_2O(l) \rightarrow SO_4^{2-}(aq) + 4H^+(aq) + 2e^-$$
$$H_2O_2(aq) + 2H^+(aq) + 2e^- \rightarrow 2H_2O(l)$$

Write the net equation for the oxidation of sulfur dioxide.

b 0.370 g of barium sulfate was collected following the precipitation reaction. Calculate the percentage by mass of sulfur dioxide in the dried apricots.

Ensure like terms cancel in the net oxidation–reduction equation

Answer:

a The electrons and water molecules cancel out on adding the two half-equations:

$$SO_2(aq) + H_2O_2(aq) + 2H^+(aq) \rightarrow SO_4^{2-}(aq) + 4H^+(aq)$$

Cancel out two hydrogen ions. Net equation:

$$SO_2(aq) + H_2O_2(aq) \rightarrow SO_4^{2-}(aq) + 2H^+(aq)$$

b $Ba^{2+}(aq) + SO_4^{2-}(aq) \rightarrow BaSO_4(s)$

$$n(BaSO_4) = m/M = \frac{0.370}{(137.3 + 32.07 + 4(16.00))} = \frac{0.370}{233.37}$$
$$= 1.585 \times 10^{-3} \text{ mol}$$

$n(SO_4^{2-}) = 1.585 \times 10^{-3}$ mol
Stoichiometry: $SO_2 : SO_4^{2-} = 1 : 1$
$n(SO_2) = 1.585 \times 10^{-3}$ mol

$$m(SO_2) = n.M = (1.585 \times 10^{2-3})(32.07 + 2(16.00))$$
$$= 0.102 \text{ g}$$

$$\%SO_2 \text{ in apricots} = \frac{(0.104)}{40.0} \times \frac{100}{1} = 0.254\%\text{w/w}$$

Precipitation titrations

- The concentration of halide ions such as bromide ions in an aqueous solution can be determined by a precipitation titration.
- The following second-hand data processing investigation will help students to understand how precipitation titrations and calculations are performed.

SECONDARY-SOURCED INVESTIGATION

Determination of bromide ion concentration via precipitation

» Students conduct investigations and/or process data involving precipitation titrations.

Read the experimental procedure and use the collected data to answer the questions.

Experiment

1. Transfer 25.00 mL of a potassium bromide solution to a conical flask using a pipette.
2. A solution of 0.100 mol/L silver nitrate solution is used to fill a burette, which is then mounted above the flask using a retort stand, bosshead and clamp.
3. Add about 1 mL of yellow potassium chromate indicator to the flask.
4. The tap of the burette is slowly opened and silver nitrate solution is added to the flask. The flask should be swirled to mix. The mixture becomes turbid as a precipitate of creamy coloured silvery bromide forms.
5. The silver nitrate is added until the indicator just changed colour to reddish-brown.
6. The volume of silver nitrate added is recorded. This first experiment will probably give a result that is too large (i.e. an outlier) and the volume is used as a guide for repeats.
7. The experiment is repeated until at least three silver nitrate volumes agree within 0.10–0.20 mL. (Note: With continued practice you could achieve an agreement of 0.05 mL.)

   Figure 14.9 shows the titration apparatus.

Sample results

Run	Volume $AgNO_3$ (mL)
1	21.35
2	20.85
3	20.80
4	20.80

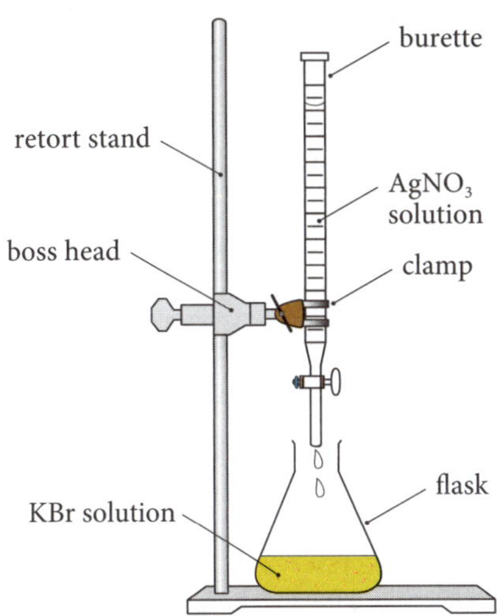

Figure 14.9 Volumetric determination of bromide concentration by precipitation

Sample calculations

1. Average volume calculation:
   The first run is a rough titration and that titre is an outlier and not used to calculate an average.
   Average volume = 20.82 mL
2. Net ionic equation for the precipitation reaction:
   $Ag^+(aq) + Br^-(aq) \longrightarrow AgBr(s)$
   1 : 1 stoichiometry
3. Calculation of bromide ion concentration:
   $n(Ag^+) = n(AgNO_3) = cV = (0.100)(0.020\,82)$
   $= 0.002\,082$ mol
   $n(Br^-) = n(Ag^+) = 0.002\,082$ mol
   $c(Br^-) = \frac{0.002\,082}{0.0250} = 0.0833$ mol/L

KEY QUESTIONS

11 Outline the procedure to quantitatively determine the percentage of sulfate ions in a soluble fertiliser.

12 Outline the procedure to quantitatively determine the concentration of bromide ions in a solution using a precipitation titration.

13 The concentration of chloride ions in solution is determined by gravimetric analysis involving a precipitation reaction. Identify a suitable reagent and explain the importance of washing and drying the precipitate that is collected on a filter paper.

Answers ➲ p. 207

4 Instrumental analytical techniques

» Students conduct investigations and/or process data to determine the concentration of coloured species and/or metal ions in aqueous solution, including but not limited to the use of colourimetry, ultraviolet-visible spectrophotometry and atomic absorption spectroscopy.

Colourimetry

➔ Colourimetry is an instrumental procedure to determine the concentration of a coloured species, such as an ion or molecule, in solution. The components of a colourimeter are:

- Light source and filter—the filter selects a narrow range of wavelengths (λ) of light emitted from the light source (e.g. a filament lamp). Modern instruments use special LED light sources. The visible wavelengths that will pass into the solution will be ones that are absorbed by the coloured species to be analysed.
- Cuvette—this is a tube of constant diameter (l) in which the coloured solution to be measured is placed.
- Reference cell—a cuvette containing the solvent.
- Detector—this is a photocell that converts light energy to an electric current. The current measured is converted to an absorbance (A) measurement on the display meter.

➔ The colourimetry analysis relies on the Beer–Lambert law, which states that the light absorbance A is directly proportional to the concentration c of the coloured species and the path length l of the solution the light passes through. This law is expressed mathematically:

$$A = \varepsilon c l$$

where ε is a constant called the molar absorptivity

This formula is supplied on the Data Sheet in the HSC exam. As long as l is constant then the absorbance A is directly proportional to concentration c.

➔ The absorbance A is determined by measuring the intensity of the light I_0 passing through a reference cell, with the light intensity I passing through the sample cell. The following equation shows that absorbance is a logarithmic function:

$$A = \log_{10}\left(\frac{I_0}{I}\right)$$

This formula is supplied on the Data Sheet in the HSC exam.

➔ The greater the concentration of the coloured species, the more light that is absorbed and the less light that is transmitted to the detector.

- If a coloured solution absorbs 60% of the light passing though it, then:
  $I_0/I = 100/60 = 1.67$
- Thus the absorbance will be:
  $A = \log_{10}(1.67) = 0.22$

➔ The first step in the calibration of the colourimeter is to place a reference sample of water in the cuvette (see Figure 14.10) and set the display to zero absorbance (100% transmittance). The next step is to determine the absorbance with at least five dilution standards. The absorbance of these dilution standards will be used to construct a calibration graph.

➔ Figure 14.10 shows the basic components of the colourimeter.

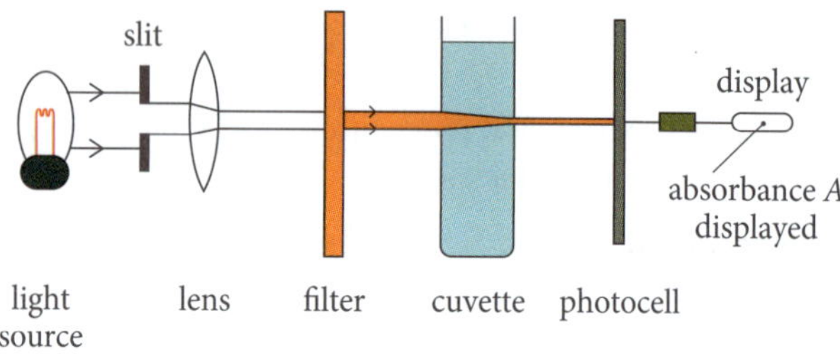

Figure 14.10 Colourimeter

➔ Chemists can measure the levels of fertilisers in soil using a variety of techniques. Analytical chemists can monitor phosphate levels using colourimetry.

➔ Phosphates form a blue complex with molybdate ions. The intensity of this blue complex is proportional to the phosphate concentration.

Experimental analysis

Colourimetric determination of phosphate

➔ A colourimeter is an instrument that measures the concentration of a coloured ion by selective light absorption. The concentration of phosphate ions can be determined using a colourimeter. The phosphate ion is colourless but it can be reacted with ammonium molybdate and ascorbic acid in the presence of sulfuric acid to form a blue complex. The colour intensity of this blue complex is directly proportional to the concentration of phosphate ions in the sample.

1. A standard aqueous solution of a pure phosphate salt is prepared. Various dilutions of this standard are accurately made. Known volumes of the diluted standards are reacted with the colouring reagent and heated to develop the colour.
2. The colour intensity (absorbance, A) of each diluted standard is determined using a colourimeter with an appropriate filter (e.g. orange filter). A standard-reference cuvette of pure water is used.
3. The phosphate in 10 g of dry soil is extracted with warm sulfuric acid. The mixture is filtered. The

filtrate is then treated with ammonium molybdate and ascorbic acid and heated to develop the blue colour. The solution is then diluted to 100 mL. The absorbance of this soil extract is measured.

4 A calibration graph of the dilution standards allows the concentration of phosphate in the soil extract to be determined.

Sample results

Phosphate concentration (ppm)	0	1	2	3	4	Soil sample
Absorbance (*A*)	0	0.020	0.041	0.062	0.080	0.052

Figure 14.11 shows the calibration graph and the interpolation of the absorbance for the soil sample.

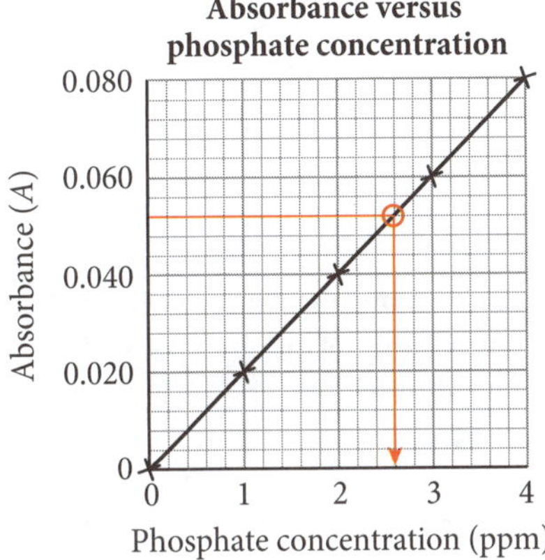

Figure 14.11 Phosphate measurements

Calculation

From Figure 14.11, soil sample extract:

$[PO_4^{3-}]$ = 2.6 ppm = 2.6 mg/L

Volume of soil extract = 100 mL

Mass PO_4^{3-} in 10 g soil = 0.26 mg

$$\text{Phosphate content of soil} = \frac{0.26\text{ mg}}{10\text{ g}} = 26\text{ mg/kg} = 26\text{ ppm}$$

➔ Further information about UV-visible spectroscopy can be found in various online videos. Enter the following titles in a search bar:

- colourimetry
- colourimeter
- Beer's law laboratory

Ultraviolet-visible spectrophotometry

➔ UV-visible spectrometers are used to measure the absorbance of ultraviolet or visible light by a solution. The wavelength range of UV radiation is 190–400 nm, while the range for visible light is 400–700 nm.

➔ The UV-visible spectrophotometer has similar components to a colourimeter.

- One common light source is a tungsten/halogen lamp that produces light in the UV and visible wavelength range.
- A monochromator is a device in which a diffraction grating splits the different wavelengths apart. The wavelength (λ_{max}) of maximum absorbance for the ion in solution is selected by the monochromator.
- The light beam of the selected wavelength is further split into two beams so that one beam passes through the test solution in a cuvette and the other beam passes through a reference solution (called *the blank*) in a second cuvette.
- Differences in light absorbance between the test and reference samples is then detected. The absorbance of light in the full UV-visible wavelength band is progressively measured and the absorbance results are graphed via a computer interface.

➔ Figure 14.12 shows the components of the UV-visible spectrometer.

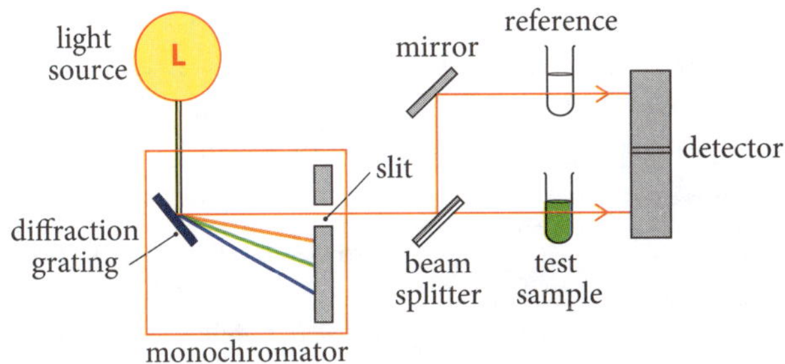

Figure 14.12 UV-visible spectrophotometer

➔ UV-visible spectrophotometry is routinely used to detect and measure the presence and concentration of dyes and organic pigments as well as the presence of specific-coloured metal complex ions. The Beer–Lambert law ($A = \varepsilon cl$, as previously mentioned) can be used to quantitatively determine concentrations.

➔ Cobalt (II) chloride solution is pink due to the presence of the hexaaquacobalt (II) ion ($Co(H_2O)_6^{2+}$). The visible spectrum (Figure 14.13) of this complex ion shows a strong absorption in the blue-green end (511 nm) of the visible spectrum. The complex ion therefore appears pink as there is little absorption of longer wavelength radiation.

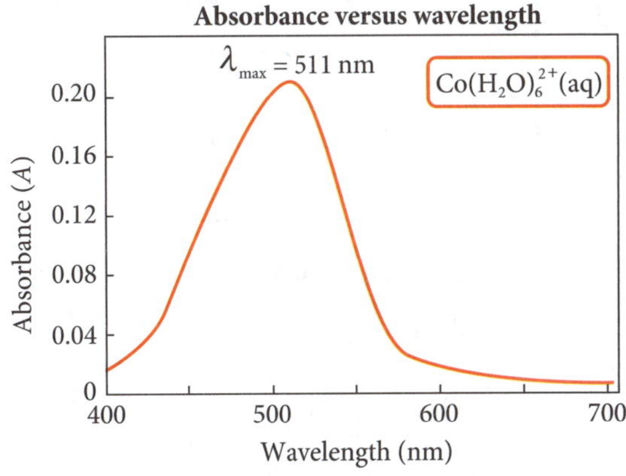

Figure 14.13 Absorption spectrum of hexaaquacobalt (II) ion

➔ Knowledge of the absorption peak maxima (λ_{max}) allows chemists to quantitatively determine the concentration of coloured ions in solution using visible absorption **spectrophotometry**. The following example demonstrates this procedure.

spectrophotometry: a laboratory method used to measure the absorbance of light as a beam of light passes through a solution

EXAMPLE 8

A standard solution of cobalt (II) chloride was prepared. The concentration of the solution was 0.150 mol/L. Three dilution standards were prepared with their concentrations listed in the table. The absorbance of each solution was measured at 511 nm wavelength using a visible absorption spectrometer. Each solution was placed in a cuvette of length 1.0 cm. A distilled-water reference blank was also included.

$Co(H_2O)_6^{2+}$ concentration (mol/L)	0	0.0600	0.0900	0.120	0.150
Absorbance (A)	0	0.321	0.489	0.645	0.818

a **Plot a calibration graph and draw the line of best fit.**

b **The absorbance of an unknown solution of hexaaquacobalt (II) ions ($Co(H_2O)_6^{2+}$) was found to be 0.560. Determine the concentration of hexaaquacobalt (II) ions in the unknown solution.**

Ensure that the graph has a title and a suitable scale for each axis is used

Answer:

a Figure 14.14 is the calibration graph and the interpolation of the absorbance of the unknown solution.

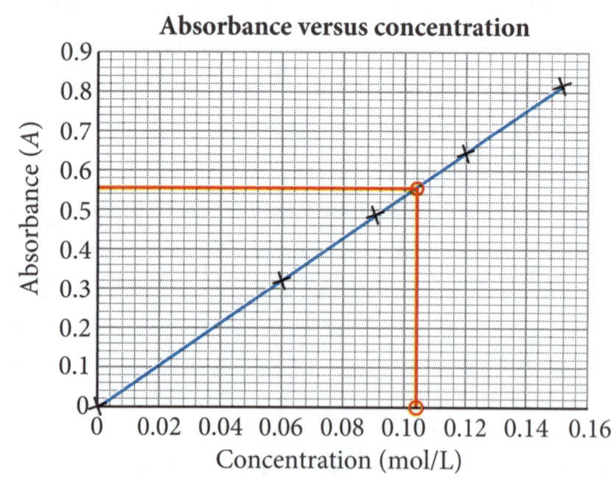

Figure 14.14 Calibration graph

b Concentration of hexaaquacobalt (II) ions in the unknown solution (interpolated from the graph) = 0.103 mol/L

Figure 14.14 has a constant slope over the range of concentrations of hexaaquacobalt (II) ions.

The gradient of the line can be determined from the graph.

(Note: In this example the line passes through all data points and so the points (0,0) and (0.150, 0.818) can be used to calculate the gradient. In examples where the line of best fit does not pass through all data points, then two widely separated points on the line should be used).

$$\text{Gradient} = \frac{0.818}{0.150} = 5.45$$

The equation for this line is:

Absorbance = (gradient)concentration

$\therefore A = 5.45c$

Therefore the concentration of the unknown solution can be calculated from the absorbance measurement.

$$c = A/5.45 = \frac{0.560}{5.45} = 0.103 \text{ mol/L}$$

EXAMPLE 9

A student measured the absorbance of standard purple solutions of potassium permanganate at a wavelength of 522 nm. The calibration graph in Figure 14.15 shows the line of best fit.

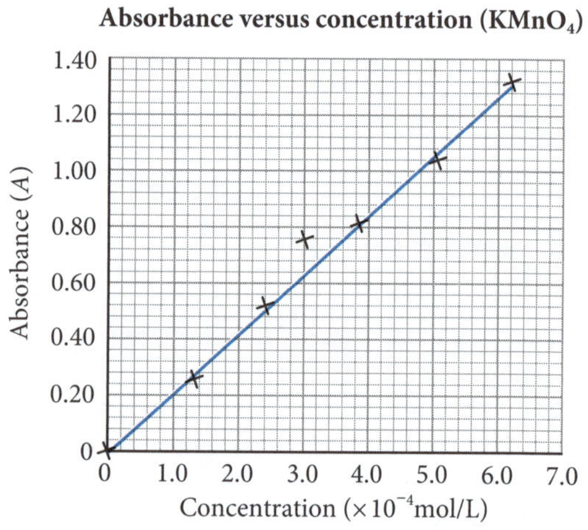

Figure 14.15 Potassium permanganate calibration graph

a **Explain why the data point at 3.0×10^{-4} mol/L was not included in drawing the line of best fit.**

b **A 1 cm cuvette was used in the experiment. Calculate the molar absorptivity (ε) for the potassium permanganate solutions.**

c **25.0 mL of an unknown solution of potassium permanganate solution was diluted to 250.0 mL in a volumetric flask. The diluted solution had an absorbance of 0.64. Calculate the concentration of $KMnO_4$ in the original undiluted solution.**

Answer:

Measure the gradient over a wide range of data points

a The data point is an outlier and should not be used to determine the line of best fit. That point indicates some human or instrumental error was made in the experiment at that time.

b From the Beer–Lambert law:

$A = \varepsilon cl$

$\therefore \varepsilon = A/cl$

A/c = gradient of line of best fit

The gradient was measured between 0 and 4.0×10^{-4} mol/L concentration.

$$\text{Gradient} = \frac{(0.84)}{(4.0 \times 10^{-4})} = 2100 \text{ and } l = 1.0 \text{ cm}$$

$$\varepsilon = \frac{2100}{1.0} = 2100 \text{ L/mol/cm}$$

c Diluted solution: $A = 0.18$

Interpolating from the graph, $c = 3.0 \times 10^{-4}$ mol/L

Using the dilution formula $c_1V_1 = c_2V_2$ (where c_1 = original concentration of undiluted solution):

$c_1(0.0250) = (3.0 \times 10^{-4})(0.2500)$

$c_1 = 3.0 \times 10^{-3}$ mol/L

➔ Organic molecules contain functional groups that strongly absorb light. Double C=C bonds and C=O bonds are examples of functional groups. These groups are called **chromophores** as they are responsible for the absorption of light and the resulting ultraviolet-visible spectrum. For example, the C=O bond in an organic molecule has a typical UV absorbance at 275 nm. Figure 14.16 shows the ultraviolet absorption spectrum for ethylene. The chromophore is the C=C double bond which has a UV absorbance at 163 nm.

In more complex organic molecules with multiple double bonds the maximum absorbance occurs at a different wavelength.

chromophore: a group of atoms in a molecule that is responsible for absorption of ultraviolet-visible light

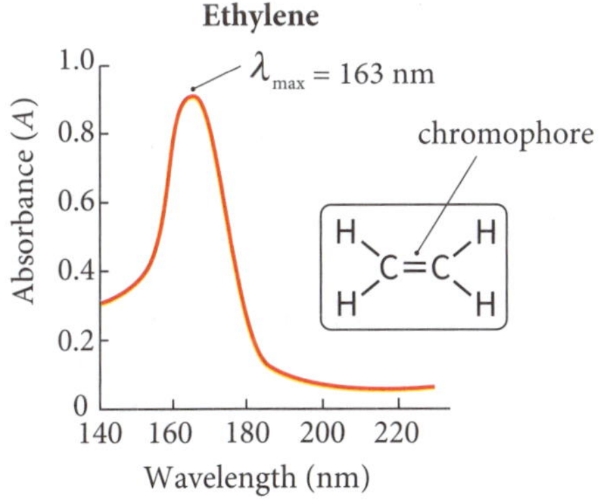

Figure 14.16 UV absorption spectrum of ethylene

➔ Figure 14.17 shows the visible absorption spectrum of chlorophyll-a and chlorophyll-b, which are two of the photosynthetic pigments in green plant leaves. Chlorophyll can be separated from other leaf pigments by a physical separation technique called *column chromatography* using a suitable solvent mixture. A green solution is obtained. The spectra of chlorophyll-a and chlorophyll-b show strong absorption of light in the blue-violet and red regions of visible light. Chlorophyll solutions appear green because of the strong absorption of the blue-violet and red components of visible light.

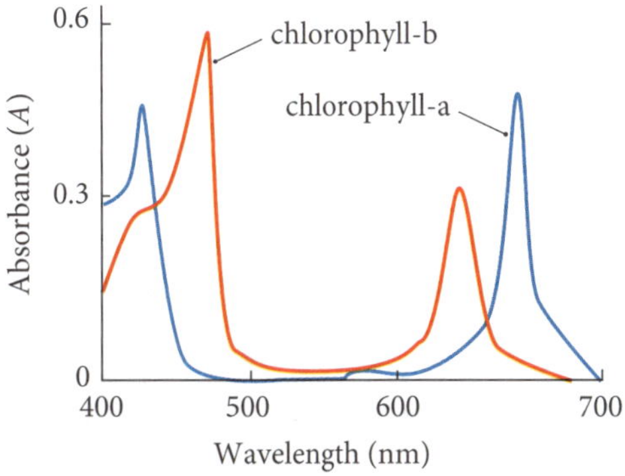

Figure 14.17 Absorption spectrum of chlorophyll-a and chlorophyll-b

➔ The concentration of chlorophyll-a in a solvent extract from green grass leaves can be measured by determining the absorbance A of light at the red wavelength maximum ($\lambda_{max} = 659$ nm). The greater the concentration of chlorophyll-a, the greater the absorbance.

Well, Holmes, was it difficult to determine on which side of the fence the grass was greener?

Elementary, I assure you. I've been instrumental in proving your grass is greeener. How absorbing is that news, my good friend?

CHLOROPHYLL LAB

WATSON HOLMES

Visible Spectrometer

Figure 14.18 Sherlock Holmes investigates chlorophyll-a absorbance using a visible light absorption spectrometer

- Further information about UV-visible spectroscopy can be found in various online videos. Enter the following titles in a search bar:
  - how a simple UV-visible spectrophotometer works
  - UV vis spectroscopy.

EXAMPLE 10

Visible spectroscopy was used to measure the spectra of two aqueous solutions containing X^{2+} and Y^{2+} ions. Figure 14.19 shows the absorbance spectrum of each solution over the same wavelength range on the same axes.

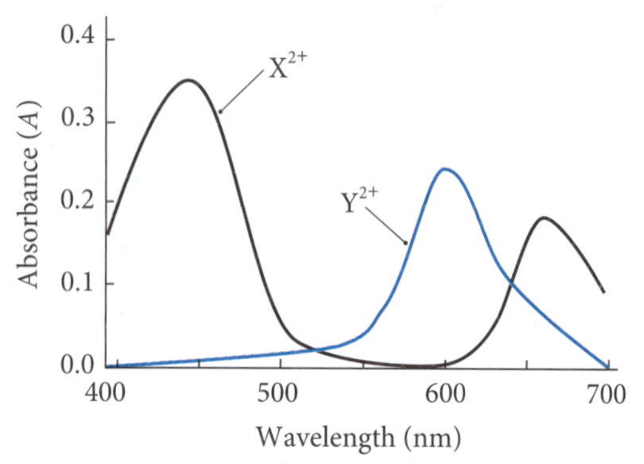

Figure 14.19 Absorbance spectra

a Compare the colours of the solutions of X^{2+} and Y^{2+}.

b 5.0 mL of an X^{2+} solution is mixed with 5.0 mL of a Y^{2+}solution. Identify a suitable wavelength that could be used to measure the absorbance of Y^{2+} in this mixture. Justify your answer.

c Explain what additional experiments would need to be performed to determine the concentration of Y^{2+} in the mixed solution.

Identify the wavelengths of maximum absorbances of both ions

Answer:

a The X^{2+} solution absorbs strongly at blue-violet and red wavelengths. Therefore its solution will appear green. The Y^{2+} solution absorbs strongly at orange-red wavelengths. Therefore its solution will appear blue-violet.

b Y^{2+} has a strong absorbance at 600 nm whereas X^{2+} has a very low absorbance. Thus 600 nm is a suitable wavelength.

c A standard solution of Y^{2+} is prepared and systematically diluted to produce at least five solutions. The absorbance of each of these dilution standards is measured at 600 nm wavelength and a calibration graph of absorbance versus concentration constructed. The absorbance of Y^{2+} in the mixed solution is then used to determine its concentration using the calibration graph.

KEY QUESTIONS

14 Identify the light source used in colourimetry and explain the purpose of the cuvette.

15 Explain the purpose of the beam splitter in a UV-visible spectrophotometer.

16 Relate the absorbance of light to the concentration of a coloured ion in colourimetry and visible spectrophotometry.

Answers ⊃ p. 207

Atomic absorption spectroscopy (AAS)

- A quantitative method of measuring the concentration of metal ions in solution was developed by the Australian scientist Alan Walsh at the CSIRO in 1954. This method measured the amount of light absorbed by a metal ion when it was vapourised in a hot flame. This technique is called *atomic absorption spectroscopy* or AAS.
- Atomic absorption spectroscopy is now a widely used tool for detecting and measuring the concentration of specific metal ions in solution. Other metal ions do not interfere with the analysis. The technique is so sensitive that trace amounts of these metal ions can be detected. Concentrations of metal ions as low as 1 mg/L (1 ppm) or 1 μg/L (1 ppb) can be measured. AAS also allows chemists to conduct rapid analyses and in multiple batches.
- The principles of the operation of a flame AAS are as follows:
  - A hollow cathode lamp made from the metal to be analysed generates specific wavelengths of light that are unique to the selected metal.
  - The light produced from the cathode lamp passes through a vapourised sample of a solution containing the atomised metal ions. A nebuliser (or atomiser) sprays the metal ion solution into the flame. The flame emerges from a slot-type burner that produces a long flame to increase the time the light spends in the flame.
  - The greater the concentration of the metal ion, the more light that is absorbed, and the less that reaches the detector. Any other metal ions that may be present in the sample do not absorb light produced from the hollow cathode. A second-reference light beam by-passes the sample.
  - The light that emerges from the flame passes through a device called a monochromator that selects one wavelength. The absorbance of that wavelength light is to be measured.
  - A photomultiplier tube (the detector) measures the light intensity or the absorbance *A* of the sample relative to the reference beam.

- Before an unknown sample is measured, a series of calibration standard solutions of known concentration are measured. From these measurements a calibration graph is constructed. The absorbance of the unknown sample can then be measured and the graph is then used to determine the concentration of metal ion in the unknown.

➔ Figure 14.20 shows an outline diagram of the flame AAS spectrometer.

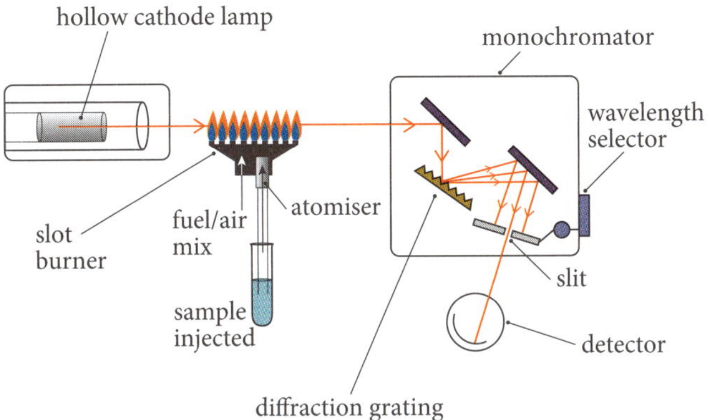

Figure 14.20 AAS components

➔ The development of AAS has allowed chemists to detect and rapidly measure the concentration of metal ions in water systems (particularly those that may be polluted with heavy metals) and in the tissues of animals and plants.

➔ Using AAS to monitor the lead concentration in the air or in soil that is suspected of being contaminated is important for the health of people because high lead concentrations in the environment are associated with nervous system disorders and brain damage. According to the NHMRC Australian Drinking Water Guidelines, drinking water must have a lead concentration of less than 0.01 mg/L (0.01 ppm).

➔ The following example outlines the use of a calibration graph to determine the concentration of lead ions in a sample of polluted water.

EXAMPLE 11

A chemist prepared a standard solution of lead (II) nitrate in which the concentration of lead ions was 100 mg/L (100 ppm). Five diluted standard solutions were prepared from this standard solution. These dilute solutions varied in lead ion concentration from 5 mg/L to 25 mg/L. Distilled water was used as a control. Each solution in turn was nebulised into a flame AAS and the absorbance of each solution was measured. The absorbance of the sample of polluted water was then measured. The results were tabulated.

Lead ion concentration (mg/L)	0	5	10	15	20	25	Polluted water
Absorbance (*A*)	0	0.18	0.36	0.54	0.72	0.90	0.68

a **Plot a calibration graph and draw the line of best fit.**

b **Use the graph to determine the concentration of lead ions in the polluted water.**

Show the interpolation to determine the concentration of the lead in the polluted water

Answer:

a The calibration graph is shown in Figure 14.21.

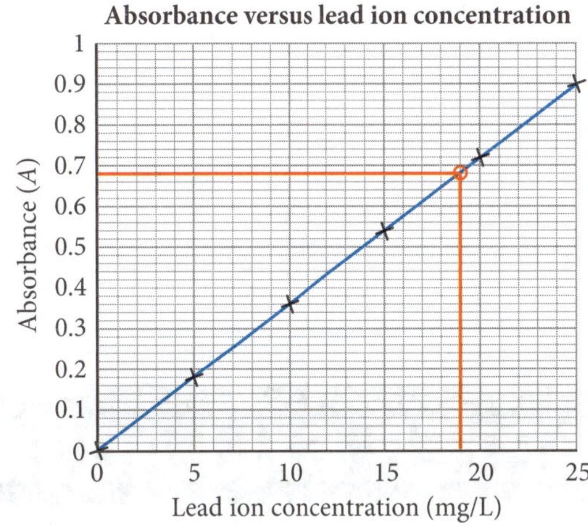

Figure 14.21 Lead calibration graph

b The polluted water has an absorbance of 0.68. This value can be used to interpolate the lead ion concentration on the calibration graph. The concentration of lead ions is 19 mg/L. This is considerably higher than the 0.01 mg/L limit for lead concentrations in drinking water. Such water is not fit for human consumption.

EXAMPLE 12

Deficiencies of trace elements in soil (e.g. Mo, Zn, Mn, Cu) leads to significant health problems for grazing farm animals such as sheep. A farmer believes the problem may be caused by poor soil fertility, leading to poor growth of grass. Explain the experimental procedure a chemist would use to determine the concentration of such trace metals as zinc in the soil on a farm. Explain how reliability is achieved.

Consider how traces of zinc could be extracted from soil

Answer:

The chemist needs to take soil samples from many sites where sheep are grazing. A minimum of five random sites should be sampled to ensure reliability.

- In the laboratory, soil samples are weighed and placed in beakers.

- Concentrated nitric acid is added and the mixture heated to extract metal ions from the soil.
- Each mixture is filtered and the filtrate transferred to a volumetric flask. Distilled water is added until the base of the meniscus is on the engraved line. The flask is stoppered and the flask shaken to mix.
- A standard stock solution of zinc is prepared by dissolving a known mass of zinc in hydrochloric acid. From this stock solution, five dilution standards are prepared.
- A zinc hollow cathode lamp is used for AAS measurements. The five dilution standards are aspirated into the flame and the absorbance of each measured. The soil extracts are then aspirated into the flame and the absorbance of each are measured.
- Using a calibration graph the zinc concentration in the soil samples can be determined. The zinc concentration in the soil is then reported in the units of mg/kg soil (1 mg/kg = 1 ppm). The ideal range of zinc concentrations in soil is 1–200 ppm.

➔ Further information about atomic absorption spectroscopy can be found in various online videos. Enter these titles in a search bar:

- AAS
- atomic absorption spectroscopy
- AAS instructional video.

➔ KEY QUESTIONS

17 **Compare the accuracy and sensitivity of gravimetric analysis with AAS analysis of metal ions in natural waters.**

18 **Explain the purpose of the monochromator in AAS.**

19 **Explain the role of AAS in measuring heavy metal pollution in waterways.**

Answers ➲ p. 207

CHAPTER SYLLABUS CHECKLIST

Are you able to answer every question from the syllabus for this chapter? Tick each question as you go through the checklist if you are able to answer it. If you cannot answer a question, turn to the relevant page in the study guide to find the answer. For NESA key word meanings, go to www.educationstandards.nsw.edu.au and search 'key words'.

	FOR A COMPLETE UNDERSTANDING OF THIS TOPIC:	PAGE NO.	✓
1	Can I give reasons for the importance of environmental monitoring?	186	
2	Can I give examples of monitoring procedures used to measure water quality?	186	
3	Can I perform mole calculations for analysis of water solutions?	187	
4	Can I describe the colours of flames produced by specific cations?	190	
5	Can I describe precipitation and complexation reactions to identify cations and anions in solution?	191	
6	Can I write equations for precipitation and complexation reactions involving cations and anions in solution?	191	
7	Can I describe experiments and perform calculations to determine the composition of compounds or the concentration of ions by gravimetric analysis?	195	
8	Can I describe experiments and perform calculations to determine the concentration of ions by a precipitation titration?	197	
9	Can I process data to determine the concentration of coloured ions using colourimetry?	198	
10	Can I process data to determine the concentration of ions using UV-visible spectrophotometry?	199	
11	Can I process data to determine the concentration of cations using atomic absorption spectroscopy?	202	

HSC EXAM-TYPE QUESTIONS

Objective-response questions (1 mark each)

1 **Select the statement that is true about atomic absorption spectroscopy (AAS).**

A The hollow cathode lamp must be made of the same metal as the metal ion to be analysed in the unknown solution.

B The solutions to be analysed are initially sprayed into the flame of a Bunsen burner.

C AAS is used to detect low concentrations of metal and non-metal ions in aqueous solutions.

D The nebuliser measures the intensity of the light that has passed through the solution of metal ions.

2 **A solution of potassium dichromate ($K_2Cr_2O_7$) is orange in colour. Its maximum absorbance occurs at a wavelength of 345 nm. A dichromate solution was diluted by a factor of 10. The diluted solution was placed in a 1.00 cm cuvette and its absorbance was measured. The absorbance was 0.552. Use the Beer–Lambert law to calculate the concentration of the undiluted solution.**

(Molar absorptivity for dichromate ions = 1044 L/mol/cm)

A 5.52×10^{-3} mol/L

B 5.29×10^{-5} mol/L

C 5.29×10^{-4} mol/L

D 5.29×10^{-3} mol/L

3 **Identify a solution that will produce a precipitate when added to a solution of ammonium sulfate.**

A potassium chloride

B sodium nitrate

C barium nitrate

D iron (II) nitrate

4 **A colourimeter is used to measure the concentration of phosphate ions. The colourless phosphate ions are reacted with ammonium molybdate and ascorbic acid to form a blue complex ion. Select the true statement about this investigation.**

A The blue solution is injected into the flame of a slot burner and the amount of light absorbed depends on the phosphate concentration.

B Light beams with wavelengths between 190 nm and 700 nm are passed through the blue solution and the absorbance is measured at each wavelength.

C The absorbance of the blue solution is measured and a calibration graph is used to determine the concentration of phosphate ions.

D A monochromator selects a specific wavelength of light to pass through the blue solution in a cuvette.

5 **Identify a qualitative test that could be used to identify the presence of iron (III) ions in an aqueous solution.**

A A deep-red colour is produced when potassium thiocyanate is added to the solution.

B A green precipitate forms when sodium hydroxide is added.

C The solution effervesces when hydrochloric acid is added.

D A white precipitate forms when sodium sulfate solution is added.

Extended-response questions

6 **25.00 mL of hard water was placed in a conical flask. The pH of the water was adjusted and it was then titrated with a standard solution of 0.010 mol/L EDTA using a suitable indicator:**

$$Ca^{2+}(aq) + EDTA^{4-}(aq) \rightarrow CaEDTA^{2-}(aq)$$

7.80 mL of the EDTA solution was added until the indicator changed colour. Calculate the total hardness of the water. Express your answer in the units mg($CaCO_3$)/L (ppm). Assume the hardness is totally due to calcium ions. (5 marks)

7 **Atomic absorption spectroscopy (AAS) is used to measure the concentration of cadmium ions in a river. The water in the river also contains sodium ions and calcium ions.**

a **Identify the metal that will form the hollow cathode lamp.** (1 mark)

b **Explain why solutions of known cadmium ion concentration are sprayed into the flame prior to the river water sample.** (1 mark)

c **Distilled water is used as a control. What absorbance will this control sample register?** (1 mark)

d **Explain why the sodium ions and calcium ions do not interfere with the absorbance measurements.** (1 mark)

8 **Figure 14.22 shows an AAS absorbance calibration graph for mercury ions in water.**

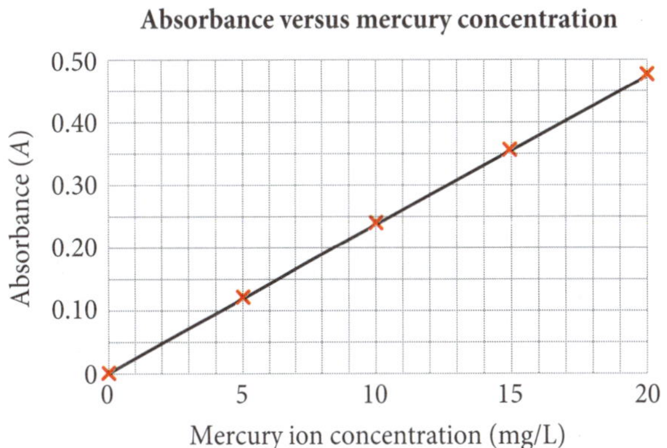

Figure 14.22 Mercury calibration graph

a A sample of polluted water was analysed using AAS. 20.0 mL of the water sample was diluted to 200.0 mL with distilled water. The absorbance of the diluted sample was 0.30. Determine the concentration of mercury ions in the polluted water. (2 marks)

b A sample of polluted water from another river was analysed using AAS. The absorbance of this sample was 0.640, which was outside the calibration range. What should the chemist now do to determine the mercury concentration in this other river? (1 mark)

9 An insoluble white solid X is dissolved in nitric acid. An odourless gas is evolved and a colourless solution Y forms.

Some of the gas is passed into a test tube containing universal indicator solution. The indicator turned pink. Some of the gas was passed into a solution of barium hydroxide and a faint white precipitate formed.

The colourless solution Y was divided into three test tubes and tested with three reagents. The results are:

- Sodium iodide solution forms a bright-yellow precipitate.
- Sulfuric acid forms a thick white precipitate.
- Sodium chloride solution forms a faint white precipitate that dissolves when the mixture is heated.

Analyse the information and identify the white solid X. Use equations as part of your explanation. (6 marks)

10 The phosphate content of a detergent is to be determined gravimetrically. The phosphate ions are precipitated as magnesium ammonium phosphate hexahydrate. 1.00 g of detergent is dissolved in 100 mL of acidified warm water. Magnesium chloride solution, ammonium chloride solution and concentrated ammonia were added and the mixture stirred. Magnesium ammonium phosphate hexahydrate formed slowly as a precipitate. The crystals were filtered, washed with ethanol and the filter paper and crystals placed in a Petri dish in a desiccator (Figure 14.23).

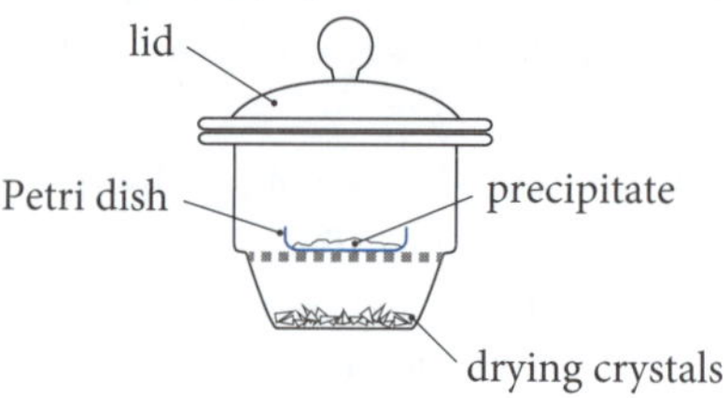

Figure 14.23 Drying precipitate in a desiccator

The equation for the precipitation reaction is:

$$Mg^{2+}(aq) + NH_4^+(aq) + PO_4^{3-}(aq) + 6H_2O(l) \rightarrow MgNH_4PO_4{\cdot}6H_2O(s)$$

a Explain the purpose of:
   i washing the precipitate with ethanol. (1 mark)
   ii the desiccator in the experiment. (1 mark)

b The following results were collected on weighing the crystalline precipitate:
   - mass of dry filter paper/Petri dish = 23.362 g
   - mass of filter paper/Petri dish + dried crystals = 23.682 g

   Determine the percentage by mass of phosphate in the detergent. (4 marks)

c Explain how the reliability of this experiment can be improved. (1 mark)

ANSWERS

KEY QUESTIONS

Key questions ⊃ p. 189

1 Water that is fit to drink should be free of pesticides, heavy metals and microbes. Water also needs to be used for crop irrigation. Chemists need to regularly monitor water to ensure it will not cause disease.

2 Acidity can be caused by illegal discharge of factory waste and acid drainage from mines. Acid rain from air pollution can also acidify natural bodies of water.

3 Dissolved calcium carbonate. Calcium and magnesium ions prevent soap from lathering. An EDTA titration is used to determine the concentration of ions causing hardness.

4 A calibrated conductivity electrode and meter is used to measure the conductivity of the river water. The TDS of the water is then determined from the calibration graph based on known standards.

Key questions ⊃ p. 193

5 Silver ions precipitate as AgCl when HCl is added. Lead (II) ions also precipitate as $PbCl_2$ when HCl is added. The lead (II) chloride dissolves on heating and can be removed via filtration. If KI solution is added to the lead ion filtrate, yellow lead (II) iodide precipitate forms. The AgCl will dissolve in ammonia solution to form a colourless complex ion.

6 To one sample of the solution add a solution of $K_3Fe(CN)_6$. The iron (II) ions form a deep-blue complex. To another sample add a solution of KSCN. A blood-red ion forms if iron (III) ions are present.

7 Add sodium fluoride solution and a white precipitate of CaF_2 forms. Filter and test the filtrate.

Add sulfuric acid to the filtrate and a white precipitate of $BaSO_4$ forms. Filter the mixture and retain the filtrate. Add excess NaOH solution and a white precipitate of $Mg(OH)_2$ forms.

Key questions ⊃ p. 195

8 Add dilute nitric acid. An effervescence indicates carbonate ions are present.

9 Acidify the solution with nitric acid and add barium nitrate solution. A white precipitate of $BaSO_4$ demonstrates that sulfate ions are present. Filter and make the filtrate alkaline with ammonia and add more barium nitrate. A white precipitate of barium phosphate forms.

10 Add $AgNO_3$ solution and produce a mixed precipitate of white AgCl and pale-yellow AgI. Add ammonia solution and the AgCl dissolves. Filter the mixture and add HNO_3 to re-acidify the filtrate. White AgCl reforms.

Key questions ⊃ p. 197

11 Weigh the fertiliser. Dissolve the fertiliser in water. Acidify with nitric acid. Add an excess of barium nitrate to precipitate the sulfate as $BaSO_4$. Filter, wash, dry and weigh the precipitate. Calculate the mass of sulfate in the $BaSO_4$. Express this mass as a percentage of the mass of the fertiliser.

12 Pipette the bromide solution into a flask. Add four drops of sodium chromate indicator and titrate with standard silver nitrate until the indicator turns reddish-brown. Use the average titre to calculate the bromide ion concentration.

13 Add excess silver nitrate to precipitate the AgCl. Filter, wash, dry and weigh the AgCl. Washing is important to remove any soluble ions adsorbed onto the precipitate. Repeated drying to the constant weight ensures the precipitate is perfectly dry.

Key questions ⊃ p. 202

14 The light source is a filament lamp and filter or an LED light. The cuvette is a glass tube that holds the solution to be investigated.

15 The beam-splitter splits the light beam so that half the beam passes through the test sample and the other half through the distilled-water reference tube. The absorbance of light by the sample is then determined.

16 The absorbance is directly proportional to the concentration of the coloured ion.

Key questions ⊃ p. 204

17 Gravimetric analysis is accurate for solutions of molar concentration but not suitable for low level metal ions in natural waters. AAS is very sensitive to very low levels of ions in ppm concentrations. Careful preparation of standards also ensures accuracy.

18 The monochromator contains a diffraction grating that diffracts the light beam emerging out of the flame and a slit selects one wavelength to determine its intensity.

19 Heavy metals (e.g. Hg, Cd, Pb) may be present in polluted waterways at concentration levels in parts per million. AAS is suitable to measure such low concentrations as it is a very sensitive technology.

HSC EXAM-TYPE QUESTIONS

Objective-response questions

1 **A.** The light emitted by the metal cathode will have the appropriate wavelength to excited electrons in the atomised sample. **B** is incorrect as a slot burner is used. **C** is incorrect as non-metal ions cannot be determined using AAS. **D** is incorrect as the nebuliser is used to spray the solution into the flame.

2 **D.** $A = \varepsilon cl$

$0.552 = (1044)c(1.00)$

$c = 5.29 \times 10^{-4}$ mol/L

The undiluted solution is 10× more concentrated. Therefore:

$c = 10 \times (5.29 \times 10^{-4})$

$= 5.29 \times 10^{-3}$ mol/L

Therefore **A**, **B** and **C** are incorrect.

3 **C.** Barium sulfate is insoluble. **A** is incorrect as potassium sulfate and ammonium chloride are soluble. **B** is incorrect as all sodium and nitrate salts are soluble. **D** is incorrect as iron (II) sulfate is soluble.

4 **C.** Dilution standards are used to create the calibration graph and then the measured absorbance of the unknown is used to interpolate a concentration on the graph. **A** is incorrect as a slot burner is used in AAS. **B** is incorrect as this method is used in UV-visible spectrophotometry. **D** is incorrect as a filter rather than a monochromator is used.

5 **A.** The red iron (III) thiocyanate ion forms. **B** is incorrect as a brown precipitate would form. **C** is incorrect as carbonates effervesce with acids. **D** is incorrect as iron (III) sulfate is soluble.

Extended-response questions

6 EM Students demonstrate their skill in calculations involving titrations. The stoichiometry should be stated and the molar concentration must be converted to parts per million to score full marks.

Calculate the number of moles of EDTA.

$n(\text{EDTA}) = c.V = (0.010)(0.0078) = 7.80 \times 10^{-5}$ mol

Reaction stoichiometry = 1 : 1

$n(Ca^{2+}) = n(\text{EDTA}) = 7.80 \times 10^{-5}$ mol ✓

As $n(Ca^{2+}) = n(CaCO_3)$

$n(CaCO_3) = 7.80 \times 10^{-5}$ mol ✓

Calculate the concentration of $CaCO_3$ dissolved.

$c(CaCO_3 = n/V = \frac{7.80 \times 10^{-5}}{0.02500} = 3.12 \times 10^{-3}$ mol/L ✓

Calculate the molar mass (M) of $CaCO_3$.

$M(CaCO_3) = 40.08 + 12.01 + 3(16.00) = 100.09$ g/mol ✓

Therefore the $CaCO_3$ concentration $= 3.12 \times 10^{-3} \times 100.09$ g/L
$= 0.312$ g/L $= 312$ mg/L

Thus the hardness of the water is 312 ppm. ✓

7 EM Students need to recall the components and purpose of each component of an atomic absorption spectrometer.

a cadmium ✓

b The solutions are calibration standards and from the absorbance data collected for these standards a calibration graph can be drawn. This graph will be used to determine the concentration of the cadmium ions in the river sample. ✓

c Zero absorbance as no cadmium ions are present. ✓

d These ions do not absorb the light as the light is generated by the cadmium cathode, which produces light waves of wavelengths specific to cadmium ions. ✓

8 EM Students need to show how to extract data from a graph and to explain the experimental procedure required if unknown solutions do not have an absorbance in the range of the standards.

a Concentration of mercury in diluted sample (from calibration graph) = 13 mg/L ✓

Thus, the concentration of mercury in the undiluted sample $= \frac{13 \times 200}{20} = 130$ mg/L ✓

b The river water should be systematically diluted until the absorbance is in the range of the standards. ✓

9 EM Students should recall the precipitation reactions of lead (II) ions and the acid–base behaviour of carbonates. Students should demonstrate logical conclusions and be able to write net ionic equations.

The indicator turns pink because the gas is producing an acidic solution. The gas is probably carbon dioxide, which dissolves in water to form carbonic acid:

$CO_2(g) + H_2O(l) \rightarrow H_2CO_3(aq)$ ✓

Carbon dioxide forms when a carbonate reacts with an acid. Therefore X is a carbonate compound.

Carbon dioxide reacts with barium hydroxide to form barium carbonate:

$Ba(OH)_2(aq) + CO_2(g) \rightarrow BaCO_3(s) + H_2O(l)$ ✓

The experiments on solution Y are consistent with the presence of lead (II) ions.

The yellow precipitate is lead (II) iodide:

$Pb^{2+}(aq) + 2I^-(aq) \rightarrow PbI_2(s)$ ✓

The thick white precipitate is lead (II) sulfate:

$Pb^{2+}(aq) + SO_4^{2-}(aq) \rightarrow PbSO_4(s)$ ✓

The faint white precipitate is lead (II) chloride, which is soluble in hot water:

$Pb^{2+}(aq) + 2Cl^-(aq) \rightarrow PbCl_2(s)$ ✓

Thus X is lead (II) carbonate. ✓

10 EM Students should recall the use of laboratory equipment and be able to show the steps of a gravimetric calculation. Students need to recall the %w/w concentration unit.

a i The ethanol helps to remove traces of water from the precipitate. Water is quite soluble in ethanol. ✓

ii The desiccator is a glass container containing drying crystals that remove traces of free water from the precipitate prior to weighing. ✓

b m = mass of crystals = 23.682 – 23.362 = 0.320 g ✓

Molar mass (M) of magnesium ammonium phosphate hexahydrate = 245.418 g/mol

Percentage by weight of PO_4 in magnesium ammonium phosphate hexahydrate

$\frac{94.97}{245.418} \times \frac{100}{1} = 38.70\%$ ✓

$M(PO_4)/M \times 100/1 = \frac{94.97}{245.418} \times \frac{100}{1} = 38.70\%$ ✓

Mass of phosphate in the crystals collected $= \frac{38.70}{100} \times 0.320$
$= 0.124$ g

% phosphate in detergent $= \frac{0.124}{1.00} \times \frac{100}{1} = 12.4\%$w/w ✓

c Repeat the experiment a minimum of five times and take an average. ✓

CHAPTER 15 ANALYSIS OF ORGANIC SUBSTANCES

MODULE 8 APPLYING CHEMICAL IDEAS

INQUIRY QUESTION:

How is information about the reactivity and structure of organic compounds obtained?

Qualitative and quantitative experiments have led chemists to an understanding of chemistry of organic compounds. Technologies such as mass spectroscopy and infrared spectroscopy have enabled a deeper understanding of functional groups and molecular structure. This chapter investigates both qualitative tests for functional groups and technologies that extend our understanding of organic structures.

1 Qualitative investigations of functional groups

» Students conduct qualitative investigations to test for the presence in organic molecules of the following functional groups: carbon–carbon double bonds, hydroxyl groups and carboxylic acids.

➔ Qualitative tests can be used to demonstrate the presence of various functional groups in organic compounds.

Carbon–carbon double bonds

➔ Carbon–carbon double bonds are present in alkenes. These bonds are very reactive. A common reaction involving double bonds is an addition reaction and this type of reaction can be used to qualitatively identify unsaturation in organic compounds. Halogens readily react with double bonds due to their high electronegativity. Unsaturated hydrocarbons also include alkynes that have carbon–carbon triple bonds.

➔ The decolourisation of bromine water is a simple experimental method used to identify **unsaturated hydrocarbons**.

unsaturated hydrocarbons: hydrocarbons containing either double or triple carbon–carbon bonds

Example: Bromine water decolourisation

- Bromine water is prepared by dissolving red-brown liquid bromine in water.
- The colour of bromine water depends on the degree of dilution. Its colour can vary from orange-brown to orange or yellow. Diluted orange-yellow bromine water is much safer to use. The experiment should be conducted in a fume cupboard.
- Bromine water contains bromine molecules (Br_2), which colour the water an orange-yellow and colourless hypobromous acid (HOBr).
- When bromine water is shaken with excess cyclohexene in a stoppered tube, the bromine water decolourises (i.e. turns from orange-yellow to colourless). When cyclohexane is shaken with bromine water, no decolourisation occurs (see Chapter 10). Figure 15.1 shows the result of the experiment with cyclohexene.

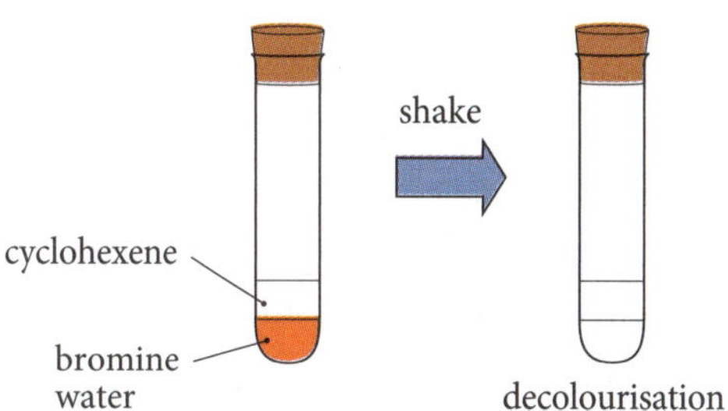

Figure 15.1 Decolourisation of bromine water

- The equations for the addition reactions with cyclohexene are:

  $C_6H_{10}(l) + Br_2(aq) \rightarrow C_6H_{10}Br_2(l)$

  $C_6H_{10}(l) + HOBr(aq) \rightarrow C_6H_{10}BrOH(l)$

- Figure 15.2 shows the structural formulas for reactants and products. Decolourisation occurs as long as excess cyclohexene is present to ensure all the bromine molecules react.

➔ Alkenes (but not alkanes) can be oxidised by an **oxidant** such as an acidified solution of potassium permanganate ($KMnO_4$). The permanganate solution turns from pink-purple to colourless in the reaction. In this reaction the double bond oxidises and an alkandiol intermediate forms. This molecule further oxidises to form **alkanal** intermediates and finally alkanoic acids.

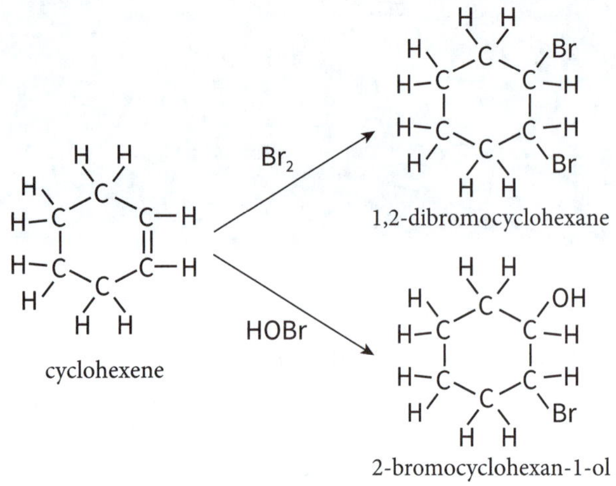

cyclohexene

Br_2

1,2-dibromocyclohexane

HOBr

2-bromocyclohexan-1-ol

Figure 15.2 Addition reactions of bromine water

oxidant: a substance that oxidises another substance by removal of 1 or more electrons
alkanal: an aldehyde formed from an alkane; aldehydes contain the CHO functional group

Thus pent-2-ene is oxidised to form a mixture of acetic acid and propanoic acid:

$$CH_3CHCHCH_2CH_3 \rightarrow CH_3CH(OH)CH(OH)CH_2CH_3$$
$$\rightarrow CH_3CHO + CH_3CH_2CHO$$
$$\rightarrow CH_3COOH + CH_3CH_2COOH$$

Hydroxyl groups

- Hydroxyl groups (OH) are functional groups that give alcohols their unique properties.
- Alcohols can be oxidised and reduced and these reactions can be used to identify these compounds.
- Primary and secondary alkanols can be oxidised using an oxidant such as an acidified solution of potassium permanganate ($H^+/KMnO_4$). **Primary alkanols** are oxidised to form alkanoic acids whereas **secondary alkanols** are oxidised to form alkanones:
  - propan-1-ol is oxidised to form propanoic acid
  - propan-2-ol is oxidised to form propanone.

  The pink-purple permanganate solution decolourises during these reactions.

primary alkanol: an alkanol that has the general formula RCH_2OH
secondary alkanol: an alkanol that has the general formula R_1R_2CHOH

- Bromine water is a much weaker oxidant and the reaction with alkanols would be very slow under lab conditions. Essentially no reaction occurs when compared to the rapid decolourisation with alkenes.
- Short chain alkanols can be reduced by metallic sodium to form ionic compounds called *alkoxides*. Bubbles of hydrogen gas are produced, as shown in Figure 15.3.

  Ethanol is reduced by sodium to sodium ethoxide ($NaOC_2H_5$):

  $$2C_2H_5OH(l) + 2Na(s) \rightarrow 2NaOC_2H_5(s) + H_2(g)$$
- This reaction is not performed at schools due to safety issues related to the high reactivity of sodium. However, students can view these reactions using online videos. Enter the following titles in a browser's search bar:
  - ethanol reacting with sodium
  - reaction of sodium with ethanol
  - sodium reacting with water, ethanoic acid and ethanol.
- Alkanes and alkenes do not react with sodium.

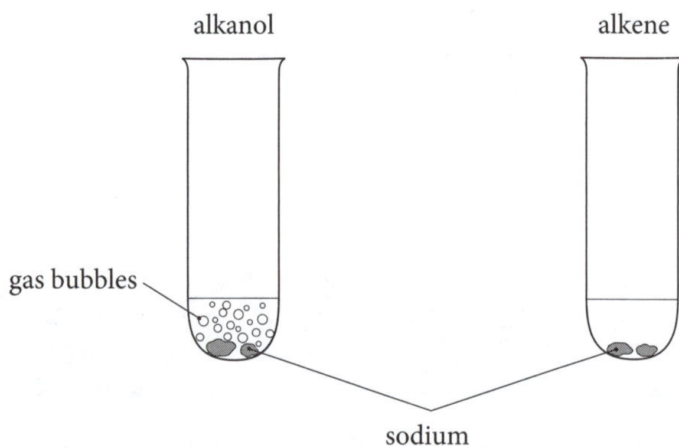

Figure 15.3 Testing an alkanol and an alkene with sodium

Carboxyl groups

- Carboxyl groups (COOH) are the functional groups present in carboxylic acids. Alkanoic acids are carboxylic acids formed from alkanes.
- Short chain alkanoic acids dissolve in water to form weakly acidic solutions.
- Alkanoic acids are neutralised by bases to form salts and water. Sodium carbonate solution reacts with short chain alkanoic acids, with the release of carbon dioxide gas. An effervescence is observed:

  $$2CH_3COOH(l) + Na_2CO_3(aq) \rightarrow 2NaCH_3COO(aq) + H_2O(l) + CO_2(g)$$
- Sodium metal also reacts with short chain alkanoic acids to produce hydrogen gas. The reaction is much slower than the reaction of sodium, with a strong acid:

  $$2Na(s) + 2CH_3COOH(l) \rightarrow 2NaCH_3COO(s) + H_2(g)$$
- Table 15.1 summarises qualitative tests for alkenes, alkanols and alkanoic acids.

Table 15.1 Qualitative tests for organic compounds

Organic compound	Bromine water	Reaction with sodium metal	Reaction with sodium carbonate solution
Alkanes	no reaction	no reaction	no reaction
Alkenes	rapid decolourisation	no reaction	no reaction
Alkanols	no reaction	effervescence	no reaction
Alkanoic acids	no reaction	effervescence	effervescence

EXAMPLE 1

An unknown, colourless liquid is known to be either an alkanol, an alkene or an alkanoic acid. The following tests were performed by a chemist on samples of the unknown. Use this information to classify the unknown. Justify your answer.

- **Test 1: A dry piece of sodium metal was added to the unknown liquid in a test tube. An effervescence occurred.**
- **Test 2: Bromine water was added dropwise to the unknown in a test tube. The mixture was agitated. The orange-coloured bromine water did not decolourise.**
- **Test 3: Drops of sodium carbonate solution were added to the unknown in a test tube. An effervescence occurred.**

Figure 15.4 shows the results of the experiment.

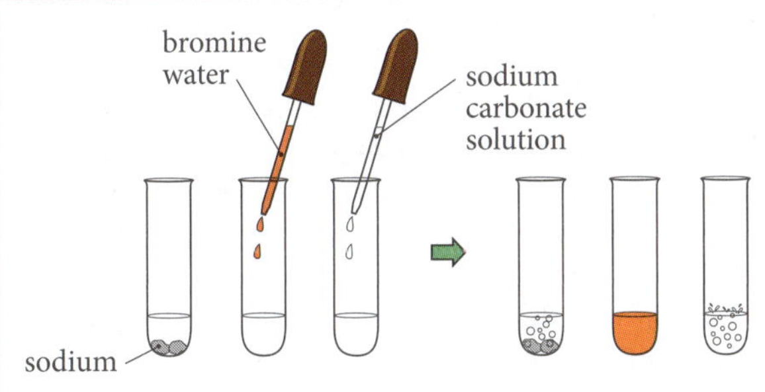

Figure 15.4 Testing an unknown

Review Table 15.1 and match this data to the observations recorded

Answer:

- Test 1. This positive test confirms the presence of an alkanol or an alkanoic acid. They react with sodium to release hydrogen gas.
- Test 2. This negative test confirms the absence of an alkene. Bromine solutions are decolourised by alkenes.
- Test 3. This positive test confirms the presence of an alkanoic acid. An alkanoic acid would have produced an effervescence with the sodium carbonate.

Thus the unknown liquid is an alkanoic acid.

→ KEY QUESTIONS

1. **Identify a simple laboratory test that can be used to distinguish pentane and pent-2-ene.**
2. **Explain how butan-1-ol can be distinguished from butanone using laboratory tests.**
3. **Write a balanced equation for the reaction of formic acid with potassium carbonate solution.**
4. **Write a balanced equation for the reaction of sodium with methanol.**

Answers ➲ p. 222

2 Instrumental analysis of organic compounds

» Students investigate the processes used to analyse the structure of simple organic compounds addressed in the course, including but not limited to proton and carbon-13 NMR, mass spectrometry and infrared spectroscopy.

→ Chemists use a wide variety of instrumental techniques to identify organic compounds. Specific functional groups can be identified using these technologies.

Carbon-13 NMR

→ Carbon consists of two stable **isotopes**, C-12 (99%) and C-13 (1%). Other carbon isotopes are radioactive.

→ Carbon-13 atoms have six protons and seven neutrons in their nuclei. This nucleus has a property called *spin*. Spin is a quantum property of subatomic particles. The **spin quantum number** can take values of 1/2, 1, 3/2, etc., for different nuclei. The C-13 nucleus belongs to nuclei groups that have a spin quantum number equal to 1/2.

isotope: elements with the same atomic number (Z) and different nucleon numbers (A)

spin quantum number: a number that describes the allowed rotational or angular momentum of a particle in the quantum theory of the atom

→ Spinning C-13 nuclei behave like tiny magnets as they create their own magnetic fields. When placed in an external magnetic field, the two fields interact. In the lowest energy state these tiny magnets align parallel with the external magnetic field lines. The higher energy state is the antiparallel state in which the tiny magnets align in the opposite direction, as shown in Figure 15.5.

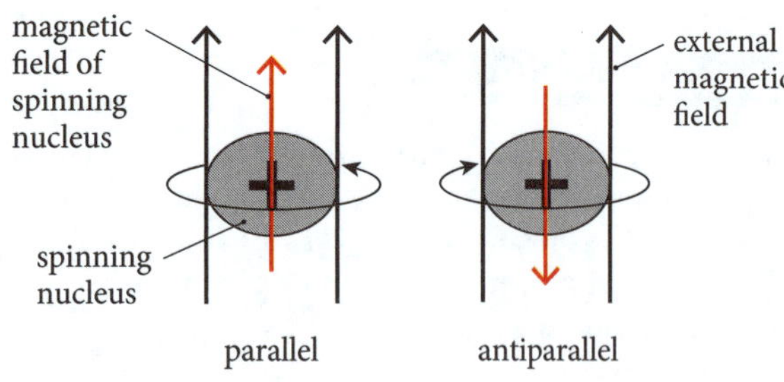

Figure 15.5 Spin states

- The spinning C-13 nucleus can absorb a **photon** of radio frequency radiation (in the range 25–100 MHz) and move from a low energy state (spin-up or parallel) to a high energy state (spin-down or antiparallel). The frequency at which this flipping of the spin state occurs is called the *resonance frequency*.

photon: a quantum of light energy

- In organic compounds the carbon atoms usually exist in different chemical environments due to different functional groups. These different chemical environments affect the frequencies at which radio frequencies are absorbed by the C-13 nuclei to achieve resonance.
- Nuclear magnetic resonance spectroscopy (NMR) measures differences in the chemical environments of C-13 atoms in organic compounds in terms of a quantity called the **chemical shift** (δ). A standard carbon compound (tetramethylsilane, $Si(CH_3)_4$) is used as a reference standard and it has a chemical shift of zero. The NMR spectra of organic compounds shows resonance peaks at different chemical shift values. The unit of chemical shift is parts per million (ppm). A chemical shift of 30 ppm (for a C-13 atom in the compound) means the frequency required to achieve resonance is 30 million times less than that required to produce resonance in the standard. High values of chemical shifts are described as being further downfield.

chemical shift: the extent to which radio frequency resonance peaks are shifted downfield from the standard

- The chemical shift of the peaks in the NMR spectra of organic molecules depends on various factors, including:
  - the presence of electronegative atoms (e.g. chlorine, oxygen, nitrogen)
  - the degree of hybridisation of carbon orbitals
  - isomerism.
- Some C-13 NMR spectra will now be compared.

Alkanes

- Figure 15.6 shows the C-13 NMR spectra of ethane (C_2H_6) and propane (C_3H_8). One peak is present in the ethane spectrum, but two peaks are present in the propane spectrum.
- Each peak is numbered and the carbon atoms in each molecule are correspondingly numbered.
- The ethane spectrum has only one peak as the two carbon atoms have equivalent chemical environments.
- In the propane molecule there are two different chemical environments. The central methylene (CH_2) carbon atom (numbered 2) has a different chemical environment to the two methyl (CH_3) carbon atoms (numbered 1).
- The vertical axis is a measure of the absorbance of specific radio frequencies.

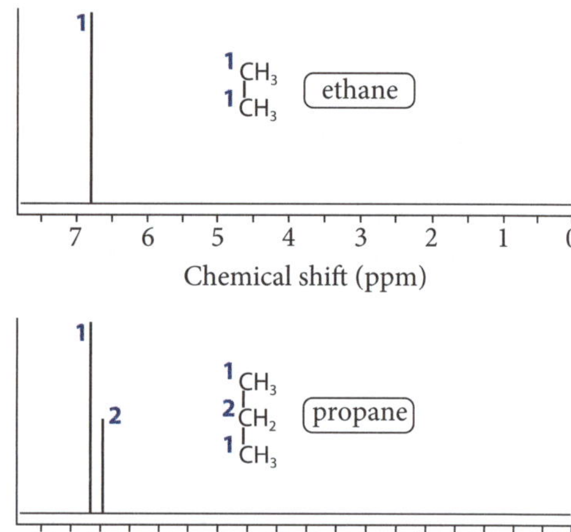

Figure 15.6 C-13 NMR spectra of ethane and propane

Alkenes

- Figure 15.7 shows the C-13 NMR spectrum of but-1-ene (C_4H_8).
- The double bond between carbons 1 and 2 is a region of high electron density involving sp^2 hybrid orbitals. The carbon atoms forming the C=C double bond in alkenes commonly have chemical shifts between 100–150 ppm. The carbon of the methylene (CH_2) group has a greater chemical shift than the carbon of the methyl group, which is furthest from the double bond.

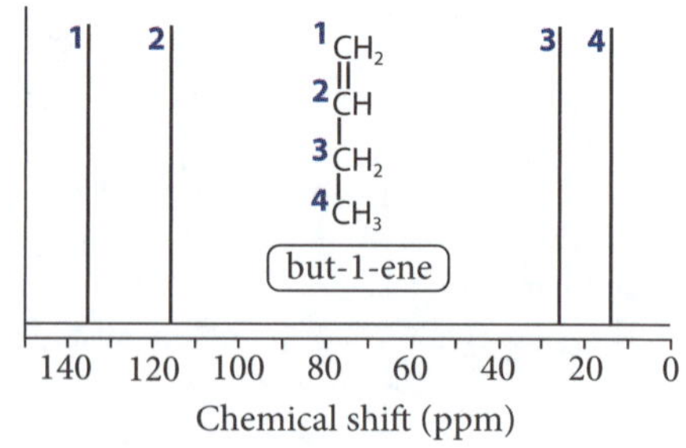

Figure 15.7 C-13 NMR spectrum of but-1-ene

Alkanols

- Figure 15.8 shows the C-13 NMR spectra of ethanol (CH_3CH_2OH) and propan-1-ol ($CH_3CH_2CH_2OH$).

- The presence of the highly electronegative oxygen atom in the hydroxyl (OH) group significantly changes the chemical environment of carbon-1 in each molecule. The electron pair in the C–O bond is shared unequally as the oxygen atom pulls electrons away from the carbon atom. This change makes it easier for the external magnetic field to achieve resonance. Therefore the chemical shift increases. This change to carbon-1 affects the electron pairs in subsequent C–C bonds. This change in the chemical environment is lower for carbon-2 in ethanol and carbon-3 in propan-1-ol.
- Generally the carbon of a primary alkanol has a chemical shift between 50–65 ppm.

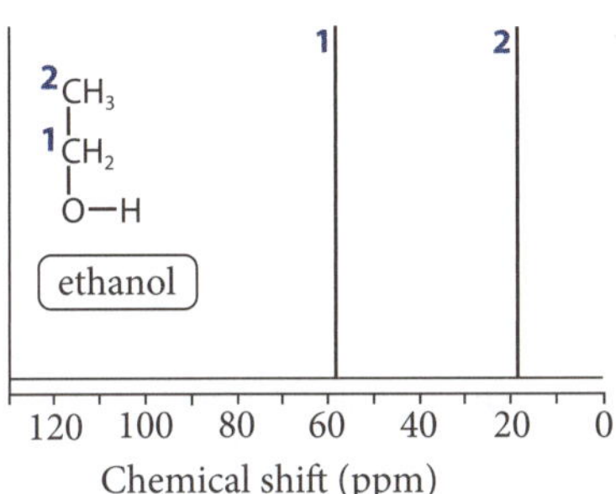

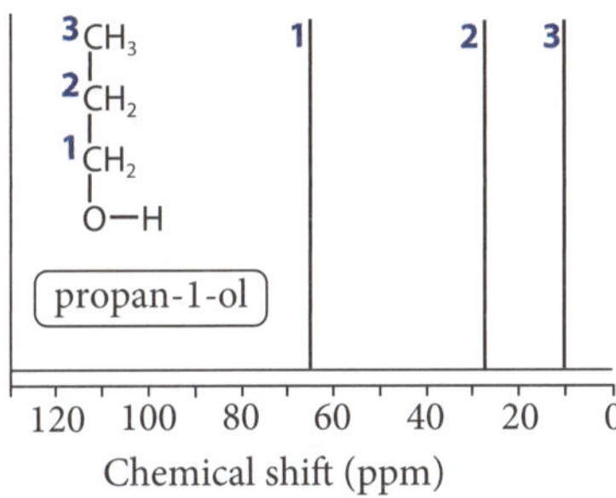

Figure 15.8 C-13 NMR spectra of ethanol and propan-1-ol

Alkanoic acid and esters

- Figure 15.9 shows the C-13 NMR spectra of two isomers, propanoic acid (CH_3CH_2COOH) and methyl acetate (CH_3COOCH_3).
- These spectra show that the carbonyl carbon has a chemical shift far downfield for both **isomers**. In methyl acetate the methyl carbon (number 2) attached to the C–O group is further downfield than carbon-2 in propanoic acid due to the electronegativity of the single-bonded oxygen atom in the ester.

isomers: molecules with the same molecular formula but different structural formulas

- Table 15.2 summarises the typical chemical shift bands for different functional groups in C-13 NMR spectra.

  In the HSC exam students can access this data in the Data Sheet provided. There is no need to memorise this data.

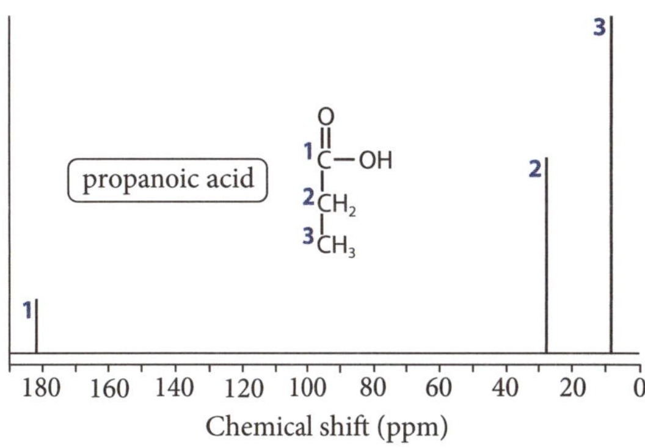

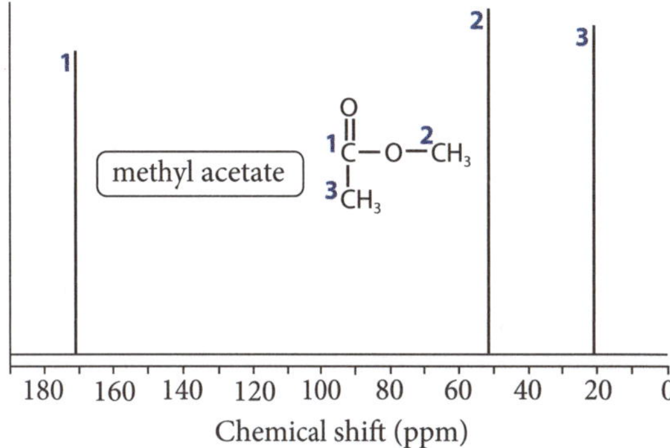

Figure 15.9 C-13 NMR spectra of propanoic acid and methyl acetate

Table 15.2 Chemical shift bands (C-13 NMR)

Carbon environment	Chemical shift δ (ppm)
C-C (alkanes)	5–40
C=C (alkenes)	90–150
C-O (alkanols/esters)	50–90
C=O (alkanoic acids and esters)	160–185
C=O (ketones)	205–220
C=O (aldehydes)	190–200
C-NH_2 (amines)	25–60
C-Cl or C-Br (haloalkanes)	10–70

EXAMPLE 2

Figure 15.10 shows the C-13 NMR spectrum of an organic molecule.

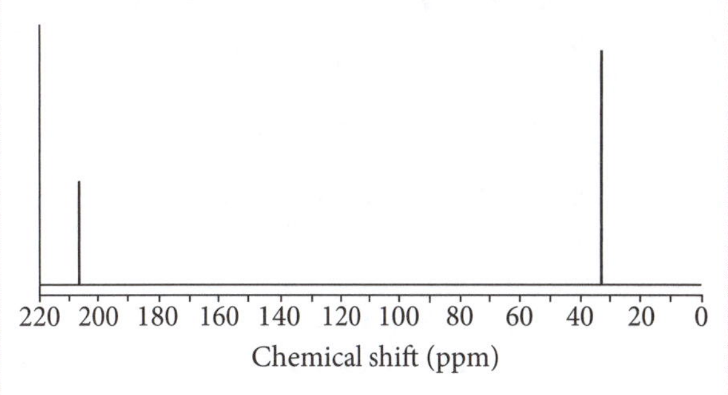

Figure 15.10 C-13 NMR spectrum

Use Table 15.2 to determine which of the following molecules could produce this spectrum:

A $CH_3CH_2CH_2COOH$

B $CH_3CH(OH)CH_3$

C CH_3COCH_3

Determine the number of different chemical environments for carbon atoms in each molecule

Answer:

Molecule A would have a C-13 NMR spectrum with four peaks as the four carbon atoms are in different chemical environments.

Molecule B would have two peaks. The central carbon to which the hydroxyl (OH) group is attached would not have a chemical shift so far downfield. Alkanols chemical shifts are typically in the 50–65 ppm range.

Molecule C would have two peaks. The central carbon to which the carbonyl group (CO) is attached would have a chemical shift downfield in the range 205–220 ppm.

Thus the spectrum is that of molecule C. This is propanone (CH_3COCH_3).

EXAMPLE 3

The following C-13 NMR chemical shift information was collected for three organic molecules (X, Y and Z):

- **Compound X: 45 ppm, 33 ppm, 29 ppm; 22 ppm, 14 ppm**
- **Compound Y: 35 ppm, 24 ppm, 15 ppm**
- **Compound Z: 140 ppm, 115 ppm, 36 ppm; 22 ppm, 14 ppm**

The three organic molecules are, in random order, pentane, pent-1-ene and 1-chloropentane.

Determine the identities of X, Y and Z. Justify your conclusions.

Use data from Table 15.2 to identify the compounds

Answer:

Compound Z must be pent-1-ene ($CH_2CHCH_2CH_2CH_3$) as the 140 ppm chemical shift is consistent with carbon–carbon double bonds (see Table15.2).

Compound Y is pentane ($CH_3CH_2CH_2CH_2CH_3$). There are only three peaks, which is consistent with molecular symmetry. The two methyl groups (C-1 and C-5) at the ends of the molecule have the same chemical environment and so only one peak will form. Similarly C-2 and C-4 are in the same chemical environment and only one peak will form. The central carbon (C-3) will have its own peak.

Compound X is 1-chloropentane ($CH_2ClCH_2CH_2CH_2CH_3$). The 45 ppm signal is consistent with a chlorine atom attached to C-1 (see Table 15.2).

➔ Further information about C-13 NMR spectroscopy can be found in various online videos. Enter the following titles in a search bar:
- how to understand carbon 13 NMR spectra
- how2 interpret a carbon-13 NMR spectrum
- CNMR spectrometry in organic chemistry.

➔ KEY QUESTIONS

5 **Explain the term *chemical environment* in relation to C-13 NMR spectroscopy.**

6 **Explain why the C-13 NMR spectrum of butane has only two peaks.**

7 **Explain why the C-13 NMR spectrum of propan-1-ol has three peaks whereas propan-2-ol has two peaks.**

8 **Identify which of the following molecules will have a chemical shift peak furthest downfield: propanal, propanone, and propanoic acid.**

Answers ➲ p. 222

Proton NMR

➔ Proton NMR (or H-1 NMR) is also used to analyse organic compounds.

➔ Like C-13 nuclei, spinning H-1 nuclei also act as tiny magnets, creating their own magnetic fields that can interact with an external magnetic field.

➔ In organic compounds the hydrogen atoms usually exist in different chemical environments due to the presence of different functional groups. These different chemical environments affect the frequencies at which radio frequency photons are absorbed by the H-1 nuclei to achieve resonance.

➔ Figure 15.11 shows low-resolution diagrams of the proton-NMR spectra of ethanol and ethyl acetate. The three resonance peaks in each spectrum demonstrate three different chemical environments of the hydrogen atoms. In ethanol the hydrogen atom of the hydroxyl group has a different chemical shift to the hydrogen atoms of the methylene (CH_2) and methyl (CH_3) groups. In ethyl acetate the hydrogen atoms of the methylene group have the greatest chemical shift as they are closer to the carboxyl group with two highly electronegative oxygen atoms.

➔ Table 15.3 summarises the typical chemical shift bands for different functional groups in H-1 NMR spectra of organic compounds. These chemical shifts deviate from the normal range if various functional groups are located nearby in the molecule.

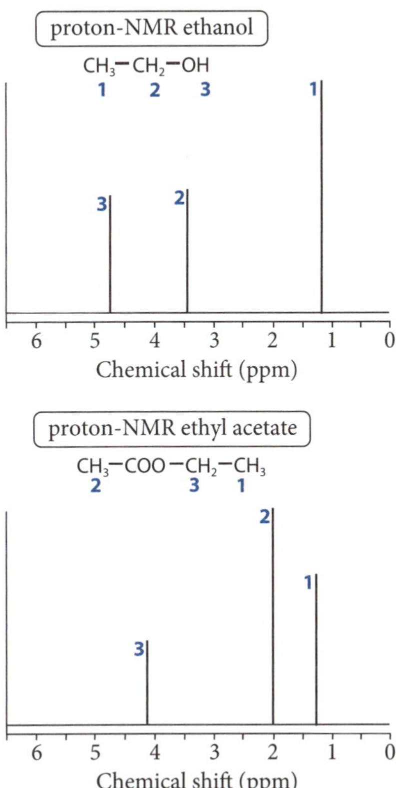

Figure 15.11 proton NMR spectra of ethanol and ethyl acetate

Table 15.3 Typical chemical shift bands (H-1 NMR)

Hydrogen environment	Chemical shift δ (ppm)
alkanes	0.8–1.9
alkenes	4.5–6.5
alkanols (ROH)	3.5–6.0
esters ($RCOOCH_2R$)	3.5–5.5
alkanoic acids (RCOOH)	10.0–13.0
aldehydes (RCHO)	9.0–10.0
amines (RNH_2)	1.0–5.0
chlorinated alkanes (RCH_2Cl)	3.0–5.0

EXAMPLE 4

The following H-1NMR chemical shift information was collected for acetic acid and 1,1-dichloroethane:

- **acetic acid: 1.6 ppm; 11.1 ppm**
- **1,1-dichloroethane: 2.1 ppm; 5.9 ppm**

Justify the presence of two spectral peaks at different chemical shifts in each spectrum.

Use the data from Table 15.3 to justify each answer

Answer:

The chemical shift of 11.1 ppm in acetic acid is typical of the proton of the carboxyl group (–COOH) in alkanoic acids. The peak at 1.6 ppm is typical of the protons of an alkyl group ($-CH_3$), which are close to an electronegative functional group such as COOH.

The chemical shift of 5.9 ppm in 1,1-dichloroethane is typical of a proton bonded to a carbon atom that also has chlorine functional groups ($-CHCl_2$). Chlorine atoms are very electronegative and two of these atoms has caused the proton's chemical shift to move downfield. The peak at 2.1 ppm is typical of the protons of an alkyl group that are close to an electronegative functional group such as chlorine.

Mass spectrometry

→ Mass spectrometry is a technology used to analyse the composition of different materials.

→ The existence of different isotopic forms of an element can be shown using a mass spectrometer.

→ This device can be used to measure the relative abundance of the isotopes of an element in a given sample as a function of their **mass-to-charge (m/z) ratio**.

mass-to-charge (m/z) ratio: a physical quantity that measures the ratio of the mass (m) of a charged particle to its electric charge (z)

- The sample is first vapourised before the atoms are charged to form positive ions by the loss of one electron.
- The **ionisation** of the gaseous atoms can be achieved in a number of ways, including the use of electron beams.

ionisation: the loss of electron(s) from a gaseous atom

- The positive ions that are formed are then accelerated and focused by electric fields.
- The accelerated ions then move through a magnetic field. The magnetic field causes the different mass ions to bend in circular arcs. The heavier the ion, the greater the radius of the arc.
- Figure 15.12 shows the basic structure of a mass spectrograph. In this example the element being analysed consists of a mixture of four isotopes.

→ Figure 15.13 shows the mass spectrum of natural chlorine gas. Two peaks are observed. The larger peak is formed by the Cl-35 ($^{35}_{17}Cl^+$) ion and the smaller peak by the Cl-37 ($^{37}_{17}Cl^+$) ion. Analysis of the mass spectrum shows that 75.8% of the chlorine is Cl-35 and 24.2% is Cl-37.

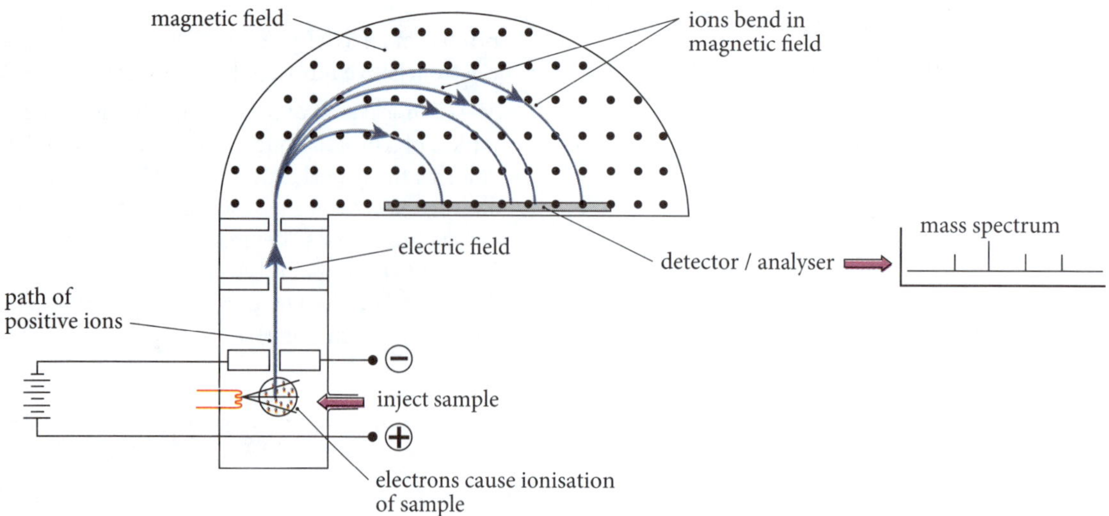

Figure 15.12 Mass spectrometer

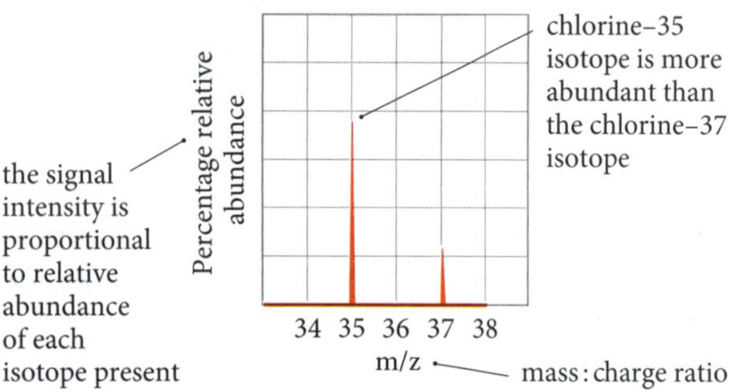

Figure 15.13 Chlorine isotopes

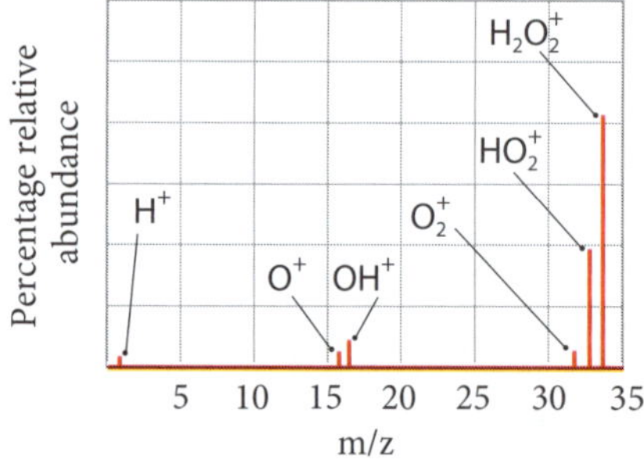

Figure 15.14 Hydrogen peroxide mass spectrum

- Mass spectrometers can also be used to determine the elemental composition of compounds.
  - The samples are vapourised into an evacuated chamber where the molecules are bombarded with electrons to ionise them.
  - These molecular ions then fragment into various smaller ions which are then accelerated in electric and magnetic fields.
  - The mass spectrum of a sample of hydrogen peroxide (H_2O_2) is shown in Figure 15.14. Six peaks are shown as well as the identity of the atomic or molecular ions producing these peaks. The largest peak is the molecular ion ($H_2O_2^+$), which has an m/z value of 34. This is the relative molecular mass of the parent molecular ion. This mass is the sum of the average isotopic masses of the component hydrogen and oxygen atoms:

    $2(1.008) + 2(16.00) = 34.02$ u $= 34$ u (2 significant figures)
  - The other peaks are formed as the molecular ion undergoes fragmentation inside the mass spectrometer.
- The purity of materials can be analysed with mass spectroscopy. Figure 15.15 compares the mass spectrum of pure methanol with a methanol (CH_3OH) sample contaminated with a small amount of ethanol (C_2H_5OH). The additional peaks in the second mass spectrum show the methanol sample is not pure.

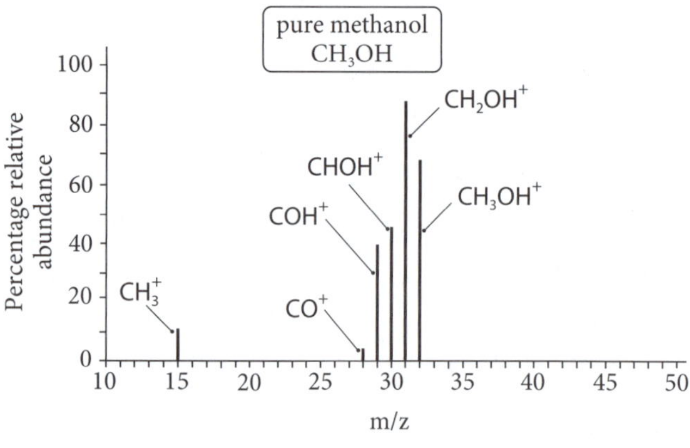

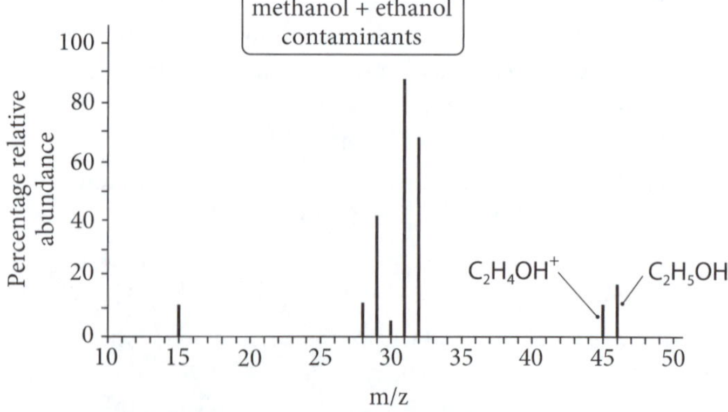

Figure 15.15 Mass spectra of pure and impure compounds

EXAMPLE 5

Figure 15.16 shows the mass spectrum of an organic amine. Some minor peaks have been omitted. Identify each spectral peak (A to H) and name the molecule.

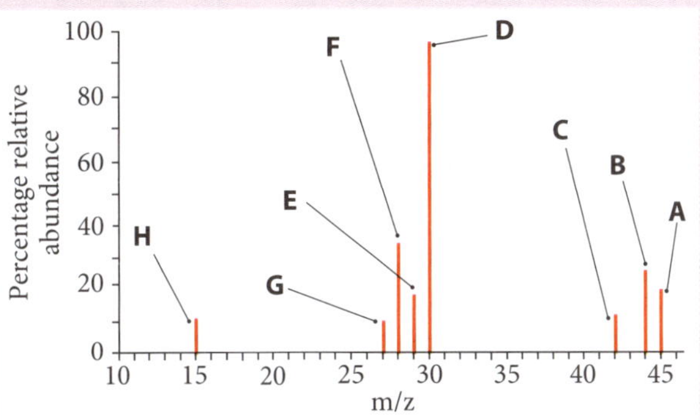

Figure 15.16 Mass spectrum

Organic amines have an NH_2 group

Answer:

Calculate possible m/z values, assuming z = 1 for each 1+ ion.

Use rounded atomic weights: C = 12 u, H = 1 u, N = 14 u

A = $CH_3CH_2NH_2^+$ ($m/z = \frac{45}{1} = 45$)

B = $CH_3CH_2NH^+$ or $CH_2CH_2NH_2^+$ ($m/z = \frac{44}{1} = 44$)

C = $CH_2CH_2N^+$($m/z = \frac{42}{1} = 42$)

D = $CH_2NH_2^+$($m/z = \frac{30}{1} = 30$)

E = CH_2NH^+ ($m/z = \frac{29}{1} = 29$)

F = CH_2N^+ ($m/z = \frac{28}{1} = 28$)

G = CHN^+ ($m/z = \frac{27}{1} = 27$)

H = CH_3^+ ($m/z = \frac{15}{1} = 15$)

The organic molecule is ethanamine ($CH_3CH_2NH_2$).

➔ Further information about mass spectroscopy can be found in various online videos. Enter the following titles in a search bar:

- mass spectrometer
- simple explanation of mass spectrometer
- how2 interpret a mass spectrum.

➔ KEY QUESTIONS

9 **Explain why the mass spectrum of chlorine has two peaks.**

10 **Identify the formula of the parent molecular ion in the mass spectrum of butane.**

11 **The two major peaks in the mass spectrum of ethanol have m/z values of 45 and 31. Identify these two peaks.**

Answers ➲ p. 222

Infrared spectroscopy

➔ Infrared spectroscopy is a technology used to identify organic molecules and their functional groups.

➔ Infrared (IR) radiation can be absorbed by organic molecules and their bonds. The absorption of this radiation makes the chemical bonds stretch and bend and the molecules bend and rotate. Figure 15.17 shows a simplified block diagram of an IR spectrometer.

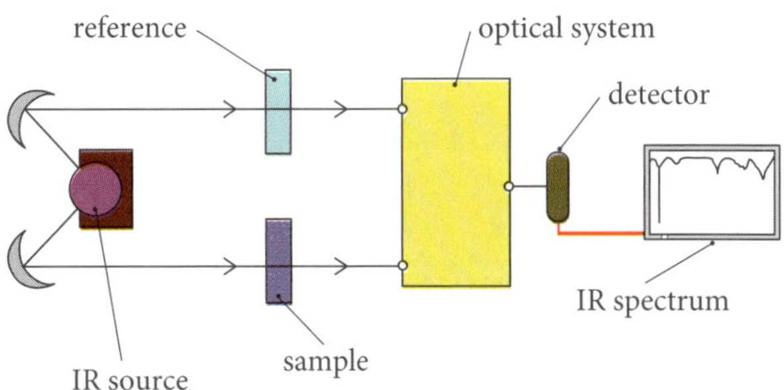

Figure 15.17 Simplified block diagram of IR spectrometer

➔ Specific functional groups have characteristic bond **vibrational frequencies** as well as molecular rotational frequencies. Table 15.4 lists some common functional groups and their characteristic infrared absorption frequencies measured in wave numbers (cm^{-1}).

vibrational frequency: frequencies where symmetric and antisymmetric, stretching, bending and twisting vibrations occur in bonds

Table 15.4 Infrared absorption frequencies

Compound type	Bond vibration	Frequency range (cm^{-1})
alkane	C–H stretch	2850–2960
alkane	C–H bend	1370 1480
alkene	C–H stretch	3010–3090
alkene	C=C stretch	1620–1680
alkene	C–H bend	900–995
alcohol	O–H stretch	3610–3640
alcohol	O–H stretch (H-bonded)	3200–3600
alcohol	C–O stretch	1050–1150
carboxylic acid	C–O stretch	1210–1320
carboxylic acid	O–H stretch	2500–3000
carboxylic acid	C=O stretch	1700–1725
amide	C=O stretch	1640–1690
amide	N–H stretch	3100–3500
ester	C=O stretch	1735–1750
alkanal	C=O stretch	1720–1740
alkanone	C=O stretch	1705–1725
amine	N–H stretch	3300–3500
amine	C–N stretch	1080–1360

→ Figure 15.18 shows that the IR spectrum is for methanol (CH_3OH). The various stretching and bending frequencies are labelled. The spectrum identifies the presence of the hydroxyl (OH) functional group. Note the O–H stretch is typical of a hydrogen-bonded molecule.

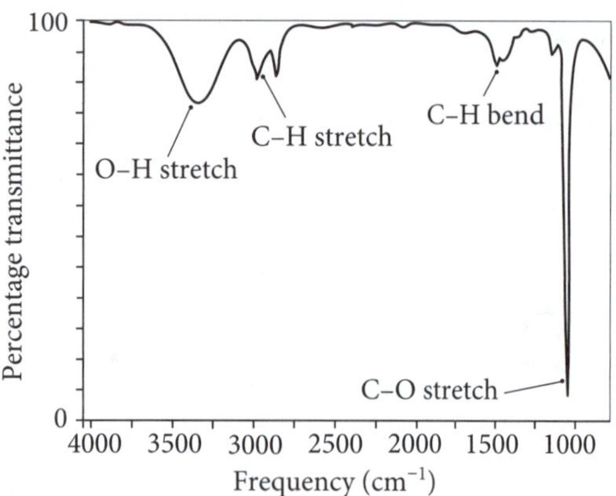

Figure 15.18 IR spectrum of methanol

EXAMPLE 6

Figure 15.19 shows the infrared spectrum of butanoic acid. Five peaks have been labelled (A to E). Use Table 15.4 to identify the three peaks that show the C=O, C–O and O–H stretching frequencies present in the IR spectrum of butanoic acid.

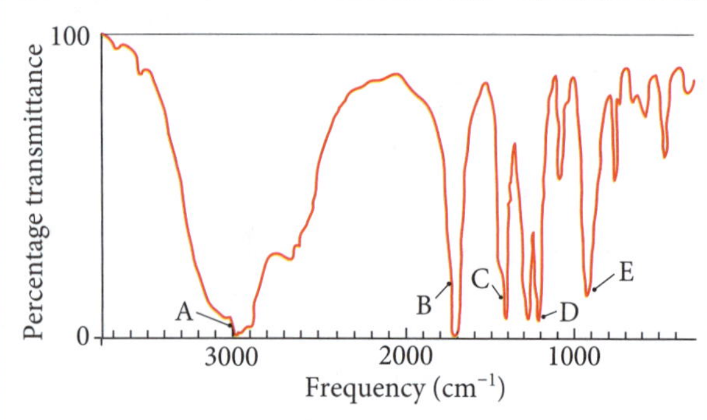

Figure 15.19 Infrared spectrum of butanoic acid

Consult Table 15.4

Answer:

Peak A is an O–H stretching frequency for a carboxylic acid.

Peak B is a C=O stretching frequency for a carboxylic acid.

Peak D is a C–O stretching frequency for a carboxylic acid.

(Peaks C and E are typical of O–H bending frequencies.)

→ Further information about mass spectroscopy can be found in various online videos. Enter the following titles in a search bar:

- interpreting IR spectra organic chemistry
- how to read IR spectroscopy—organic chemistry tutorials.

→ KEY QUESTIONS

12 **Infrared radiation is absorbed by an organic molecule. Explain what occurs within the molecule when the IR is absorbed.**

13 **The IR stretching frequency for hydroxyl groups in alkanols is lower in short chain alkanols that hydrogen bond than in long chain alkanols that have very little hydrogen bonding. Explain this observation.**

Answers ➲ p. 222

CHAPTER SYLLABUS CHECKLIST

Are you able to answer every question from the syllabus for this chapter? Tick each question as you go through the checklist if you are able to answer it. If you cannot answer a question, turn to the relevant page in the study guide to find the answer. For NESA key word meanings, go to www.educationstandards.nsw.edu.au and search 'key words'.

FOR A COMPLETE UNDERSTANDING OF THIS TOPIC:		PAGE NO.	✓
1	Can I describe qualitative tests that distinguish alkenes and alkanes?	209	
2	Can I describe qualitative tests that distinguish primary and secondary alkanols?	210	
3	Can I describe qualitative tests that distinguish alkanols and alkanoic acids?	210	
4	Can I explain how C-13 NMR spectra can be used to identify different molecules?	211	
5	Can I explain how H-1 NMR spectra can be used to identify molecules?	214	
6	Can I explain how mass spectrometry can be used to identify different molecules?	215	
7	Can I interpret mass spectra and identify the molecule that produces the spectrum?	216	
8	Can I interpret infrared spectra and identify the molecule that produces the spectrum?	217	

EXTENSION BOX **High-resolution proton NMR spectra**

A high-resolution proton NMR spectrum shows that the peaks are split compared with the low-resolution spectrum. The extent of splitting of the peaks gives additional information about the molecular structure.

The n+1 rule

The amount of splitting tells you about the number of hydrogens attached to the carbon atom or atoms **next door** to the one you are currently interested in. The number of sub-peaks in a cluster is **one more** than the number of hydrogens attached to the next door carbon(s).

Singlet peaks are next to a carbon atom with no H atoms; **doublet peaks** are next to a CH group; **triplet peaks** are next to a CH_2 group; and **quartet peaks** are next to a CH_3 group.

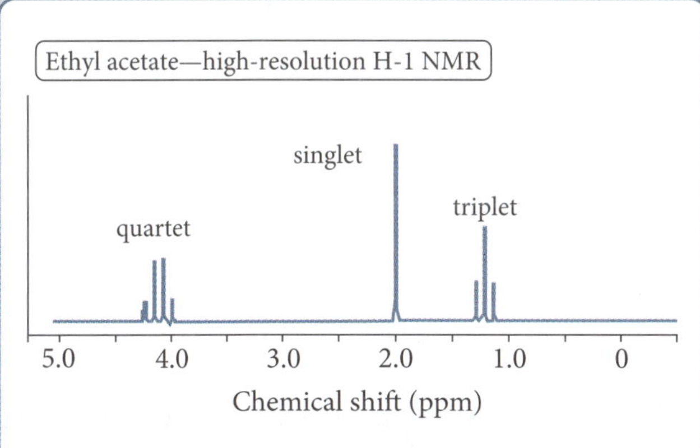

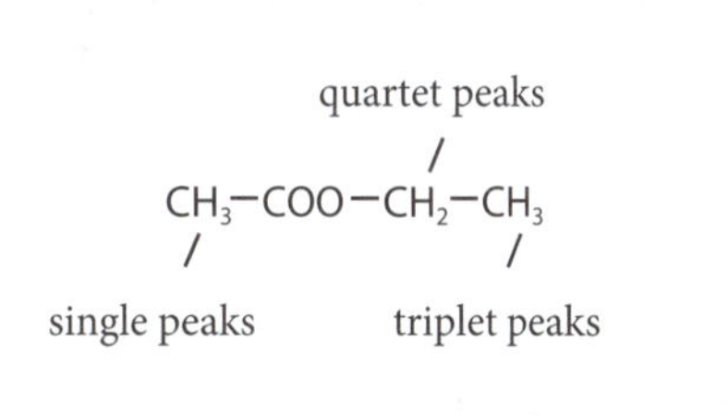

The ethyl acetate spectrum illustrates the n+1 rule. Thus the H atoms in the left CH_3 group are next to the carbon of the COO group that has no H atoms and so n+1 = 0+1 = 1(singlet).

HSC EXAM-TYPE QUESTIONS

Objective-response questions (1 mark each)

1 A small piece of sodium metal was added to a colourless liquid in a test tube. A slow effervescence of a colourless gas was observed. Identify which of the following compounds could have been in the test tube.

A butan-1-ol

B hex-1-ene

C hexane

D ethyl ethanoate

2 Select the true statement about infrared spectroscopy.

A In the IR spectrum of a carboxylic acid molecule there is no peak for an O–H bond.

B Infrared radiation ionises the organic molecules.

C IR spectroscopy is used to measure the chemical shift of functional groups.

D The absorption of infrared radiation makes the chemical bonds in organic molecules stretch and bend.

3 Select the true statement about mass spectrometry.

A The vapourised sample is converted to negative ions by bombardment with neutrons.

B The positive ions are accelerated using only magnetic fields.

C The ions are deflected in a magnetic field and, the greater the mass of the ions, the greater the arc through which they are bent.

D The mass spectrum of methane shows only one spectral line.

4 Select the statement that is true about C-13 NMR spectroscopy.

A The standard used is dimethylpropane.

B The chemical shift of the NMR peaks is affected strongly by electronegative atoms in functional groups.

C The C-13 NMR spectrum of methane contains five peaks.

D Carbon–carbon double bonds have chemical shifts between 10–50 ppm.

5 Select the correct statements about qualitative tests for functional groups in hydrocarbons.

A Bromine water is rapidly decolourised by alkanes and alkanoic acids.

B Sodium metal reacts with alkanones, releasing hydrogen gas.

C An acidified solution of potassium permanganate is decolourised when reacted with propan-2-ol.

D When sodium hydroxide solution reacts with propanoic acid there is an effervescence of carbon dioxide gas.

Extended-response questions

6 An alkanol X has the molecular formula $C_4H_{10}O$. The alkanol was mixed with acidified potassium permanganate and the mixture turned from pink-purple to colourless on heating. The liquid product Y of this reaction was isolated from the reaction mixture and tested with sodium carbonate solution. An effervescence of a colourless gas occurred.

a Identify the product Y. (2 marks)

b Identify alkanol X. (2 marks)

7 Figure 15.20 shows the mass spectrum of an element X.

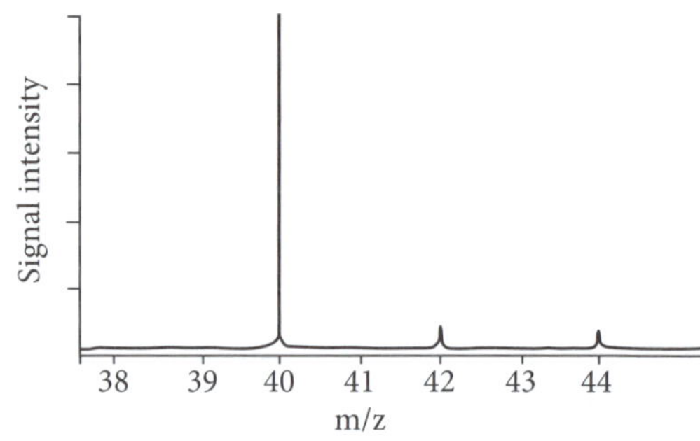

Figure 15.20 Mass spectrum of an element

a Account for the presence of one large peak and two smaller peaks. (1 mark)

b Explain why the atomic weight of this element is 40.08 u. (2 marks)

8 Figure 15.21 is the mass spectrum is of a compound that contains only carbon and hydrogen atoms.

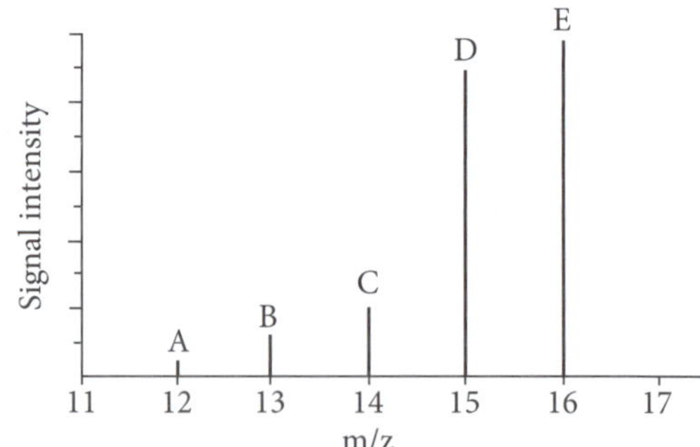

Figure 15.21 Mass spectrum of a hydrocarbon

Five peaks in the spectrum are labelled A, B, C, D and E. Each peak represents a positive ion that has been formed as the original molecule has broken down inside the mass spectrometer. If peak E is formed by the CH_4^+ molecular ion, determine the likely identity of the ions that form the remaining peaks. Justify your answer. (6 marks)

9 The empirical formula of a liquid organic compound is C_2H_4O. Samples of this compound were tested with sodium, bromine water and sodium carbonate solution.

All tests were negative. The mass spectrum of this compound showed that its molar mass was 88.1 g/mol. The infrared spectrum and the C-13 NMR spectrum of this compound is shown in Figure 15.22. Refer to the Data Sheet for infrared absorption data and C-13 chemical shift data.

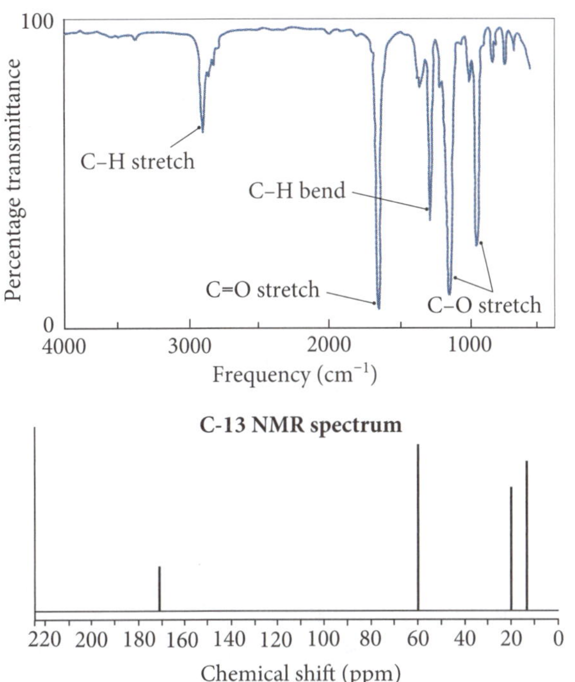

Figure 15.22 Organic compound IR and C-13 NMR spectra

Use this information to determine the possible names and structural formulas of the compound. (6 marks)

10 Figure 15.23 shows the mass spectrum and infrared spectrum are for an organic compound X. Use this information to determine the name and chemical formula of X. (3 marks)

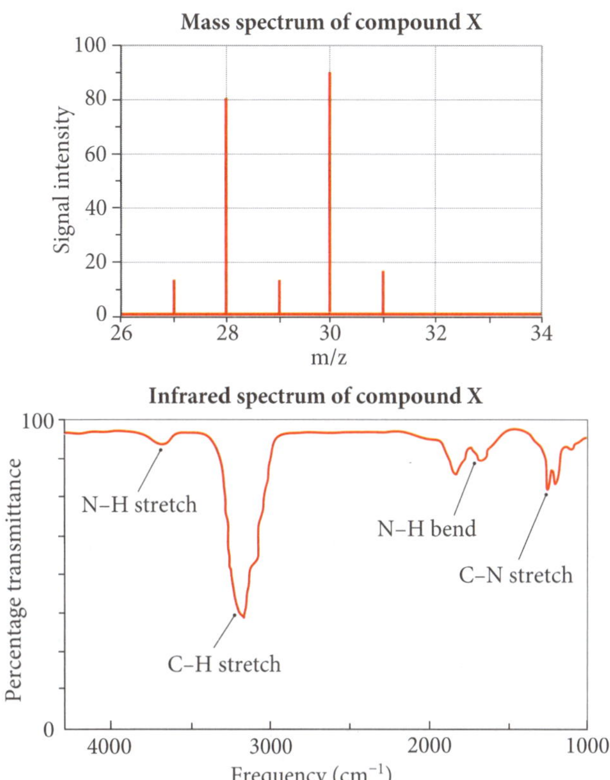

Figure 15.23 Mass spectrum and IR spectrum

ANSWERS

KEY QUESTIONS

Key questions ⊃ p. 211

1 The bromine water decolourisation test will be positive for pent-2-ene but negative for pentane.

2 Add drops of purple-pink acidified potassium permanganate to each chemical and gently heat. Decolourisation will occur with butan-1-ol as it will be oxidised. Butanone is not oxidised and so no decolourisation occurs.

3 $2HCOOH(aq) + K_2CO_3(aq) \rightarrow 2KHCOO(aq) + H_2O(l) + CO_2(g)$

4 $2Na(s) + 2CH_3OH(l) \rightarrow 2NaCH_3O(s) + H_2(g)$

Key questions ⊃ p. 214

5 The chemical environment of the carbon atoms refers to how different functional groups affect the sharing of electrons with carbon atoms. Highly electronegative atoms (e.g. O) attract electron pairs strongly and shift the resonance peaks downfield.

6 Butane is a symmetrical molecule and so instead of four peaks there are only two peaks.

7 Propan-1-ol is not symmetrical whereas propan-2-ol is symmetrical. Thus the three carbons in propan-1-ol will produce three peaks, whereas propan-2-ol has two peaks because the end carbon atoms are in identical chemical environments.

8 Propanone has a chemical shift further downfield. The C=O group in these three molecules has different chemical shift bands: C=O (ketones) = 205–220 ppm; C=O (aldehydes) = 190–200 ppm; C=O (alkanoic acids) = 170–185 ppm

Key questions ⊃ p. 217

9 Chlorine has two stable isotopes (Cl-37; Cl-35), which form the two peaks.

10 $C_4H_{10}^+$

11 $C_2H_5O^+$ (peak 45), CH_2OH^+ (peak 31)

Key questions ⊃ p. 218

12 The absorbed energy causes bond stretching and bond bending.

13 When hydrogen bonding occurs between neighbouring alkanols the O–H bond is weakened and less energy needs to be absorbed to cause bond stretching. In non-hydrogen-bonding alkanols, the O–H bond is stronger and a higher infrared frequency is required for stretching.

HSC EXAM-TYPE QUESTIONS

Objective-response questions

1 **A.** Short chain alkanols react with sodium, releasing hydrogen gas. **B**, **C** and **D** are incorrect as these molecules have no reaction with sodium.

2 **D.** The IR energy is only sufficient to bend or stretch bonds. **A** is incorrect as OH peaks are common in carboxylic acids. **B** is incorrect as the IR radiation has insufficient energy to ionise a molecule. **C** is incorrect as chemical shifts are recorded in NMR spectra.

3 **C.** Ions have different masses and they experience different forces in a magnetic field and this separates the fragments. **A** is incorrect as positive ions are formed and neutrons are not involved. **B** is incorrect as the ions are initially accelerated in an electric field. **D** is incorrect as methane (CH_4) would produce multiple ions due to fragmentation of the molecule.

4 **B.** Electronegative atoms such as oxygen and nitrogen cause a large chemical shift in the C-13 atoms to which they are bonded. **A** is incorrect as the standard is tetramethyl silane. **C** is incorrect as there is only one carbon atom. **D** is incorrect as double bonds have chemical shifts in the band 115–140 ppm.

5 **C.** A secondary alkanol can be oxidised by acidified permanganate, which is decolourised. **A** is incorrect as bromine water is too weak to oxidise alkanes and acids. **B** is incorrect as alkanones have no reaction with sodium. **D** is incorrect as carbon dioxide will be produced only if sodium carbonate solution is used.

Extended-response questions

6 EM Students are required to recall the reactions of acidified potassium permanganate and sodium carbonate solution. They then need to apply this knowledge to identify the classification of the alkanol.

a Y is an alkanoic acid as the sodium carbonate solution effervesced. ✓
Y has four carbon atoms per molecule. It is butanoic acid ($CH_3CH_2CH_2COOH$). ✓

b X is a primary alkanol because primary alkanols can be oxidised to alkanoic acids. ✓ Thus X is butan-1-ol. ✓

7 EM Students need to recall that elements consist of isotopic forms with different masses and that the peak heights are a measure of relative abundance.

a The natural sample of this element contains three isotopic forms. ✓

b The much greater abundance of the X-40 isotope means that the average atomic weight of the natural mixture will be very close to 40 u. ✓ The average atomic weight will be a little greater than 40 u as the other two isotopes are heavier. ✓ Element X is calcium (see periodic table).

8 EM Students use the mass spectrum to predict which fragments form each peak. m/z values should be calculated for possible fragments. A justification is needed to score full marks.

$A = C^+$; ✓ $B = CH^+$; ✓ $C = CH_2^+$; ✓ $D = CH_3^+$ ✓

Justification: Assume all ions have a +1 (z = 1) charge. Carbon has an atomic weight of approximately 12 u and hydrogen has an approximate atomic weight of 1 u. Thus peak A at m/z = 12 is likely to be C^+, which would have an m/z value of $\frac{12}{1} = 12$. The peak at B is probably CH^+ as m = 12 + 1 = 13. ✓

The peak at C is probably CH_2^+ as m = 12 + 1 + 1 = 14 and the peak at D is probably CH_3^+ as m = 12 + 1 + 1 + 1 = 15. ✓

9 EM Students use the IR spectrum to identify key functional groups. Using the known molar mass and empirical formula can be determined. Use the C-13 NMR spectrum to determine key functional groups and the number of chemical environments of the carbon atoms. Clear structural formula diagrams and correct IUPAC naming is required for full marks.

EM (C_2H_4O) = 44.05 g/mol. The empirical molar mass is half the molar mass (88.1 g/mol). Thus the molecular formula is $C_4H_8O_2$. ✓
The negative tests show the compound is not an alkanol, alkene or alkanoic acid. ✓

The IR spectrum shows the presence of a carbonyl group (C=O) as well as a single-bonded O atom (C–O–C).

The C-13 NMR spectrum shows characteristic chemical shifts for a carbonyl group (171 ppm) and a C–O group (59 ppm). The carbon atoms are in four different chemical environments.

The compound is most likely to be an ester. ✓

Figure A15.1 shows possible structures and names of esters with a molar mass of 88.1 g/mol. In these examples there are four different chemical environments of each of the carbon atoms.

a

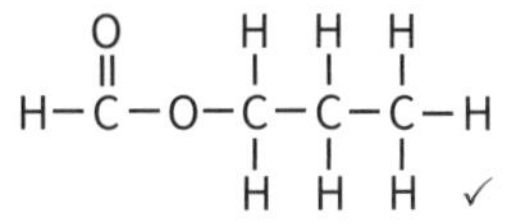

```
     O      H  H  H
     ||     |  |  |
  H—C—O—C—C—C—H
            |  |  |
            H  H  H  ✓
```

1-propyl methanoate

b

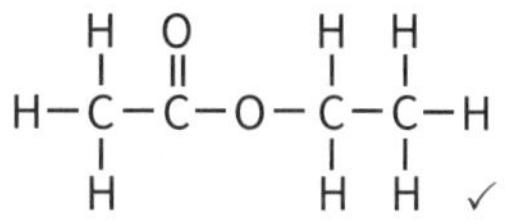

```
     H  O      H  H
     |  ||     |  |
  H—C—C—O—C—C—H
     |         |  |
     H         H  H  ✓
```

ethyl ethanoate

c

```
     H  H  O      H
     |  |  ||     |
  H—C—C—C—O—C—H
     |  |         |
     H  H         H  ✓
```

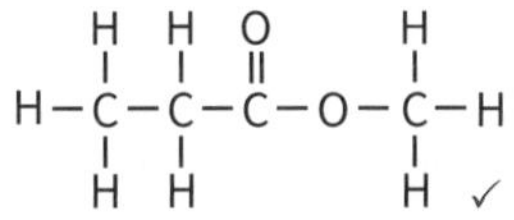

methyl propanoate

Figure A15.1 Ester structures

10 EM Students need to identify the key information in both spectra to determine the molar mass and then the chemical formula and name.

The IR spectrum shows the presence of C–N and N–H bonds and this is consistent with an amine compound. ✓ The mass spectrum has a m/z value of 31. ✓ This is consistent with methanamine (CH_3NH_2), which has a molar mass of 31.1 g/mol. ✓

CHAPTER 16 CHEMICAL SYNTHESIS AND DESIGN

INQUIRY QUESTION:

What are the implications for society of chemical synthesis and design?

There are huge numbers of different compounds that have been synthesised from raw materials. Many of these are polymeric or simple compounds that have replaced natural materials that can no longer meet demand. For example, natural fertilisers had to be replaced by synthetic fertilisers due to the rapid increase in the human population. Chemists continue to experiment to develop new products to meet the demands of society as well as replace past materials that caused environmental pollution. This chapter investigates the chemical synthesis process from an industrial perspective.

1 Chemical synthesis

» Students evaluate the factors that need to be considered when designing a chemical synthesis process, including but not limited to availability of reagents; reaction conditions; yield and purity; industrial uses (e.g. pharmaceutical, cosmetics, cleaning products, fuels); and environmental, social and economic issues.

The chemical industry

➔ Chemical industries produce a wide variety of inorganic and organic compounds. The reactants used, the steps of the reaction sequence and the conditions employed vary from one chemical industry to another. The conditions that are selected usually represent a compromise between equilibrium, kinetic and economic factors.

- The yield of product in a chemical process is dependent upon the extent of the equilibrium reactions involved. In chemical industry, changes in temperature and concentration affect the yield of the product at equilibrium.
- The rate of production of a product depends on the activation energy for the reaction. Raising the temperature and the use of suitable catalysts increases the rate of the reaction.
- The environmental and social issues of the production of chemicals need to be investigated during the planning stages of chemical factory construction.
- The cost of providing the conditions required to maximise rate and yield must also be considered. If high temperatures are required then the cost of providing heat energy to raise the temperature becomes an economic issue for industrial plants. Products that are manufactured via continuous processing on a production line can only be economic as long as large amounts of products are produced for sale.

➔ Chemical industries must choose a suitable site to undertake the recovery of raw materials or to construct the processing factory. Figure 16.1 summarises the key issues.

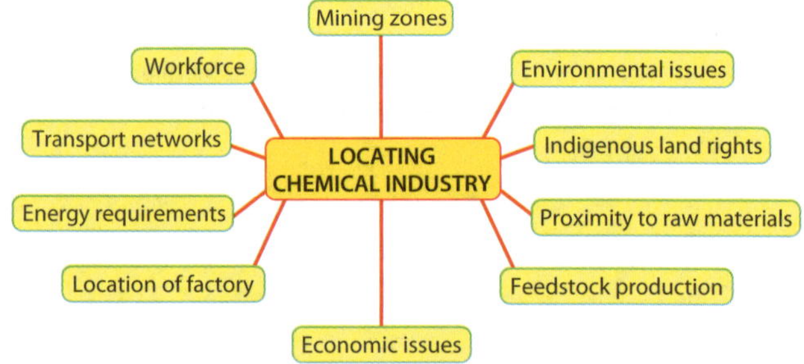

Figure 16.1 Issues concerning the location of chemical industries

➔ The following factors should be considered when siting chemical industries. Many of these factors are determined by economic issues.

Location of mining zones

Some regions of Australia are protected for environmental reasons or because Indigenous Australians hold land rights over the land. The mining of raw materials for a chemical industry cannot be easily conducted if the deposit is located within a high population zone.

Raw materials and feedstocks

Some factory sites need to be as close as possible to the site of **raw material** production or **feedstock** availability. The production of ammonia and sodium carbonate illustrate this requirement.

raw materials: unprocessed natural materials obtained from the lithosphere, hydrosphere, atmosphere or biosphere
feedstocks: materials produced from the raw materials which become the reactants in the chemical process

Economic issues

Companies need to determine the cost of mining, milling and extracting the required feedstocks from the raw materials. Some feedstocks are more expensive to extract than others from ore bodies due to the higher energy costs involved. Aluminium is extracted from bauxite ore bodies and it is more expensive to extract and manufacture than copper due to the high costs of the electrolytic process used to make aluminium.

Co-location of chemical industries

In many cases a new factory could be sited in an urban industrial zone where other industries supplying the chemicals and equipment needed to run the factory are located. The markets for the products produced will be close by in large urban areas. Where possible the chemical factory should be sited near rural sites for raw materials or near power plants if electrical energy is required for the chemical process. In such cases the company may need to provide housing for qualified workers.

Energy

Electrical power must be readily available for many chemical industries such as the electrolytic production of sodium hydroxide and the production of aluminium. Energy companies often provide lower power cost incentives to new industries.

Transport networks

Efficient transport networks need to be available to move materials to and from the industrial site. Transport costs need to be minimised to maximise the economics of the industry. Transportation hubs for product distribution locally or overseas are essential.

Workforce

A large workforce may be required and many chemical jobs require staff with tertiary qualifications in chemistry as well as skilled technicians. Siting a chemical factory near a city or large town will provide this skilled labour force. If a remote location must be used then large companies will build houses and shops for staff.

➔ In the industrial production of chemicals it is important to maximise the final yield of products in order to maximise the profits for the company. Consider the industrial production of methanol (CH_3OH). Methanol is an important fuel, solvent and precursor for the production of esters and resins.

- It is manufactured by reacting carbon monoxide and hydrogen gas using a mixed catalyst comprising copper, zinc oxide and aluminium oxide at ~250 °C and 5–10 MPa. A 97% yield is obtained under these conditions:

  $$CO(g) + 2H_2(g) \leftrightarrows CH_3OH(g);\ \Delta H = -90\ \text{kJ/mol}$$

- The high pressure used maximises the yield of methanol as the equilibrium shifts to the product side (according to Le Chatelier's principle) to counteract the imposed pressure by reducing the number of particles. As the reaction is exothermic, the temperatures used are a compromise so that a reasonable yield is obtained and the rate of the reaction is maintained. The catalyst also ensures a higher reaction rate.

EXAMPLE 1

Carbon monoxide gas and hydrogen gas react to form methanol vapour:

$$CO(g) + 2H_2(g) \leftrightarrows CH_3OH(g)$$

a Use the following thermochemical data to show that the reaction is spontaneous under standard conditions.

Data:
ΔG^o (kJ/mol): $CO(g)$ (−137); $H_2(g)$ (0); $CH_3OH(g)$ (−163)

b 1.00×10^4 tonnes of carbon monoxide and the stoichiometric amount of hydrogen are reacted to form methanol at 250 °C and 10 MPa.

i Calculate the theoretical yield of methanol.

ii Calculate the percentage yield by weight of methanol if 1.11×10^4 tonnes of methanol are formed in the reaction.

Consider the reaction stoichiometry

Answer:

a $\Delta G^o = \Delta G^o(CH_3OH) - \Delta G^o(CO) - 2\Delta G^o(H_2)$
$= (-163) - (-137) - 2(0) = -26\ \text{kJ/mol} < 0$

As the Gibbs free energy change is negative, the reaction is spontaneous under standard conditions.

b i $CO(g) + 2H_2(g) \leftrightarrows CH_3OH(g)$

1.00×10^4 t of CO $= 1 \times 10^7$ kg $= 1 \times 10^{10}$ g

$n(CO) = m/M = \dfrac{1 \times 10^{10}}{28.01} = 3.57 \times 10^8$ mol

Stoichiometry $CO : CH_3OH = 1 : 1$

$n(CH_3OH) = n(CO) = 3.57 \times 10^8$ mol

$m(CH_3OH) = n.M = (3.57 \times 10^8)(32.042)$
$= 1.14 \times 10^{10}\ \text{g} = 1.14 \times 10^4\ \text{t}$

Theoretical yield $= 1.14 \times 10^4$ t

ii Percentage yield $= \dfrac{1.11 \times 10^4}{1.14 \times 10^4} \times \dfrac{100}{1} = 97.4\%$w/w

➔ An emerging technology in chemical industries is computer-aided molecular design (CAMD). This technology is helping polymer chemists to select the most appropriate functional groups to be included in monomers that will produce polymers with pre-determined physical and chemical characteristics.

- Traditionally molecular design has relied on experimental data for synthesising possible products and evaluating their performance. This older approach is more costly and time consuming than CAMD. The pharmaceutical industry requires plastic containers for their products that seal properly so that the pharmaceutical contents have an extended shelf life prior to sale and the entry of air, moisture and microorganisms is prevented.

- CAMD principles can be used to design polymeric materials that provide the best seal for such products. The first step is to determine the structural and chemical property restraints of the final polymer products. For example, the elasticity and permeability limits of the final polymer need to be determined. The polymer's repeating units should not contain polar or hydrogen-bonding functional groups as this will decrease flexibility of the final polymer seal. The CAMD program analyses the chemical and physical data supplied and selects the best monomers to produce the desired properties in the polymer.

2 Chemical industries

» Students evaluate the factors that need to be considered when designing a chemical synthesis process, including but not limited to availability of reagents; reaction conditions; yield and purity; industrial uses (e.g. pharmaceutical, cosmetics, cleaning products, fuels); and environmental, social and economic issues.

➔ This section investigates a range of chemical industries.

Production of ammonia: The Haber process

➔ The production of ammonia illustrates the issues involved in many chemical industries. The Haber process is used to manufacture ammonia. Ammonia is an important industrial chemical. Solutions of ammonia in water are used domestically as cleaning agents to dissolve and remove grease and dirt from floors and windows. The production of fertiliser accounts for about 85% of the worldwide use of ammonia.

- Ammonia gas is made by the direct reaction of hydrogen gas and nitrogen gas over a catalyst at elevated temperatures and pressures. The reaction is an exothermic equilibrium:

$$N_2(g) + 3H_2(g) \leftrightharpoons 2NH_3(g); \Delta H = -92 \text{ kJ/mol}$$

- The nitrogen and hydrogen feedstocks are derived from such common raw materials as air, water and natural gas. These raw materials are in plentiful supply.
- Hydrogen is produced by reacting the natural gas with steam over a nickel catalyst at 750 °C in a furnace. Carbon monoxide is also formed. Air is then introduced and the carbon monoxide and some of the hydrogen undergo combustion to form carbon dioxide and water. The unreacted nitrogen from the air together with the remaining hydrogen is removed and purified.
- It is essential that no carbon monoxide remains in the hydrogen gas as it will poison (deactivate) the catalyst used in the subsequent Haber process. The final gaseous mixture contains nitrogen and hydrogen in the volume ratio of 1 : 3, as required for the reaction stoichiometry of the Haber process.
- The conditions employed in the Haber process vary from one manufacturing plant to another. The selected set of conditions represents a compromise between equilibrium, kinetic and economic factors. The Haber process conditions are summarised as follows:
  - Reactants: N_2 and H_2 (volume ratio 1 : 3).
  - Pressure: 20–25 MPa. Kinetic and equilibrium considerations favour greater yields of ammonia at high pressures.
  - Temperature: 400–550 °C. Equilibrium factors favour a low temperature and kinetic factors favour a high temperature. The compromise-temperature conditions allow the reactants to have sufficient kinetic energy to overcome the activation energy barrier as well as achieve a reasonable yield per cycle through the reaction vessel.
  - Catalyst: Finely powdered magnetite (Fe_3O_4) fused with K_2O, Al_2O_3 and CaO and the magnetite is then partly reduced (by hydrogen) to porous iron on the surface. The catalyst has a high surface area. The catalyst increases the rate of the reaction by lowering the activation energy in the first stage of the reaction. Figure 16.2 shows the activation energy profile for the Haber process.

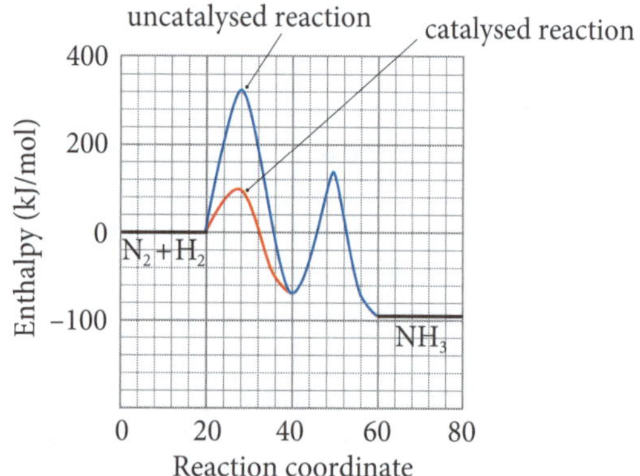

Figure 16.2 Enthalpy profile for the Haber process

  - Yield: At 20 MPa/400 °C the yield is ~39%. At 20 MPa/500 °C the yield is ~18%. After 5–6 cycles about 98% of the reactants are converted to ammonia.
  - Continuous removal of ammonia: The ammonia is removed after each cycle through the reactor by liquefaction (under pressure). This removal increases the net yield of ammonia due to an equilibrium shift to the right, as predicted using Le Chatelier's principle. Unreacted gases are

- recycled many times to achieve a high conversion of reactants into ammonia over many cycles.
  - Economic issues: As high pressures are required the pipes and vessels need to be very strong and such equipment is expensive to buy and maintain.
  - Location of Haber plants: In Australia, Haber plants are located near the supply of natural gas that is used to generate the hydrogen feedstock. The North West Shelf of Western Australia is the location of large reserves of natural gas in Australia. This coastal site is also near transport hubs to move the liquefied ammonia and manufactured fertiliser interstate and overseas.
- ➔ Haber plants provide ammonia for the production of fertilisers for use in agriculture.
- ➔ In the production of ammonia, environmental pollution must be minimised. Environmental pollution has a social impact and can result in the EPA fining the company. The high temperatures required for the Haber process require heat energy, which commonly is produced from the combustion of fossil fuels. The carbon dioxide released in the production of electricity via fossil fuel combustion leads to increased global warming.
- ➔ The fertilisers produced from ammonia have allowed a large expansion of agriculture to produce food for an increasing world population. This is a major benefit to society. However, some fertiliser runs off agricultural land during rain and drains into creeks and rivers, where it can cause **eutrophication**. Eutrophication is the abundant growth of algae and water plants due to the presence of high levels of nitrogenous and phosphate nutrients in the waterway. The surface algal blooms block sunlight and oxygen from entering the water. On the death of the algae the dissolved oxygen levels in the water become severely depleted as **aerobic decomposers** break down their remains. Once the dissolved oxygen levels drop to near zero, anaerobic decomposers become active and foul-smelling gases are released in the stagnant water.

eutrophication: the abundant growth of algae in a waterway that ultimately leads to the depletion of dissolved oxygen and the death of aquatic organisms

aerobic decomposers: organisms such as bacteria and moulds that break down dead remains and require oxygen to carry out the process

EXAMPLE 2

The equilibrium concentration of ammonia, nitrogen and hydrogen were measured at 500 °C in a sample of an equilibrium reaction mixture removed from the Haber reactor. The equilibrium concentrations were:

$[NH_3]$ = 1.24 mol/L; $[N_2]$ = 0.97 mol/L;
$[H_2]$ = 2.62 mol/L

a **Calculate the equilibrium constant for the ammonia equilibrium at 500 °C.**

b **The temperature of the vessel was changed and the mixture allowed to come to equilibrium. The new concentrations were:**
$[NH_3]$ = 0.93 mol/L; $[N_2]$ = 1.16 mol/L;
$[H_2]$ = 3.49 mol/L

- **i** **Calculate the equilibrium constant at the new temperature.**
- **ii** **If the ammonia synthesis is an exothermic equilibrium, determine whether the new reaction temperature is higher or lower than 500 °C.**

Review the concept of equilibrium constants in Chapter 3

Answer:

a $K_{eq} = [NH_3]^2/[N_2][H_2]^3 = \dfrac{(1.24)^2}{(0.97)(2.62)^3} = 0.088$

b **i** $K_{eq} = \dfrac{(0.93)^2}{(1.16)(3.49)^3} = 0.0175$

ii The value of K_{eq} has decreased. Thus the new temperature is higher as the equilibrium has shifted to the left to reduce the concentration of ammonia and raise the concentration of nitrogen and hydrogen.

➔ Further information about the Haber process can be found in various online videos. Enter the following title in a search bar: synthetic ammonia—Fritz Haber and Carl Bosch

➔ KEY QUESTIONS

1. **Identify factors that influence the yield of a product in an industrial chemical synthesis.**
2. **Identify the feedstocks in the Haber process and identify the raw materials from which the feedstocks are sourced.**
3. **Use an equation to explain why ammonia is continually removed from the reaction vessel in the Haber process.**

Answers ➲ p. 240

Production of sodium carbonate: The Solvay process

➔ Sodium carbonate has many major uses in society. Two major uses of sodium carbonate are:

- Glass manufacturing—the sodium carbonate acts as a **flux** because it lowers the melting point of the mixture of silicon dioxide and calcium carbonate.

- Soap and detergents—sodium carbonate can be used as a base in soap and detergent manufacture. It is also used as a **water softener** in laundry powders.

flux: a chemical that softens a substance and reduces its melting point, making it flow more readily
water softener: hard water contains calcium ions that prevent soap lathering; water softeners precipitate the calcium ions and soften the water

- Sodium carbonate is manufactured from sea salt, limestone (calcium carbonate) and ammonia via the Solvay process. The sea salt and limestone are readily available and ammonia is acquired from a Haber factory.
- The stages of the Solvay process are summarised in the following equations:
  A $CaCO_3(s) \rightarrow CaO(s) + CO_2(g)$
  B $CO_2(g) + H_2O(l) \leftrightharpoons H_2CO_3(aq)$
  C $H_2CO_3(aq) + NH_3(aq) \leftrightharpoons NH_4HCO_3(aq)$
  D $NaCl(aq) + NH_4HCO_3(aq) \rightarrow NaHCO_3(s) + NH_4Cl(aq)$
  E $2NaHCO_3(s) \rightarrow Na_2CO_3(s) + H_2O(l) + CO_2(g)$
- A 30%w/w **brine** solution is prepared from purified sea salt, which is obtained in Australia by the evaporation of seawater in coastal salt farms close to the factory site. The salt must first be treated to remove calcium and magnesium ions that will interfere in the Solvay process.
- Limestone is transported from a local limestone mine by rail to the factory site. Crushed limestone is thermally decomposed in a **coke**-fired kiln to produce calcium oxide and carbon dioxide gas (reaction A). The carbon dioxide released is compressed and cooled. The calcium oxide is then used to produce calcium hydroxide suspension (called *milk of lime*).

brine: salty water
coke: a product produced by the carbonisation of coal at high temperatures to enrich the carbon content

- The ammonia (manufactured via the Haber process) is transported as liquefied ammonia via rail to the factory. The purified brine is saturated with ammonia in the ammonia absorber tower. Ammonia is recycled so little is lost from cycle to cycle. The ammoniating process is exothermic and the heat released is removed via cold water pipes. This ensures that the brine is fully ammoniated.
- The ammoniated brine enters the top of the Solvay towers (carbonators). Carbon dioxide gas is pumped under pressure (~300 kPa) into the base of the Solvay towers. The ammoniated brine trickles downward over the surfaces of curved and perforated baffle plates inside the towers. These plates provide a large surface area for the absorption of carbon dioxide. Carbonic acid forms (reaction B) when the rising carbon dioxide dissolves in the ammoniated brine. Figure 16.3 shows a typical Solvay tower.

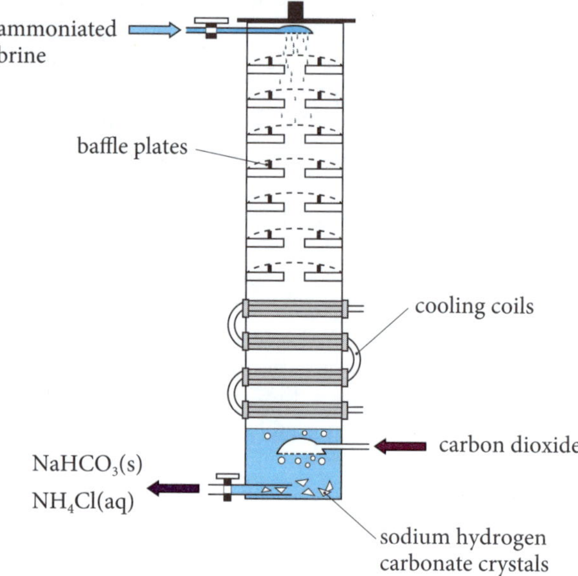

Figure 16.3 Solvay tower

- An acid–base reaction occurs between the ammonia and the carbonic acid in which ammonium hydrogen carbonate forms (reaction C).
- The low temperature and high salt concentration promotes the crystallisation of sodium hydrogen carbonate (reaction D). The crystals of sodium hydrogen carbonate are washed downwards to the base of the Solvay towers.
- The sodium hydrogen carbonate crystals are recovered from the filtrate using a rotary vacuum filter. The filtrate (containing ammonium chloride) is returned to the ammonia recovery plant, where it is mixed with the milk of lime and then distilled to recover the ammonia, which is then recycled.
- The sodium hydrogen carbonate crystals are washed and dried. Any adsorbed ammonia that is released on drying is recycled. The dried crystals are then decomposed in a rotary kiln to produce anhydrous sodium carbonate, water vapour and carbon dioxide (reaction E). Heat is removed from the hot carbon dioxide before it too is recycled back to the carbonator.
- This waste solution is cooled using heat exchangers and discharged safely into the ocean as the calcium and chloride ions in the solution do not substantially change the salinity of the ocean. However, such waste solutions could not be discharged into rivers as it would increase the salinity of the fresh water. Ammonia is always recycled and does not pose a pollution problem.
- Figure 16.4 shows a flowchart of the Solvay process.

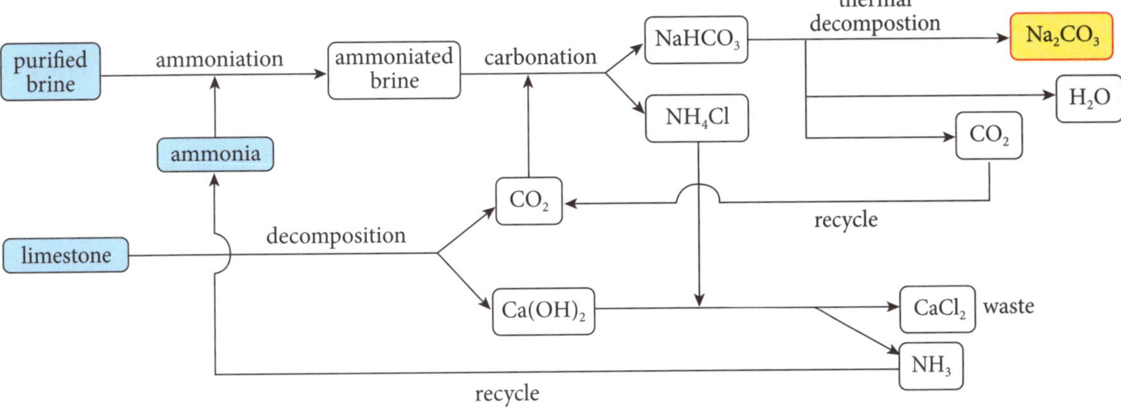

Figure 16.4 Solvay process flowchart

EXAMPLE 3

Figure 16.5 shows a map of possible locations for various chemical industries. The locations are labelled A, B, C and D.

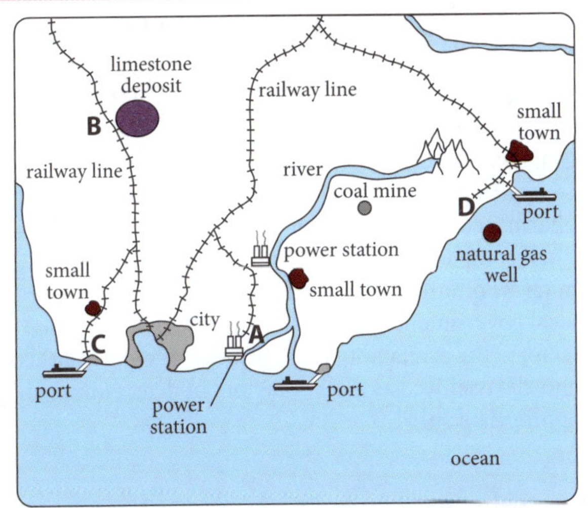

Figure 16.5 Hypothetical locations of chemical industries

Select the most appropriate location (A, B, C or D) for the production of:

a sodium carbonate

b ammonia.

Justify your response.

Consider the location of raw materials required for each industry

Answer:

a C is the best site for the production of sodium carbonate because:
- the salt required is locally available from salt fields located on the coast.
- the limestone mine is connected to the factory site by rail. This railway line is used to transport the raw material to the factory.
- a port is nearby to transport the product to overseas markets.
- the city is nearby to market the product and also a site for staff to live and easily commute to the factory.

b D is the best site for the production of ammonia because:
- the natural gas well is near the coast and the natural gas required for hydrogen feedstock production can be transferred by undersea pipes to the factory.
- a coal mine is nearby so that coking coal can be obtained to use as a fuel for heating the reactor.
- rail tankers can transport the liquefied ammonia and fertiliser to the local port for transport to overseas markets or interstate coastal ports.
- a town is nearby for staff to live and easily commute to the factory.

➔ KEY QUESTIONS

4 **Explain the purpose of the carbonation step in the Solvay process.**

5 **Identify the conditions that promote the formation of crystals of sodium hydrogen carbonate in the Solvay tower.**

6 **Explain why waste calcium chloride solutions cannot safely be discharged into a river.**

Answers ➲ p. 240

Production of pharmaceuticals: Aspirin manufacture

➔ Many pharmaceuticals are manufactured throughout the world on large scales. One of these pharmaceuticals is aspirin.

➔ Aspirin is an **analgesic** as it relieves mild pain. It is also an antipyretic as it reduces fever and is an anti-inflammatory as it reduces swelling. It also acts as an **anticoagulant** in the blood and thus is helpful for preventing strokes and heart attacks. Aspirin is therefore of great social importance.

> **analgesic:** relieves pain
> **anticoagulant:** prevents blood cells clotting and blocking blood vessels

➔ The active ingredient in aspirin tablets is acetylsalicylic acid. It is manufactured by a one-step reaction between salicylic acid and acetic anhydride.

The feedstocks required for aspirin production are manufactured from other organic compounds.

- Acetic anhydride is industrially manufactured from acetic acid. The acetic acid is produced by a reaction between methanol and carbon monoxide on a commercial scale. A condensation reaction between two acetic acid molecules produces acetic anhydride.
- Salicylic acid is manufactured from sodium phenoxide and carbon dioxide.

➔ In the laboratory preparation concentrated sulfuric acid is also added as a catalyst. The mixture is heated and the acetylsalicylic acid forms. The mixture is then cooled to precipitate the product. Figure 16.6 shows the structural equation for this reaction. The reactants and products are in equilibrium.

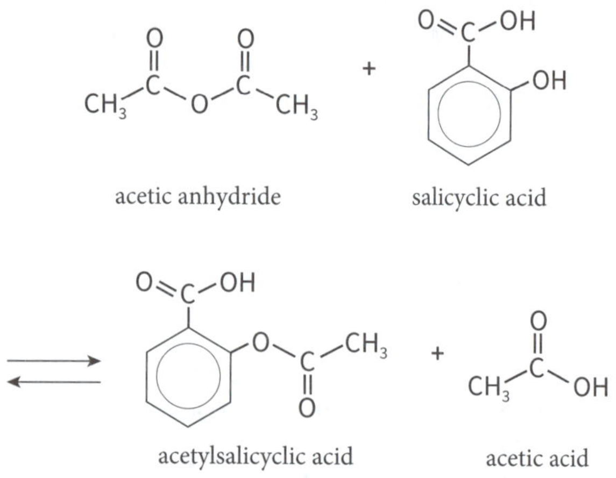

acetic anhydride + salicyclic acid ⇌ acetylsalicyclic acid + acetic acid

Figure 16.6 Manufacture of acetylsalicylic acid

➔ The methods used in the commercial manufacture of acetylsalicylic acid vary. One common method is described as follows:

- In the industrial process a large glass-lined vessel (~7 ML) is used. The acetic anhydride is dissolved in liquid toluene and then added to the vessel. The salicylic acid is then added and the mixture and heated under reflux for about 20 hours at 90 °C. The heating increases the reaction rate.
- The acetic anhydride is added in excess and so the reaction equilibrium is shifted to increase the yield of products. The reaction mixture is then transferred to another vessel to cool down over three days.
- Large crystals of acetylsalicylic acid form during this time. Some acetylsalicylic acid remains dissolved and also has to be recovered. The unreacted salicylic acid and acetic anhydride are recovered from the solvent by centrifugation and recycled. The acetic acid is also recovered as it is a by-product of the industrial process.
- The crystals of acetylsalicylic acid are washed with distilled water and dried using warm air. The final yield of acetylsalicylic acid is about 98%.

➔ Aspirin tablets are created by mixing corn starch and water with the correct amount of the acetylsalicylic acid. The corn starch acts as a binding agent to hold the ingredients together and produce a tablet of appropriate size for swallowing.

➔ In recent times the increased use of pharmaceuticals by an increasing world population has raised some concerns about water pollution. Some analgesic ingredients are not fully metabolised before they are excreted. Sewage waste finds its way into rivers and oceans and consequently some types of pharmaceuticals enter the hydrosphere and may affect the biosphere. Their potential impact on beneficial organisms in the environment has yet to be determined.

➔ Chemical factories that produce pharmaceuticals are typically located in industrial areas of cities. Pharmaceuticals are produced by very large as well as small companies. They have no special requirements in terms of infrastructure and transport. They require local government approval for the site chosen for the factory. The availability of feedstock, potential markets and electricity costs are key factors in determining an appropriate site.

EXAMPLE 4

Aspirin (acetylsalicylic acid, $C_6H_4OOCCH_3COOH$) is manufactured from salicylic acid ($C_6H_4OHCOOH$) and acetic anhydride ($C_4H_6O_2$) according to the following reaction:

$$C_6H_4OHCOOH + C_4H_6O_2 \leftrightharpoons C_6H_4OOCCH_3COOH + CH_3COOH$$

The reaction was performed by a student in the laboratory. 22.00 g of salicylic acid and 22.00 g of acetic anhydride were heated together for 30 minutes in a conical flask. The flask was then cooled back to room temperature and crystals of aspirin formed on the base of the flask. The crystals were filtered, dried and weighed. The crystals weighed 21.50 g.

a Determine which reactant is in excess.

b Determine the theoretical yield of aspirin.

c Calculate the percentage yield of aspirin in this experiment.

> Calculate the molar masses of each reactant

Answer:

a M(salicylic acid) = 122.12 g/mol

n(salicylic acid) = $m/M = \frac{22.00}{122.12}$ = 0.180 mol

M(acetic anhydride) = 86.088 g/mol

n(acetic anhydride) = $m/M = \frac{22.00}{86.088}$ = 0.256 mol

The reaction stoichiometry is 1 : 1.

Thus the excess reagent is acetic anhydride.

b Salicylic acid is the limiting reagent. Therefore:

n(salicylic acid) = n(aspirin)

n(aspirin) = 0.180 mol

M(aspirin) = 180.154 g/mol

m(aspirin) = $n.M$(aspirin) = (0.180)(180.154)

= 32.43 g = theoretical yield

c Percentage yield = $\frac{21.50}{32.43} \times \frac{100}{1}$ = 66.3%

Production of cosmetics: Hair conditioner manufacture

- Hair conditioners are examples of **emulsions**.
- There are two types of emulsions: oil-in-water and water-in-oil. Oil-in-water emulsions are composed of **colloidal** droplets of oil dispersed in water (the continuous phase). Water-in-oil emulsions are composed of colloidal droplets of water dispersed in an oil (the continuous phase).
- Emulsions are stabilised by **surfactants.** A surfactant molecule is an organic molecule with a polar group that interacts with water and a non-polar end that interacts with non-aqueous liquids. The polar group is called the *head* and it is hydrophilic (or *water loving*). The non-polar hydrocarbon chain is called the *tail* and it is hydrophobic (or *water hating*).

> **emulsion:** a colloidal suspension of one liquid in another liquid
> **colloid:** particles in the size range 1–1000 nm that remain suspended in a liquid
> **surfactant:** a surface-active molecule that has a hydrophilic end and a hydrophobic end

- Alkyl quaternary ammonium salts are common surfactants used in hair conditioners. Figure 16.7 shows the typical structure of hexadecyltrimethyl ammonium chloride, which is an example of this type of surfactant. This surfactant is classified as a cationic surfactant as it has a positive-head group. Such surfactants are commonly manufactured by petrochemical industries.

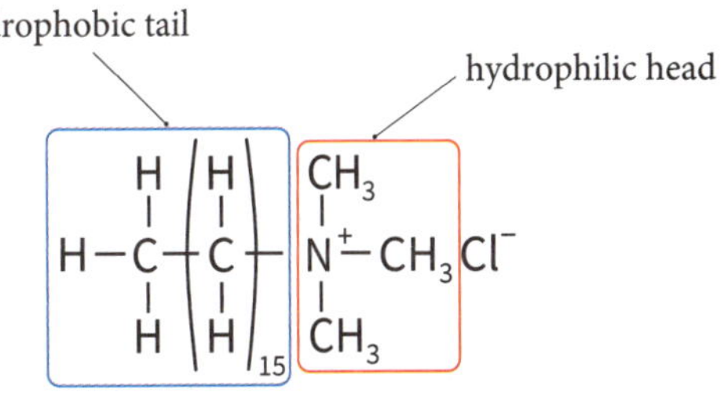

Figure 16.7 Hexadecyltrimethyl ammonium chloride surfactant

- Cationic surfactants can be manufactured from long chain carboxylic acids. These fatty acids are reacted with ammonia and the product reduced and alkylated to form the cationic surfactant. Cationic surfactants are strongly foaming in water and this makes them useful in hair conditioners as well as mild antiseptics.
- Figure 16.8 shows the interaction of the cationic surfactant with the oil and water in an oil-in-water emulsion.

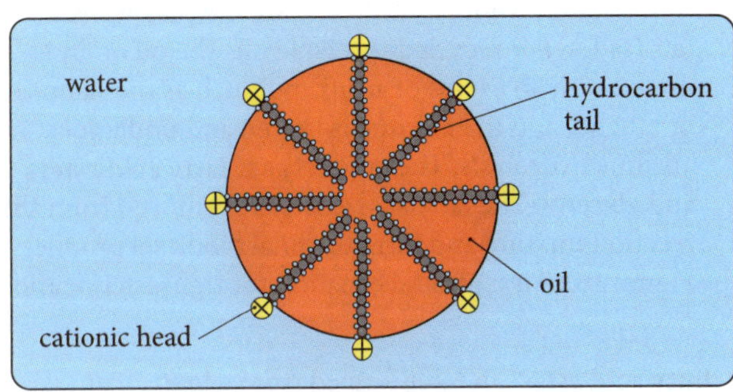

Figure 16.8 Cationic surfactant stabilises an oil-in-water emulsion

- Hair conditioners help to make hair shiny and prevent tangling. They are examples of oil-in-water emulsions.
- The water continuous phase is compatible with the skin and hair and provides a medium to transfer the oily component (e.g. rose oil) onto the hairs.
- The positively charged head groups of the cationic surfactants are attracted to the negative charges on the protein surfaces of the hair strands.
- The long hydrocarbon tails of the surfactant cling to the hair, make it feel soft and smooth and give it weight (Figure 16.9).
- By forming a stabilised colloidal emulsion with water, these materials can be diluted to an appropriate level and be deposited evenly on all the hair.

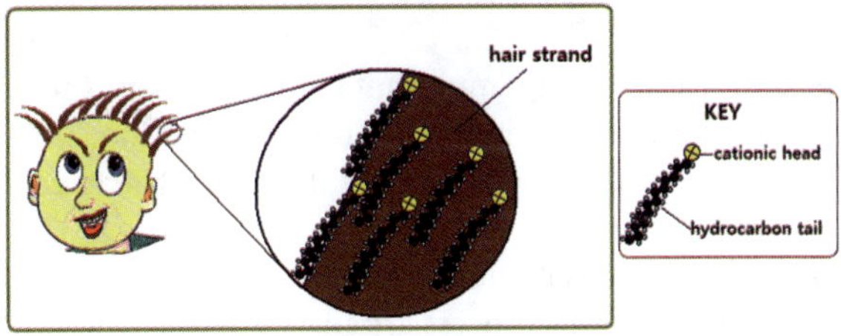

Figure 16.9 Cationic surfactants cling to hair strand

- Hair conditioners are manufactured by large companies as well as small businesses. The ingredients (natural plant oils, water and surfactants) are non-toxic. Once the ingredients are added in the correct quantities they are warmed and mixed to form the stable emulsion.
- Other personal care products such as skin lotions and face creams are also emulsions and manufactured from natural or synthetic materials derived from crude oil.

Production of fuels: Biodiesel manufacture

- Alternative renewable biofuels such as biodiesel are under development.
- Biodiesel is a general term referring to many types of fatty esters that can be used as a fuel in diesel vehicles, although some engine modification may be required in some vehicles. Biodiesel can also be blended with petroleum-derived diesel (petrodiesel).
- The chemical process for manufacturing biodiesel is transesterification. The feedstock is typically beef fats (tallow) or recycled vegetable oil. The oil or fat is mixed with an excess of an alkanol such as methanol or ethanol and the mixture is heated and undergoes alkaline hydrolysis. The products are **fatty acid esters** and glycerol. The fatty acid esters are removed from the reaction mixture and form the final biodiesel product. Figure 16.10 is a general equation for transesterification.

fatty acid ester: an ester formed from a long chain carboxylic acid (fatty acid) and an alkanol

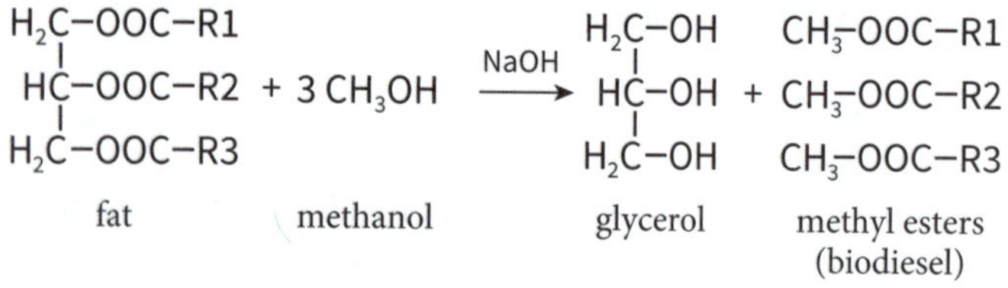

$$\begin{array}{l} H_2C-OOC-R1 \\ HC-OOC-R2 \\ H_2C-OOC-R3 \end{array} + 3\,CH_3OH \xrightarrow{NaOH} \begin{array}{l} H_2C-OH \\ HC-OH \\ H_2C-OH \end{array} + \begin{array}{l} CH_3-OOC-R1 \\ CH_3-OOC-R2 \\ CH_3-OOC-R3 \end{array}$$

fat — methanol — glycerol — methyl esters (biodiesel)

Figure 16.10 Transesterification—production of biodiesel

- Biodiesel has superior lubricating properties and a higher octane rating than petrodiesel.
- Biodiesel is a renewable fuel. There is a social and environmental benefit in using waste fats and oils instead of dumping them. Biodiesel fuel use reduces net greenhouse gas emissions.

Production of sulfuric acid: The Contact process

- Sulfuric acid is an important industrial chemical. It has many uses including fertiliser production, anionic detergent manufacture and as the electrolyte in car batteries.
- Sulfuric acid is manufactured using the Contact process. The steps of this process are all exothermic and they are summarised below:

KEY QUESTIONS

7 **In the industrial manufacture of aspirin, explain why the acetylsalicylic acid is washed with distilled water prior to drying.**

8 **Explain why cationic surfactants rather than anionic surfactants are used in hair conditioners.**

9 **Explain how a cationic surfactant can stabilise an oil-in-water emulsion.**

10 **Identify the common feedstocks used in the production of biodiesel fuel.**

Answers p. 240

- In Australia the most common feedstock of sulfuric acid is sulfur dioxide, which is a by-product of the roasting and **smelting** of sulfide ores such as chalcopyrite ($CuFeS_2$) at 1000–1200 °C during the production of copper. The net reaction is:

$$2CuFeS_2(s) + 5O_2(g) \rightarrow 2Cu(l) + 2FeO(s) + 4SO_2(g)$$

smelting: extracting a metal from an ore by heating and melting it at high temperatures

- The sulfur dioxide is collected and oxidised by an air stream (100–200 kPa) to form sulfur trioxide using a vanadium (V) oxide catalyst. This reaction occurs in the catalytic tower. The moderate temperature used is in the range 400–550 °C. The gas stream makes four cycles to maximise yield. Unreacted gases are recycled. The sulfur trioxide is cooled using a heat exchanger:

$$2SO_2(g) + O_2(g) \leftrightarrows 2SO_3(g)$$

The equilibrium is shifted to the right to increase the yield of sulfur trioxide. This is achieved by:
  - increasing the partial pressure of oxygen. (Le Chatelier's principle predicts that the addition of more reactants will shift the equilibrium to the product side to counteract the change.)
  - increasing the total pressure (100–200 kPa). (Le Chatelier's principle predicts that the increase in total pressure will shift the equilibrium to the side of lower numbers of molecules to counteract the change.)
  - also controlling the rate and equilibrium position by adjusting the temperature in each catalyst bed. In the first catalyst bed, the heat liberated raises the temperature from 400 to 550 °C. The rate increases and a 70% yield is achieved. In subsequent catalyst beds the temperature is lowered to 400–425 °C, which increases the yield of sulfur trioxide to 97%.
- The sulfur trioxide is absorbed into previously concentrated sulfuric acid to form oleum ($H_2S_2O_7$)

in the absorber tower. This process generates less heat and avoids the production of hot acidic mists that form when sulfur trioxide is added to water:

$SO_3(g) + H_2SO_4(l) \rightarrow H_2S_2O_7(l)$

- Small quantities of water are added to the oleum to form concentrated sulfuric acid (98%) in the diluter vessel:

$H_2S_2O_7(l) + H_2O(l) \rightarrow 2H_2SO_4(l)$

➔ High yields (~99.7%) of sulfuric acid are achieved with very low toxic emission of sulfur dioxide (~0.3%) to the atmosphere. Government regulations require sulfur dioxide emissions to be this low.

➔ Figure 16.11 shows a flowchart for the Contact process.

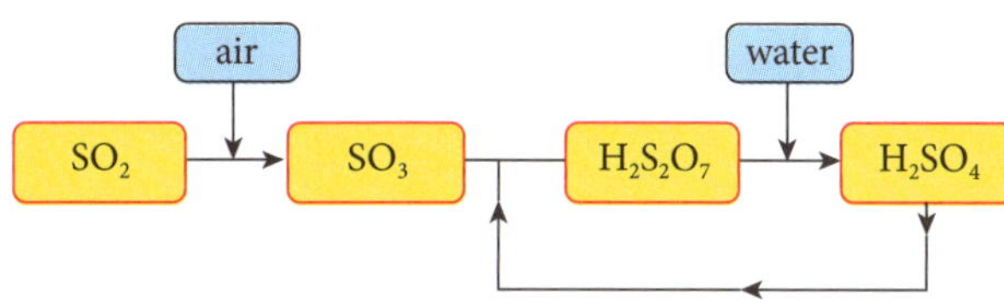

Figure 16.11 Contact process flowchart

➔ By locating the sulfuric acid plant near the sulfide smelter and chalcopyrite mine the process is more economic. Australia's largest sulfide mineral mines are located in South Australia and Queensland.

EXAMPLE 5

The conversion of sulfur dioxide to sulfur trioxide in the Contact process is an equilibrium reaction:

$2SO_2(g) + O_2(g) \leftrightharpoons 2SO_3(g)$

The gas mixture is passed sequentially through four catalysts beds.

a The temperature in the first catalyst bed is 550 °C. How will this high temperature affect the rate of the reaction?

The value of the equilibrium constant (K_{eq}) is temperature dependent. The following table shows the values of K_{eq} over a range of temperatures.

Temperature (°C)	400	450	650
K_{eq}	100	5	1

b In the remaining three catalyst beds, the temperature is lowered to 400–425 °C. Explain how this temperature range will affect the equilibrium yield of sulfur trioxide.

Relate the equilibrium constant value to the equilibrium position

Answer:

a The reaction rate increases as the temperature increases as more reactant molecules have sufficient energy to overcome the activation energy barrier.

b The lower temperature in the remaining catalyst beds will result in an increased yield of sulfur trioxide as the K_{eq} value is greater at these lower temperatures. If the temperature had remained at 550 °C, the yield at equilibrium would be significantly reduced. (Experimental data from the industrial plant shows a 97% conversion at the lower temperature down to 70% conversion at the higher temperature.)

➔ Further information about the Contact process can be found in various online videos. Enter the following title in a search bar: sulphuric acid—the contact process.

Steel production

➔ Steel manufacture is an important chemical industry. Port Kembla in New South Wales has a long history of steel production. Steel products have a wide variety of uses, including car bodies, rail tracks, tools and building construction.

➔ A blast furnace is used to extract iron from iron ore (Fe_2O_3). Coal is an important raw material used to produce coke. Coke is made by heating the coal at high temperature to remove the volatile components and raise the carbon content. In the blast furnace the coke burns to form carbon monoxide. Carbon monoxide reduces the iron (III) oxide to iron. The molten iron that forms is run off into moulds and is called pig iron. Pig iron contains about 4%w/w of carbon, which makes it brittle. Pig iron can be converted to a more useful iron alloy called steel using the basic oxygen steel-making (BOS) process. Pure oxygen is then blasted onto the surface of the melted iron to oxidise the carbon, silicon and phosphorus impurities. The steel that forms has less than 0.2%w/w carbon:

$Fe_2O_3(s) + 3CO(g) \rightarrow 2Fe(l) + 3CO_2(g)$

➔ The Port Kembla blast furnace is located at a coastal port south of Sydney. This region provides local coal for the production of coke used to produce carbon monoxide in the furnace. The iron ore is transported to Port Kembla via cargo ships from mines in Western Australia and South Australia.

➔ The location of Port Kembla provides multiple transport networks. Raw materials can be transported by rail and sea to Port Kembla. The steel produced can then be shipped across Australia and overseas. The city of Wollongong and other local towns nearby provides local housing for steel workers.

Electrolytic industries

➔ Many chemical industries use **electrolysis** to produce elements or compounds from raw materials.

electrolysis: an electrochemical process that uses a direct electric current to drive a non-spontaneous redox reaction

- Aluminium is extracted from bauxite (hydrated aluminium oxide) ore using electrolysis. The largest bauxite mine in Australia is at Weipa on the Gulf of Carpentaria in northern Queensland. The large size of the ore body and its high concentration of bauxite ensures that profits can eventually be made by the mining company. The ore is transported by ship to an electrolytic refinery at Yarwun on the eastern Queensland coast. Following removal of impurities from the ore the aluminium oxide is dissolved in molten cryolite (Na_3AlF_6) at 1000 °C and the mixture is electrolysed using graphite anodes and a graphite cathode lining in the steel tank. Molten aluminium forms at the cathode and is regularly tapped off and run into moulds. Figure 16.12 shows an electrolytic cell.

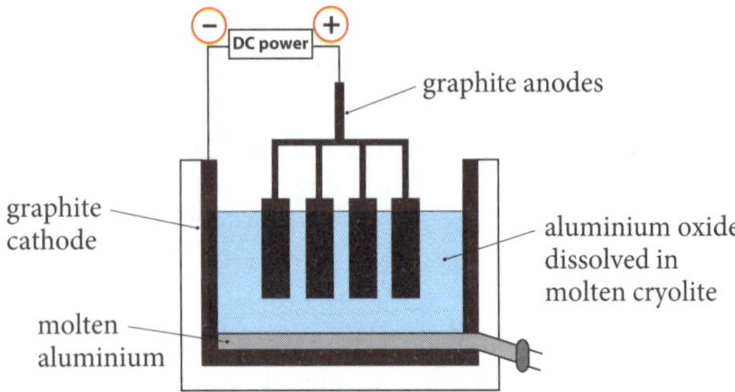

Figure 16.12 Aluminium electrolytic cell

Graphite cathode (reduction): $Al^{3+}(l) + 3e^- \rightarrow Al(l)$

Graphite anode (oxidation): $2O^{2-}(l) \rightarrow O_2(g) + 4e^-$

The electrolytic refinery must be located near electrical power stations to supply the large amounts of electricity used in the refining process.

Figure 16.13 is a flowchart summary of the production of aluminium.

- Electroplating is another important application of the electrolytic process. Electroplating is an electrolytic process in which a thin film or layer of metal is deposited on the surface of another metal. Electroplating may be done for decorative reasons. Some pieces of jewellery may be made out of a less expensive metal and then coated with a thin layer of a more expensive and decorative metal. In general the electroplating cell uses an electrolyte that contains an ion of the metal to be plated.

EXAMPLE 6

The mineral in an ore body must be sufficiently concentrated to make it economical to mine. This requirement is called the *grade* of the ore. If the concentration is too low then it may be currently non-economic, although future improvements in technology can mean that a non-economic ore can become economic. At Weipa the typical grade of the bauxite ore is 51%w/w Al_2O_3. Calculate the theoretical yield of aluminium that could be extracted from 50 000 tonnes of bauxite.

Express your answer in tonnes

Answer:

$m(Al_2O_3) = \left(\frac{51}{100}\right)(50\,000) = 25\,500$ t

% Al in $Al_2O_3 = (M(Al)/M(Al_2O_3)) \times 100/1$

$$= \frac{(2 \times 26.98)}{(2 \times 26.98 + 3 \times 16.00)} \times \frac{100}{1}$$

$$= \frac{53.96}{101.96} \times \frac{100}{1} = 52.92\%$$

Theoretical yield of aluminium $= \frac{52.92}{100} \times 25\,500$

$= 1.3 \times 10^4$ tonnes (2 s.f.)

The anode is usually composed of the plating metal. For example, if a spoon is to be silver-plated the spoon is made the cathode. A solution of silver cyanide is used as the electrolyte and a pure rod of silver is the anode.

Cathode: $Ag^+(aq) + e^- \rightarrow Ag(s); E^\circ = +0.80$ V

Anode: $Ag(s) \rightarrow Ag^+(aq) + e^-; E^\circ = -0.80$ V

The net cell potential is equal to zero volts (E°cell = (+0.8) + (−0.8) = 0 V). Thus only a small positive voltage is required to make the process spontaneous and increase the rate of the electrolysis. Electroplating factories are normally small-scale chemical industries located in industrial zones of cities. Electroplating does not require the high currents used in electrolytic refining.

- Electroforming is a type of electroplating process used to create a metallic copy of the external shape and patterning of the original object (called the *mandrel*). The original object may be removed or it may be left inside the copy.
  - A piece of jewellery can be created out of glass, wax or plastic. The surface can be then coated with a

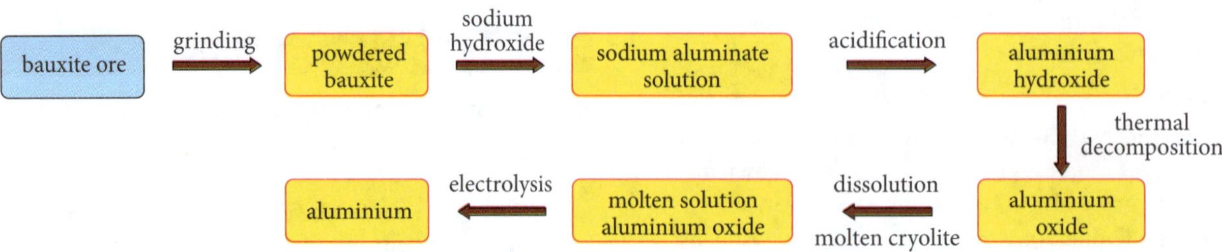

Figure 16.13 Extraction of aluminium flowchart

conductive layer such as graphite paint. The graphite-coated object is made the cathode of an electrolytic cell, and the metal to coat the object (e.g. copper) is made the anode. Figure 16.14 shows an example of this process.

- The electrolysis cell is set up and a suitable electrolyte (e.g. copper (II) sulfate) is used. A voltage is applied to electroplate the jewellery object. If the original object was made of wax then this can be removed by drilling a tiny hole and heating the object to melt the wax and allowing the melted wax to drain out.
- Electroforming factories are also located in industrial zones in cities as high electrical currents are not required.

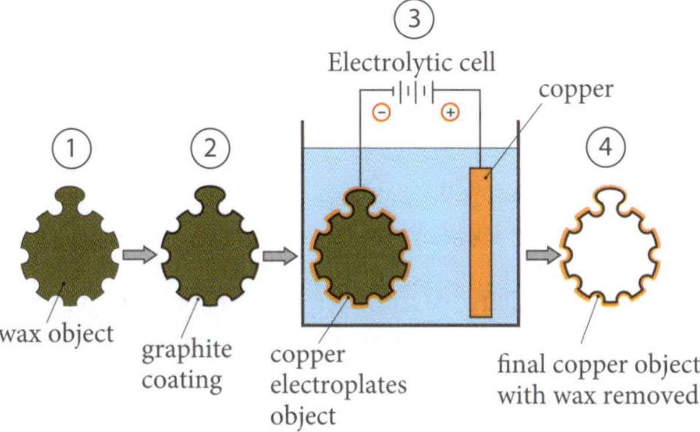

Figure 16.14 Electroforming

> **→ KEY QUESTIONS**
>
> **11** **Describe the conditions used to maximise the yield of sulfur trioxide in the Contact process.**
>
> **12** **Explain why sulfur trioxide is not directly dissolved in water to produce sulfuric acid in the Contact process.**
>
> **13** **Explain why Port Kembla in New South Wales is a suitable location for the production of steel in a blast furnace.**
>
> **14** **Explain why aluminium refineries need to be located near electrical power stations.**
>
> Answers ⊃ p. 240

Specialised products

Polypeptides

→ Modern chemical methods are able to create specialised products for medical applications. Chemists can synthesise polypeptides with specific amino acid sequences. A polypeptide consists of chains of amino acids, as shown in Figure 16.15. The synthesis is performed on a solid supporting structure such as polystyrene gel.

→ The first step is to bind the first amino acid to the supporting structure. This leaves a reactive carboxylic acid group, which then condenses via a peptide bond with the amine group of the next amino acid. The peptide chain grows longer as subsequent condensation reactions occur in a specified sequence. This procedure can form polypeptide chains that are about 130 amino acids long. Polypeptides are short chain polyamides.

→ The polypeptides are being scientifically investigated for a variety of medical uses. These include reduction of high blood pressure in patients with cardiovascular disease and stimulation of the immune response.

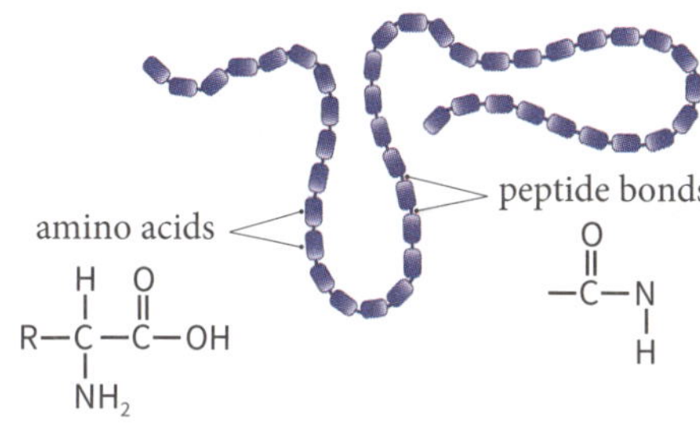

Figure 16.15 Polypeptide

Carbon nanotubes

→ Carbon nanotubes are allotropes of carbon. They are being studied extensively for their use in structural composite materials, electronic circuits as well as in medicine. Nanotubes are stronger than steel and they can be used in carbon composite materials that require high tensile strength. Figure 16.16 shows the structure of a carbon nanotube.

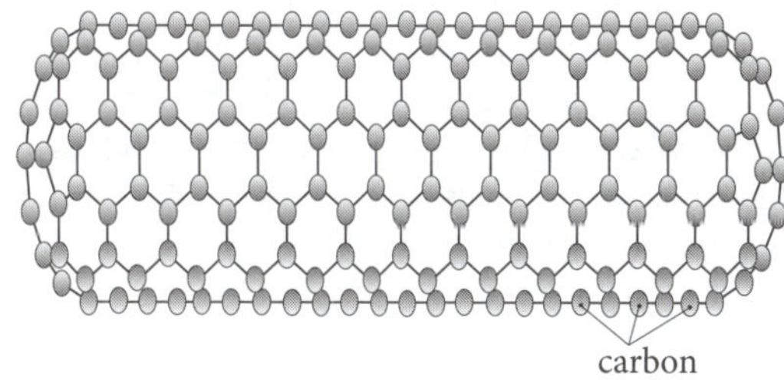

Figure 16.16 Carbon nanotube

→ An important area of nanotube research is the design and production of biosensors that have medical applications. A biosensor is an electrochemical device that detects changes in the local chemical environment by the production of an electrical signal. A biosensor utilising nanotubes and artificial DNA fragments has been developed to detect low levels of pathogenic bacteria.

→ Biosensors that are used to measure the concentration of molecules such as glucose and cholesterol in the blood must be sensitive enough to detect the normal range of concentrations found in human blood. A glucose sensor can be used to detect diabetes. A glucose level of 6 mmol/L or less in the blood is considered normal whereas a level of 7 mmol/L or higher suggests diabetes may be present.

- A cholesterol sensor must have a sensitivity in the range of 2.5–10 mmol/L. Blood cholesterol levels of less than 5 mmol/L are considered to be satisfactory, whereas cholesterol levels of more than 6 mmol/L are considered to be dangerous for the cardiovascular system.

Nanorobots

- Nanorobotics is a technology in the research and development stage. Nanorobots are small microscopic devices designed to deliver pharmaceuticals (e.g. anticancer drugs) to specific sites in the body via the bloodstream without affecting other tissues. Trials on rabbits have shown that anticancer drugs mixed with magnetic nanoparticles can be delivered to cancers in the liver via the use of nanorobots and magnetic resonance imaging (MRI). These nanorobots can be made from biodegradable polymers.

Organochlorine pesticides

- Some specialised organic products have led to environmental damage. Organochlorine pesticides such as DDT (dichlorodiphenyltrichloroethane) were developed in World War II to control mosquitos that transmitted malaria and typhus to troops and civilians. After the war DDT became widely used to kill insect pests that damaged crops. By the 1960s it had become apparent that insects were becoming resistant to the DDT and that the compound was causing widespread ecological damage, particularly to birdlife. In the 1970s and 1980s DDT was banned in most Western countries.
- In more modern times pyrethrum insecticides have been developed. Pyrethrum is a natural plant product produced by Pyrethrum daisies. The active ingredients are a mixture of esters known as pyrethrins that work together to kill insects. Studies have shown that pyrethrins do not harm animals and humans. They also break down quickly, especially in sunlight, and do not persist in the environment. Figure 16.17 shows the structures of DDT and a pyrethrin.

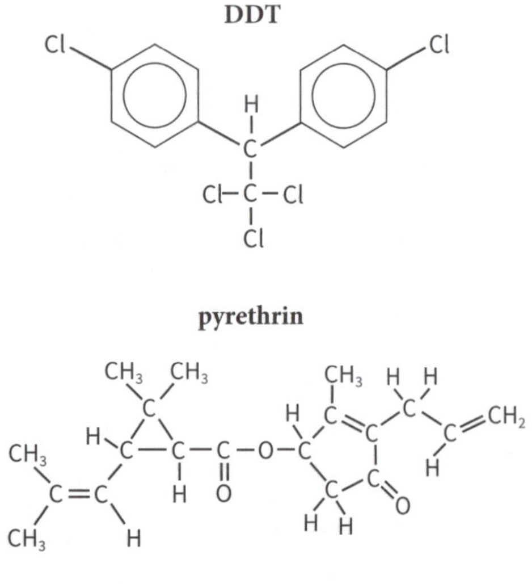

DDT

pyrethrin

Figure 16.17 DDT and pyrethrin structures

Chlorofluorocarbons

- Chlorofluorocarbons (CFCs) were specialised organic compounds developed in the 1950s for use as refrigerants in air conditioners and refrigerators, as well as blowing agents for foam plastics, propellants in aerosol cans and as solvents in industry. CFCs are hydrocarbon molecules in which all the hydrogen atoms have been replaced by chlorine and fluorine functional groups.
- By the 1980s it was discovered that the CFC compounds were destroying the ozone (O_3) layer in the stratosphere. The ozone layer is important in absorbing some of the more harmful UV rays from the Sun. It therefore acts as a UV shield. Research showed that the CFC molecules such as trichlorofluoromethane ($CFCl_3$) absorb UV radiation in the upper atmosphere and **photodissociate** to produce reactive chlorine radicals (Cl•). Chlorine radicals were responsible for destroying ozone molecules:

$$CFCl_3 + UV \rightarrow CFCl_2\cdot + Cl\cdot$$
$$Cl\cdot + O_3 \rightarrow ClO\cdot + O_2$$
$$2ClO\cdot \rightarrow Cl_2O_2 \rightarrow 2Cl\cdot + O_2$$

photodissociate: decomposition of a compound by high energy radiation

Figure 16.18 shows the photodissociation of trichlorofluoromethane by UV light.

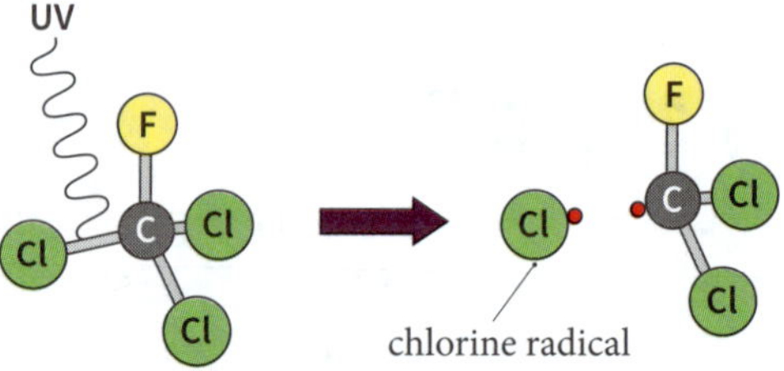

Figure 16.18 Photodissociation of trichlorofluoromethane

- The chlorine radicals reform after each reaction with ozone, so they have a long lifetime in the stratosphere. Consequently significant damage was caused to the ozone layer. The United Nations established the Montreal Protocol in 1987. Its aim was to ban the production of CFCs and replace them initially by hydrochlorofluorocarbons (HCFCs) that would largely be broken down in the troposphere and then finally by hydrofluorocarbons (HFCs) that contained no chlorine that could damage the ozone layer. By 2050 it is hoped that ozone levels will return to 1970s levels.

KEY QUESTIONS

15 **Identify specialised products developed via carbon nanotube research.**

16 **Identify an environmental issue resulting from the development of chlorofluorocarbons in the 1950s.**

Answers p. 240

CHAPTER SYLLABUS CHECKLIST

Are you able to answer every question from the syllabus for this chapter? Tick each question as you go through the checklist if you are able to answer it. If you cannot answer a question, turn to the relevant page in the study guide to find the answer. For NESA key word meanings, go to www.educationstandards.nsw.edu.au and search 'key words'.

	FOR A COMPLETE UNDERSTANDING OF THIS TOPIC:	PAGE NO.	✓
1	Can I explain how the rate and yield of a chemical synthesis are important in industry?	224	
2	Can I explain why environmental and social issues need to be considered in establishing a chemical factory?	224	
3	Can I list the factors that should be considered in choosing a site for a chemical industry?	224	
4	Can I describe how the production of ammonia demonstrates the issues confronting chemical industries?	226	
5	Can I describe how the production of sodium carbonate demonstrates the issues confronting chemical industries?	227	
6	Can I describe how the production of aspirin, hair conditioner and biodiesel fuel illustrate issues in chemical synthesis?	229	
7	Can I describe how the production of sulfuric acid demonstrates the issues confronting chemical industries?	232	
8	Can I explain why steel blast furnaces should be located at coastal ports?	233	
9	Can I describe examples of chemical industries that use electrolysis?	233	
10	Can I describe examples of the development of specialised products?	235	

HSC EXAM-TYPE QUESTIONS

Objective-response questions (1 mark each)

1 **In the industrial production of aspirin the reactants are heated under reflux for about 20 hours. The reflux process is used to:**

- A allow the system to reach equilibrium without the loss of volatile reactants and products.
- B increase the rate of reaction between the reactant molecules.
- C increase the pressure and therefore the yield of aspirin.
- D catalyse the reaction.

2 **Lime (calcium oxide) is prepared from the thermal decomposition of limestone:**

$$CaCO_3(s) \leftrightharpoons CaO(s) + CO_2(g); \Delta H = +178 \text{ kJ/mol}$$

Identify a suitable industrial method for the production of lime.

- A Raise the temperature from 20 °C to 40 °C.
- B Use an open kiln operating at a high temperature.
- C Use large pieces of limestone.
- D Keep the reaction vessel closed to increase the gas pressure during decomposition.

3 **Sodium carbonate produced via the Solvay process has many uses in society.**
Select the correct statement about sodium carbonate production and its uses.

- A Sodium chloride, ammonia and calcium carbonate are the feedstocks used to produce sodium carbonate.
- B Sodium carbonate is used in the production of synthetic fertilisers.
- C In the Solvay process, ammonium chloride is the waste product that is released into the ocean.
- D Sodium carbonate is used in the production of sulfuric acid.

4 **Select the statement that is true about the production of hair conditioner.**

- A Hair conditioner is manufactured by the alkaline hydrolysis of fats.
- B Hair conditioners contain anionic surfactants that are attracted to the positive charges on hair strands.
- C Hair conditioner is an example of a colloidal oil-in-water emulsion.
- D Hair conditioners potentially cause environmental issues when waste shower water is released into the hydrosphere.

5 **Identify which environmental pollution issue would lead to prosecution in the nominated chemical industry?**

- A release of calcium chloride solution into the ocean from the Solvay process
- B emissions of carbon dioxide into the atmosphere from fossil fuel combustion used to heat the reaction vessel in the Haber process
- C heat pollution of the hydrosphere during the production of analgesics such as aspirin
- D emissions of sulfur dioxide into the atmosphere in the Contact process

Extended-response questions

6 **The conversion of sulfur dioxide to sulfur trioxide in the Contact process is an exothermic equilibrium.**

- a **Write a balanced equation for this conversion process.** (1 mark)
- b **Predict the effect of a large reduction of the volume of the reaction vessel at constant temperature on the yield of sulfur trioxide.** (1 mark)
- c **In practice only a slight increase in pressure of air (above atmospheric pressure) is used rather than high pressure apparatus. Account for this practice.** (1 mark)

7 **An industrial chemist investigated the effect of temperature and pressure on the equilibrium yield of ammonia in the Haber process. Nitrogen and hydrogen in a 1 : 3 volume ratio were mixed and introduced into the reaction vessel containing an iron catalyst at a specific temperature and pressure. The percentage by volume of ammonia at equilibrium was measured and the results shown in the table.**

Percentage v/v ammonia in equilibrium mixture

Temperature (°C)	1 MPa	10 MPa	20 MPa
200	51	82	89
300	15	53	67
500	1	11	18

- a **Use this data to describe the effect on the equilibrium yield of ammonia of an increase in:**
  - i **temperature.** (1 mark)
  - ii **pressure.** (1 mark)
- b **Explain why the commercial Haber process normally operates at temperatures between 450–500 °C and pressures of 15–25 MPa.** (2 marks)
- c **Explain why the iron catalyst is powdered for use in the industrial synthesis of ammonia.** (1 mark)

8 **Aspirin (acetylsalicylic acid, $C_6H_4OOCCH_3COOH$) is manufactured from salicylic acid ($C_6H_4OHCOOH$) and acetic anhydride ($C_4H_6O_2$), according to the following reaction:**

$$C_6H_4OHCOOH + C_4H_6O_2 \leftrightharpoons C_6H_4OOCCH_3COOH + CH_3COOH$$

20 moles of salicylic acid were reacted with an excess of acetic anhydride. The yield of aspirin was 98% by mass.

a Explain why the acetic anhydride was added in excess. (1 mark)

b Calculate the mass of aspirin produced. (3 marks)

9 A new factory is to be established to manufacture sodium hydroxide and chlorine from sodium chloride. Electrolysis is to be used to decompose purified salt water to produce sodium hydroxide, chlorine and hydrogen. A polymer membrane is used to separate the chlorine gas produced at the anode and the hydrogen gas and sodium hydroxide solution at the cathode. A suitable industrial site is to be found. Discuss the issues associated with the choice of a suitable site. (5 marks)

10 The Ostwald process is used to manufacture nitric acid. The first step in the process is the oxidation of ammonia using hot air at ~230 °C and 0.4–1.0 MPa to form nitric oxide and water vapour using a platinum gauze catalyst. The ammonia and oxygen are mixed in a 1 : 8 volume ratio and this reaction is an exothermic equilibrium (ΔH = –905 kJ/mol). Under these conditions 95% of the ammonia is converted to nitric oxide gas. In the second step the hot nitric oxide is removed to a new chamber where it is cooled to 50 °C and reacted with oxygen to form nitrogen dioxide (ΔH = –114 kJ/mol). The nitrogen dioxide is pumped into a third chamber called the absorption tower where it reacts in the third step with a fine shower of water to form dilute nitric acid, and nitric oxide is released and recycled (ΔH = –117 kJ/mol). By recycling the acid solution many times through the absorption tower the concentration of the acid increases and after further dehydration the concentration of the nitric acid rises to about 15 mol/L.

a Write a balanced equation for the first step. (1 mark)

b Justify the reaction conditions used for the first step. (3 marks)

c Write equations for steps 2 and 3. (2 marks)

d Explain why a fine shower of water is used in the absorption tower. (1 mark)

e Draw a flowchart for the Ostwald process. (4 marks)

ANSWERS

KEY QUESTIONS

Key questions ➲ p. 227

1 temperature, concentration (and gas pressure involving gaseous equilibria)

2 Feedstocks = nitrogen and hydrogen; raw materials = air used to produce nitrogen; natural gas and water to produce hydrogen

3 $N_2(g) + 3H_2(g) \leftrightharpoons 2NH_3(g)$. The constant ammonia removal shifts the equilibrium towards the right (according to Le Chatelier's principle) and leads to a yield increase.

Key questions ➲ p. 229

4 The carbonation step allows the carbonic acid formed to react with the ammoniated brine to produce sodium hydrogen carbonate and ammonium chloride.

5 very low temperatures and high salt concentrations

6 Rivers are fresh water. Dumping calcium chloride solutions will raise the salinity of the water, leading to extreme stress to aquatic organisms.

Key questions ➲ p. 232

7 The washing removes adhering acetic acid and acetic anhydride from the crystal surfaces.

8 Cationic surfactants have positive head groups, which are attracted to negative charges on protein molecules on the hair strands. The hydrocarbon tails then hang on the hair and help to make the hair shine.

9 The positive head group attracts polar water molecules and the hydrocarbon tails interact with the oil via dispersion forces. These combined effects stabilise the colloidal droplets in the water.

10 animal or plant fats and an alkanol such as methanol under alkaline conditions

Key questions ➲ p. 235

11 A compromise temperature maintains a suitable yield and a reasonable reaction rate. The vanadium (V) oxide catalyst increases the rate but not the yield. A slightly raised air pressure increases the oxygen concentration, which shifts the equilibrium to the right to increase the yield of sulfur trioxide.

12 Too much heat is liberated when sulfur trioxide dissolves in water. This creates acidic mists that are difficult to control in a chemical factory. Consequently the sulfur trioxide is dissolved in pre-made sulfuric acid to form oleum, which is then reacted with water to form sulfuric acid.

13 Port Kembla is near local coal mines. Coal is converted to coke, which is used to reduce the iron ore. Port Kembla is a coastal port. Ships can transport iron ore from Western Australia and South Australia. The port can also be used to ship the manufactured steel to overseas markets. The workforce can be sourced from local towns and the city of Wollongong.

14 Aluminium oxide is converted to aluminium via electrolysis, which requires high current densities. Nearby electrical power stations are required to provide the electricity.

Key questions ➲ p. 236

15 Carbon nanotube sensors for the measurement of glucose and cholesterol in the blood.

16 CFCs caused destruction of ozone in the stratosphere. The ozone layer acts as a UV shield to reduce excessive levels of UV reaching the Earth's surface.

HSC EXAM-TYPE QUESTIONS

Objective-response questions

1 **A.** The reactants need to be heated for a long time and, to avoid vapour pressure build up, the mixture is refluxed in an open vessel. The cooling water in the condenser returns volatile components to the reaction vessel. **B** is incorrect as the reflux condenser does not change the rate. **C** is incorrect as the reactants are in solution and pressure has no effect. **D** is incorrect as refluxing is not a catalytic process.

2 **B.** By using an open kiln the carbon dioxide can continuously escape and drive the equilibrium to the right. The reaction is endothermic so a high temperature is required for decomposition to occur. **A** is incorrect as this small rise in temperature is insufficient for decomposition. **C** is incorrect as large pieces have a low surface area and the rate is too low. Crushing into a powder increases the surface area and the rate. **D** is incorrect as an equilibrium will be achieved in a closed vessel and the calcium carbonate will not completely decompose.

3 **A.** NaCl, NH_3 and $CaCO_3$ are the feedstocks. **B** is incorrect as sodium carbonate is a base and not used in fertiliser production. **C** is incorrect as calcium chloride is the waste. **D** is incorrect as sulfuric acid is manufactured from sulfur dioxide and water.

4 **C.** Colloidal oil particles are emulsified in water and stabilised by a cationic surfactant. **A** is incorrect as alkaline hydrolysis is used in the manufacture of soap. **B** is incorrect as cationic surfactants are used as negative charges are present on hair strands. **D** is incorrect as the components are biodegradable.

5 **D.** Sulfur dioxide is toxic and released into the atmosphere above 0.3% can lead to prosecution. **A** is incorrect as the release of calcium chloride into the ocean does not cause pollution issues. **B** is incorrect as carbon dioxide from fossil fuel combustion occurs whenever fossil fuels or biofuels undergo combustion. **C** is incorrect as there is very little heat generated in this process and hot water is not released into the hydrosphere. The reaction vessel cools and heat is released into the atmosphere.

Extended-response questions

6 EM Students need to demonstrate a high level of understanding of the conditions employed in the first equilibrium stage of the Contact process.

- a $2SO_2(g) + O_2(g) \leftrightharpoons 2SO_3(g)$ ✓
- b Increased pressure results from a reduction in volume and this shifts the equilibrium to the right, producing a greater yield of sulfur trioxide. ✓
- c The extra costs of high pressure equipment are not warranted as high yields can be produced by increasing the concentration of oxygen using a slightly raised air pressure. ✓

7 EM Students interpret tabulated data and relate this to the ammonia equilibrium. Students need to demonstrate that catalyst particle size relates only to reaction rate.

a i A temperature increase leads to a decrease in yield as the reaction is exothermic. ✓

ii A pressure increase increases the yield as the equilibrium shifts to the right to counteract the imposed pressure. ✓

b This temperature range is a compromise because too high a temperature shifts the equilibrium towards the reactants as the equilibrium is exothermic. If the temperature is too low the rate is too slow. ✓ High pressures are used to push the equilibrium to the right to reduce the number of molecules (four molecules to two molecules). ✓

c Powdering the catalyst increases the surface area, leading to a rate increase. ✓

8 EM Students demonstrate the importance of increasing yield by adding one reactant in excess. They need to demonstrate proficiency in mole calculations to score full marks.

a Adding excess acetic anhydride causes the equilibrium to shift to the right and a greater yield of aspirin is achieved. ✓

b $n(C_6H_4OHCOOH) = 20$ mol
Reaction stoichiometry = 1 : 1
$n(C_6H_4OOCCH_3COOH) = 20$ mol ✓
$m(C_6H_4OOCCH_3COOH) = n.M = (20)(180.154) = 3603.08$ g ✓
Yield = 98/100 × 3603.08 = 3531 g ✓

9 EM Students demonstrate the variety of factors that chemical industry must consider before deciding on a suitable site for the factory.

Considerations when siting factories can include:

- Energy requirements: Large amounts of electricity are needed for electrolysis of the salt water. Therefore the factory should be sited near a power station. Special rates are usually available to such industries. ✓
- Raw materials: Large amounts of salt water are needed. Therefore locate the factory on the coast. ✓
- Waste disposal: Waste brine must be able to be discharged safely into the ocean but not too close to public beaches. Factory design should ensure that no chlorine can be emitted into the atmosphere. ✓
- Transport: The site must be near transport networks such as road, rail and shipping to minimise transport costs of raw materials to the site and for products to be sent off to markets in Australia or overseas. ✓
- Staff: The factory should be relatively close to towns or cities where staff live and commute quickly to work. ✓

10 EM Students need to read and interpret supplied information and relate this to a new chemical industry. Skills in drawing flowcharts must be demonstrated to score full marks.

a $4NH_3(g) + 5O_2(g) \leftrightharpoons 4NO(g) + 6H_2O(g)$ ✓

b The temperature is raised to a moderate level to increase the rate and still achieve a reasonable yield. ✓ The catalyst increases the rate ✓ and the excess oxygen shifts the equilibrium to the right, increasing the yield of nitric oxide gas. ✓

c $2NO(g) + O_2(g) \leftrightharpoons 2NO_2(g)$ ✓
$3NO_2(g) + H_2O(l) \longrightarrow 2HNO_3(aq) + NO(g)$ ✓

d The fine droplets of water have a high surface area and this increases the rate of gas absorption. ✓

e Figure A16.1 shows a flowchart for the production of nitric acid.

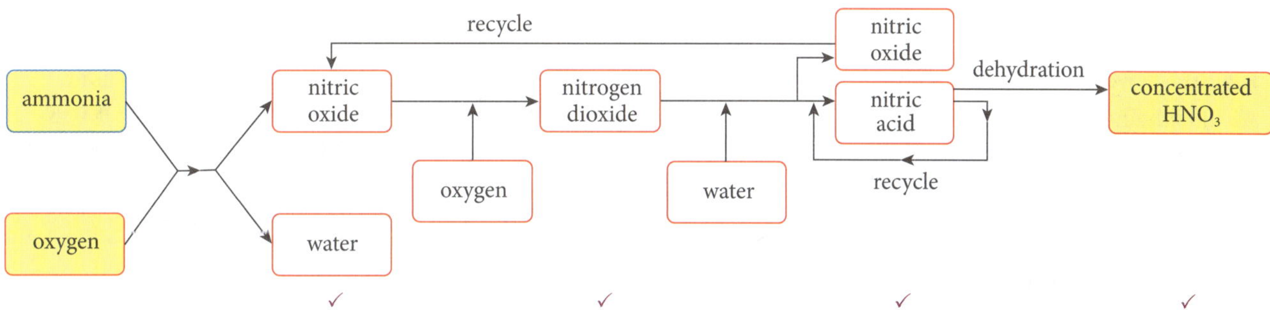

Figure A16.1 Nitric acid flowchart

SAMPLE HSC EXAMINATION 1

Try to complete these papers as if they are the real thing. These are the instructions you need to follow in the HSC Exam:

General instructions

- Reading time: 5 minutes
- Working time: 3 hours
- Write using black pen.
- NESA approved calculators may be used.
- Use the Data Sheet and Periodic Table in this book.

Total marks: 100

Section I: 20 marks

Section II: 80 marks

- Attempt all questions.

Section I: 20 marks

Attempt Questions 1–20.
Allow about 35 minutes for this section.

1 **A solution of barium chloride is sprayed into a blue Bunsen burner flame. Identify the colour that the flame will turn.**

A orange-red
B yellow
C bright green
D pale yellow-green

2 **Name the hydrocarbon shown in Figure E1.1.**

```
  H  CH3 H   H           H
  |   |   |   |          |
H-C - C - C - C - C = C - C-H
  |   |   |   |   |   |  |
  H   H   H  CH3  H   H  H
```

Figure E1.1 Hydrocarbon

A dimethylheptene
B hept-2-ene
C 4,6-dimethylhept-2-ene
D 2,4-dimethylhept-5-ene

3 **Identify the weak acid.**

A sulfuric acid
B hydrochloric acid
C sulfurous acid
D nitric acid

4 **Calculate the pH of a 0.000 50 mol/L solution of sulfuric acid.**

A 5.00
B 3.30
C 11.00
D 3.00

5 **Identify which of the following aqueous mixtures would act as a buffer.**

A propanoic acid + potassium propanoate
B phosphoric acid + sodium carbonate
C sulfurous acid + sodium sulfate
D formic acid + potassium hydroxide

6 **Three solutions X, Y and Z were tested with universal indicator. The indicator changed colour as recorded below:**

X = violet; Y = yellow; Z = red

Use this information to classify the acidity or basicity of the three solutions.

A X = acidic; Y = basic; Z = basic
B X = basic; Y = acidic; Z = acidic
C X = basic; Y = basic; Z = acidic
D X= acidic; Y = acidic; Z = basic

7 **Consider the following equilibrium involving copper (II) ions in water:**

$$Cu(H_2O)_4^{2+}(aq) + 4Cl^-(aq) \leftrightarrows CuCl_4^{2-}(aq) + 4H_2O(l)$$

blue **green**

The equilibrium solution is a blue-green colour. Identify which of the following statements is true.

A The solution will become more green if additional water is added.
B The solution will become more green if crystals of NaCl are dissolved in the solution.
C The solution will become more blue if concentrated sodium nitrate is added to the solution.
D The colour of the solution will not change if hydrochloric acid is added.

8 **The pH of soda water in an unopened bottle is 4. The pH of the soda water rises when the bottle is opened. Select the correct statement about this change.**

A The pH rises to 7 as all the gas is lost from the water.
B Additional carbonic acid forms on opening the bottle.
C The rise in pH is due to a temperature increase.
D The concentration of hydrogen ions decreases as carbonic acid decomposes due to the loss of carbon dioxide from the bottle.

9 **Identify the true statement about nitrous acid (HNO_2).**

A The conjugate base of nitrous acid is the nitrite ion.
B Nitrous acid is a strong acid.

C Nitrous acid is completely dissociated in water.

D The pH of 0.0010 mol/L nitrous acid is 3.0.

10 Ammonia gas is manufactured from hydrogen and nitrogen gas:

$N_2(g) + 3H_2(g) \leftrightarrows 2NH_3(g)$

At 500 °C, the equilibrium constant (K_{eq}) is 0.060. The reaction is exothermic.

Select the correct statement about this equilibrium.

A The equilibrium constant will increase if a catalyst is used.

B The yield of ammonia at 500 °C is not changed if helium gas is injected into an equilibrium mixture in the constant volume reaction vessel.

C The equilibrium constant will increase if the temperature is raised.

D The yield of ammonia will increase if the volume of the reaction vessel is increased.

11 A solution of a weak monoprotic acid, HA, is 8% dissociated at 25 °C. Its pH is 4.10. Calculate the concentration of the weak acid (HA) in the solution.

A 0.080 mol/L

B 2.13 mol/L

C 9.93×10^{-3} mol/L

D 9.93×10^{-5} mol/L

12 Identify which of the following solutions will cause precipitation when added to a potassium chloride solution.

A barium nitrate

B silver nitrate

C copper (II) nitrate

D zinc bromide

13 The solubility product (K_{sp}) for lead (II) bromide is 6.6×10^{-6}. Determine the concentration of bromide ions in a saturated solution.

A 0.0236 mol/L

B 0.0118 mol/L

C 2.57×10^{-3} mol/L

D 0.0189 mol/L

14 Name the molecule shown in Figure E1.2.

```
    H   H   H   H   H   H   H
    |   |   |   |   |   |   |
H — C — C — C — C — C — C — C — H
    |   |   |   |   |   |   |
    H  CH3  H  OH   H   H   H
```

Figure E1.2 Alkanol

A 2-methylheptan-4-ol

B 6-methylheptan-4-ol

C methylheptanol

D 2-methylheptan-4-one

15 A flame atomic absorption spectrometer (AAS) uses a slot burner. The purpose of the flame is to:

A generate light of specific wavelengths to be absorbed by metal ions.

B atomise the material.

C diffract the light waves.

D absorb light from the hollow cathode lamp.

16 The equilibrium constant for an equilibrium reaction in the gaseous state is:

$K_{eq} = [CH_4][H_2O]/[CO][H_2]^3$

Select the equilibrium equation that is consistent with the equilibrium constant.

A $CH_4(g) + H_2O(g) \leftrightarrows CO(g) + H_2(g)$

B $CH_4(g) + H_2O(g) \leftrightarrows CO(g) + 3H_2(g)$

C $CO(g) + 3H_2(g) \leftrightarrows CH_4(g) + H_2O(g)$

D $CO(g) + 3H_2(g) \leftrightarrows CH_4(g) + H_2O(l)$

17 Hexan-2-ol burns with a yellow, smoky flame. Identify which equation is consistent with that observation.

A $C_6H_{13}OH(l) + 9O_2(g) \rightarrow 6CO_2(g) + 7H_2O(l)$

B $C_6H_{15}OH(l) + 9O_2(g) \rightarrow 5CO_2(g) + CO(g) + 8H_2O(l)$

C $2C_6H_{13}OH(l) + 15O_2(g) \rightarrow 6CO_2(g) + 6CO(g) + 14H_2O(l)$

D $C_6H_{13}OH(l) + 6O_2(g) \rightarrow 2CO_2(g) + 2CO(g) + 2C(s) + 7H_2O(l)$

18 Saturated solutions of the following salts were prepared. The solubility product constant for each salt is included. Identify the solution that has the lowest sulfate ion concentration.

A $SrSO_4$ ($K_{sp} = 3.44 \times 10^{-7}$)

B $CaSO_4$ ($K_{sp} = 4.93 \times 10^{-5}$)

C $PbSO_4$ ($K_{sp} = 2.53 \times 10^{-8}$)

D $BaSO_4$ ($K_{sp} - 1.08 \times 10^{-10}$)

19 A student stirred a mixture of red cobalt (II) chloride crystals and water to produce a saturated solution. Identify a procedure that could be used to determine when the system had reached equilibrium.

A Use a colourimeter to analyse the colour of the solution.

B Titrate the solution with sodium hydroxide.

C Use an electronic balance to measure the mass of the system.

D Evaporate samples of the solution and measure the mass of solid formed.

20 Select the true statement about C-13 NMR spectra of organic molecules.

A One peak will be observed for ethanol.

B Methyl acetate and propanoic acid are both three-carbon molecules and their C-13 NMR spectra will be identical.

C The chemical shift for the C=O bond in ketones is typically in the range 205–220 ppm.

D Two peaks will be observed for ethane.

Section II: 80 Marks

Attempt Questions 21–36.

Allow about 2 hours and 25 minutes for this section.

Instructions
- In the HSC Exam you will answer the questions in the spaces provided. These spaces provide guidance for the expected length of response.
- Show all relevant working in questions involving calculations.

---

21 An analytical chemist determines the concentration of zinc minerals in the soil on a farm. Five soil samples were collected for analysis. Each sample was heated with nitric acid to dissolve the zinc minerals.

a Discuss the use of gravimetric analysis, volumetric analysis and AAS in determining the zinc levels in the soil. (2 marks)

b Describe the steps of the method and calculation used to determine the zinc concentration in the soil. (5 marks)

22 You are provided with a punnet of blueberries and asked to evaluate the use of the blueberry extract as an acid–base indicator. Describe your investigations and the procedures you will use to evaluate the effectiveness of the indicator. (6 marks)

23 Write balanced equations for the following reactions and name the organic reaction product(s):

a esterification of pentanoic acid and ethanol (2 marks)

b addition reaction of HI and pent-2-ene (4 marks)

24 An alkanol has a molar mass of 102.172 g/mol. Draw structural formulas for isomeric molecules that have this molar mass. Name these molecules systematically. (7 marks)

25 Figure E1.3 shows the structural formula of a halogenated alkene.

```
F     F
 \   /
  C=C
 /   \
F     F
```

Figure E1.3 Halogenated alkene

a Name this molecule. (1 mark)

b This molecule is polymerised using a catalyst. The product is called Teflon or PTFE.

- **i** What type of polymerisation occurs? (1 mark)
- **ii** Draw the repeating structure of Teflon. (1 mark)

26 A common polyester used to make clothing is PET (polyethylene terephthalate). This polymer is manufactured from two monomers. Figure E1.4 shows the PET polymer structure.

```
 ┌ O          O     H   H    ┐
 │ ‖          ‖     |   |    │
-│-C-(C6H4)-C-O-C-C-O-│-
 │                  |   |    │
 └                  H   H    ┘n
```

Figure E1.4 PET

a Identify the type of polymerisation that produces this polymer. (1 mark)

b Draw structural formulas of the monomers. (2 marks)

c What property of PET makes it useful in creating textile products? (1 mark)

27 25.00 mL of hard water from a lake was pipetted into a conical flask. The solution was titrated with a standard solution of 0.0100 mol/L EDTA using Eriochrome Black T indicator. At the end point the titre was 13.75 mL:

$$Ca^{2+}(aq) + EDTA^{4-}(aq) \rightarrow CaEDTA^{2-}(aq)$$

a Explain how lake water can become hard. (1 mark)

b How can hard water be used to clean dirty clothing if soap will not lather? (2 marks)

c Calculate the total hardness of the lake water. Assume the hardness is totally due to calcium ions. (5 marks)

28 A colourless solution contains lead (II) nitrate and calcium nitrate. The solution was given to a student as an unknown in a practical exam. The student is told that two cations from the following list are present in the unknown:

Ca^{2+}, Ba^{2+}, Fe^{3+}, Pb^{2+}, Cu^{2+}

Describe the steps required to ensure that the student correctly identifies the two cations in the mixture. (4 marks)

29 Solutions of sodium hydrogen carbonate exhibit both acidic and basic properties. The hydrogen carbonate ion is amphiprotic. Write net ionic equations to demonstrate that it is amphiprotic. (2 marks)

30 Polymers can be foamed to reduce their density. One method of introducing gas bubbles into the polymer is to add a 'blowing agent' such as sodium hydrogen carbonate, which decomposes to form sodium carbonate on heating to form bubbles in the foam.

a Write a balanced equation for the thermal decomposition of sodium hydrogen carbonate. (1 mark)

b Calculate the volume of carbon dioxide (at 25 °C/100 kPa) produced by the decomposition of 8.41 g of sodium hydrogen carbonate. (Molar volume of a gas = 24.79 L/mol) (3 marks)

c Identify the change in property that results by introducing gas bubbles into a plastic such as polystyrene or polypropene. (1 mark)

31 A student performed a simple fermentation experiment in the school laboratory. Dried yeast and a glucose solution were placed in a stoppered side arm flask. The flask was kept warm in a water bath at 37 °C. The carbon dioxide evolved was collected in a weighed U-tube containing powdered calcium oxide.

a Explain why the flask is kept at 37 °C. (1 mark)

b Explain why calcium oxide was used to collect the carbon dioxide. (1 mark)

c Write a balanced equation for the fermentation of glucose. (1 mark)

d Calculate the maximum mass of carbon dioxide that can be formed by the complete fermentation of 18.0 g of glucose. (2 marks)

e In one experiment, 18.0 g of glucose was dissolved in the water. After two days the calcium oxide tube had increased in mass by 7.0 g. What percentage of the theoretical yield of carbon dioxide has been reached after two days? (3 marks)

32 Octadecanoic acid ($C_{17}H_{35}COOH$) is a weak organic acid. 15.0 g of octadecanoic acid dissolves in 6.0 mol/L potassium hydroxide solution to form potassium octadecanoate and water. Potassium octadecanoate is a white insoluble solid with a soapy feel.

a Write the chemical formula of potassium octadecanoate. (1 mark)

b Write a balanced equation for this reaction. (1 mark)

c Calculate the volume of potassium hydroxide required to completely react with the octadecanoic acid. (3 marks)

33 A piece of white phosphorus (P_4) is burnt in pure oxygen. The phosphorus burns with a yellow flame to produce clouds of a white smoke containing tetraphosphorus decaoxide.

a Write a balanced equation for this combustion reaction. (1 mark)

b The white powder is dissolved in water and purple litmus added. The litmus turned red. Account for this observation, with the aid of a balanced equation. (2 marks)

c Sodium hydroxide solution is now carefully added to the solution in part b until the indicator turns purple. Write an equation for the reaction occurring and name the salt formed in this reaction. (2 marks)

34 Sulfurous acid is one of the acids present in acid rain.

a Explain how sulfurous acid is formed in rainwater, with the aid of a balanced equation. (1 mark)

b Marble is composed of calcium carbonate. The acid rain attacks a marble statue. Write a balanced equation for this reaction. (1 mark)

c A sample of the acid rain was tested and found to have a hydrogen ion concentration of 1.0×10^{-5} mol/L. Calculate the pH of the rain. (1 mark)

d What colour would each of the following indicators turn in samples of this acid rain?

i phenolphthalein ii bromothymol blue (1 mark)

35 Determine whether or not silver sulfate will precipitate when 60 mL of 0.010 mol/L silver nitrate and 30 mL of 0.010 mol/L sodium sulfate are mixed at 25 °C. (K_{sp}(silver sulfate) = 1.20×10^{-5}) (3 marks)

36 An unknown alkanol X and an unknown alkanoic acid Y have the same molar mass but different melting and boiling points as shown below:

X: m.p = –45 °C; b.p = 158 °C

Y: m.p = –34 °C; b.p = 186 °C

a Explain why the melting and boiling points of Y are greater than those of X. (1 mark)

b The molecular formula of Y is $C_5H_{10}O_2$. Name molecule Y. (1 mark)

c When X is oxidised it is converted to the next member of Y's homologous series. Suggest a name for X. (1 mark)

SAMPLE HSC EXAMINATION 2

Section I: 20 Marks

Attempt Questions 1–20.
Allow about 35 minutes for this section.

1 Figure E2.1 is a C-13 NMR spectrum of an organic molecule.

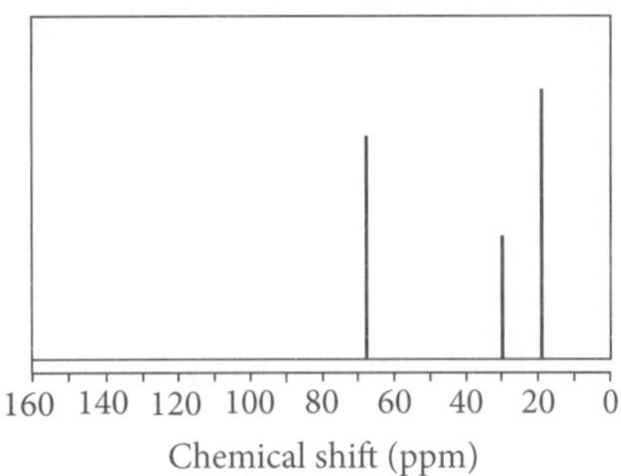

Figure E2.1 C-13 NMR spectrum

Use the Data Sheet in the inside cover of this book to identify which one of the listed molecules would produce this C-13 NMR spectrum.

A butan-1-ol
B butane
C 2-methylpropan-1-ol
D butanone

2 Figure E2.2 is an infrared spectrum of an organic molecule.

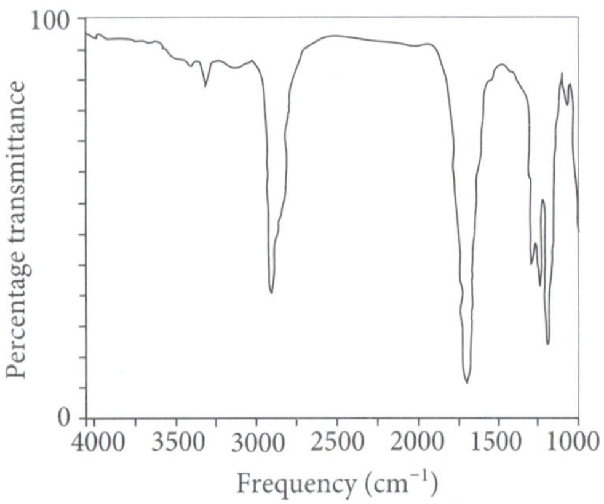

Figure E2.2 IR spectrum

Use the Data Sheet on the inside cover of this book to identify which one of the listed molecules would produce this infrared spectrum.

A butan-1-ol
B butan-1-amine
C 2-methylpropan-1-ol
D butanone

3 Figure E2.3 is a simplified mass spectrum of an organic molecule.

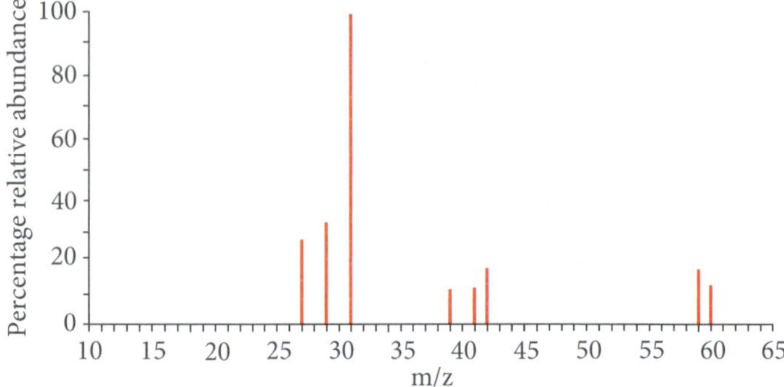

Figure E2.3 Mass spectrum

Identify which one of the listed molecules would produce this mass spectrum.

A propan-1-ol
B propane
C propanoic acid
D but-2-ene

4 Three code-labelled bottles X, Y and Z are known to contain an alkanol, an ester and an alkanoic acid with the same molecular weights. The boiling points of each liquid were determined. Use the following results to identify the component in each bottle:

X (141 °C); Y (118 °C); Z (57 °C)

	X	Y	Z
A	alkanol	ester	alkanoic acid
B	alkanoic acid	ester	alkanol
C	alkanoic acid	alkanol	ester
D	ester	alkanol	alkanoic acid

5 A sample of high density polyethylene (HDPE) has a molar mass of 1.00×10^6 g/mol. Determine the approximate number of monomers that have undergone addition polymerisation to form this polymer.

A 33 258
B 36 977
C 17 824
D 35 648

6 The condensed structural formulas of three molecules X, Y and Z are:

$X = CH_3CHFCHF_2$
$Y = CHCl_2CH_2CHBr_2$
$Z = CH_3CH_2CH_2CHBrCH_2CBr_3$

Identify the correct IUPAC names of these molecules.

	X	Y	Z
A	1,1,2-trifluoropropane	1,1-dibromo-3,3-dichloropropane	1,1,1,3-tetrabromohexane
B	trifluoropropane	dibromodichloropropane	tetrabromohexane
C	2,3,3-trifluoropropane	1,1-dichloro-3,3-dibromopropane	4,6,6,6-tetrabromohexane
D	1,1,2-trifluoropropane	1,1-dichloro-3,3-dibromopropane	1,1,1,3-tetrabromohexane

7 Three bottles X, Y and Z contain the following solutions:

X = ammonia cleanser; Y= sea water; Z = lemonade

The solutions were tested with universal indicator and a pH meter.

Identify which answer correctly identifies the colour of the indicator and the pH.

	X	Y	Z
A	violet (pH 11.5)	green-blue (pH 7.9)	red (pH 2.5)
B	violet (pH 9.0)	green (pH 7.0)	orange (pH 4.8)
C	blue (pH 10.0)	green (pH 7.0)	orange-red (pH 4.0)
D	violet (pH 10.0)	yellow (pH 6.0)	red (pH 3.8)

8 30.0 mL of 0.0250 mol/L HNO_3 is added to 220.0 mL of water to form 250.0 mL of solution.

Identify the response that has correctly calculated the pH and pOH of the solution.

	pH	pOH
A	1.60	12.40
B	2.52	11.48
C	11.48	2.52
D	2.47	11.53

9 Acetic acid dissociates according to the following equation:

$$CH_3COOH(aq) + H_2O(l) \leftrightharpoons CH_3COO^-(aq) + H_3O^+(aq)$$

The K_a value for acetic acid is 1.77×10^{-5}. Determine the pH of a 0.15 mol/L solution.

A 4.8

B 2.8

C 5.6

D 2.0

10 Esters have many uses in society. 1-propyl acetate is a volatile solvent used in cosmetics and aerosol sprays. 1-pentyl propanoate is a slow-evaporating solvent used in enamels and car refinishing. Identify the response that correctly identifies the condensed structural formula of each ester and the likely boiling point consistent with its use.

	1-propyl acetate	b.p (°C)	1-pentyl propanoate	b.p (°C)
A	$CH_3COOC_3H_7$	102	$CH_3CH_2COOC_5H_{11}$	165
B	$CH_3CH_2COOC_2H_5$	102	$CH_3CH_2CH_2CH_2COOC_3H_7$	165
C	$CH_3COOC_3H_7$	165	$CH_3CH_2COOC_5H_{11}$	102
D	$CH_3CH_2COOC_2H_5$	165	$CH_3CH_2CH_2CH_2COOC_3H_7$	102

11 Concentrated sulfuric acid is added to hexan-1-ol. Select the true statement about this reaction.

A Hex-1-ene is a product.

B Hexane is a product.

C The hexan-1-ol is oxidised to form hexanoic acid.

D Formic acid and pentanoic acid are formed.

12 Select the true statement about mass spectrometry.

A The vapourised sample is ionised using pulses of high frequency radio waves.

B The parent molecular ion always forms a peak that has the highest intensity.

C The mass spectrum of methane will have a peak with an m/z value of 16.

D The mass spectrum of ethylene will have a peak at a frequency of 28 cm^{-1}.

13 A solution of silver ions was tested with the following reagents:

- dilute hydrochloric acid
- dilute sodium hydroxide
- dilute sulfuric acid
- dilute potassium iodide solution

Select the response that correctly shows the observations recorded.

	HCl(aq)	NaOH(aq)	H_2SO_4(aq)	KI(aq)
A	white precipitate	brown precipitate	no reaction	no reaction
B	no reaction	white precipitate	no reaction	white precipitate
C	white precipitate	white precipitate	white precipitate	yellow precipitate
D	white precipitate	brown precipitate	faint white precipitate	yellow precipitate

14 As part of the refining of nickel, impure nickel is reacted with carbon monoxide gas to form nickel tetracarbonyl gas:

$$Ni(s) + 4CO(g) \leftrightharpoons Ni(CO)_4(g); \Delta H = -163 \text{ kJ/mol}$$

An increase in the concentration (or partial pressure) of carbon monoxide causes the equilibrium to shift towards the product. Impurities in the crude nickel do not react and are removed.

Select the response that correctly explains how pure nickel can now be obtained.

A Decompose the nickel tetracarbonyl by heating the compound and recycle the carbon monoxide.

B Decrease the partial pressure of carbon monoxide.

C Cool the gaseous product till solid nickel forms.

D Reduce the volume of the vessel to raise the total pressure.

15 In the industrial manufacture of soap the blended fats (e.g. beef tallow and coconut oil) are mixed with concentrated sodium hydroxide in large vats and steam jets are used to heat the mixture. Following saponification, hot brine is added. The soap curd separates from the aqueous layer. The aqueous layer is pumped out and processed to extract the glycerol. Water is added to the soap curd. The soap is then vacuum dried before processing into soap bars, flakes or powders.

Select the correct response about the soap-making process.

A The hot brine is added to precipitate the soap curd.

B The glycerol extracted is used to manufacture more soap.

C Saponification is an example of the acidic hydrolysis of a fat.

D Water is added to hydrate the soap before processing to form bars of soap.

16 In the industrial preparation of sodium carbonate an important step is to recover the ammonia used in the industrial process. Calcium carbonate can be used to recover ammonia gas from solutions of ammonium chloride. The calcium carbonate is first thermally decomposed to form calcium oxide solid. This solid is dissolved in water to form calcium hydroxide solution. The calcium hydroxide solution is mixed with the ammonium chloride solution and on heating ammonia is formed.

Identify the correct response concerning these industrial reactions.

A The conversion of calcium carbonate to calcium oxide is an exothermic process.

B Sodium carbonate will also form in the reaction that produces the ammonia.

C The ammonia is recovered and recycled to reduce costs and avoid pollution of the atmosphere if released.

D Calcium carbonate decomposes to form calcium hydroxide.

17 Identify the response that shows the condensed structural formulas of functional group isomers.

A $CH_3CH_2CH_2CH_2CH_3$ and $CH_3CH(CH_3)CH_2CH_3$

B $CH_3CH_2CH_2CH_2CHO$ and $CH_3COCH_2CH_2CH_3$

C $CH_3CHBrCHBrCH_2CH_3$ and $CH_2BrCH_2CHBrCH_2CH_3$

D CH_3CH_2COOH and $CH_3CH_2CH_2OH$

18 Cetyl trimethyl ammonium bromide is the common name of a surfactant molecule. Its condensed structural formula is: $CH_3(CH_2)_{12}CH_2N(CH_3)_3{}^+Br^-$

Select the true statement about such surfactants.

A The surfactant is anionic and can be used in dishwashing liquids.

B The types of surfactants are produced from biomass and readily biodegrade.

C The surfactant is weakly lathering in water and can be used to create stable emulsions in house paint.

D The surfactant is a cationic surfactant that can be used in hair conditioners.

19 Select the statement that is true about photosynthesis.

A Chlorophyll molecules selectively absorb green wavelengths of solar radiation.

B Water undergoes oxidation during the photosynthetic process.

C The Gibbs free energy change for photosynthesis is negative and therefore the reaction is spontaneous.

D The oxygen gas molecules liberated by the photosynthetic process are formed from oxygen atoms present in the carbon dioxide molecules.

20 Figure E2.4 shows a map where chemical industries may be established at sites labelled W, X, Y and Z.

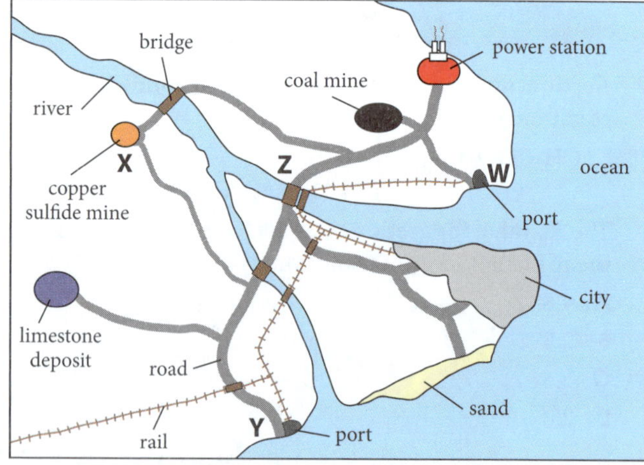

Figure E2.4 Possible sites for chemical industries

The possible industries to be established are:

- **Sodium carbonate production—the raw materials and feedstocks required are salt, limestone and ammonia; ammonia can be obtained by rail tanker from interstate.**

- Sodium hydroxide production—salt is the raw material and electrical energy is required to electrolyse salt water.
- Sulfuric acid production—sulfur dioxide feedstock can be obtained from the roasting of sulfide ores.

Select the response that shows the best locations for each chemical plant.

	Sodium carbonate	Sodium hydroxide	Sulfuric acid
A	Y	W	X
B	Y	Z	X
C	W	X	Y
D	W	Y	Z

Section II: 80 Marks

Attempt Questions 21–36.

Allow about 2 hours and 25 minutes for this section.

Instructions
- In the HSC Exam you will answer the questions in the spaces provided. These spaces provide guidance for the expected length of response.
- Show all relevant working in questions involving calculations.

21 Hydrogen iodide gas was placed in a flask at 490 °C. It was allowed to decompose to form an equilibrium mixture of hydrogen gas and iodine vapour. The concentrations of each species were measured at equilibrium:

$[HI] = 1.76 \times 10^{-2}$ mol/L; $[H_2] = 2.60 \times 10^{-3}$ mol/L; $[I_2] = 2.60 \times 10^{-3}$ mol/L

a Write an equation for the equilibrium. (1 mark)

b Write an expression for K_{eq}. (1 mark)

c Calculate the value of K_{eq} at 490 °C and state the position of this equilibrium. (2 marks)

d The value of K_{eq} at 423 °C is 0.018. What can we conclude about this equilibrium from the K_{eq} values at two different temperatures? (1 mark)

22 Figure E2.5 is the pH titration graph for the reaction between 25.00 mL of a standardised 0.00245 mol/L sulfuric acid solution and an ammonia solution of unknown concentration.

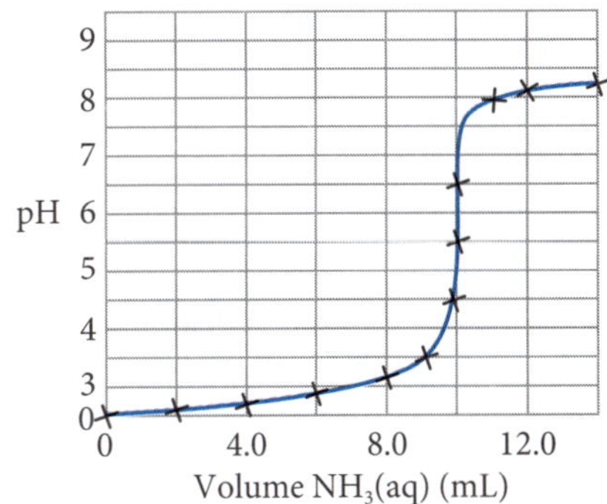

Figure E2.5 Acid–base pH titration

a Write a balanced whole formula equation for the neutralisation reaction. (1 mark)

b Identify which solution is placed in the burette. (1 mark)

c Determine the approximate pH at the equivalence point. (1 mark)

d Determine the titre for the titration. (1 mark)

e Calculate the concentration of the ammonia solution. (2 marks)

f Name an appropriate indicator if the titration had been performed using an indicator rather than a pH meter. (1 mark)

23 An insoluble white solid X is dissolved in nitric acid. An odourless gas is evolved and a colourless solution forms. Some of the gas is passed into a test tube containing universal indicator solution. The indicator turned pink. Some of the gas was passed into a solution of barium hydroxide and a faint white precipitate forms. The colourless solution was divided into three test tubes and tested with three reagents. The results are:

- Sodium iodide solution forms a bright yellow precipitate.
- Sulfuric acid forms a thick white precipitate.
- Sodium chloride solution forms a faint white precipitate that dissolves when the mixture is heated.

Analyse the information and identify the white solid X. Use equations as part of the explanation. (5 marks)

24 Atomic absorption spectroscopy (AAS) was used to measure the concentration of mercury in a sample of polluted water.

a Describe how a chemist could prepare a dilution standard with a mercury concentration of 20 ppm using crystalline mercury (II) nitrate that contains 61.80% mercury by weight. (3 marks)

b Further dilution standards were prepared and used to calibrate the atomic absorption spectrometer. Explain why this calibration was performed. (1 mark)

c The absorbance of the dilution standards was measured. The results are tabulated.

$[Hg^{2+}]$ (ppm)	0	5	10	15	20
Absorbance	0	0.095	0.190	0.290	0.385

The polluted water sample was diluted by a factor of 1 in 10 and its absorbance measured. The absorbance of the diluted sample was 0.120.

i Draw a calibration graph. (5 marks)

ii Use this data to determine the concentration of mercury in the original polluted water sample. (1 mark)

25 Bomb calorimeters are used to accurately measure enthalpies of combustion. An analytical chemist placed a 5.00 g sample of ethanol in a steel bomb calorimeter containing an excess of pure oxygen gas. The ethanol was vapourised and ignited. The heat released by the combustion was absorbed by the stainless steel calorimeter and its water bath that rose in temperature by 23.68 Celsius degrees. Calculate the molar enthalpy of combustion of ethanol from this data. Assume that the steel calorimeter and water bath were equivalent to 1.500 kg of water with a specific heat capacity of 4.18×10^3 J/kg/K. (3 marks)

26 Hydrogen sulfate ions are acidic in water solution due to the following equilibrium:

$$HSO_4^-(aq) + H_2O(l) \leftrightharpoons H_3O^+(aq) + SO_4^{2-}(aq)$$

a A 0.100 mol/L solution of sodium hydrogen sulfate has a pH of 1.54 at 25 °C. Calculate the acid dissociation constant for the above equilibrium. (3 marks)

b Write a whole formula equation and a net ionic equation for the reaction of sodium hydrogen sulfate solution with potassium hydroxide solution. (2 marks)

27 Phosgene gas ($COCl_2$) is formed by the reaction of carbon monoxide and chlorine.

a Write an equilibrium equation for this reaction. (1 mark)

b The enthalpy change for the reaction is −108 kJ/mol. How will the position of the equilibrium change if the temperature is increased? (1 mark)

c One mole of CO and 1 mole of Cl_2 are mixed in a 2-litre vessel at a fixed temperature. The system is allowed to come to equilibrium. The amount of phosgene present at equilibrium was 0.30 mol.

i Write an expression for the equilibrium constant. (1 mark)

ii Calculate a value for the equilibrium constant. (3 marks)

28 The Solvay process is used industrially to manufacture sodium carbonate. One of the steps in the process is the carbonation of ammoniated salt water (brine). Ammonia gas is first passed through saturated salt water to saturate the solution in ammonia. Inside the carbonation tower the ammoniated brine sprayed onto serrated plates arranged vertically in the tower. As the solution trickles over the plates the carbon dioxide is then pumped in at the base of the tower. The rising carbon dioxide gas is absorbed by the ammoniated brine and a solution of ammonium chloride and sodium hydrogen carbonate form. The carbonation tower is cooled during this process.

a Explain how the use of serrated plates increases the rate of reaction between the carbon dioxide and ammoniated brine solution. (1 mark)

b Write a balanced equation for the reaction between the carbon dioxide and the ammoniated brine solution. (1 mark)

c Sodium hydrogen carbonate crystals form under the conditions used. Explain what conditions lead to the crystallisation of sodium hydrogen carbonate. (2 marks)

29 25.00 mL of a nitric acid solution was pipetted into a beaker and a conductivity probe was clamped into position in the acid solution. A burette was cleaned and filled with a standardised 0.100 mol/L KOH solution. The conductivity of the solution was recorded as the titration was performed. The conductivity was plotted as a function of the volume of added KOH. The graph is shown in Figure E2.6. Determine the concentration of the nitric acid solution. (3 marks)

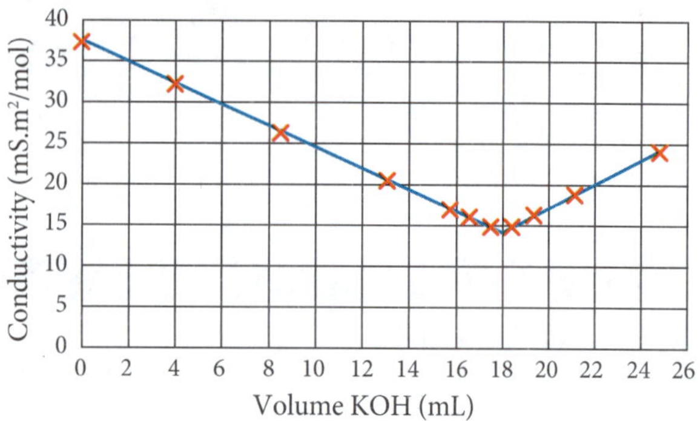

Figure E2.6 Conductivity graph versus volume KOH

30 One white salt and one green salt were mixed together and dissolved in water. A pale-green solution was formed. The tests in the table were performed on samples of this solution to identify the ions that may be present. Use the observations to identify the ions in the mixture. (4 marks)

Test	Observation
1 Dilute nitric acid was added.	no change
2 Barium nitrate solution was added to the solution from test 1.	no change
3 Silver nitrate solution was added to the solution from test 2.	white precipitate
4 A flame test was conducted on the original solution.	green flame
5 HCl was added to original solution.	no change
6 Sulfuric acid was added to the original solution.	white precipitate
7 Sodium hydroxide was added to the original solution.	pale-blue precipitate
8 Sodium fluoride was added to the original solution.	no change

31 Lauric acid is a straight chain alkanoic acid. Its molar mass is 200.312 g/mol.

a Determine the condensed structural formula of lauric acid. (2 marks)

b Lauric acid is a waxy solid that is insoluble in water. Explain why the acid is insoluble despite the presence of the carboxyl functional group. (1 mark)

c Lauric acid will dissolve in a hot solution of potassium hydroxide. Write an equation for this reaction. (1 mark)

32 Seawater contains a high concentration of chloride ions. A gravimetric analysis was performed by a student using a 100 mL sample of seawater. Excess silver nitrate was added to seawater and the precipitate produced weighed 7.56 g.

- **a** The student assumed that the precipitate was silver chloride. Is this a valid assumption? Explain. (2 marks)
- **b** Assuming that only chloride ions were precipitated, calculate the chloride ion concentration in the seawater in parts per million. (2 marks)

33 The molecule 6-aminohexanoic acid can undergo condensation polymerisation to form a polyamide called nylon-6.

- **a** Draw the structural formula of 6-aminohexanoic acid. (1 mark)
- **b** Draw a three-monomer section of the polymer chain to show the amide linkage. (1 mark)

34 A 100 mL bottle was completely filled with a sample of fresh, aerated water from a waterfall in the mountains. The water temperature was measured and the bottle was sealed. Back in the lab a Winkler titration was performed. A solution of manganese (II) ions reacted with the oxygen in the water and then an alkaline solution of potassium iodide was added. The mixture was acidified and the iodine that formed was titrated with a standard solution of 0.0100 mol/L sodium thiosulfate ($Na_2S_2O_3$). The titre was 12.35 mL.

- **a** The net reaction in the titration is:

$$O_2(aq) + 4S_2O_3^{2-}(aq) + 4H^+(aq) \rightarrow 2H_2O(l) + 2S_4O_6^{2-}(aq)$$

  Calculate the dissolved oxygen (DO) level in the water sample in the units of parts per million. (3 marks)
- **b** Use the tabulated data to determine the temperature of the water sample, given that it was saturated in oxygen. (1 mark)

Temperature (°C)	10	15	20	25
DO (ppm)	11.1	9.9	9.0	8.2

35 The colours of thymol blue indicator at different concentrations of hydrogen ion are shown in the table.

$[H^+]$ mol/L	Colour	$[H^+]$ mol/L	Colour	$[H^+]$ mol/L	Colour
10^{-2}	red	10^{-5}	yellow	10^{-8}	yellow-green
10^{-3}	yellow	10^{-6}	yellow	10^{-9}	green
10^{-4}	yellow	10^{-7}	yellow	10^{-10}	blue

- **a** What colour will thymol blue indicator turn if it is added to:
  - **i** 1 mol/L HCl? (1 mark)
  - **ii** 1 mol/L NaOH? (1 mark)
- **b** Thymol blue is added to a 1 mol/L sodium hydroxide solution and the solution slowly neutralised using sulfuric acid. Describe the colour changes of the indicator as neutralisation occurs. (3 marks)
- **c** An unknown solution X was tested with bromothymol blue and thymol blue indicators. Bromothymol blue turned green. What colour will thymol blue turn? (1 mark)

36 The solubility of lead (II) carbonate was found by experiment to be 7.27×10^{-6} g/100 g water at 25 °C. Determine the solubility product constant for lead (II) carbonate. (Density of water = 1 g/mL) (3 marks)

37 An equimolar mixture of ammonium chloride and ammonia solution is prepared. The pH was measured and found to be 9.3.

- **a** Write an equilibrium equation between ammonia and ammonium ions in water solution. (1 mark)
- **b** The original ammonia solution had a pH greater than 9.3. Explain why the addition of ammonium chloride has caused the pH to drop to 9.3. (2 marks)
- **c** Explain how this mixture can act as a buffer. (2 marks)

ANSWERS

SAMPLE HSC EXAMINATION 1

Section I

1 **D.** Barium ions produce a yellow-green flame. **A** is incorrect as calcium ions produce an orange-red flame. **B** is incorrect as sodium ions produce yellow flames. **C** is incorrect as copper (II) ions produce bright-green flames.

2 **C.** The double bond has priority in numbering the chain. **A** is incorrect as no locants are shown. **B** is incorrect as the methyl groups are not named. **D** is incorrect as the double bond has not been given priority.

3 **C.** Sulfurous acid is a weak acid. Thus **A**, **B** and **D** are all incorrect as they are all strong acids.

4 **D.** Sulfuric acid is a strong acid that ionises completely when diluted:

$H_2SO_4(aq) \rightarrow 2H^+(aq) + SO_4^{2-}(aq)$

$[H^+] = 2 \times [H_2SO_4] = 2 \times 0.00050 = 0.0010$ mol/L

$pH = -\log_{10}[H^+] = -\log_{10}(0.0010) = 3.00$

Thus **A**, **B** and **C** are incorrect.

5 **A.** A buffer is a mixture of a weak acid and its conjugate base. **B** is incorrect as the carbonate ion is not the conjugate base of phosphoric acid. **C** is incorrect as sulfate is not the conjugate base of sulfurous acid. **D** is incorrect as KOH is a strong base.

6 **B.** Universal indicator is violet in strongly basic solutions but yellow or red in acidic solutions. **A** is incorrect because X is basic as the indicator turns violet. **C** is incorrect because Y is weakly acidic as the indicator is yellow. **D** is incorrect because Z is acidic as the indicator is red.

7 **B.** The chloride ion concentration increases and shifts the equilibrium to the right. **A** is incorrect as water will shift the equilibrium to the left and the solution becomes more blue. **C** is incorrect as the nitrate ion will have no effect. **D** is incorrect as the HCl will cause an increase in chloride concentration and the solution turns more green.

8 **D.** The system is open and carbon dioxide escapes and this leads to less carbonic acid in the water. **A** is incorrect as the water will be slightly acidic due to carbon dioxide partially dissolving in the water. **B** is incorrect as the loss of carbon dioxide will reduce the concentration of carbonic acid. **C** is incorrect as no temperature change has occurred.

9 **A.** The nitrite ion (NO_2^-) is the conjugate base of HNO_2. **B** is incorrect as nitrous acid is a weak acid. **C** is incorrect as weak acids are partially dissociated in water. **D** is incorrect as 3.0 would be the pH of a strong acid with this concentration.

10 **B.** Helium is neither a reactant nor product and cannot react with any other gas as it is a noble gas. It cannot alter the position of the equilibrium as the concentration of reactants or products have not changed as the vessel volume is constant. **A** is incorrect as the value of K_{eq} depends only on temperature. **C** is incorrect as the equilibrium constant would decrease as the reaction is exothermic. **D** is incorrect as an increase in yield is caused by a total pressure increase as the result of reducing the volume of the reaction vessel.

11 **C.** $[H_3O^+] = 10^{-pH} = 10^{-4.10}$ mol/L

Degree of dissociation = 8 = $[H_3O^+]/[HA] \times 100$

$[HA] = \left(\frac{10^{-4.10}}{8}\right) \times 100 = \left(\frac{7.94 \times 10^{-5}}{8}\right) \times 100 = 9.93 \times 10^{-3}$mol/L

Thus **A**, **B** and **D** are incorrect.

12 **B.** Silver chloride is insoluble. **A** is incorrect as all products are soluble. **C** is incorrect as copper (II) chloride is soluble. **D** is incorrect as zinc chloride is soluble.

13 **A.** $PbBr_2(s) \leftrightarrows Pb^{2+}(aq) + 2Br^-(aq)$

$K_{sp} = [Pb^{2+}][Br^-]^2$

Let x be the mol of $PbBr_2$ dissolved per litre.

At equilibrium: $[Pb^{2+}] = x$; $[Br^-] = 2x$

$K_{sp} = (x)(2x)^2 = 4x^3 = 6.6 \times 10^{-6}$

Solve for x.

$x = 0.0118$ mol/L

$[Br^-] = 2x = 0.0236$ mol/L

Thus **B**, **C** and **D** are incorrect.

14 **A.** The hydroxyl functional group has numbering priority. **B** is incorrect as numbering is from left to right to give the lowest number to the methyl functional group. **C** is incorrect as no locants are shown. **D** is incorrect as the molecule is not an alkanone.

15 **B.** The metal ion solution is atomised in the flame. **A** is incorrect as the hollow cathode lamp generates the light. **C** is incorrect as the diffraction grating diffracts the light. **D** is incorrect as the flame does not absorb the light. The ions in the flame absorb the light.

16 **C.** The numerator of the equilibrium constant is the product of the methane and water vapour concentrations. **A** is incorrect as the equation is not balanced. **B** is incorrect as the quotient is inverted. **D** is incorrect as the water formed is in the gaseous state.

17 **D.** The yellow smoky flame is caused by carbon particles. **A** is incorrect as this is a complete combustion equation that would have produced a blue flame. **B** and **C** are incorrect as no carbon particles are present.

18 **D.** The smallest solubility product indicates the sulfate salt of lowest solubility and therefore the lower sulfate concentration. Thus **A**, **B** and **C** are incorrect as their K_{sp} values are larger.

19 **A.** The colourimeter can measure the colour of the solution and, when the equilibrium is attained, then the light absorbed by the solution will no longer change in intensity. **B** is incorrect because titrations are too long a procedure as samples would need to be removed and this would affect the attainment of equilibrium. **C** is incorrect as the mass remains constant. **D** is incorrect as the procedure is time consuming and could disturb the time to reach saturation equilibrium.

20 **C.** Ketones typically have chemical shifts at large downfield positions compared with the standard. **A** is incorrect as two peaks would be seen as the two carbon atoms are in different chemical environments. **B** is incorrect as the carbon atoms are in different chemical environments. **D** is incorrect as only one peak is observed because each carbon atom is in the same chemical environment.

Section II

21 **EM** Students need to determine which method can accurately measure low levels of zinc ions in soil. Students score full marks if they demonstrate the full experimental method.

a AAS is the only method that is sufficiently sensitive to detect the low concentrations of zinc minerals in a typical farm soil. ✓ Gravimetric and volumetric analysis are used in cases where concentrations are at molar levels. ✓

b **1** Weigh each soil sample. Add nitric acid to each sample and heat to dissolve the zinc minerals. ✓

2 Filter and collect each filtrate. Transfer each filtrate to 100 mL volumetric flasks and add distilled water to the engraved mark. Stopper and mix. ✓

3 Prepare a stock zinc standard from a known mass of pure zinc dissolved in nitric acid. Prepare dilution standards from this stock standard. ✓

4 Use a zinc hollow cathode lamp. Aspirate the standards in the AAS flame and measure the absorbance of each. Create a calibration graph. Measure the absorbance of the soil extracts and use the calibration graph to determine the zinc concentrations. ✓

5 Calculate the zinc concentration of zinc in each soil sample (in ppm = mg/kg soil). Average the results of the five samples. ✓

22 **EM** Students must accurately describe the experimental method as well as explain how effective the indicator is for different types of titrations.

Preparation of indicator:

1 Use a mortar and pestle to squash the blueberries. Add the squashed berries to a beaker, cover with water and heat gently to extract the dyes. ✓

2 Filter the mixture and collect the filtrate. Use this filtrate as an indicator. ✓

Testing the indicator:

1 Prepare acid and base solutions of different pH levels between 0 and 14 using a pH meter and 1 mol/L HCl, 1 mol/L NaOH and water. ✓

2 Place the different pH solutions in test tubes and add drops of the blueberry indicator.

Record the colour of the indicator at each pH. ✓

3 Use these results to determine the pH range(s) of the blueberry indicator. ✓

4 Determine whether or not the indicator ranges would be suitable to be used in titrations involving strong/weak acids and bases. ✓

23 **EM** Students must demonstrate equation writing skills as well as a knowledge of possible reaction products. In part **b** students need to realise that two possible isomers can form.

a $CH_3CH_2CH_2CH_2COOH + CH_3CH_2OH \rightleftharpoons CH_3CH_2CH_2CH_2COOCH_2CH_3 + H_2O$ ✓

Ester = ethyl pentanoate ✓

b $CH_3CHHCH_2CH_3 + HI \rightarrow CH_3CH_2CHICH_2CH_3$ ✓

3-iodopentane ✓

$CH_3CHHCH_2CH_3 + HI \rightarrow CH_3CHICH_2CH_2CH_3$ ✓

2-iodopentane ✓

24 **EM** Students need to determine that a vast number of isomers can form. They also need to show that each isomer has a unique name.

These alkanols have six carbon atoms.

MF = $C_6H_{13}OH$

M = (6 × 12.01) + (14 × 1.008) + 16.00 = 102.172 g/mol

Figure AE1.1 shows the structural formulas and names of the isomers.

25 **EM** Students must demonstrate skills in naming alkenes. Students should show an understanding of the addition polymerisation process and the structure of the repeating unit.

a tetrafluoroethylene ✓

b **i** addition polymerisation ✓

ii Figure AE1.2 shows the structural formula of the polymer's repeating unit.

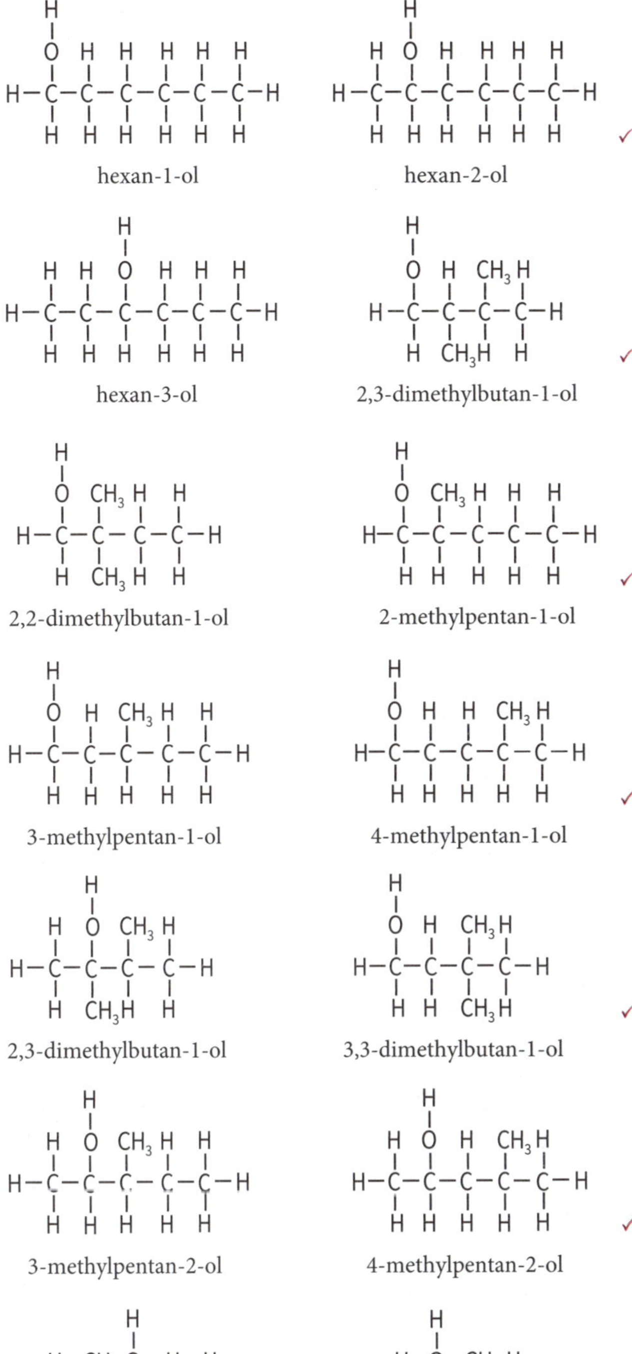

Figure AE1.1 Isomers

Figure AE1.2 Teflon repeating structure

26 **EM Students must demonstrate an understanding of condensation polymerisation and how functional groups interact. Full marks are given to students who have a wide understanding of the social use of such polymers.**

a condensation polymerisation ✓

b Figure AE1.3 shows the structural formulas of the two monomers.

$$H{-}O{-}\overset{O}{\overset{\|}{C}}{-}C_6H_4{-}\overset{O}{\overset{\|}{C}}{-}O{-}H \quad ✓ \qquad H{-}O{-}\underset{H}{\overset{H}{C}}{-}\underset{H}{\overset{H}{C}}{-}O{-}H \quad ✓$$

Figure AE1.3 Monomers

c PET has low water absorption and minimal shrinkage after washing. ✓

27 **EM Students relate the social/environmental issues of hard water to the analysis of the water using an EDTA titration. Full marks require students to express hardness in the correct units.**

a Limestone rocks bordering the lake are leached over time and calcium carbonate dissolves into the water. ✓

b Synthetic detergents (such as anionic detergents) are used. ✓ They lather and foam in hard water and can clean dirty clothes. ✓

c $n(\text{EDTA}) = cV = (0.0100)(0.010.75) = 1.075 \times 10^{-4}$ mol ✓
Reaction stoichiometry = 1 : 1
$n(Ca^{2+}) = 1.075 \times 10^{-4}$ mol ✓
The calcium ions are present due to calcium carbonate undergoing dissolution in the water.
$n(CaCO_3) = 1.075 \times 10^{-4}$ mol ✓
Calculate the mass of $CaCO_3$.
$M(CaCO_3) = 40.08 + 12.01 + 3(16.00) = 100.09$ g/mol
$m(CaCO_3) = n.M = (1.075 \times 10^{-4})(100.09) = 0.0108 \text{ g} = 10.8$ mg ✓
$c(CaCO_3) = n/V = \dfrac{10.8}{0.0250} = 432 \text{ mg/L} = 432$ ppm ✓

28 **EM Students need to recall the order of elimination tests and describe confirmatory tests that confirm the presence of various cations.**

1 Add excess HCl to a sample of the solution. If a white precipitate forms then lead (II) ions are indicated. This can be confirmed by adding sodium iodide to another sample and a bright yellow precipitate forms. ✓

2 If a precipitate formed when excess HCl was added, filter the mixture and keep the filtrate for testing. If no precipitate formed, then the next test can be conducted on the acidified solution. Add sulfuric acid to a sample of the filtrate or solution. A white precipitate indicates that calcium and/or barium ions are present. To another sample of filtrate, add sodium fluoride solution. If a white precipitate formed, then calcium ions are present. If no precipitate formed, then barium ions are present. ✓

3 Use a new sample of the unknown solution. Test the solution with potassium thiocyanate solution. A deep-red colouration demonstrates that iron (III) ions are present. ✓

4 Use a new sample of the unknown solution. Add ammonia solution in excess. A blue precipitate forms and then the precipitate dissolves to form a deep-blue solution if copper (II) ions are present. ✓

29 **EM Students show an understanding of amphiprotic behaviour as well as choosing a strong base and a strong acid to demonstrate the amphiprotic property of the hydrogen carbonate ions.**

HCO_3^- as an acid: $HCO_3^-(aq) + OH^-(aq) \rightarrow CO_3^{2-}(aq) + H_2O(l)$ ✓
HCO_3^- as a base: $HCO_3^-(aq) + H_3O^+(aq) \rightarrow CO_2(g) + 2H_2O(l)$ ✓

30 **EM Students are to use the stoichiometry of an equation to calculate the volume of a gaseous product. Students must then relate the use of the gas to produce a polymer foam and explain the properties of the polymer foam that makes it useful in a social context.**

a $2NaHCO_3(s) \rightarrow Na_2CO_3(s) + H_2O(l) + CO_2(g)$ ✓

b $M(NaHCO_3) = 22.99 + 1.008 + 12.01 + 3(16.00) = 84.008$ g/mol
$n(NaHCO_3) = m/M = \dfrac{8.41}{84.008} = 0.100$ mol ✓
$n(CO_2) = n(NaHCO_3)/2 = 0.0500$ mol ✓
$V(CO_2) = (0.0500)(24.79) = 1.24$L ✓

c The gas holes formed in the foam make the plastic foam a good thermal insulator. ✓

31 **EM Students must show a deep understanding of the conditions required for fermentation and use supplied information to explain why calcium oxide can absorb carbon dioxide. For full marks students should demonstrate skills in mole calculations.**

a At this temperature the yeast's biochemical processes are optimised. Enzymic processes occur at their optimal rate near 'blood temperature'. ✓

b Calcium oxide is a base and carbon dioxide is an acidic oxide. The calcium oxide reacts with the carbon dioxide to form solid calcium carbonate. Thus the carbon dioxide is effectively trapped. ✓

c $C_6H_{12}O_6(aq) \rightarrow 2C_2H_5OH(aq) + 2CO_2(g)$ ✓

d $M(\text{glucose}) = 6(12.01) + 12(1.008) = 6(16.00) = 180.156$ g/mol
$n(\text{glucose}) = m/M = \dfrac{18.0}{180.156} = 0.0999$ mol
$n(CO_2) = 2n(\text{glucose}) = 0.200$ mol
$M(CO_2) = 44.01$ g/mol
$m(CO_2) = 0.200 \times 44.01 = 8.80$ g ✓

e $m(CO_2)$ evolved in two days = 7.0 g
Percentage $= \left(\dfrac{7.0}{8.80}\right) \times \dfrac{100}{1} = 79.5\%$ ✓

32 **EM Students should understand that the reaction involves neutralisation and be able to predict the formula of the fatty acid salt as well as perform a mole calculation to determine the volume of base for the complete reaction.**

a $KC_{17}H_{35}COO$ ✓

b $C_{17}H_{35}COOH(s) + KOH(aq) \rightarrow KC_{17}H_{35}COO(s) + H_2O(l)$ ✓

c $M(C_{17}H_{35}COOH) = 18 \times 12.01 + 36 \times 1.008 + (2 \times 16.00)$
$= 284.468$ g/mol
$n(C_{17}H_{35}COOH) = m/M = \dfrac{15.0}{284.468} = 0.0527$ mol ✓
Stoichiometry = 1 : 1
$n(KOH) = 0.0527$ mol ✓
$V(KOH) = n/c = \dfrac{0.0527}{6.0} = 8.78 \times 10^{-3} \text{ L} = 8.78$ mL ✓

33 **EM Students need to demonstrate the acidic properties of phosphorus oxides and write balanced equation for a neutralisation reaction.**

a $P_4(s) + 5O_2(g) \rightarrow P_4O_{10}(s)$ ✓

b The solution is acidic due to the production of phosphoric acid: ✓
$P_4O_{10}(s) + 6H_2O(l) \rightarrow 4H_3PO_4(aq)$ ✓

c $H_3PO_4(aq) + 3NaOH(aq) \rightarrow Na_3PO_4(aq) + 3H_2O(l)$ ✓
Salt = sodium phosphate ✓

34 **EM Students should recall the acidic oxides that can pollute the atmosphere and how acid rain can damage natural and built structures. Students need to calculate the pH of solutions and relate pH to colours of different indicators.**

a SO_2 pollution of the atmosphere combines with water to form sulfurous acid:
$SO_2(g) + H_2O(l) \rightarrow H_2SO_3(aq)$ ✓

b $CaCO_3(s) + H_2SO_3(aq) \rightarrow CaSO_3(s) + H_2O(l) + CO_2(g)$ ✓
c $pH = -\log_{10}[H^+] = -\log_{10}[1.0 \times 10^{-5}] = 5.0$ ✓
d i colourless ii yellow ✓

35 EM Students must show mathematical skills in determining the concentration of ions on mixing solutions and determine whether the ionic product is greater or lesser than the solubility product constant.

Calculate the concentration of silver and sulfate ions on mixing.

$n(Ag^+) = cV\ (0.010)(0.060) = 6.0 \times 10^{-4}$ mol

$n(SO_4^{2-}) = cV = (0.010)(0.030) = 3.0 \times 10^{-4}$ mol

Total volume on mixing = 60 + 30 = 90 mL = 0.090 L

$[Ag^+] = n/V = \left(\frac{6.0 \times 10^{-4}}{0.090}\right) = 6.67 \times 10^{-3}$ mol/L

$[SO_4^{2-}] = n/V = \left(\frac{3.0 \times 10^{-4}}{0.090}\right) = 3.33 \times 10^{-3}$ mol/L ✓

Calculate the ionic product (*IP*). The ionic product is the product of concentration of ions of the electrolyte, each raised to the power of their coefficients in the balanced chemical equation in the dissolution equilibrium.

$IP = [Ag^+]^2[SO_4^{2-}] = (6.67 \times 10^{-3})^2(3.33 \times 10^{-3}) = 1.48 \times 10^{-7}$ ✓

The ionic product is much less than the K_{sp}.

$IP = 1.48 \times 10^{-7} << 1.20 \times 10^{-5}$

Therefore no precipitate will form on mixing the solutions. ✓

36 EM Students should recall different types of intermolecular forces to explain differences in the properties of organic compounds. They also need to demonstrate an understanding of homologous series of organic compounds to score full marks.

a Alkanoic acids are more polar than alkanols and the hydrogen-bonding in alkanoic acids is stronger in short chain alkanoic acids than in similar alkanols. These stronger intermolecular forces result in higher melting and boiling points for alkanoic acids than alkanols. ✓

b pentanoic acid ✓

c The next member of Y's alkanoic acid series is hexanoic acid. Hexan-1-ol is oxidised to hexanoic acid. X = hexan-1-ol ✓

SAMPLE HSC EXAMINATION 2

Section I

1 **C.** 2-methylpropan-1-ol has three carbon atoms in different chemical environments. (Carbon 1 is attached to the OH group; C-2 is a CH_2 (methylene group); the remaining carbons are both methyl groups attached to C-2. **A** is incorrect as the spectrum would have four peaks. **B** is incorrect as butane would have two peaks. **D** is incorrect as butanone would have four peaks.

2 **D.** Butanone contains a carbonyl (C=O) group, which is consistent with the peak at 1705–1725 cm^{-1}. The other large peak is a C–H stretching frequency (typically 2850–2960 cm^{-1}). **A** is incorrect as butan-1-ol would have an O–H stretch at 3610–3640 cm^{-1}. **B** is incorrect as butan-1-amine would have an N–H stretch at 3300–3500 cm^{-1}. **C** is incorrect as 2-methylpropan-1-ol would have an O–H stretch at 3610–3640 cm^{-1}.

3 **A.** Propan-1-ol has a molar mass of 60.1 g/mol. This is consistent with the small peak at an m/z value of 60. The large peak at m/z = 31 is a fragment (CH_2OH^+). **B** is incorrect as propane would have a parent ion peak at m/z = 44. **C** is incorrect as propanoic acid would have a parent ion peak at m/z = 74. **D** is incorrect as but-2-ene would have a parent ion peak at m/z = 56.

4 **C.** Alkanoic acids have the highest boiling points as they are the most polar of the three molecules, and the ester has the lowest boiling point as it is the least polar. Therefore **A**, **B** and **D** are incorrect as the order is incorrect.

5 **D.** $M(CH_2CH_2) = (2 \times 12.01) + (4 \times 1.008) = 28.052$ g/mol

$n = \left(\frac{1.00 \times 10^6}{28.052}\right) = 35\ 648$ mol

Therefore **A**, **B** and **C** are mathematically incorrect.

6 **A.** In X the numbering is from right to left to give the lowest locant set. In Y the numbering is also right to left because the locant set numbers are the same from left to right as right to left, and so the alphabetical order of functional groups decides the numbering order. Z is also numbered from right to left to give the lowest locant set for the four bromo groups. **B** is incorrect as no locant positions are shown for the functional groups. **C** is incorrect as X and Z have been incorrectly numbered from left to right. **D** is incorrect as Y has been incorrectly numbered from left to right and priority has incorrectly been given to chloro groups instead of bromo groups.

7 **A.** Ammonia cleansers typically have a pH of 11.5 and universal indicator is violet at this pH. Sea water has a pH range of about 7.5–8.5 and universal indicator would be green-blue. Lemonades typically have a pH between 2 and 3 and the universal indicator would be red. Therefore **B**, **C** and **D** are incorrect.

8 **B.**

$n(HNO_3) = cV = (0.0250)(0.0300) = 7.50 \times 10^{-4}$ mol

After dilution:

$c(HNO_3) = n/V = \left(\frac{7.50 \times 10^{-4}}{0.250}\right) = 3.00 \times 10^{-3}$ mol/L

$[H^+] = c(HNO_3) = 3.00 \times 10^{-3}$ mol/L

$pH = -\log_{10}[H^+] = -\log_{10}(3.00 \times 10^{-3}) = 2.52$

$pOH = 14 - pH = 14 - 2.52 = 11.45$

Thus **A**, **C** and **D** are mathematically incorrect.

9 **B.** $K_a = [H_3O^+][CH_3COO^-]/[CH_3COOH]$

Hydronium ions and acetate ions have the same concentration according to the balanced equation. Let this concentration = x.

At equilibrium: $[H_3O^+] = [CH_3COO^-] = x$

$[CH_3COOH] = 0.150 - x$

$[H_3O^+][CH_3COO^-]/[CH_3COOH] = (x)(x)/(0.150 - x) = 1.77 \times 10^{-5}$

Solve for x.

$x = 1.6 \times 10^{-3}$ mol/L $= [H_3O^+]$

$pH = -\log_{10}[1.6 \times 10^{-3}] = 2.8$

Thus **A**, **C** and **D** are mathematically incorrect.

10 **A.** 1-propyl acetate is more volatile and will have the lower boiling point; the condensed structural formulas are correct. **B** is incorrect as the formulas are incorrect. **C** is wrong as the boiling points are incorrect. **D** is incorrect as both the formulas and boiling points are incorrect.

11 **A.** Hex-1-ene is produced in the dehydration reaction. **B** is incorrect as alkenes and not alkanes form on dehydration of alkanols. **C** and **D** are incorrect as oxidation does not occur.

12 **C.** Methane has an m/z value of 16 as the molar mass is 16.042 g/mol. **A** is incorrect as radio waves are not used in this instrument. **B** is incorrect as parent peaks are not necessarily the largest peak. **D** is incorrect as mass spectra do not have units of cm^{-1}.

13 **C.** HCl, NaOH and H_2SO_4 all form white precipitates, whereas KI produces a yellow precipitate. Therefore **A**, **B** and **D** are incorrect.

14 **A.** The reverse reaction is endothermic and therefore heating the nickel tetracarbonyl will decompose it to form pure nickel. **B** is incorrect as the reverse reaction will not go to completion. **C** is incorrect as the reverse reaction requires heat. **D** is incorrect because increasing the total pressure will promote the formation of nickel tetracarbonyl.

15 **A.** The high salt concentration causes the soap to precipitate. **B** is incorrect as the glycerol is used to manufacture other products but not soap. **C** is incorrect as the hydrolysis occurs under alkaline conditions. **D** is incorrect as water is used to wash alkali and salt out of the soap curd.

16 **C.** Ammonia is recycled to reduce costs. **A** is incorrect as this is an endothermic process. **B** is incorrect as calcium chloride forms. **D** is incorrect as the calcium carbonate decomposes to form calcium oxide.

17 **B.** These molecules are alkanals and alkanones, which have different functional groups. **A** is incorrect as these molecules are chain isomers. **C** is incorrect as these are position isomers. **D** is incorrect as these molecules are not isomers because they have different molecular formulas.

18 **D.** The positive head group is present in cationic surfactants and these are used in hair conditioners. **A** is incorrect as the surfactant is cationic. **B** is incorrect are mainly produced from the petroleum by-products. **C** is incorrect as these cationic surfactants are strongly lathering.

19 **B.** The oxygen atoms in water are converted to oxygen gas. **A** is incorrect as chlorophyll absorbs red and violet light. **C** is incorrect as the Gibbs free energy change is positive and the reaction is not spontaneous. **D** is incorrect as oxygen is produced by water oxidation.

20 **A.** Sodium carbonate requires salt and limestone, which are closest at site Y. Sodium hydroxide production requires salt and large amounts of electricity, which are close to site W. Sulfuric acid uses sulfur dioxide as a feedstock and so site X is best as a sulfide mine is nearby. The sulfide can be roasted to generate the sulfur dioxide. Therefore **B**, **C** and **D** are incorrect.

Section II

21 EM Students need to recall the format of the equilibrium constant. In the calculation the student should show the substitution of values into the formula.

a $2HI(g) \rightleftarrows H_2(g) + I_2(g)$ ✓

b $K_{eq} = [H_2][I_2]/[HI]^2$ ✓

c $K_{eq} = \dfrac{(2.60 \times 10^{-3})(2.60 \times 10^{-3})}{(1.76 \times 10^{-2})^2}$

$= 0.0218$ ✓

The equilibrium lies to the left (reactants). ✓

d The equilibrium constant decreased as the temperature decreased. Therefore the reaction is endothermic in the forward direction. ✓

22 EM This question requires graphical interpretation as well as a deep understanding of titrations and mole calculations. Students need to recall the colours of indicators at different pH levels.

a $H_2SO_4(aq) + 2NH_3(aq) \rightarrow (NH_4)_2SO_4(aq)$ ✓

b ammonia solution ✓

c pH = 5.5 ✓

d $V = 10.0$ mL ✓

e $n(H_2SO_4) = cV = (0.00245)(0.0250) = 6.125 \times 10^{-5}$ mol ✓

Reaction stoichiometry = 1 : 2

$n(NH_3) = 2(6.125 \times 10^{-5}) = 1.225 \times 10^{-4}$ mol

$c(NH_3) = n/V = \left(\dfrac{1.225 \times 10^{-4}}{0.0100}\right) = 0.0123$ mol/L ✓

f methyl orange ✓

23 EM Students need to recall characteristic reactions of cation and anions and then write appropriate equations for the observed reactions.

The gas is acidic as the indicator turned pink. The gas is probably carbon dioxide because it forms carbonic acid when dissolved in water. Therefore X is a carbonate. ✓

Carbon dioxide reacts with barium hydroxide solution to form barium carbonate precipitate:

$Ba(OH)_2(aq) + CO_2(g) \rightarrow BaCO_3(s) + H_2O(l)$ ✓

The cation in the solution is lead (II) ions. Lead (II) ions form a yellow precipitate of PbI_2 when sodium iodide solution is added:

$Pb^{2+}(aq) + 2I^-(aq) \rightarrow PbI_2(s)$ ✓

Lead (II) ions form a white precipitate of $PbSO_4$ when sulfuric acid is added:

$Pb^{2+}(aq) + SO_4^{2-}(aq) \rightarrow PbSO_4(s)$ ✓

Lead (II) ions form a faint white precipitate of $PbCl_2$ when sodium chloride is added. $PbCl_2$ dissolves when the mixture is heated:

$Pb^{2+}(aq) + 2Cl^-(aq) \rightarrow PbCl_2(s)$ ✓

24 EM Students must demonstrate high level understanding of AAS and how dilution standards are prepared. Students need to demonstrate graphical skills to achieve full marks.

a The following steps are taken to prepare these standards:

1 Prepare a stock solution of higher concentration and dilute this stock solution systematically. ✓

2 A stock solution containing 2000 ppm (= 2000 mg Hg^{2+}/L = 2.000 g/L) of mercury (II) ions is prepared by weighing out 3.236 g of mercury (II) nitrate and dissolving it in distilled water. The solution made up to 1.00 L in a volumetric flask.

$(m = 2.000 \times \left(\dfrac{100}{61.80}\right) = 3.236\text{ g})$ ✓

3 To prepare the 20 ppm solution, use a 10 mL pipette to transfer 10.00 mL of the stock solution to a 1 L volumetric flask and add distilled water to the mark and mix. ✓

b The AAS needs to be calibrated so that absorbance measurements can be related to known mercury ion concentrations. Usually dilution levels are selected so that there is a linear relationship between absorbance and concentration. ✓

c i Figure AE2.1 shows the absorbance versus concentration graph for mercury ions.

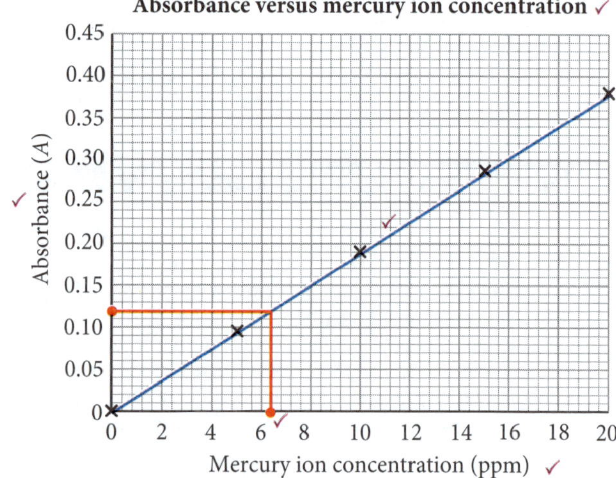

Figure AE2.1 Absorbance graph

ii $A = 0.120$

From the graph, $c(Hg^{2+}) = 6.2$ ppm

Original concentration of mercury $= 6.2 \times \left(\dfrac{10}{1}\right) = 62$ ppm ✓

25 EM Students need to use the calorimetry formula and show correct substitution and ensuring the correct units are used. The relationship between heat absorbed and the molar enthalpy change should be shown in the equation.

$q = mc\Delta T = (1.500)(4.18 \times 10^3)(23.68) = 148\,474\text{ J} = 148.474\text{ kJ}$ ✓

$M(C_2H_5OH) = 2 \times 12.01 + 6 \times 1.008 + 16.00 = 46.068$ g/mol

$n(C_2H_5OH) = m/M = \dfrac{5.00}{46.068} = 0.1085$ mol ✓

$\Delta H = -q/n = \dfrac{-148.474}{0.1085} = -1368$ kJ/mol ✓

26 EM Students need to recall the K_a formula for a weak acid and relate p*Ka* to pH. Substitution of data should be shown into the equations to obtain full marks.

a $[H_3O^+] = 10^{-pH}$
$= 10^{-1.54} = 0.0288$ mol/L ✓
$= [SO_4^{2-}]$ ✓
$K_a = [H_3O^+][SO_4^{2-}]/[HSO_4^-]$
$= \dfrac{(0.0288)(0.0288)}{(0.100 - 0.0288)}$
$= 1.16 \times 10^{-2}$ ✓

b $NaHSO_4(aq) + KOH(aq) \rightarrow KNaSO_4(aq) + H_2O(l)$ ✓
$HSO_4^-(aq) + OH^-(aq) \rightarrow SO_4^{2-}(aq) + H_2O(l)$ ✓

27 EM Students need to determine the initial concentrations, change in concentrations and equilibrium concentrations prior to substitution into the equilibrium constant expression.

a $CO(g) + Cl_2(g) \rightleftharpoons COCl_2(g)$ ✓

b The equilibrium will shift to the left to counteract the addition of heat. More CO and Cl_2 will form. ✓

c **i** $K_{eq} = [COCl_2]/[CO]\cdot[Cl_2]$ ✓

ii

Concentration	CO	Cl_2	$COCl_2$	
I: initial concentration (mol/L)	$\frac{1}{2} = 0.5$	$\frac{1}{2} = 0.5$	0	
C: change (mol/L)	−0.15	−0.15	$\frac{+30}{2} = 0.15$	✓
E: equilibrium concentration (mol/L)	0.35	0.35	0.15	✓

$K_{eq} = [COCl_2]/[CO][Cl_2] = \dfrac{(0.15)}{(0.35)(0.35)} = 1.2$ ✓

28 EM Students use the supplied information to explain an industrial synthesis reaction. Students need to demonstrate an understanding of rates of reaction in relation to surface areas as well as conditions that promote product formation.

a The serrated plates increase the surface area of contact between the gas and the solution. This surface area increase causes the rate of the reaction to increase. ✓

b $NH_3(g) + H_2O(l) + NaCl(aq) + CO_2(aq) \rightarrow NH_4Cl(aq) + NaHCO_3(s)$ ✓

c The low temperatures ✓and high sodium ion concentration promote crystallisation of the sodium hydrogen carbonate. ✓

29 EM Students are required to use the graphical information and reaction stoichiometry to determine the moles of nitric acid and its concentration.

From graph: Volume (KOH) = 18.0 mL at equivalence point
$HNO_3(aq) + KOH(aq) \rightarrow KNO_3(aq) + H_2O(l)$ ✓
$n(KOH) = cV = (0.100)(0.0180) = 0.001\ 80$ mol
Stoichiometry = 1 : 1
$n(HNO_3) = 0.001\ 80$ mol ✓
$c(HNO_3) = n/V = \dfrac{(0.001\ 80)}{(0.025\ 00)} = 0.0720$ mol/L ✓

30 EM Students must be able to relate the observations in each test to conclusions that can be made about the possible ions that would give a positive test. A final conclusion should be stated.

Copper (II) ions are present as test 4 gives a green flame characteristic of copper (II) ions. This observation is supported by test 7, which results in a pale-blue precipitate consistent with copper (II) hydroxide.
Chloride ions are present as test 3 produces a white precipitate of silver chloride. ✓
There are no carbonate ions as test 1 was negative.
There are no sulfate ions as test 2 is negative.
There are no silver or lead (II) ions as test 5 is negative. ✓
Calcium and/or barium ions are present as test 6 produces a white precipitate.
Test 8 is negative and this eliminates calcium ions. Therefore barium ions are present. ✓
Therefore copper (II) ions, barium ions and chloride ions are present. ✓

31 EM Students need to recall the general structure of an alkanoic acid and then use the molar mass data to determine the molecular formula. The physical properties of fatty acids and their neutralisation by a strong base need to be demonstrated.

a Alkanoic acid = $CH_3(CH_2)_nCOOH$. (n = integer)
Molar mass = $(n + 2)(12.01) + (2n + 4)(1.008) + 2(16.00) = 200$ ✓
Solve for n.
$n = 10$
Condensed structural formula = $CH_3(CH_2)_{10}COOH$ ✓

b The long hydrocarbon tail is non-polar and has little affinity with the polar water solvent. The polar carboxyl head's interaction with water is insufficient to assist dissolution of the solid in water. ✓

c $CH_3(CH_2)_{10}COOH(s) + KOH(aq) \rightarrow KCH_3(CH_2)_{10}COO(aq) + H_2O(l)$ ✓

32 EM Students need to show they understand that sea water is not just a solution of sodium chloride. They need to show that they can convert concentrations in mol/L units to ppm units.

a The assumption is invalid. ✓ Seawater contains other anions such as bromide and sulfate ions that can be precipitated by silver ions. ✓

b $Ag^+(aq) + Cl^-(aq) \rightarrow AgCl(s)$
$m(AgCl) = 7.56$ g
$n(AgCl) = m/M = \dfrac{7.56}{(107.9 + 35.45)} = 0.0527$ mol ✓
$n(Cl^-) = 0.0527$ mol
$m(Cl^-) = n.M = (0.0527)(35.45) = 1.87$ g = 1870 mg
$c(Cl^-) = 1870$ mg/0.100 L = 18700 ppm ✓

33 EM Students should demonstrate that they can draw structural formula given the name of a molecule. They also need to demonstrate knowledge of condensation polymerisation in order to draw the repeating structure of a polymer.

a Figure AE2.2 shows the structural formula of 6-aminohexanoic acid.

```
      H H H H H
H\    | | | | |  //O
  N-C-C-C-C-C-C
H/    | | | | |  \O-H
      H H H H H
```
✓

Figure AE2.2

b Figure AE2.3 shows a section of the polymer chain with monomers linked via amide functional groups.

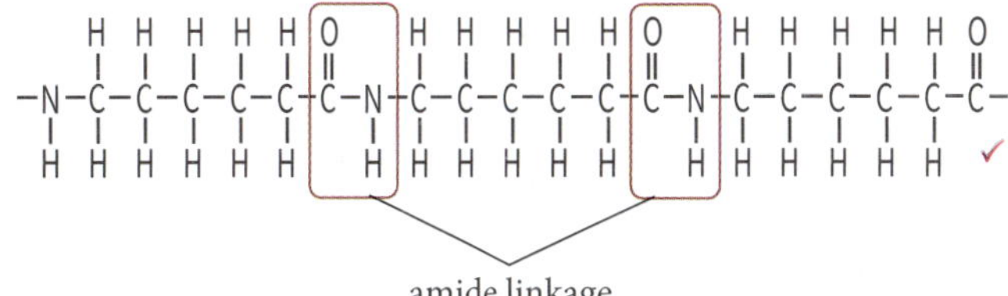

Figure AE2.3

34 EM Students need to use the reaction stoichiometry to perform the mole calculation and be able to convert concentration data to ppm units. Students then need to use the tabulated data to determine the temperature of the water sample.

a $O_2(aq) + 4S_2O_3^{2-}(aq) + 4H^+(aq) \rightarrow 2H_2O(l) + 2S_4O_6^{2-}(aq)$

$O_2 : S_2O_3^{2-}$ stoichiometry = 1 : 4

$n(S_2O_3^{2-}) = cV = (0.0100)(0.01235) = 1.235 \times 10^{-4}$ mol ✓

$n(O_2) = \frac{1}{4} \times 1.235 \times 10^{-3} = 3.089 \times 10^{-5}$ mol ✓

$m(O_2) = n.M = (3.089 \times 10^{-5})(32.00) = 9.885 \times 10^{-4}$ g = 0.989 mg

$c(O_2) = m/V = \frac{0.989 \text{ mg}}{0.100 \text{ L}} = 9.89$ ppm ✓

b Temperature = 15 °C ✓

35 EM Recall of indicator colours in acidic and basic solutions is required. The student needs to relate pH to the hydrogen ion concentration.

a i red ✓ ii blue ✓

b The indicator changes from blue to green ✓ then to green-yellow ✓ and then yellow ✓ in the last drop of acid as neutralisation is just achieved. (Note: The indicator will stay yellow if excess acid is added.)

c Bromothymol blue is green in the pH range 6.0–7.6 ($[H^+] = 10^{-6} - 10^{-7.6}$ mol/L). Thymol blue would be yellow in this pH range. ✓

36 EM Students need to write an appropriate solubility equation and using the supplied data to determine the solubility product constant. Significant figures need to be considered.

Calculate the molar mass of lead (II) carbonate.

$M(PbCO_3) = 207.2 + 12.01 + (3 \times 16.00) = 267.21$ g/mol

$n(PbCO_3) = m/M = \frac{(7.27 \times 10^{-6})}{267.21} = 2.72 \times 10^{-8}$ mol ✓

Write the equation for the dissolution equilibrium:

$PbCO_3(s) \leftrightarrows Pb^{2+}(aq + CO_3^{2-}(aq)$

Reaction stoichiometry: $Pb^{2+} : CO_3^{2-} = 1 : 1$

$n(Pb^{2+}) = n(CO_3^{2-}) = 2.72 \times 10^{-8}$ mol

100 g water = 100 mL = 0.100 L

Calculate the ion concentrations at equilibrium.

$[Pb^{2+}] = [CO_3^{2-}] = \frac{2.72 \times 10^{-8}}{0.100} = 2.72 \times 10^{-7}$ mol/L ✓

Calculate K_{sp}.

$K_{sp} = [Pb^{2+}] \cdot [CO_3^{2-}] = (2.72 \times 10^{-7})^2 = 7.40 \times 10^{-14}$ ✓

37 EM Students need to understand the composition of a buffer and how buffering occurs when small amounts of strong acids and bases are added to the buffer solution.

a $NH_3(aq) + H_2O(l) \leftrightarrows NH_4^+(aq) + OH^-(aq)$ ✓

b The ammonium ion is a Brønsted–Lowry acid and it reacts with hydroxide ions. The equilibrium shifts to the left ✓ and the pH decreases as the hydroxide ion concentration decreases. ✓

c The mixture contains a Brønsted–Lowry base (NH_3) and its conjugate acid (NH_4^+). ✓ As the solution is equimolar in the base and its conjugate acid, then the solution acts as a buffer because the addition of small quantities of a strong acid or a strong base leads to a very small pH change. The strong acid will be neutralised by NH_3 and the strong base will be neutralised by NH_4^+. ✓

INDEX

A

acetic acid (ethanoic acid) 58
acetylsalicylic acid 229
acid dissociation constants 40, 97–100
acid ionisation constant 40
acid rain 60
acidic buffer 104
acidic salts 80
acidic substances 57
acidity, adjusting/neutralising
  soil 64–5
  stomach 64
  swimming pool 64
acids *see also by name*
  acid–base analysis, applications of 100–3
  acid–base indicators 59–60, 81
  acid–base reactions 62–3
  acid–base theories 66–9
  acid–carbonate reactions 63
  acid–metal reactions 63–4
  common laboratory 58
  corrosive 66
  diprotic 78, 97
  dissociation of 39–40
  inorganic 57, 58
  modelling in solution 84–5
  monoprotic 78, 97
  primary standard, preparation of 92
  properties of 58–9
  strengths, investigation of 83–4
  strong 39, 67
  strong acid/strong base titrations 94
  strong acid/weak base titrations 94
  strong base/weak acid titrations 94–5
  triprotic 78, 98
  weak 39, 67, 78, 103
addition polymerisation 173
addition reactions 136
  halogen 136
  hydrogen halide 137
aerobic decomposers 227
air quality
  testing and monitoring 188
alcohols 114–15, 143, 158 *see also* alkanols
  boiling points 144
  classification 143–5
  combustion 145–7
  dehydration 147
  fermentation 150–1
  oxidation 148
  production of 150–2
  properties of 143–4
  reactions 147–50
  substitution reaction 147, 152
aldehydes 115, 159
alkaline solutions 57
alkalis 57
alkanes 111–12, 124
  boiling point 127
  homologous series 111
  hybridisation of orbitals 126
  instrumental analysis 212
alkanoic acids 115
  instrumental analysis 213
alkanols 114, 158, 209, 210 *see also* alcohols
  boiling points 144
  combustion 145–7
  instrumental analysis 212–13
  oxidation 147–50
  physical properties 144–5
  primary 114, 147, 158, 210
  secondary 114, 150, 158, 210
  tertiary 114, 158
alkenes 112–13, 124
  hybridisation of orbitals 126
  instrumental analysis 212
alkyl alkanoates *see* esters
alkyl functional groups 112
alkynes 113, 124
  hybridisation of orbitals 126
aluminium 234
amides 117
  amide functional group 160
amines 116–17
  amine functional group 159–60
ammonia 4, 40, 228–9
  production of 226–7
ammonia solution 58
amphiprotic substances 79
analgesics 229, 230
anion analysis 193–5
  confirmatory tests 193
  identification of anions 194
  precipitation tests 193
anionic detergent 167
answers, sample exam 252–8
anthracite coal 153
anticoagulants 230
Arrhenius, Svante (1884) 67, 69, 73
artesian water 186
aspirin 229
  manufacturing 229–31
association 46
atomic absorption spectroscopy (AAS) 202–4
auto-ionisation 74, 75

B

back titrations 92–3
bases 57, 68
  acid–base analysis, applications of 100–3
  acid–base indicators 59–60, 81
  acid–base reactions 62–3
  acid–base theories 66–9
  common laboratory 58
  diprotic 78
  dissociation of 40
  inorganic 57
  modelling in solution 84–5
  monoprotic 78
  primary standard, preparation of 90–2
  properties of 58–9
  strengths, investigation of 83–4
  strong 67
  strong acid/strong base titrations 94
  strong acid/weak base titrations 94
  strong base/weak acid titrations 94–5
  triprotic 78
  weak 67, 78
basic buffer 104
basic salts 80
basic substances 57
basicity, monitoring
  window cleaner 65
beta-glucose 179
biocompatibility 180
biodegradable condensation polymers 179–80
biodiesel manufacture 232
bioethanol
  cellulosic biomass, production from 168
biofuels 153–4
biomass 168
black coal (bituminous coal) 153
boiling chips 162
bonding
  intermolecular 126
  intramolecular 126
brine 228
bromides
  ion concentration 197
  solubility rules 49
bromine water test 136–7, 209
bromothymol blue 59, 60, 81, 82
  titrations, and 94
Brønsted, Johannes (1923) 68
Brønsted–Lowry (B/L) theory 68, 69, 73, 78–85
  acid–base pairs 78
  hydrolysis of salts 80–1
  indicators, and 81–5
brown coal (lignite) 153
buffers 104
  laboratory 105
  natural systems, in 104–5
  preparation and testing 105
  properties of 104–6
butanoic acid 99

C

calcium carbonate, decomposition of 6
calcium hydroxide 58
calibration 74
  graph 37

carbon-13 NMR 211–14
carbon–carbon double bonds 209–10
carbon nanotubes 234–5
carbonates
acid–carbonate reactions 63
carbonate ion 78–9
solubility rules 49
carbonyl functional group (CO) 159
carboxyl functional group (COOH) 160–1, 210–11
carboxylic acids 115
carcinogens 49
catalysts, effect of 21, 23
cation analysis 189–93
confirmatory tests 190
identification of cations 191–2
precipitation tests 189
cationic detergent 167
cellulose 179
biopolymers 180
cellulosic biomass 168
chain isomers 118
chemical industries 223, 225–35
co-location 225
economic issues 225
energy 225
transport networks 225
workforce 225–6
chemical reactions
failure to complete 4
reversibility, investigating 7–9
chemical shift 212
chemical synthesis 168–9, 223–5
chlorides
solubility rules 49
chlorofluorocarbons (CFCs) 236
chlorophyll 10
chromophores 201
closed systems 6
coal 129, 153
coke 228
colloids 231
colourimetry 198–9
column chromatography 201
combustion 9–10, 147
computer-aided molecular design (CAMD) 225–6
concentration changes 19
condensation polymers 178–81
biodegradable 179–80
condensation reaction 161, 178
conductivity titrations 95–7
conjugate 78
Contact process 23, 232–3
coordinate bond 57, 68
corrosives 66, 67
cosmetics, production of 231–2
cycasin 49

D

Davy, Humphry (1810) 67
dehydration reaction 147
detergent 167–8, 228
anionic 167
cationic 167
non-ionic 167
dichlorodiphenyltrichloroethane (DDT) 236
dihydrogen phosphate ion 79
dimers 160
diprotic
acids 78
bases 78
dissociation 39, 46, 97
acid dissociation constants 40, 97–100
dissolution 46–8
endothermic 47, 48
exothermic 47, 48
steps 47
drain cleaner, analysis of 100–1
dynamic balance 4
dynamic equilibrium 5
dissolution, and 46
modelling 4–6

E

effervescence 195
efflorescence 25
electroforming 234–5
electrolysis 233
electrolytes 38, 59
electrolytic industries 233–5
electron pair acceptor/electron pair donor theory 68
electroplating 234
elimination testing 189
emulsion 166
emulsions 231
end point 91
endothermic dissolution 47, 48
endothermic reactions 10–11, 18
energy
activation 10
enthalpy of combustion 145–6
enthalpy of formation 9
enthalpy of neutralisation 66
entropy 9
environmental monitoring 186–9
equilibrium
decomposition 33
disturbances in 22–7
dynamic *see* dynamic equilibrium
factors affecting 17–27
heterogeneous, disturbing a 22–3
homogeneous, disturbing a 23–7
position 21, 32
pressure, effects 35
reaction quotient 36
static *see* static equilibrium
temperature change, and 18–19, 23, 35
equilibrium constant 11, 32–3
acids, dissociation of 39–40
bases, dissociation of 40
calculations 33–5
determination of 37–8
homogeneous systems 32–3
ionic compounds, dissociation of 38–9
values, comparison of 33
equilibrium vapour pressure 128
equivalence point 91
esterification 162
esters 161, 162, 231
instrumental analysis 213
nomenclature rules 162
preparation of 161–2, 163
properties and uses 163
ethanol 153
combustion of 4
production of 150–1
ethylene 173
eutrophication 167, 227
exams, sample 240–9
answers 250–6
exothermic dissolution 47, 48
exothermic reactions 9, 11, 18

F

fatty acids 165, 232
feedstock 224
fermentation 150–1
flowchart 151
glucose 151
fertiliser, sulfate ion content 195–6
flammability 128
flash points 128–9
flux 227
forward rate constant 11
forward reaction 6
fossil fuels 152–3
fractions 128
crude oil 130
free radicals 173, 174
fuel production 152–4, 231
functional groups 111, 117
amide 160
amine 159–60
carbonyl 159
carboxyl 160–1, 210–11
hydroxyl 114, 143, 158–9, 210
isomers 119
qualitative investigations of 209–11

G

gas, natural 129
gas pressure changes 19–21, 23
Gibbs free energy 9
gravimetric analysis 195–6
gravitational force 4

H

Haber process 36, 226–7
hair conditioner, manufacture of 231–2
halogen addition reactions 136
halogenated organic compounds 117–18
halogens 117
heterogeneous equilibrium
dissolution, and 46
disturbing a 22–3
high density polyethylene (HDPE) 174–5, 177

homogeneous equilibrium
    disturbing a 23–7
homogeneous systems
    equilibrium constants 32–3
homologous series 111, 143
hybridisation 125
    orbitals 125–6
hydrated cobalt (II) chloride equilibrium 25
hydrated cobalt (II) chloride reactions 7–8
hydration 137
hydrocarbons 111–13, 124–31
    boiling point 127
    hybridisation of orbitals 125–6
    melting point 127
    models 124–5
    nomenclature 111–13
    physical properties 126–8
    production of 129–31
    reactions 136–40
    saturated 111–12, 124
    sources 129–31
    unsaturated 112–13, 209
    uses of 129–31
    volatility 128–9
hydrochloric acid 58, 67
hydrochlorofluorocarbons (HCFCs) 236
hydrocyanic acid 99
hydrofluoric acid 39, 99
hydrogen halide addition reactions 137
hydrogenation 137
hydrolysis of salts 80–1
hydrous ethanol 150
hydroxides
    solubility rules 49
hydroxyl functional group (OH) 114, 143, 210
    properties 158–9

I

indicators 59–62
    acid–base 59–60, 81
    Brønsted Lowry theory, and 81–5
    natural, preparation and testing of 60–1
Indigenous Australian food sources 49
    acidic and basic materials, use of 102–3
infrared absorption frequencies 217
infrared spectroscopy 217–18
instrumental analytical techniques 198–204
intermolecular forces 126
intramolecular forces 126
iodides, solubility rules 49
ionic compounds, dissociation of 38–9
ionic product (IP) 39
ionisation 39, 97, 215
    acid ionisation constant 97
ions
    qualitative analysis 189–95
    quantitative analysis 195–7
iron (III) thiocyanate equilibrium 8, 26–7
irreversible reactions 4
isomers 118–20, 213
    chain 118
    functional group 119
    position 119
isotopes 211

K

ketones 115–16, 159

L

Lavoisier, Antoine (1776) 67
Le Chatelier's principle 17, 24, 105
Lewis, Gilbert (1923) 68
lignite (brown coal) 153
limiting reagent 77
litmus 59, 82
    paper 58
locant 112
low density polyethylene (LDPE) 173–4, 177
    polylactic acid, comparison with 180
Lowry, Thomas (1923) 68

M

macroscopic properties 11
magnesium/steel wool combustion 8–9
mandrel 234
mass spectrometry 215–17
mass-to-charge (m/z) ratio 215
metals
    acid–metal reactions 63–4
methane 143
methyl orange 59, 81, 82
    titrations, and 94
micelles 165
mining zones, location of 223
molecular shapes 125–6
monochromator 199
monomers 168, 173
monoprotic
    acids 78, 97
    bases 78

N

nanorobots 236
neutral salts 80
neutral substances 57
neutralisation 62
    enthalpy 66
    modelling 96
    pH calculations, and 77
    reactions, applications of 64–5
    Solvay process 65
nitrates
    solubility rules 49
nitric acid 58, 78
nitrogen dioxide/dinitrogen tetroxide equilibrium 26
nomenclature
    alcohols 114–15
    aldehydes 115
    amides 117
    amines 116–17
    carboxylic acids 115
    chain isomers 118
    functional group isomers 119
    halogenated organic compounds 117–18
    hydrocarbons 111–13
    isomers 118–20
    ketones 115–16
    position isomers 119
non-equilibrium systems 9–10
non-ionic detergent 167
nylon 178

O

open systems 5
orbitals 125
    hybridisation of 125–6
organic compounds, instrumental analysis 211–18
organochlorine pesticides 235
oxidants 63, 209, 210
oxidation 148
    primary alkanols 148–50
    secondary alkanols 150
oxygen theory of acidity 67

P

paraffins 138
peat 153
pesticides
    organochlorine 236
petroleum 129, 153, 173
    economic issues 131
    environmental issues 131
    sociocultural issues 131
pH
    blood 104–5
    calculations 74, 76–7
    investigating 74
    neutralisation reactions 77
    scale 73–4
pharmaceuticals, production of 229–31
phenolphthalein 59, 81, 82
    titrations, and 94
phosphate
    colourimetric determination 198–9
phosphoric acid 58, 78, 98
photodissociation 236
photon 212
photosynthesis 10
pOH
    calculations 74, 76
polar functional groups 48
polar molecules 48
    solubility 48–9
polyamides 178
polyesters 178
polyethylene (PE) 173–5
    high density polyethylene (HDPE) 174–5, 177
    low density polyethylene (LDPE) 173–4, 177, 180
polylactic acid (PLA) 180
    low density polyethylene, comparison with 180

polymers 173
biodegradable 173
catalytic method 164–5
condensation 178–81
initiator method 163–4
petrochemical 173
properties 177
structure 177
uses 177
polypeptides 235
polystyrene (PS) 175–6, 177
polytetrafluoroethylene (PTFE) 176–7
polyvinyl acetate 173
production of 168–9
polyvinyl chloride (PVC) 173, 175, 177
position isomers 119
potassium hydroxide 58
precipitation titrations 197
pressure changes 19–21, 23
equilibrium, effect on 35
primary standard 90
acid, preparation of 92
base, preparation of 90–2
proton NMR 214–15

R

radical initiator 173, 174
raw materials 224
reaction quotient 36
reactions
acid–base 62–3
acid–carbonate 63
acid–metal 63–4
addition 136–7
alcohols 147–50
condensation 161
dehydration 147
endothermic 10–11, 18
exothermic 9, 11, 18
hydrocarbons 136–40
irreversible 4
modelling 138–40
neutralisation, applications of 64–5
oxidation 148–50
pH calculations, and 77
reaction profiles 93–5
reversible 4, 10–12
substitution 138, 147, 152
reductant 63
refluxing 162
reverse rate constant 11
reversible reactions 4
energy 10–12
rate 10–12
rough titre 91

S

safety issues 128
salt 62
salts
acidic 80
basic 80
hydrolysis of 80–1
neutral 80
saponification 165
saturated hydrocarbons 111–12, 124
saturated solution 51
secondary standard 91
smelting 231
soap 165–7, 227
sodium carbonate 58
production of 227–9
sodium hydrogen carbonate (baking soda) 58
sodium hydroxide 58
acid–base analysis, applications of 100–1
soil acidity
decreasing 65
increasing 64
soil salinity
testing and monitoring 188–9
solubility
precipitation reactions, and 50–1
product constant 51–3
rules 49–51
Solvay process 65, 227–9
neutralisation, and 65
spectrophotometry 200
ultraviolet-visible 199–202
spectroscopy
atomic absorption (AAS) 202–4
infrared 217–18
mass 215–17
spin 211
spin quantum number 211
standard solution 90
static balance 4
static equilibrium 4
modelling 4–6
steel, production of 233
stomach acidity, neutralising 64
strong acid 39
strong acid/strong base titrations 94
strong acid/weak base titrations 94
strong base 67
strong base/weak acid titrations 94–5
substitution reactions 138, 147, 152
sulfates
solubility rules 49
sulfur dioxide, oxidation of 33–4
sulfuric acid 58, 78, 231
production of 232–3
sulfurous acid 97
surfactants 231
swimming pool acidity, adjusting 64

T

temperature change 18–19, 23
dissolution equilibria, and 47
equilibrium, effect on 35
thermal (or steam) cracking 130
thermoplastics 180
titrant 91
titrations 90–3
acid–base analysis, commercial applications of 100–1
back 92–3
conductivity 95–7
modelling 96–7
precipitation 197
steps 90–2
strong acid/strong base 94
strong acid/weak base 94
strong base/weak acid titrations 94–5
titration graphs 93–7
titre 91
total dissolved solids (TDS) 186–7
translational motion 4
triprotic
acids 78
bases 78

U

ultraviolet-visible spectrophotometry 199–202
universal indicator (UI) 60
colours 60, 73
unsaturated hydrocarbons 112–13, 209
unsaturated solution 51
upwards reaction force 4

V

vibrational frequencies 217
vibrational motion 4
vitamin C, analysis of 102
volatile substances 128
volume changes 19–21
volumetric analysis 90

W

water
acidity 187
hardness 187–8
softener 228
testing and monitoring 186
weak acid 39, 67, 78
household, analysis of 103
titrations 94–5
weak base 67, 78
titrations 94–5
wine
acid content of 101–2
white, titration of 103

Z

zeolites 167

Reprinted 2019, 2022, 2025

ISBN 978 1 74125 676 5

Pascal Press
PO Box 250
Glebe NSW 2037
www.pascalpress.com.au

Publisher: Vivienne Joannou
Project editor: Mark Dixon
Edited by Karen Enkelaar
Reviewed and answers checked by Janette Ellis
Indexed by Puddingburn Publishing Services
Cover and page design by Sonia Woo
Typeset by Kim Webber
Printed by Vivar Printing/Green Giant Press

Students
All care has been taken in the preparation of this study guide, but please check with your teacher or the NSW Education Standards Authority about the exact requirements of the course you are studying as these can change from year to year.

The validity and appropriateness of the internet addresses (URLs) in this book were checked at the time of publication. Due to the dynamic nature of the internet, the publisher cannot accept responsibility for the continued validity or content of these web addresses.

DATA SHEET

Chemical shift data (^{13}C-NMR)

Carbon environment	Chemical shift (δ) (ppm)	Carbon environment	Chemical shift (δ) (ppm)
C–C (alkanes)	5–50	C=O (ketones)	205–220
C=C (alkenes)	90–150	C=O (aldehydes)	190–200
C–O (alkanols/esters)	50–90	C–NH_2 (amines)	25–60
C=O (alkanoic acids and esters)	160–185	C–Cl or C–Br (haloalkanes)	10–70

Infrared absorption data

Compound type	Bond vibration	Frequency range (cm^{-1})	Compound type	Bond vibration	Frequency range (cm^{-1})
alkane	C–H stretch	2850–2960	carboxylic acid	O–H stretch	2500–3000
alkane	C–H bend	1370–1480	carboxylic acid	C=O stretch	1700–1725
alkene	C–H stretch	3010–3090	amide	C=O stretch	1640–1690
alkene	C=C stretch	1620–1680	amide	N–H stretch	3100–3500
alkene	C–H bend	900–995	ester	C=O stretch	1735–1750
alcohol	O–H stretch	3610–3640	alkanal	C=O stretch	1720–1740
alcohol	O–H stretch (H-bonded)	3200–3600	alkanone	C=O stretch	1705–1725
alcohol	C–O stretch	1050–1150	amine	N–H stretch	3300–3500
carboxylic acid	C–O stretch	1210–1320	amine	C–N stretch	1080–1360

Solubility product data (K_{sp})

Compound	K_{sp}	Compound	K_{sp}
barium carbonate	2.58×10^{-9}	lead (II) iodide	9.8×10^{-9}
barium sulfate	1.08×10^{-10}	lead (II) sulfate	2.53×10^{-8}
calcium carbonate	3.36×10^{-9}	magnesium hydroxide	5.61×10^{-12}
calcium sulfate	4.93×10^{-5}	silver bromide	5.35×10^{-13}
copper (II) carbonate	1.4×10^{-10}	silver carbonate	8.46×10^{-12}
lead (II) bromide	6.6×10^{-6}	silver chloride	1.77×10^{-10}
lead (II) chloride	1.7×10^{-5}	silver iodide	8.52×10^{-17}

UV absorption (numerical data values vary with chemical environment; a range for C≡C is provided)

chromophore	C–H	C–C	C=C	C≡C	C=O	C–Cl	C–Br
λ_{max} (nm)	122	135	180	173–222	275	173	208

Acid dissociation constant data (K_a)

Acid	K_a	Acid	K_a
formic acid (HCOOH)	1.8×10^{-4}	hydrofluoric acid (HF)	6.8×10^{-4}
acetic acid (CH_3COOH)	1.7×10^{-5}	nitrous acid (HNO_2)	7.1×10^{-4}

EXCEL YEAR 12 CHEMISTRY PASS CARDS

This set of cards is designed for you to use as the final step in your revision program. The author has carefully selected the most important facts of the course for you to focus on just before your exam or test. You can use your ***Excel*** Year 12 Pass Cards:

YOUR PATH TO SUCCESS!

- ✓ **ON THE GO.** If you are by yourself, read over each card again and again until you completely master its content. If you're with a friend, revise as a team by turning the bullet points into questions and quizzing each other on key points. Your answers will be there on the cards.
- ✓ **AT HOME.** Read each card thoroughly and make sure you understand all the points. You should also know more detailed information on each topic—if you are not completely sure of a topic, revise it in your ***Excel*** Year 12 study guide.
- ✓ **ON YOUR DIGITAL DEVICE.** Download a **FREE** digital copy of these cards at **www.pascalpress.com.au/free**

CONTENTS (2)

37 Hydrolysis of salts
38 Indicators and the Brønsted–Lowry theory

QUANTITATIVE ANALYSIS
39 Preparation of a primary standard base
40 Titration
41 Titration calculation
42 Indicator selection for titrations
43 Acid dissociation constant (K_a)
44 Applications of acid–base analysis
45 Buffers
46 Natural buffers

MODULE 7: ORGANIC CHEMISTRY

NOMENCLATURE
47 Alkanes
48 Alkenes and alkynes
49 Alcohols
50 Alkanol classification
51 Carboxylic acids
52 Aldehydes and ketones
53 Amines and amides
54 Halogenated hydrocarbons
55 Isomers

HYDROCARBONS
56 Structural formulas of hydrocarbons
57 Modelling hydrocarbons
58 Physical properties of hydrocarbons
59 Sources and uses of hydrocarbons

PRODUCTS OF REACTIONS INVOLVING HYDROCARBONS
60 Addition reactions
61 Substitution reactions

ALCOHOLS
62 Physical properties of alcohols
63 Enthalpy of combustion of alcohols
64 Reactions of alcohols
65 Oxidation of alcohols
66 Ethanol production by fermentation
67 Types of fuels

REACTIONS OF ORGANIC ACIDS AND BASES
68 Alkanols, alkanals and alkanones
69 Amines and amides
70 Carboxylic acids and esters
71 Esterification
72 Soap
73 Detergents
74 Bioethanol production
75 Polyvinyl acetate production

STATIC AND DYNAMIC EQUILIBRIUM — 1

Equilibria and closed systems

- → In **chemical systems** the particles are always in constant motion at a fixed temperature.
- → In a **static chemical equilibrium**, both forward and backward reactions have come to a halt as no changes occur at the molecular level.
- → In a **dynamic chemical equilibrium** the forward and reverse reactions occur at the same rate.
- → In **chemical systems** reactions may occur in the forward and reverse directions. They are said to be reversible.
- → The **attainment of equilibrium** depends on whether the system is open or closed.
- → A **closed system** is a system in which components cannot escape. Chemical reactions are therefore reversible in closed systems.

See ➲ *Excel* Year 12 Chemistry p. 4

STATIC AND DYNAMIC EQUILIBRIUM — 3

Combustion—a non-equilibrium system

- → When **hydrocarbons undergo complete combustion** in an excess of oxygen, carbon dioxide and water are formed. These reactions are strongly exothermic:

  $CH_4(g) + 2O_2(g) \rightarrow CO_2(g) + 2H_2O(l)$
- → The **release of heat energy in this exothermic reaction** drives the reaction in the forward direction. The reaction is spontaneous as the Gibbs free energy change is negative ($\Delta G < 0$).
- → The **high activation energy for the reverse reaction** reduces the chance that products will recombine to form reactants.
- → As **combustion reactions occur in open systems**, the products can escape into the surroundings and therefore they cannot recombine to produce reactant molecules.

See ➲ *Excel* Year 12 Chemistry p. 9

STATIC AND DYNAMIC EQUILIBRIUM — 5

Rate and energy changes in chemical equilibrium

- → In an **endothermic process** the activation energy for the forward reaction is much greater than the activation energy for the reverse reaction. Heating the system increases the forward rate at which reactants are turned into products. The equilibrium shifts to the right as more products form.
- → In an **exothermic process** the activation energy for the reverse reaction is much greater than for the forward reaction. If the system is heated the rate of the reverse reaction increases and more reactants form. The equilibrium shifts to the left.
- → The **rate of a reaction** is dependent on the frequency of collision between reacting particles. The greater concentration of the reacting particles, the greater the collision frequency. Over time the reacting particles are consumed and their concentration decreases.
- → **Equilibrium** is achieved when the rate of the forward reaction equals the rate of the reverse reaction.

See ➲ *Excel* Year 12 Chemistry p. 11

FACTORS THAT AFFECT EQUILIBRIUM — 7

Gaseous equilibria

- → In **homogeneous gaseous systems**, changes in pressure (caused by changes in the volume of the vessel) will lead to changes in concentrations of reactants or products.
- → In general an **increase in total pressure by volume reduction** causes a gaseous equilibrium to shift to the side of fewer gas molecules. If the number of particles is the same on each side of the equilibrium, then a change in total pressure has no effect.

Example: $N_2O_4(g) \leftrightarrows 2NO_2(g)$

- N_2O_4 is colourless and NO_2 is brown.
- If the volume of the reaction vessel is increased, the total pressure decreases (Boyle's law).
- According to Le Chatelier's principle, the system counteracts the change by shifting to the right to increase the number of particles. The system will become more brown.

See ➲ *Excel* Year 12 Chemistry p. 19

FACTORS THAT AFFECT EQUILIBRIUM — 9

Heterogeneous equilibria 2

- → The **thermal decomposition** of solid calcium carbonate in a closed system is an example of a heterogeneous reaction because the carbon dioxide produced is gaseous:

  $CaCO_3(s) \leftrightarrows CaO(s) + CO_2(g)$
- → The **equilibrium** is endothermic in the forward direction. If the system is heated to a higher temperature, the equilibrium shifts to the right (according to Le Chatelier's principle) to counteract the change and the partial pressure of carbon dioxide increases.
- → The **addition of more solid calcium oxide** or solid calcium carbonate causes no change to the equilibrium position.
- → The **addition of more carbon dioxide** gas raises the gas pressure and causes the equilibrium to shift to the left to counteract the change. The partial pressure of carbon dioxide will decrease until the new equilibrium is established.

See ➲ *Excel* Year 12 Chemistry p. 22

CALCULATING THE EQUILIBRIUM CONSTANT (K_{eq}) — 11

Determining the equilibrium constant

- → The **decomposition equilibrium for ammonia** is:

  $2NH_3(g) \leftrightarrows N_2(g) + 3H_2(g)$
- → The **K_{eq} expression** is $K_{eq} = [N_2][H_2]^3/[NH_3]^2$
- → The following **equilibrium data** was collected at 30 MPa and 427 °C:

$[NH_3]$ (mol/L)	$[N_2]$ (mol/L)	$[H_2]$ (mol/L)
1.42	0.71	2.14

- → The **value of K_{eq}** can now be calculated:

$$K_{eq} = \frac{[0.71][2.14]^3}{[1.42]^2} = 3.5$$

- → **$K_{eq} > 1$.** The equilibrium lies slightly to the right.

See ➲ *Excel* Year 12 Chemistry p. 33

CALCULATING THE EQUILIBRIUM CONSTANT (K_{eq}) — 13

Dissociation of ionic compounds

- → **Ionic compounds** have different solubilities in water. At 25 °C a saturation equilibrium is achieved when the salt reaches its maximum solubility.
- → The **product of the ion concentrations** of the salt is called the *ionic product (IP)*.
- → *IP* increases as more salt dissolves and reaches a constant value once saturation is achieved.
- → For **CaI_2, the dissociation equilibrium** is:

  $CaI_2(s) \leftrightarrows Ca^{2+}(aq) + 2I^-(aq)$

  For a saturated solution of CaI_2, $IP = [Ca^{2+}][I^-]^2 = 0.34$ mol^3 L^{-3}

See ➲ *Excel* Year 12 Chemistry p. 38

STATIC AND DYNAMIC EQUILIBRIUM

Reversibility in chemical reactions

2

- The following **equilibrium** exists in a solution of cobalt (II) chloride:

$$Co(H_2O)_6^{2+}(aq) + 4Cl^-(aq) \leftrightarrows CoCl_4^{2-}(aq) + 6H_2O(l)$$

The $Co(H_2O)_6^{2+}$ ion is pink and the $CoCl_4^{2-}$ ion is blue.

- The **addition of HCl or NaCl solutions** increases the chloride ion concentration so that the equilibrium shifts to the right and the system becomes more blue. A new equilibrium position is established.
- The **addition of more water** causes the equilibrium to shift to the left and the system becomes more pink. A new equilibrium position is established.
- **Repeated changes to the concentration of chloride** ions or water show that this reaction is reversible.

See ↓ *Excel* Year 12 Chemistry p. 7

CONTENTS (3)

POLYMERS
76 Polyethylene
77 Addition polymers—polystyrene
78 Addition polymers—PVC, PS and PTFE
79 Condensation polymers
80 Polylactic acid

MODULE 8: APPLYING CHEMICAL IDEAS

ANALYSIS OF INORGANIC SUBSTANCES
81 Environmental monitoring
82 Cation analysis
83 Flame tests
84 Anion analysis
85 Gravimetric analysis
86 Precipitation titration
87 Colourimetry
88 The Beer–Lambert law
89 UV-visible spectrophotometry 1
90 UV-visible spectrophotometry 2
91 Atomic absorption spectroscopy (AAS) 1
92 Atomic absorption spectroscopy (AAS) 2

ANALYSIS OF ORGANIC SUBSTANCES
93 Functional group analysis
94 Carbon-13 and proton NMR
95 Mass spectrometry
96 Infrared spectroscopy

CHEMICAL SYNTHESIS AND DESIGN
97 The chemical industry
98 The Haber process—ammonia manufacture
99 The Solvay process—sodium carbonate manufacture
100 Aspirin manufacture
101 Personal care products—hair conditioners
102 Renewable fuels—production of biodiesel
103 The Contact process—production of H_2SO_4
104 Electrolytic industries

CONTENTS (1)

MODULE 5: EQUILIBRIUM AND ACID REACTIONS

STATIC AND DYNAMIC EQUILIBRIUM
1 Equilibria and closed systems
2 Reversibility in chemical reactions
3 Combustion—a non-equilibrium system
4 Photosynthesis
5 Rate and energy changes in chemical equilibrium

FACTORS THAT AFFECT EQUILIBRIUM
6 Le Chatelier's principle
7 Gaseous equilibria
8 Heterogeneous equilibria 1
9 Heterogeneous equilibria 2

CALCULATING THE EQUILIBRIUM CONSTANT (K_{eq})
10 The equilibrium constant expression
11 Determining the equilibrium constant
12 Temperature and the equilibrium constant
13 Dissociation of ionic compounds
14 Weak acid dissociation

SOLUTION EQUILIBRIA
15 Solubility
16 Solubility rules
17 Solubility product constant
18 Predicting precipitation

MODULE 6: ACID–BASE REACTIONS

PROPERTIES OF ACIDS AND BASES
19 Common acids and bases
20 Properties of acids and bases
21 Acid–base indicators
22 Acid–base reactions
23 Acid–carbonate reactions
24 Acid–metal reactions
25 Neutralisation applications 1
26 Neutralisation applications 2
27 Enthalpy of neutralisation
28 Theories of acids 1
29 Theories of acids 2
30 Theories of acids 3

USING BRØNSTED–LOWRY THEORY
31 pH
32 pOH
33 Neutralisation and pH
34 Acid–base conjugate pairs
35 Equilibria and the Brønsted–Lowry theory
36 Amphiprotism

FACTORS THAT AFFECT EQUILIBRIUM

Heterogeneous equilibria 1

8

- A **saturated sodium chloride solution** is an example of a heterogeneous equilibrium. Solid salt crystals are in equilibrium with ions in solution:

$$NaCl(s) \leftrightarrows Na^+(aq) + Cl^-(aq)$$

- If drops of **concentrated hydrochloric acid are added** to the saturated solution, the concentration of chloride ions increases. This disturbance shifts the equilibrium to the left to counteract the change. More NaCl crystals form.
- If **additional NaCl crystals are added** to the saturated solution, no change is observed as the sodium and chloride ions are already at their maximum concentration in the water.
- **Adding water** will dilute the concentration of the ions. The equilibrium shifts to the right as more salt crystals dissolve to restore the system to saturation.

See ↓ *Excel* Year 12 Chemistry p. 22

FACTORS THAT AFFECT EQUILIBRIUM

Le Chatelier's principle

6

- **Le Chatelier's principle** states:
  When a change is made to an equilibrium system, the system moves to counteract the imposed change and restore the system to a new equilibrium.
- The **new equilibrium position differs** from the old equilibrium position. The position of an equilibrium refers to the comparative concentration of reactants and products.

Example: $2CrO_4^{2-}(aq) + 2H^+(aq) \leftrightarrows Cr_2O_7^{2-}(aq) + H_2O(l)$

- The chromate ion (CrO_4^{2-}) is yellow and the dichromate ion ($Cr_2O_7^{2-}$) is orange.
- Increasing the concentration of hydrogen ions is a change in the system.
- According to Le Chatelier's principle, to counteract the change the hydrogen ion concentration is decreased by chromate ions reacting with some of the hydrogen ions to produce more dichromate ions. Thus the system becomes more orange. A new equilibrium position is established.

See ↓ *Excel* Year 12 Chemistry p. 17

STATIC AND DYNAMIC EQUILIBRIUM

Photosynthesis

4

- **Photosynthesis** occurs in the chloroplasts of green plant cells. Solar energy is absorbed by chlorophyll molecules inside the chloroplasts.
- **Chlorophyll** selectively absorbs red and violet light in the visible spectrum.
- This energy is used to convert carbon dioxide and water into glucose and oxygen:

$$6CO_2(g) + 6H_2O(l) \rightarrow C_6H_{12}O_6(s) + 6O_2(g)$$

- The **Gibbs free energy** change for photosynthesis is positive ($\Delta G > 0$) and therefore the reaction is non-spontaneous. This reaction requires solar energy to occur.
- The reaction products are lost from the system and this prevents the reverse reaction occurring.

See ↓ *Excel* Year 12 Chemistry p. 10

CALCULATING THE EQUILIBRIUM CONSTANT (K_{eq})

Weak acid dissociation

14

- The **degree of dissociation of a weak acid** can be calculated by measuring the concentration of hydronium ions formed from a solution of the acid of known concentration.
- Consider a weak acid, HA. The **dissociation equilibrium equation** is:

$$HA(aq) + H_2O(l) \leftrightarrows H_3O^+(aq) + A^-(aq)$$

The degree of dissociation can be determined using the equation:

Degree of dissociation = $[H_3O^+]/[HA] \times 100/1\%$

- A **solution of hydrofluoric acid at 25 °C** has a hydronium ion concentration of 0.004 12 mol/L and a HF concentration of 0.0250 mol/L.

The degree of dissociation is $\frac{[0.004\,12]}{[0.0250]} \times \frac{100}{1} = 16.5\%$

See ↓ *Excel* Year 12 Chemistry p. 39

CALCULATING THE EQUILIBRIUM CONSTANT (K_{eq})

Temperature and the equilibrium constant

12

- The **dimerisation equilibrium for nitrogen dioxide** is:

$$2NO_2(g) \leftrightarrows N_2O_4(g)$$

- The following experimental equilibrium data was collected at two temperatures:

Temperature (°C)	NO_2 (mol/L)	N_2O_4(mol/L)	K_{eq}
0	0.370	9.815	71.7
43	2.58	8.71	1.31

- The **equilibrium lies further** to the right at 0 °C.
- The **equilibrium is exothermic** in the forward direction because higher temperatures reduce the value of the equilibrium constant as the equilibrium shifts to the left.

See ↓ *Excel* Year 12 Chemistry p. 35

CALCULATING THE EQUILIBRIUM CONSTANT (K_{eq})

The equilibrium constant expression

10

- The **position or extent of a chemical equilibrium** can be expressed quantitatively using the equilibrium constant (K_{eq}). If the value of K_{eq} is large then the equilibrium lies to the product side. If K_{eq} is small then the equilibrium lies to the reactant side.
- Consider the following **homogeneous gaseous equilibrium**:

$$2A_2(g) + 3B_2(g) \leftrightarrows 2A_2B_3(g)$$

- The **equilibrium constant** for this equilibrium is:

$$K_{eq} = [A_2B_3]^2/[A_2]^2[B_2]^3$$

- If $K_{eq} >> 1$, then the equilibrium lies to the right.
- If $K_{eq} << 1$, then the equilibrium lies to the left.

See ↓ *Excel* Year 12 Chemistry p. 33

Solubility

15

- Ionic and molecular compounds vary in their water solubility. In general solubility increases with increasing temperature.
- For ionic compounds undergoing dissolution the ions are stabilised in the water by attraction between the ions and water dipoles.
- Ionic sulfides, oxides and carbonates are generally insoluble in water, whereas nitrates, chlorides and sulfates are generally soluble.
- Polar organic molecules contain polar functional groups (e.g. OH, COOH, NH_2) that improve solubility, particularly if they can hydrogen bond with water molecules.
- Toxic polar organic compounds found in some plants and fruits can slowly be removed by continued washing of the crushed flesh with water. This technique was used by Indigenous Australians to make various native fruits edible.

See Excel Year 12 Chemistry p. 46

Solubility product constant

17

- For any saturated solution of an ionic solid M_xN_y, the dissolution equilibrium equation is:

  $M_xN_y(s) \leftrightarrows xM^{y+}(aq) + yN^{x-}(aq)$

- The equilibrium constant expression for this solubility equilibrium is called the *solubility product constant*, symbol K_{sp}.

$$K_{sp} = [M^{y+}]^x[N^{x-}]^y$$

Examples:

1 $PbCl_2(s) \leftrightarrows Pb^{2+}(aq) + 2Cl^-(aq)$; $K_{sp} = [Pb^{2+}][Cl^-]^2 = 1.7 \times 10^{-5}$

2 $PbI_2(s) \leftrightarrows Pb^{2+}(aq) + 2I^-(aq)$; $K_{sp} = [Pb^{2+}][I^-]^2 = 9.8 \times 10^{-9}$

PbI_2 is more insoluble than $PbCl_2$ as the K_{sp} is smaller for PbI_2.

See Excel Year 12 Chemistry p. 51

Common acids and bases

19

- Common laboratory acids include:

  hydrochloric acid (HCl); nitric acid (HNO_3); sulfuric acid (H_2SO_4)
- Common laboratory bases include:

  sodium hydroxide (NaOH); potassium hydroxide (KOH);

  ammonia solution ($NH_3(aq)$)
- Common household acids include:

  acetic acid (in vinegar); citric acid in oranges/lemons;

  carbonic acid (in soda water)
- Common household bases include:

  baking soda ($NaHCO_3$); washing soda (Na_2CO_3) in laundry powders; ammonia solution in floor/window cleaners

See Excel Year 12 Chemistry p. 56

Acid–base indicators

21

- Acid–base indicators change colour as the acidity of the surroundings change.
- The following table summarises the changes.

Indicator	High acidity	Low acidity	Neutral	Low alkalinity	High alkalinity
methyl orange	red	orange to yellow	yellow	yellow	yellow
litmus	red	red	purple	blue	blue
bromothymol blue	yellow	yellow	green	blue	blue
phenolphthalein	colourless	colourless	colourless	pink	crimson

See Excel Year 12 Chemistry p. 59

Acid–carbonate reactions

23

- When an acidic solution reacts with a metal carbonate or a metal hydrogen carbonate, a neutralisation reaction occurs. The products of such reactions are carbon dioxide, water and a salt.

Examples:

1 Copper (II) carbonate is neutralised by the nitric acid to form copper (II) nitrate, carbon dioxide and water. The green solid gradually dissolves and a solution of blue copper (II) nitrate forms:

$CuCO_3(s) + 2HNO_3(aq) \rightarrow Cu(NO_3)_2(aq) + H_2O(l) + CO_2(g)$

2 Baking soda (sodium hydrogen carbonate) is a white solid that is neutralised by a solution of citric acid to form a colourless solution of sodium citrate, carbon dioxide and water. This reaction is used in cooking as a rising agent because of the evolved carbon dioxide:

$3NaHCO_3(s) + C_3H_5O(COOH)_3(aq) \rightarrow Na_3(C_3H_5O(COO)_3)(aq) + 3H_2O(l) + 3CO_2(g)$

See Excel Year 12 Chemistry p. 63

Neutralisation applications 1

25

- Neutralisation can be used to reduce the symptoms of gastric acid reflux.
- Commercial antacids are used to neutralise excess acid. These antacids contain weak bases such as magnesium hydroxide, aluminium hydroxide and sodium hydrogen carbonate.
- The neutralisation reactions remove the excess hydrochloric acid that burns the oesophagus during gastric reflux:

  $Mg(OH)_2(s) + 2H_3O^+(aq) \rightarrow Mg^{2+}(aq) + 4H_2O(l)$

  $Al(OH)_3(s) + 3H_3O^+(aq) \rightarrow Al^{3+}(aq) + 6H_2O(l)$

  $NaHCO_3(s) + H_3O^+(aq) \rightarrow Na^+(aq) + 2H_2O(l) + CO_2(g)$

See Excel Year 12 Chemistry p. 64

Enthalpy of neutralisation

27

Example: 50.0 mL of 1.0 mol/L potassium hydroxide solution and 50.0 mL of 1.0 mol/L hydrochloric acid were mixed in a polystyrene foam calorimeter. The mixture had a mass of 0.1038 kg and the temperature increased by 6.5 °C during the neutralisation. (Assume $c = 4.2 \times 10^3$ J/kg/K)

Answer:

$HCl(aq) + KOH(aq) \rightarrow KCl(aq) + H_2O(l)$

$q = mc\Delta T = (0.1038)(4.2 \times 10^3)(6.5) = 2834$ J

$n(HCl) = (1.0)(0.050) = 0.050$ mol;

$n(KOH) = (1.0)(0.050) = 0.050$ mol

$\Delta H = -q/n = \frac{-2834}{0.050} = -5668$ J/mol $= -56.7$ kJ/mol

See Excel Year 12 Chemistry p. 66

Theories of acids 2

29

- Svante Arrhenius (1884)—the hydrogen ion theory
  - Arrhenius observed that acidic solutions conducted electricity just like salt solutions. He proposed that acids released hydrogen ions when they dissolved in water.
  - Arrhenius proposed that alkalinity was due to the presence of hydroxide ions in solution.
  - Arrhenius explained how acid solutions neutralised alkaline solutions in terms of the interaction of the aqueous hydrogen ions and hydroxide ions:

    $OH^-(aq) + H^+(aq) \rightarrow H_2O(l)$
  - Arrhenius's theory also explained the difference between strong acids such as hydrochloric acid (HCl) and weak acids, such as hydrofluoric acid (HF) as being due to their different degrees of dissociation in water.

See Excel Year 12 Chemistry p. 67

pH

31

- The pH of a solution is defined as:

$$pH = -\log_{10}[H^+] = -\log_{10}[H_3O^+]$$

Acidic solutions have a pH < 7. Alkaline solutions have a pH > 7.

Example: Calculate the pH of a 0.00250 mol/L nitric acid solution.

Answer: Nitric acid is a strong monoprotic acid and completely dissociates in water:

$HNO_3(aq) + H_2O(l) \rightarrow H_3O^+(aq) + NO_3^-(aq)$

$[H_3O^+] = [HNO_3] = 0.002\,50$ mol/L

$pH = -\log_{10}[0.002\,50] = 2.60$

See Excel Year 12 Chemistry p. 73

PROPERTIES OF ACIDS AND BASES

Properties of acids and bases

20

Acids	Bases
weak diluted acids taste sour	weak diluted bases taste bitter
turn blue litmus red	turn red litmus blue
neutralise bases	neutralise acids
react with and dissolve active metals with the release of a gas	generally do not react with active metals
react with carbonates with the evolution of carbon dioxide gas	do not react with carbonates

See ➲ *Excel* Year 12 Chemistry p. 59

SOLUTION EQUILIBRIA

Predicting precipitation

18

Question: 20 mL of 0.0010 mol/L magnesium nitrate is added to 20 mL of 0.0010 mol/L sodium carbonate. Determine whether or not precipitation of magnesium carbonate will occur.

$K_{sp}(MgCO_3) = 2 \times 10^{-5}$

Answer:

$n(Mg^{2+}) = (0.0010)(0.020) = 2.00 \times 10^{-5}$ mol

$n(CO_3^{2-}) = (0.0010)(0.020) = 2.0 \times 10^{-5}$ mol

Volume on mixing = 20 + 20 = 40 mL = 0.040 L

$[Mg^{2+}] = \frac{2.0 \times 10^{-5}}{0.040} = 5.0 \times 10^{-4}$ mol/L

$[CO_3^{2-}] = \frac{2.0 \times 10^{-5}}{0.040} = 5.0 \times 10^{-4}$ mol/L

$IP = [Mg^{2+}][CO_3^{2-}] = (5.0 \times 10^{-4})(5.0 \times 10^{-4}) = 2.5 \times 10^{-7}$

$IP < K_{sp}$. Therefore no precipitate of $MgCO_3$ will form.

See ➲ *Excel* Year 12 Chemistry p. 52

SOLUTION EQUILIBRIA

Solubility rules

16

➔ The **solubility rules** are used to identify ions in solution.

Anion	Solubility rule
NO_3^-	All nitrates are soluble.
Cl^-	Most chlorides are soluble except AgCl and $PbCl_2$.
Br^-	Most bromides are soluble except AgBr and $PbBr_2$.
I^-	Most iodides are soluble except AgI and PbI_2.
SO_4^{2-}	Most sulfates are soluble except $BaSO_4$, $PbSO_4$.
CO_3^{2-}	Most carbonates are insoluble except Group 1 carbonates.
OH^-	Most hydroxides are insoluble except Group 1 hydroxides and $Ba(OH)_2$.

➔ All sodium and potassium salts are soluble.

See ➲ *Excel* Year 12 Chemistry p. 49

PROPERTIES OF ACIDS AND BASES

Neutralisation applications 2

26

➔ **Optimal plant growth** depends on the acidity or pH of the soil. Camellias grow best in moderately acidic soils (pH 4.5–5.5) whereas chrysanthemums grow best in mildly acidic soils (pH 6.0–6.3).

➔ **Decreasing soil acidity** can be achieved by adding weakly basic substances that neutralise excess soil acidity. Adding powdered limestone ($CaCO_3$) to an acidic soil reduces the acidity of the soil water as calcium carbonate is a weak base and a neutralisation reaction occurs:

$$CaCO_3(s) + 2H_3O^+(aq) \rightarrow Ca^{2+}(aq) + 3H_2O(l) + CO_2(g)$$

➔ To **monitor the acidity of soil water** during these pH adjustments a wet soil sample is placed in a Petri dish and white barium sulfate is applied to the wet soil surface. The soil water absorbs into the $BaSO_4$ layer. When universal indicator is added to the white layer the colour change reveals the pH of the soil water.

See ➲ *Excel* Year 12 Chemistry p. 64

PROPERTIES OF ACIDS AND BASES

Acid–metal reactions

24

➔ **The reaction of active metals with dilute hydrochloric or sulfuric acid** is an example of an **oxidation–reduction** reaction rather than a neutralisation reaction.

➔ The **metal is oxidised to form a metal ion**. The hydronium ions in the acidic solution are reduced to hydrogen gas. Water is also formed.

Examples:

1 Zinc solid dissolves in dilute hydrochloric acid and a colourless solution of zinc chloride and hydrogen gas are formed:

$$Zn(s) + 2H_3O^+(aq) \rightarrow Zn^{2+}(aq) + H_2(g) + 2H_2O(l)$$

2 Iron solid dissolves slowly in dilute sulfuric acid to form a pale-green solution of iron (II) sulfate. Hydrogen gas is evolved:

$$Fe(s) + 2H_3O^+(aq) \rightarrow Fe^{2+}(aq) + H_2(g) + 2H_2O(l)$$

See ➲ *Excel* Year 12 Chemistry p. 63

PROPERTIES OF ACIDS AND BASES

Acid–base reactions

22

➔ When an **acidic solution reacts with a base**, such as a metal oxide or metal hydroxide, a neutralisation reaction occurs. The products of such acid–base reactions are water and a salt.

Examples:

1 Copper (II) oxide is neutralised by dilute sulfuric acid to form copper (II) sulfate and water. The black copper (II) oxide solid gradually dissolves and the blue copper (II) sulfate product is water soluble:

$$CuO(s) + H_2SO_4(aq) \rightarrow CuSO_4(aq) + H_2O(l)$$

2 Cobalt (II) hydroxide is neutralised by the hydrochloric acid to form cobalt (II) chloride and water. The purple cobalt (II) hydroxide solid gradually dissolves and a solution of pink-red cobalt (II) chloride forms:

$$Co(OH)_2(s) + 2HCl(aq) \rightarrow CoCl_2(aq) + 2H_2O(l)$$

See ➲ *Excel* Year 12 Chemistry p. 62

USING BRØNSTED–LOWRY THEORY

pOH

32

➔ The **pOH of a solution** is defined as:

$$pOH = -\log_{10}[OH^-]$$

Acidic solutions have a pOH > 7. Alkaline solutions have a pOH < 7.

➔ **$[H^+]$, $[OH^-]$ and K_w** (water ionisation constant) are related:

$K_w = 1.0 \times 10^{-14} = [H^+][OH^-]$

➔ **pH and pOH are related:** pH + pOH = 14

Example: Calculate the pH of a 0.008 50 mol/L NaOH solution.

Answer: NaOH is a strong base and fully dissociated in water:

$[OH^-] = [NaOH] = 0.008\,50$ mol/L

$pOH = -\log_{10}[OH^-] = -\log_{10}(0.008\,50) = 2.07$

$pH = 14 - pOH = 14 - 2.07 = 11.93$

See ➲ *Excel* Year 12 Chemistry p. 76

PROPERTIES OF ACIDS AND BASES

Theories of acids 3

30

➔ **Johannes Brønsted and Thomas Lowry (1923)—the proton theory**

- Hydrogen ions are protons. These protons associate with water molecules to form hydronium ions (H_3O^+). B/L acids donate protons and B/L bases accept protons.
- Bases include substances that neutralise acids but which are not classified as alkalis as no OH^- ions are present. In the following equation the hydronium ion donates a proton to the carbonate ion:

$$H_3O^+(aq) + CO_3^{2-}(aq) \rightarrow H_2O(l) + HCO_3^-(aq)$$

➔ **Gilbert Lewis (1923)—the electron pair theory**

- A Lewis acid accepts an electron pair and a Lewis base donates an electron pair.
- When hydrogen ions and hydroxide ions undergo neutralisation, the H^+ ion accepts a pair of non-bonding electrons from the oxygen atom of the OH^- ion to form a covalent bond in the resulting water molecule.

See ➲ *Excel* Year 12 Chemistry p. 68

PROPERTIES OF ACIDS AND BASES

Theories of acids 1

28

➔ **Antoine Lavoisier (1776)—the oxygen theory**

- Lavoisier demonstrated that non-metallic oxides (e.g. sulfur trioxide) dissolved in water made the water acidic.
- Lavoisier proposed that oxygen was the cause of acidity. Lavoisier's theory could not explain why metal oxides were not acidic.

➔ **Humphry Davy (1810)—the hydrogen theory**

- Davy observed that when HCl gas and H_2S gas dissolved in water they produced acidic solutions. These molecules contained no oxygen atoms. Davy's theory attributed acidity to the presence of replaceable hydrogen. These results were not consistent with Lavoisier's oxygen theory.
- Davy's theory could not explain why not all hydrogen compounds were acidic.

See ➲ *Excel* Year 12 Chemistry p. 67

USING BRØNSTED–LOWRY THEORY

Neutralisation and pH

33

- When a **solution of a strong acid is mixed with a strong base solution** in the correct mole ratio the final solution will have a pH of 7. If the acid and base solutions are not mixed in the correct mole ratios, then the final mixture will be either acidic or alkaline.

Example:

Calculate the final pH of a solution formed by mixing 200 mL of 0.100 mol/L HNO_3 and 200 mL of 0.100 mol/L $Ba(OH)_2$.

Answer:

$2HNO_3(aq) + Ba(OH)_2(aq) \rightarrow Ba(NO_3)_2(aq) + 2H_2O(l)$

Stoichiometry = $HNO_3 : Ba(OH)_2 = 2:1$

$n(HNO_3) = (0.100)(0.200) = 0.0200$ mol; $n(Ba(OH)_2) = (0.100)(0.200) = 0.0200$ mol

The nitric acid is limiting and is completely neutralised, leaving an excess of $Ba(OH)_2$.

$n(Ba(OH)_2(\text{excess})) = 0.0100$ mol; $n(OH^-) = 2 \times 0.0100 = 0.0200$ mol

$[OH^-] = \frac{0.0200}{0.400} = 0.0500$ mol/L; pOH = $-\log_{10}(0.0500) = 1.30$; pH = 14 – 1.3 = 12.7

See ➲ *Excel* Year 12 Chemistry p. 77

USING BRØNSTED–LOWRY THEORY

Equilibria and the Brønsted–Lowry theory

35

- When **strong acids and strong bases react** the products are weak conjugate acids and bases. These neutralisation reactions go to completion. The hydronium ion is a strong acid and the hydroxide ion is a strong base. Water is produced when these ions react:

  $H_3O^+(aq) + OH^-(aq) \rightarrow H_2O(l) + H_2O(l)$

- When a **strong acid reacts with a weak base, or a strong base reacts with a weak acid**, the reaction also goes to completion:

  $H_3O^+(aq) + NH_3(aq) \rightarrow H_2O(l) + NH_4^+(aq)$

  $OH^-(aq) + HF(aq) \rightarrow H_2O(l) + F^-(aq)$

- When **weak acids and weak bases react** an equilibrium is established. These reactions do not proceed to completion:

  $NH_3(aq) + HF(aq) \leftrightarrows NH_4^+(aq) + F^-(aq)$

See ➲ *Excel* Year 12 Chemistry p. 78

USING BRØNSTED–LOWRY THEORY

Hydrolysis of salts

37

- **Water** is a common amphiprotic substance. Water can act as a proton donor or a proton acceptor.
- The **salts of weak bases** (e.g. NH_4Cl) hydrolyse in water to form acidic solutions as hydronium ions are formed. The ammonium ion acts as a B/L acid and water acts as a B/L base:

  $NH_4^+(aq) + H_2O(l) \leftrightarrows NH_3(aq) + H_3O^+(aq)$

- The **salts of weak acids** (e.g. $NaCH_3COO$) hydrolyse in water to form alkaline solutions as hydroxide ions are formed. The acetate ion acts as a B/L base and water acts as a B/L acid:

  $CH_3COO^-(aq) + H_2O(l) \leftrightarrows CH_3COOH(aq) + OH^-(aq)$

See ➲ *Excel* Year 12 Chemistry p. 80

QUANTITATIVE ANALYSIS

Preparation of a primary standard base

39

- **Volumetric analysis** is a quantitative laboratory technique used to determine the concentration of a solution by reacting it with another solution of known concentration.
- The **first step in acid–base volumetric analysis** involves the preparation of a solution of known concentrations called the *primary standard*.
- **Anhydrous sodium carbonate** is one of the most commonly used primary standard bases. It is stable, pure and readily dried so that it can be weighed easily and accurately.
- **To prepare 250.0 mL of a 0.100 mol/L solution of sodium carbonate**, 2.65 g of dried Na_2CO_3 is weighed out, dissolved in water and the solution is quantitatively transferred to a 250 mL volumetric flask. Water is added until the base of the meniscus is on the engraved mark in the volumetric flask. The solution is mixed thoroughly.

See ➲ *Excel* Year 12 Chemistry p. 90

QUANTITATIVE ANALYSIS

Titration calculation

41

Example: 25.00 mL of the standard 0.0500 mol/L Na_2CO_3 was titrated with a solution of HNO_3 of unknown concentration using methyl orange indicator. The average of three titres was 23.45 mL. Calculate the concentration of the nitric acid solution.

Answer:

$Na_2CO_3(aq) + 2HNO_3(aq) \rightarrow 2NaNO_3(aq) + H_2O(l) + CO_2(g)$

$n(Na_2CO_3) = c.V = (0.0500)(0.02500) = 1.25 \times 10^{-3}$ mol

Reaction stoichiometry: $Na_2CO_3 : HNO_3 = 1:2$

$n(HNO_3) = 2 \times 1.25 \times 10^{-3} = 2.50 \times 10^{-3}$ mol

$c(HNO_3) = n/V = \frac{(2.50 \times 10^{-3})}{0.02345} = 0.107$ mol/L

See ➲ *Excel* Year 12 Chemistry p. 92

QUANTITATIVE ANALYSIS

Acid dissociation constant (K_a)

43

- The **equation for the dissociation of a weak acid** (HA) in water is:

  $HA(aq) + H_2O(l) \leftrightarrows H_3O^+(aq) + A^-(aq)$

- The equilibrium constant is called the *acid dissociation constant* (K_a):

  $K_a = [H_3O^+][A^-]/[HA]$

- The **smaller the value of K_a**, the weaker the acid:
  - Formic acid (HCOOH): $K_a = 1.8 \times 10^{-4}$
  - Acetic acid (CH_3COOH): $K_a = 1.7 \times 10^{-5}$
- pK_a is defined as $pK_a = -\log_{10}K_a$. The larger the pK_a, the weaker the acid.

See ➲ *Excel* Year 12 Chemistry p. 97

QUANTITATIVE ANALYSIS

Buffers

45

- **Buffers** are solutions that resist large changes in pH on the addition of small quantities of strong acids or bases. Buffer solutions contain a weak Brønsted–Lowry acid and its conjugate base, or a weak Brønsted–Lowry base and its conjugate acid.
- The **greatest buffering effect** is obtained if the buffer solution contains equimolar concentrations of the weak acid and its conjugate base.

  Example:

  Buffer = $CH_3COOH(aq) + NaCH_3COO(aq)$

  Changes: 1 Add drops of NaOH solution.
  2 Add drops of HCl solution.

  Buffering action: 1 The hydroxide ions are neutralised by the acetic acid molecules.
  2 The hydronium ions are neutralised by the acetate ions.

See ➲ *Excel* Year 12 Chemistry p. 104

NOMENCLATURE

Alkanes

47

- **Alkanes** are saturated hydrocarbons that contain only carbon–carbon single bonds. The general formula of alkanes is:

  C_nH_{2n+2} $(n = 1,2,3 \ldots)$

- All alkanes are named using the suffix *–ane*.
- The **first four alkanes** are:

  methane (n = 1) (CH_4); ethane (n = 2) (C_2H_6);
  propane (n = 3) (C_3H_8); butane (n = 4) (C_4H_{10})

- **Methyl groups (CH_3) and ethyl groups (CH_2CH_3)** are common alkyl groups that form branches along an alkane chain.

  Example: $CH_3CH_2CH(CH_3)CH_2CH_2CH_3$ = 3-methylhexane

See ➲ *Excel* Year 12 Chemistry p. 111

NOMENCLATURE

Alcohols

49

- **Alcohols** are organic compounds that contain the hydroxyl functional group (OH).
- **Alkanols** are derivatives of alkanes and their general formula is:

  $C_nH_{2n+1}OH$

- **Alkanols are named** using the *–ol* suffix.
- The *–ol* suffix has priority over alkyl groups in allocating locant positions of functional groups.

Examples:

$CH_3CH(OH)CH_3$: name = propan-2-ol

$CH_3CH(OH)CH_2CH(CH_3)CH_3$: name = 4-methylpentan-2-ol

See ➲ *Excel* Year 12 Chemistry p. 114

USING BRØNSTED–LOWRY THEORY 38
Indicators and the Brønsted–Lowry theory

→ **Acid–base indicators** are weak acids. Acid–base indicators exist in two forms, the unionised Brønsted–Lowry acid (HIn) and its conjugate base (In^-):

$$HIn(aq) \rightleftarrows H^+(aq) + In^-(aq)$$

→ **The HIn indicator molecule** has a different colour to the In^- ion.

Indicator	HIn	In^-	pH range
methyl orange	red	yellow	3.1 – 4.4
bromothymol blue	yellow	blue	6.0 – 7.6
phenolphthalein	colourless	crimson	8.3 – 10.0

Example: Methyl orange is red when the pH < 3.1 and yellow when the pH > 4.4. In the pH range 3.1–4.4 the indicator is orange.

See ⮌ *Excel* Year 12 Chemistry p. 81

USING BRØNSTED–LOWRY THEORY 36
Amphiprotism

→ In the **Brønsted–Lowry theory** some **species are amphiprotic**. This means they can behave as either a Brønsted–Lowry acid or a Brønsted–Lowry base, depending on the other chemical substances present.

Example: The hydrogen phosphate ion (HPO_4^{2-})

- HPO_4^{2-} is a B/L acid:

  $HPO_4^{2-}(aq) + OH^-(aq) \rightarrow PO_4^{3-}(aq) + H_2O(l)$
- HPO_4^{2-} is a B/L base:

  $HPO_4^{2-}(aq) + H_3O^+(aq) \rightarrow H_2PO_4^-(aq) + H_2O(l)$

See ⮌ *Excel* Year 12 Chemistry p. 79

USING BRØNSTED–LOWRY THEORY 34
Acid–base conjugate pairs

→ **In the Brønsted–Lowry theory** each acid has a corresponding base called the **conjugate base**. Each Brønsted–Lowry (B/L) base has a corresponding conjugate acid.

→ A **strong Brønsted–Lowry (B/L)** acid has a weak conjugate base (e.g. HCl/Cl^-).

→ A **strong Brønsted–Lowry base** has a weak conjugate acid (e.g. OH^-/H_2O).

Examples:

B/L acid	Conjugate base	B/L base	Conjugate acid
H_2SO_4	HSO_4^-	SO_4^{2-}	HSO_4^-
H_3O^+	H_2O	NH_3	NH_4^+
H_3PO_4	$H_2PO_4^-$	PO_4^{3-}	HPO_4^{2-}

See ⮌ *Excel* Year 12 Chemistry p. 78

QUANTITATIVE ANALYSIS 44
Applications of acid–base analysis

→ **Acid–base titrations** can be used to determine the concentration of sodium hydroxide in a liquid drain cleaner.

→ A **pH meter and glass electrode** can be used to determine the acidity of wine.

→ The **ascorbic acid content of Vitamin C tablets** can be determined by titrating a solution of the Vitamin C tablet with standardised NaOH using phenolphthalein indicator. Chemists can also determine the Vitamin C content of traditional foods used by Indigenous Australians (e.g. Quandong fruits).

→ A **pH titration** can be performed on samples of diluted vinegar to determine the acetic acid content.

See ⮌ *Excel* Year 12 Chemistry p. 100

QUANTITATIVE ANALYSIS 42
Indicator selection for titrations

→ To select the **most appropriate indicator for the titration** the pH at the equivalence point must be matched to the pH range in which the indicator changes colour. The end point of a titration occurs when the indicator just changes colour.

→ The **indicator must change colour** when the pH has its greatest rate of change during the titration.

Titration	Suitable indicator(s)
strong acid–strong base	methyl orange; phenolphthalein
strong acid–weak base	methyl orange
weak acid–strong base	phenolphthalein

See ⮌ *Excel* Year 12 Chemistry p. 93

QUANTITATIVE ANALYSIS 40
Titration

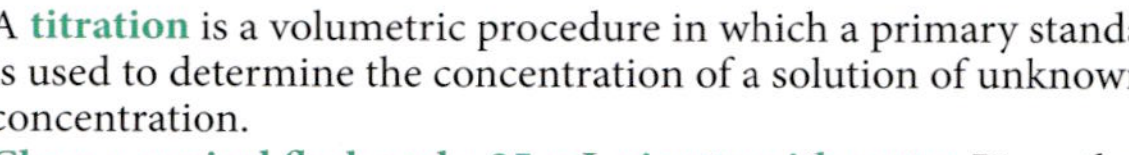

→ A **titration** is a volumetric procedure in which a primary standard is used to determine the concentration of a solution of unknown concentration.

→ **Clean a conical flask and a 25 mL pipette with water.** Rinse the pipette with the standard sodium carbonate solution. Fill the pipette with the sodium carbonate solution and transfer the solution to the conical flask.

→ **Clean a 50 mL burette with water** and then rinse the burette with the acid solution. Discard the washings and fill the burette with the acid (e.g. HCl) solution.

→ **Add four drops of the appropriate indicator** (e.g. methyl orange) to the flask and add the acid to the flask until the indicator turns from yellow to orange. The volume of acid is recorded. This volume (called the *titre*) is used to calculate the acid concentration.

See ⮌ *Excel* Year 12 Chemistry p. 90

NOMENCLATURE 50
Alkanol classification

→ **Alkanols** can be classified as primary, secondary or tertiary. This classification is based on the position of the hydroxyl functional group in the alkanol molecule (R_1, R_2, R_3 = alkyl groups).

- Primary alkanol: R_1CH_2OH
- Secondary alkanol: R_1R_2CHOH
- Tertiary alkanol: $R_1R_2R_3COH$

Examples: Figure PC.1 shows the structural formulas of three alkanols.

Primary	Secondary	Tertiary
propan-1-ol	butan-2-ol	2-methylbutan-2-ol

Figure PC.1

See ⮌ *Excel* Year 12 Chemistry p. 114

NOMENCLATURE 48
Alkenes and alkynes

→ **Alkenes** have a double carbon–carbon bond and alkynes have a triple carbon–carbon bond. They are unsaturated hydrocarbons.

→ **Alkene general formula** = C_nH_{2n};

alkyne general formula = C_nH_{2n-2}

→ **Alkenes are named** using the *–ene* suffix.

Alkynes use the *–yne* suffix.

Examples:

$CH_3CH_2CHCHCH_2CH_2CH_3$: name = hept-3-ene

$CH_3CH_2C(CH_3)CHCH_2CH_3$: name = 3-methylhex-3-ene

$CHCCH_2CH_2CH_3$: name = pent-1-yne

See ⮌ *Excel* Year 12 Chemistry p. 112

QUANTITATIVE ANALYSIS 46
Natural buffers

→ The **pH of our blood** must remain in the range 7.35 to 7.45. Buffers are used to maintain blood pH. One important buffer system in the blood is the carbonic acid–hydrogen carbonate buffer.

→ The **buffering equations** are:

1 $CO_2(g) + H_2O(l) \rightleftarrows H_2CO_3(aq)$

2 $H_2CO_3(aq) + H_2O(l) \rightleftarrows HCO_3^-(aq) + H_3O^+(aq)$

→ By **breathing deeper and faster** the carbon dioxide is removed from the blood and the pH rises as both equilibria shift to the left and the hydronium ion concentration decreases.

→ The **kidneys excrete HCO_3^- into the blood**. By altering the level of this conjugate base the pH of the blood can be stabilised over a narrow range.

See ⮌ *Excel* Year 12 Chemistry p. 104

NOMENCLATURE

Carboxylic acids

51

- **Carboxylic acids** are organic molecules containing the carboxyl (COOH) functional group. This group is always on the first carbon atom in the chain.
- **Alkanoic acids are carboxylic acids** formed from alkanes.
- The **general formula of alkanoic acids** is:

  $C_nH_{2n+1}COOH \quad (n = 0,1,2,3 \ldots)$

- **Alkanoic acids are named** using the *–oic acid* suffix. The preferred IUPAC name for the first two alkanoic acids is their common name rather than their systematic name.

Examples:
HCOOH: name = formic acid (or methanoic acid)
CH_3COOH: name = acetic acid (or ethanoic acid)
$CH_3CH_2CH_2CH_2COOH$: name = pentanoic acid

See Excel Year 12 Chemistry p. 115

NOMENCLATURE

Amines and amides

53

- **Amines** are organic molecules containing the amino (NH_2) functional group.
  The general formula for primary amino alkanes is:

  $C_nH_{2n+1}NH_2 \quad (n = 1,2,3 \ldots)$

  Primary amino alkanes are named using the *–amine* suffix.
  Example: $CH_3CH_2CH_2CH_2NH_2$: name = butan-1-amine
- **Amides** are derivatives of carboxylic acids. The OH group of the carboxylic acid is replaced by an amine (NH_2) group. The general formula for amides is:

  $C_nH_{2n+1}CONH_2 \quad (n = 0,1,2 \ldots)$

- Amides are named using the *–amide* suffix.
  Example: $CH_3CH_2CH_2CONH_2$: name = butanamide

See Excel Year 12 Chemistry p. 116

NOMENCLATURE

Isomers

55

- **Isomers** are molecules with the same molecular formula but different structural formulas.
- 1 **Chain isomers**—these isomers differ due to chain branching:
  $CH_3CH_2CH_2CH_2CH_2CH_3$ (hexane)
  $CH_3CH_2CH(CH_3)CH_2CH_3$ (3-methylpentane)
- 2 **Position isomers**—these isomers differ due to different positions of the functional groups:
  $CH_3CHFCH_2CH_2F$ (1,3-difluorobutane)
  $CH_2FCH_2CH_2CH_2F$ (1,4-difluorobutane)
- 3 **Functional group isomers**—these isomers differ due to different functional groups:
  $CH_3CH_2COCH_3$ (butanone); $CH_3CH_2CH_2CHO$ (butanal)

See Excel Year 12 Chemistry p. 118

HYDROCARBONS

Modelling hydrocarbons

57

- **Hydrocarbons** are three-dimensional molecules (see Figure PC.3).
- In **alkanes,** carbon–carbon single bonds are formed by the overlap of sp^3 hybrid orbitals. This results in a tetrahedral arrangement of atoms around each carbon.
- In **alkenes,** carbon–carbon double bonds are formed by the overlap of sp^2 hybrid orbitals and p orbitals. This results in a planar arrangement of atoms around the double bond.
- In **alkynes,** carbon–carbon triple bonds are formed by the overlap of sp hybrid orbitals and two sets of p orbitals. This results in a linear arrangement of atoms around the triple bond.

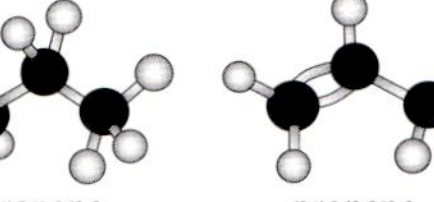

Figure PC.3

See Excel Year 12 Chemistry p. 125

HYDROCARBONS

Sources and uses of hydrocarbons

59

- The **majority of hydrocarbons** are extracted from organic sources such as petroleum and natural gas located in geological structures. Petroleum is a fossil fuel formed by the decay of the remains of plankton and algae over long periods of geological time.
- The **petroleum is** a mixture of crude oil with dissolved hydrocarbon gases. The components of crude oil are separated into fractions containing mixtures of various hydrocarbons.
- The **petrol fraction** (7–9 carbon atoms) is used as a fuel for cars. The kerosene fraction (11–16 carbon atoms) is used as aviation fuel and a solvent. Heavy fractions with greater than 20 carbon atoms are used as waxes, greases and bitumen surfacing of roads.
- **Hydrocarbons release carbon dioxide** on combustion. The increase in atmospheric levels of CO_2 over the last 200 years has led to increases in global warming.

See Excel Year 12 Chemistry p. 129

PRODUCTS OF REACTIONS INVOLVING HYDROCARBONS

Substitution reactions

61

- **Alkanes** are relatively unreactive towards many common chemical reagents.
- UV light is used to make alkanes reactive towards molecules such as chlorine. The UV creates reactive free radicals. In **substitution reactions** the halogen atoms replace hydrogen atoms:

  $CH_4(g) + Cl_2(g) + UV \rightarrow CH_3Cl(l) + HCl(g)$

- The **chlorinated products** are now more reactive and other compounds can be synthesised from them. Alcohols can be manufactured from these chlorinated products. Methanol can be manufactured from chloromethane:

  $CH_3Cl(l) + KOH(aq) \rightarrow CH_3OH(aq) + KCl(aq)$

See Excel Year 12 Chemistry p. 138

ALCOHOLS

Enthalpy of combustion of alcohols

63

- **Calorimetry experiments** can be used to determine the enthalpy of combustion of alcohols.
- The **calorimeter consists** of a Pyrex or metal beaker containing a known mass of water.
- The **alcohol is placed in a weighed spirit burner**. The burning alcohol heats the water and the temperature rise and the mass of alcohol burnt is determined.
- **The results obtained for $\Delta_c H$ using different alkanols** are much lower than the published data because considerable heat is lost to the surroundings and the combustions are incomplete.
- The **longer the hydrocarbon chain of the alkanols**, the more incomplete is the combustion. The flames become increasingly smoky due to the formation of soot:

  $C_5H_{11}OH(l) + 6O_2(g) \rightarrow 3CO_2(g) + CO(g) + C(s) + 6H_2O(l)$

See Excel Year 12 Chemistry p. 145

ALCOHOLS

Oxidation of alcohols

65

- **Primary alkanols** can be oxidised by a mild oxidant (e.g. H^+/CrO_3) to form alkanals:

  $C_3H_7OH(l) \rightarrow CH_3CH_2CHO(l)$ (product = propanal)

- **Strong oxidants** (e.g. H^+/MnO_4^-) oxidise primary alkanols to form alkanoic acids:

  $C_3H_7OH(l) \rightarrow CH_3CH_2COOH(l)$ (product = propanoic acid)

- **Secondary alkanols** can be oxidised by strong oxidants to form alkanones:

  $CH_3CH(OH)CH_2CH_2CH_3 \rightarrow CH_3COCH_2CH_2CH_3$ (product = pentan-2-one)

- **Tertiary alkanols** cannot be oxidised by these oxidants.

See Excel Year 12 Chemistry p. 148

ALCOHOLS

Types of fuels

67

- **Petrol, kerosene, diesel and natural gas** are fossil fuels because they are derived from petroleum that formed over millions of years from the fossilised remains of marine organisms.
- **Coal** is also a fossil fuel that is used to generate electricity in many power stations. Coal is the fossilised remains of swamp plants.
- **Ethanol and biodiesel** are examples of biofuels because they are derived from biological materials such as plant matter.
- Ethanol is a 10% component of E10 petrol.
- **Biodiesel** is a biofuel manufactured from organic sources such as fats and oils.

See Excel Year 12 Chemistry p. 152

HYDROCARBONS

Structural formulas of hydrocarbons

56

- The **structural formulas of hydrocarbons** represent the arrangement of atoms in the molecules in two dimensions. Figure PC.2 shows the structural formulas of an alkane, an alkene and an alkyne.

```
   H  H  H  H  H           H     H  H  H  H            H  H        H  H  H
   |  |  |  |  |           |     |  |  |  |            |  |        |  |  |
H—C—C—C—C—C—H       H—C—C=C—C—C—C—H       H—C—C—C≡C—C—C—C—H
   |  |  |  |              |  |  |  |  |               |  |        |  |
   H  H  H  CH3H           H  H  H  CH3H  H            H  H        CH3H  H
```

2-methylpentane   4-methylhex-2-ene   5-methylhept-3-yne

Figure PC.2

Download a FREE digital copy of these cards at www.pascalpress.com.au/free

See Excel Year 12 Chemistry p. 124

NOMENCLATURE

Halogenated hydrocarbons

54

- **Halogens** can bond to hydrocarbon molecules to form halohydrocarbons.
- **Halogens are functional groups.** The halogen functional groups are:
  Br = bromo; Cl = chloro; F = fluoro; I = iodo
  The halogen functional groups are named using prefixes in the name of the hydrocarbon molecule.
- **The halogens are named alphabetically** in a halogenated hydrocarbon.

Examples:

$CH_3CH_2CHI_2$: name = 1,1-diiodopropane

$CH_3CHFCBr_2CH_2CH_2CH_3$: name = 3,3-dibromo-2-fluorohexane

$CHF_2CH_2CBr_2CH_2CHCl_2$: name = 3,3-dibromo-1,1-dichloro-5, 5-difluoropentane

See Excel Year 12 Chemistry p. 117

NOMENCLATURE

Aldehydes and ketones

52

- The carbonyl group (C=O) is present in both aldehydes and ketones. Aldehydes and ketones formed from alkanes are called **alkanals** and **alkanones**.
- **Aldehydes** have the carbonyl group at the first (terminal) carbon. Thus the functional group for aldehydes is CHO. Ketones have the carbonyl group on non-terminal carbon atoms in the chain.
- **Alkanals have the general formula** RCHO. Alkanones have the general formula R_1R_2CO.
- **Alkanals are named** using the *–al* suffix. Alkanones are named using the *–one* suffix.

Examples:

$CH_3CH_2CH_2CH_2CH_2CHO$: name = hexanal

$CH_3CH_2COCH_2CH_2CH_2CH_3$: name = heptan-3-one

See Excel Year 12 Chemistry p. 115

ALCOHOLS

Physical properties of alcohols

62

- **Alcohols** are organic compounds that contain the hydroxyl functional group (OH).
- **Alkanols** are polar molecules due to the polar hydroxyl group. The boiling points of alkanols increases with increasing chain length.
- **Alkanols** are less volatile and much more water soluble than the corresponding alkane. Apart from dipole–dipole attraction, hydrogen bonding is also a significant intermolecular force for short chain alkanols.
- **Ethanol** is an important solvent. The non-polar ethyl (C_2H_5) group in ethanol assists some low-molecular weight non-polar solutes to dissolve in ethanol. Ethanol–water mixtures are used in many pharmaceutical products.

See Excel Year 12 Chemistry p. 143

PRODUCTS OF REACTIONS INVOLVING HYDROCARBONS

Addition reactions

60

- **Unsaturated hydrocarbons** have reactive double or triple carbon–carbon bonds. These bonds have high electron density and this property promotes reactions with other molecules such as halogens that have high electronegativity.
- **Alkenes** react with chlorine to form chlorinated alkanes. Alkenes react with hydrogen bromide to form brominated alkanes via addition reactions.
- Alkenes react with hydrogen (over a nickel catalyst) to form alkanes.

Examples:

$CH_2CH_2 + Cl_2 \rightarrow CH_2ClCH_2Cl$; name of product = 1,2-dichloroethane

$CH_2CH_2 + HBr \rightarrow CH_3CH_2Br$; name of product = bromoethane

$CH_2CH_2 + H_2 \rightarrow CH_3CH_3$; name of product = ethane

See Excel Year 12 Chemistry p. 136

HYDROCARBONS

Physical properties of hydrocarbons

58

- **Hydrocarbons** are non-polar molecules. The dispersion intermolecular forces between these molecules are caused by the interaction of temporary dipoles and induced dipoles.
- These dispersion forces are much weaker than the intramolecular chemical covalent bonds joining the atoms together within the molecule.
- The **boiling points of hydrocarbons** increase as the molecular weight of the molecules increases. The larger the molecule, the greater the dispersion forces.
- In general the boiling points of straight chain alkanes and alkenes with the same number of carbon atoms per molecule decrease in the order alkanes > alkenes > alkynes.
- **Low molecular weight hydrocarbons** are volatile and pose a safety risk during use and storage. Sparks can cause fires when exposed to hydrocarbon vapour–air mixtures.

See Excel Year 12 Chemistry p. 126

REACTIONS OF ORGANIC ACIDS AND BASES

Alkanols, alkanals and alkanones

68

- **Alkanols** are molecules containing the hydroxyl (OH) functional group. Alkanols are neutral, polar molecules. Low molecular weight alkanols are water soluble.
- Primary alkanols can be oxidised by strong oxidants to form alkanoic acids. Secondary alkanols can be oxidised to form alkanones.
- **Alkanals** are molecules containing the carbonyl (CO) functional group on the terminal carbon. Alkanals can be formed by mild oxidation of primary alkanols. Alkanals can also be oxidised by strong oxidants to form alkanoic acids.
- **Alkanones** are molecules containing the carbonyl (CO) functional group on the non-terminal carbon atoms. Alkanones can be formed by strong oxidation of secondary alkanols. Alkanones are not oxidised further.

See Excel Year 12 Chemistry p. 158

ALCOHOLS

Ethanol production by fermentation

66

- **Sugarcane** is grown for sugar (sucrose) production. A by-product is a dark-brown syrup called *molasses*. The sucrose sugar ($C_{12}H_{22}O_{11}$) in the molasses is first hydrolysed to produce simple sugars such as glucose ($C_6H_{12}O_6$).
- **Glucose** can then be anaerobically fermented at 37 °C by genetically modified yeast to produce ethanol (C_2H_5OH):

  $C_6H_{12}O_6(aq) \rightarrow 2C_2H_5OH(l) + 2CO_2(g)$

- **Fractional distillation of the fermented solution** produces hydrous ethanol (95% ethanol).
- **Anhydrous ethanol** is produced by removing the remaining water using absorbent solids (e.g. zeolites).

See Excel Year 12 Chemistry p. 150

ALCOHOLS

Reactions of alcohols

64

Substitution reaction:

- Alkanols undergo substitution reactions when they react with hydrogen halides. An acid catalyst is required:

  $CH_3CH_2OH(l) + HBr(g) \rightarrow CH_3CH_2Br(l) + H_2O(l)$

Dehydration reaction:

- When alkanols are dehydrated, alkenes are formed. Concentrated sulfuric acid is a common dehydrating agent:

  $C_3H_7OH(l) \rightarrow C_3H_6(g) + H_2O(l)$

Complete combustion:

- Alkanols undergo complete combustion in excess oxygen to produce CO_2 and H_2O:

  $2CH_3OH(l) + 3O_2(g) \rightarrow 2CO_2(g) + 4H_2O(l)$

See Excel Year 12 Chemistry p. 147

REACTIONS OF ORGANIC ACIDS AND BASES 69
Amines and amides

- **Amines** are polar organic molecules containing the amine (NH_2) functional group.
- **Amines** are weak bases in aqueous solution. They react with strong acids to form salts. Methanamine forms methylammonium chloride when reacted with HCl:

  $CH_3NH_2(aq) + HCl(aq) \rightarrow CH_3NH_3Cl(aq)$
- **Amides** are strongly polar organic molecules containing the amide ($CONH_2$) functional group.
- The boiling points of amides are considerably higher than the equivalent alkanol or alkanoic acids.
- Amides are very weak bases in water solution.

MODULE 7: ORGANIC CHEMISTRY See Excel Year 12 Chemistry p. 159

REACTIONS OF ORGANIC ACIDS AND BASES 71
Esterification

- **Esterification** is the procedure used to make an ester. The reaction is kinetically slow and the yield of ester is quite variable.
- The **rate of the reaction** can be increased by moderate heating (using an electric heating mantle), as well as by adding a catalyst such as concentrated sulfuric acid.
- The **equilibrium yield of ester** can be increased by adding an excess of one reactant.
- The **reaction is performed** using a reflux condenser that condenses the hot vapours and returns them to the reaction flask so that the reaction can reach equilibrium.
- The **condenser is open** to the atmosphere so dangerous vapour pressure build-up is avoided. Boiling chips in the flask prevent large vapour bubbles from forming that can cause vibrations in the apparatus.

MODULE 7: ORGANIC CHEMISTRY See Excel Year 12 Chemistry p. 162

REACTIONS OF ORGANIC ACIDS AND BASES 73
Detergents

- **Detergents** are cleaning agents. Most detergents are manufactured from petrochemicals. Like soap they have long hydrocarbon tails and a charged or polar head group.
- Detergents are classified into three main categories:
  - Anionic detergents: the head group is anionic (e.g. $-C_6H_4SO_3^-$)
  - Cationic detergents: the head group is cationic (e.g. $-N(CH_3)_3^+$)
  - Non-ionic detergents: the head group is polar (e.g. $-(CH_2CH_2O)_n$)
- **Anionic detergents** are widely used in laundry detergents, dishwashing liquids and oven cleaners as they are strongly foaming.
- **Cationic detergents** are used as fabric softeners and hair and fabric conditioners.
- **Non-ionic detergents** are widely used in automatic dishwashing detergents and cosmetics.

MODULE 7: ORGANIC CHEMISTRY See Excel Year 12 Chemistry p. 167

REACTIONS OF ORGANIC ACIDS AND BASES 75
Polyvinyl acetate production

- **Polyvinyl acetate** is a synthetic polymer used for wood glues and other adhesives. It is also used in the preparation of latex paints and in the lamination of metal foils.
- The **steps used in its industrial production** from crude oil are summarised below:
  - The naphtha fraction of crude oil is extracted by fractional distillation.
  - The naphtha fraction undergoes thermal (steam) cracking to produce smaller molecules, including reactive ethylene gas (C_2H_4).
  - The ethylene gas is reacted with acetic acid and oxygen (using a palladium catalyst) to form the vinyl acetate ($CH_3CO_2CHCH_2$) monomer.
  - The vinyl acetate monomer undergoes free radical addition polymerisation to form polyvinyl acetate.

MODULE 7: ORGANIC CHEMISTRY See Excel Year 12 Chemistry p. 168

POLYMERS 77
Addition polymers—polystyrene

- **Polystyrene** (PS) is an important commercial polymer. It is manufactured from the styrene monomer ($CH_2CHC_6H_5$). The IUPAC name for this monomer is phenylethylene.
- **Commercial PS polymers** typically vary in length from 10 000 to 40 000 monomer units. The polymer is quite rigid due to the presence of the large phenyl group. The phenyl group (C_6H_5) consists of a hexagonal ring of carbon atoms in which hybridised orbitals form a ring of electrons above and below the plane of the ring.
- Polystyrene is a transparent, hard plastic. The high stiffness of PS makes it suitable for car battery cases, plastic cutlery and handles for tools. Its high refractive index makes it suitable for clear containers such as plastic drinking glasses and packaging of CDs and DVDs.
- **Polystyrene can be converted into polystyrene foam** by heating and using blowing agents such as propane gas to expand the polymer droplets and create holes in the plastic. The polystyrene foam is white and has a very low density. It is an excellent thermal insulator.

MODULE 7: ORGANIC CHEMISTRY See Excel Year 12 Chemistry p. 175

POLYMERS 79
Condensation polymers

- **Condensation polymerisation** involves monomers condensing together with the elimination of a smaller molecule such as water.
- **Polyesters** are examples of condensation polymers. Polyesters form when an alkandiol monomer bonds (via an ester linkage (COO)) with a dialkanoic acid monomer. The monomers are linked by the ester functional group. Polyester fibres are strong and elastic and are extensively used in the production of textiles.

  A common polyester used to make textiles is PET (polyethylene terephthalate; see Figure PC.6).

$$\left[-\overset{O}{\overset{\|}{C}}-C_6H_4-\overset{O}{\overset{\|}{C}}-O-CH_2-CH_2-O- \right]_n$$

PET

Figure PC.6

MODULE 7: ORGANIC CHEMISTRY See Excel Year 12 Chemistry p. 178

ANALYSIS OF INORGANIC SUBSTANCES 81
Environmental monitoring

- **Environmental monitoring** is critical to knowing whether the quality of our environment is getting better or worse. The systematic sampling of air, water, soil and living organisms are conducted in order to observe and study the environment.
- **Water quality is monitored** to ensure the water is fit to drink. Chemists measure water acidity, total dissolved solids, hardness and dissolved oxygen levels in this monitoring process.
- **Air quality is monitored** to determine the levels of polluting oxides such as CO, NO_2 and SO_2 produced by vehicles and industries.
- **Soil quality** is tested to determine the levels of contamination of heavy metals and organic pesticides discharged by industries and farmers into the environment.

MODULE 8: APPLYING CHEMICAL IDEAS See Excel Year 12 Chemistry p. 186

ANALYSIS OF INORGANIC SUBSTANCES 83
Flame tests

- When **aqueous solutions of various metal ions are atomised** in a blue (non-luminous) Bunsen burner flame, they produce a characteristic colour in the flame. A clean platinum wire loop is dipped into a solution of a salt and then held in the flame (see Figure PC.8).

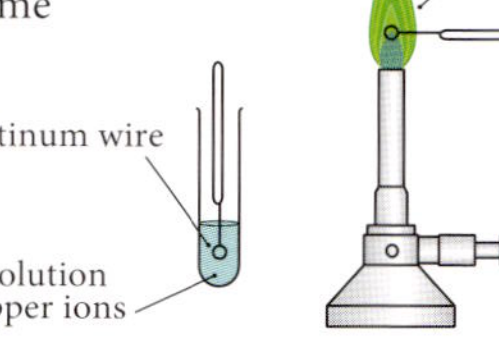

Figure PC.8

- **Flame colour examples:**

  Ba^{2+} = pale yellow-green; Cu^{2+} = green; Ca^{2+} = orange-red

MODULE 8: APPLYING CHEMICAL IDEAS See Excel Year 12 Chemistry p. 190

ANALYSIS OF INORGANIC SUBSTANCES 85
Gravimetric analysis

- **The composition of a mixture** can be determined by gravimetric analysis involving a precipitation reaction. The sulfate concentration of a fertiliser can be found this way.
- **A sample of fertiliser** is weighed and dissolved in water. After filtration the mixture is acidified and excess barium nitrate is added to precipitate $BaSO_4$. The precipitate is filtered, washed and dried to constant weight. The percentage by weight of SO_4^{2-} in the fertiliser can be calculated.

Example: mass (fertiliser) = 1.85 g; mass ($BaSO_4$) = 0.475 g

$n(BaSO_4) = \frac{0.475}{233.37} = 2.04 \times 10^{-3}$ mol; $n(SO_4^{2-})$ in fertiliser $= 2.04 \times 10^{-3}$ mol

$m(SO_4^{2-}) = (2.04 \times 10^{-3})(96.07) = 0.196$ g

$\%SO_4^{2-}$ in fertiliser $= \left(\frac{0.196}{1.85}\right) \times \frac{100}{1} = 10.6\%$w/w

MODULE 8: APPLYING CHEMICAL IDEAS See Excel Year 12 Chemistry p. 195

REACTIONS OF ORGANIC ACIDS AND BASES 74

Bioethanol production

- **Cellulosic biomass** can be used to produce bioethanol fuel.
- **The bioethanol fuel** can be produced by first converting biomass into glucose. The glucose undergoes anaerobic fermentation to produce ethanol. Figure PC.4 illustrates the process in the form of a flowchart.

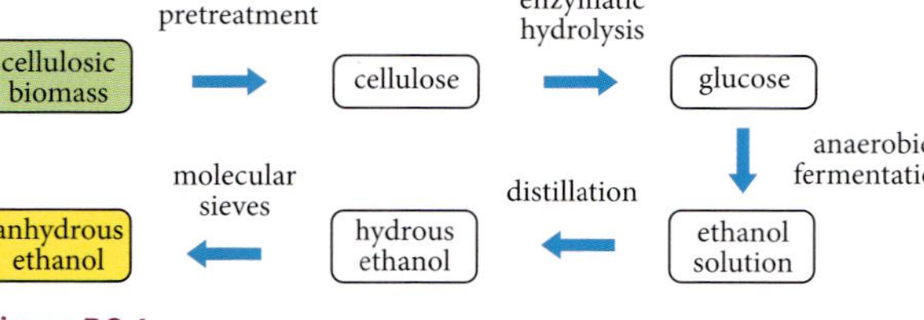

Figure PC.4

MODULE 7: ORGANIC CHEMISTRY

See ⮌ *Excel* Year 12 Chemistry p. 168

REACTIONS OF ORGANIC ACIDS AND BASES 72

Soap

- **Sodium stearate** ($NaCH_3(CH_2)_{16}COO$) is a soap. It is the salt of the fatty acid called stearic acid. The soap consists of a long hydrocarbon 'tail' and a charged 'head' group (COO^-).
- The soap is made by heating a fatty ester called *tristearin* with concentrated sodium hydroxide solution. The soap curd is precipitated by adding saturated salt water:

$$(C_3H_5O)(OOC(CH_2)_{16}CH_3)_3 + 3NaOH \rightarrow C_3H_5O(OH)_3 + 3NaCH_3(CH_2)_{16}COO$$

- **Soap** is an efficient cleaning agent. The hydrocarbon tails of the soap penetrate oily stains on clothes and form spherical structures called *micelles*, with the negative head groups on their surface. These charged groups interact with polar water molecules so the micelles lift off the clothes, dissolve in the water and are then washed away.

MODULE 7: ORGANIC CHEMISTRY

See ⮌ *Excel* Year 12 Chemistry p. 165

REACTIONS OF ORGANIC ACIDS AND BASES 70

Carboxylic acids and esters

- **Carboxylic acids** are organic molecules containing the carboxyl (COOH) functional group. Carboxylic acids are highly polar molecules. Alkanoic acids are carboxylic acids with the carboxyl group on the terminal carbon atom.
- **Alkanoic acids are weak acids** in water solution. Alkanoic acids can be neutralised by strong bases to form salts:

$$CH_3COOH(aq) + KOH(aq) \rightarrow KCH_3COO(aq) + H_2O(l)$$

- **Alkanoic acids react with alkanols** in the presence of an acid catalyst to form esters. Esters are alkyl alkanoate molecules containing the ester (COO) functional group. These reactions are condensation reactions:

$$CH_3OH + CH_3CH_2CH_2COOH \leftrightarrows CH_3CH_2CH_2COOCH_3 + H_2O$$

MODULE 7: ORGANIC CHEMISTRY

See ⮌ *Excel* Year 12 Chemistry p. 160

POLYMERS 80

Polylactic acid

- **Polylactic acid** (PLA) is an example of a synthetic biopolymer. PLA is a biodegradable condensation polymer that is strong and flexible. The monomer is lactic acid ($CH_3CHOHCOOH$), which is produced from starch waste using *Lactobacillus* bacteria. Plastic bags, composting bags, disposable plates and tableware can be made from PLA.
- **PLA is biocompatible** and can be used in medical applications inside the body (Figure PC.7).

PLA

Figure PC.7

MODULE 7: ORGANIC CHEMISTRY

See ⮌ *Excel* Year 12 Chemistry p. 180

POLYMERS 78

Addition polymers—PVC, PS and PTFE

- **Polyvinyl chloride** (PVC) is an addition polymer made from the chloroethylene (vinyl chloride) monomer. PVC is a rigid polymer used to make drainage pipes. Polystyrene (PS) is manufactured from the styrene monomer. PS is a clear, rigid polymer used to make containers and tool handles. Polytetrafluoroethylene (PTFE) is made from the tetrafluoroethylene monomer. PTFE is used as non-stick coatings in frypans.
- Figure PC.5 shows their structural formulas.

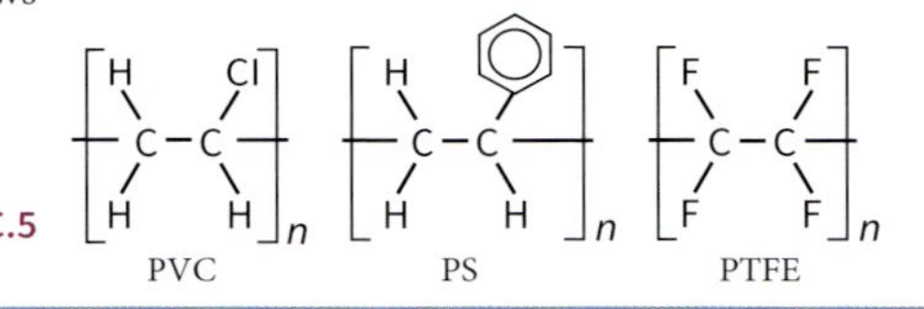

Figure PC.5

MODULE 7: ORGANIC CHEMISTRY

See ⮌ *Excel* Year 12 Chemistry p. 175

POLYMERS 76

Polyethylene

- **Ethylene monomers** can join together in long chains to form polymers. This type of polymerisation is called **addition polymerisation**.
- **Low density polyethylene** (LDPE) is produced by a free radical peroxide initiator method. The radical attacks the monomer to produce a monomer radical, which then attacks another monomer to form a dimer radical. During propagation chain growth continues via a series of addition reactions to produce a branching polymer chain.
- **The branches** cause the hydrocarbon chains to pack non-uniformly. This weakens the dispersion forces between the polymer chains.
- **LDPE** is soft and flexible and is used to manufacture plastic bags, squeeze bottles and electrical insulation.

MODULE 7: ORGANIC CHEMISTRY

See ⮌ *Excel* Year 12 Chemistry p. 173

ANALYSIS OF INORGANIC SUBSTANCES 86

Precipitation titration

- The **concentration of bromide ions in an aqueous solution** can be determined by a precipitation titration. Silver ions precipitate bromide ions in the titration.
- **A 0.100 mol/L silver nitrate solution** was used to fill a burette. 25.00 mL of NaBr solution was pipetted into a conical flask. 1 mL of yellow K_2CrO_4 indicator was added. The silver nitrate is added and the end point occurs when the indicator turns from yellow to reddish-brown.

Example:

Average titre = 21.55 mL $AgNO_3$

$n(AgNO_3) = (0.100)(0.021\,55) = 2.115 \times 10^{-3}$ mol

$Ag^+(aq) + Br^-(aq) \rightarrow AgBr(s)$; $n(NaBr) = 2.115 \times 10^{-3}$ mol

$c(Br^-) = c(NaBr) = \left(\frac{2.115 \times 10^{-3}}{0.025\,00}\right) = 0.0846$ mol/L

MODULE 8: APPLYING CHEMICAL IDEAS

See ⮌ *Excel* Year 12 Chemistry p. 197

ANALYSIS OF INORGANIC SUBSTANCES 84

Anion analysis

- **Various acid–base and precipitation reactions** can be used to identify various anions.
- **CO_3^{2-} and OH^-** ions form alkaline solutions, which can be detected using a pH meter. Nitric acid causes effervescence of CO_2 from CO_3^{2-} ions but not with OH^- ions.
- **Cl^- forms a white precipitate with Ag^+**, whereas Br^- forms a cream precipitate and I^- forms a pale-yellow precipitate. AgCl dissolves in dilute ammonia. AgBr dissolves in concentrated ammonia solution. AgI does not dissolve in concentrated ammonia solution.
- **SO_4^{2-} precipitates in an acidified solution of Ba^{2+}**, but PO_4^{3-} will not. The phosphate ion will precipitate with barium ions in a solution made alkaline with ammonia solution. Acetate (CH_3COO^-) ions do not precipitate with Ba^{2+} in acidic or alkaline solutions.

MODULE 8: APPLYING CHEMICAL IDEAS

See ⮌ *Excel* Year 12 Chemistry p. 193

ANALYSIS OF INORGANIC SUBSTANCES 82

Cation analysis

- **Qualitative tests** can be performed to identify the presence of various cations in solution.
- **Mg^{2+}, Ca^{2+} and Ba^{2+}** can be distinguished using several tests. Sulfuric acid produces white precipitates with Ca^{2+} and Ba^{2+}. Sodium hydroxide produces only a white precipitate with Mg^{2+}. Calcium ions produce an orange-red flame colour and barium ions produce a yellow-green flame colour.
- **Pb^{2+} and Ag^+** both form white precipitates with HCl. However, $PbCl_2$ is soluble in hot water.
- **Cu^{2+} ions** form a blue precipitate with NaOH solution. Fe^{2+} ions form a green precipitate and Fe^{3+} ions form a brown precipitate with NaOH solution.

MODULE 8: APPLYING CHEMICAL IDEAS

See ⮌ *Excel* Year 12 Chemistry p. 189

ANALYSIS OF INORGANIC SUBSTANCES
Colourimetry
87

- **Colourimetry** is an instrumental procedure to determine the concentration of a coloured ion, or molecule, in solution.
- **Light of a specific colour or wavelength** is passed through the coloured solution and the amount of light absorbed (absorbance, A) is directly proportional to the concentration (c) of the coloured ion or molecule.
- A **series of dilution standards** of known concentration are prepared and their absorbance measured. A calibration graph is constructed (A versus c).
- The **absorbance of solutions of unknown concentration** is measured and the calibration graph is used to determine the concentration of these solutions.

ANALYSIS OF INORGANIC SUBSTANCES
UV-visible spectrophotometry 1
89

- **UV-visible spectrometers** are used to measure the absorbance of ultra violet or visible light by a solution in the wavelength range 190–700 nm.
- **UV-visible spectrophotometry** is used to detect and measure the concentration of dyes and organic pigments as well as the presence of specific coloured-metal complex ions.

Example: The spectrum of chlorophyll-a in Figure PC.9 shows strong absorbance peaks in the violet and red regions of visible light. Chlorophyll-a is green as little green light is absorbed by this molecule.

Figure PC.9

ANALYSIS OF INORGANIC SUBSTANCES
Atomic absorption spectroscopy (AAS) 1
91

- **Atomic absorption spectroscopy** is used to measure the concentration of specific metal ions in solution.
- A hollow cathode lamp made from the metal to be analysed generates specific wavelengths of light. The light passes through a slot burner's flame containing the atomised metal ions of the sample being analysed. The greater the concentration of the metal ion, the more light that is absorbed and the less that reaches the detector.
- The **absorbance of a series of calibration standard solutions** is measured and a calibration graph is constructed. The absorbance of the unknown sample can then be measured and the graph is then used to determine the concentration of the metal ion in the unknown.

ANALYSIS OF ORGANIC SUBSTANCES
Functional group analysis
93

- **Unsaturated hydrocarbons**, alkanol and alkanoic acids can be distinguished by performing tests that are characteristic of their functional groups.
- The following table shows **observations for each test.** Alkanes give negative results for each test.

Organic compound	Bromine water	Reaction with sodium metal	Reaction with sodium carbonate solution
alkenes alkynes	rapid decolourisation	no reaction	no reaction
alkanols	no reaction	effervescence	no reaction
alkanoic acids	no reaction	effervescence	effervescence

ANALYSIS OF ORGANIC SUBSTANCES
Mass spectrometry
95

- **Mass spectrometry** is used to analyse the composition of different materials.
- The sample is vaporised and ionised. The ions are accelerated through electric and magnetic fields.
- The **mass spectrogram** in Figure PC.10 plots the intensity of each ion as a function of its **mass to charge ratio (m/z)**. The ions may fragment to produce a characteristic fragmentation pattern.

Figure PC.10

CHEMICAL SYNTHESIS AND DESIGN
The chemical industry
97

- **Chemical industries** produce a wide variety of compounds. The reaction conditions that are selected represent a compromise between equilibrium, kinetic and economic factors.
- The **yield of product** depends on the extent of the equilibrium reactions involved. In chemical industries changes in temperature and concentration affect the yield.
- Raising the temperature and the use of suitable catalysts increases the rate of the reaction.
- The **environmental and social issues of the production of chemicals** need to be investigated during the planning stages of chemical factory construction.
- **Electrical power** must be readily available for many chemical industries. In some industries the site chosen may need to be close to the site of raw materials or feedstocks.

CHEMICAL SYNTHESIS AND DESIGN
The Solvay process—sodium carbonate manufacture
99

- **Sodium carbonate** is used in glass, soap and detergent manufacture. Salt is obtained from sea water and calcium carbonate is obtained from limestone mines. The Solvay industry is located on the coast and near limestone mines.
- The **flowchart** in Figure PC.11 summarises the steps to manufacture sodium carbonate. Ammonia is recycled to save costs and to avoid atmospheric pollution. Some CO_2 is also recycled.

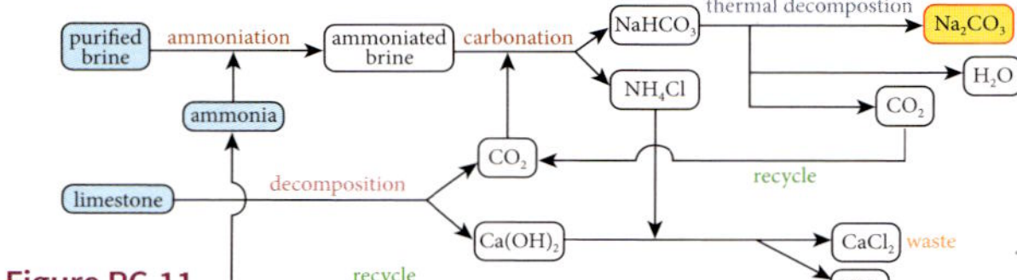

Figure PC.11

CHEMICAL SYNTHESIS AND DESIGN
Personal care products—hair conditioners
101

- **Hair conditioners** are examples of oil-in-water emulsions. The water continuous phase is compatible with the skin and hair and provides a medium to transfer the oily component onto the hairs.
- The **positively charged head groups** of the **cationic surfactants** are attracted to the negative charges on the protein surfaces of the hair strands. The long hydrocarbon tails of the surfactant cling to the hair and make it feel soft and smooth and give it weight.
- **Hair conditioners** are manufactured by large companies as well as small businesses. These factories are located in industrial zones in cities or large towns with suitable infrastructure and road networks.

CHEMICAL SYNTHESIS AND DESIGN
The Contact process—production of sulfuric acid
103

- **Sulfuric acid has many uses**, including fertiliser production, anionic detergent manufacture and as the electrolyte in car batteries.
- **Sulfuric acid plants in Australia** are usually located near sulfide smelters that generate the SO_2 feedstock. The sulfur dioxide is oxidised by an air stream (100–200 kPa) to form sulfur trioxide. The raised air pressure increases rate and yield. A vanadium (V) oxide catalyst increased the reaction rate. Because the reaction is exothermic a moderate temperature is used.
- **Recycling of unreacted SO_2** four times increases the net yield of SO_3. The SO_3 is initially dissolved in pre-prepared H_2SO_4 to form $H_2S_2O_7$, which is diluted carefully to form H_2SO_4.
- High yields of sulfuric acid are achieved with very low toxic emission of sulfur dioxide to the atmosphere as required by government regulations.

ANALYSIS OF INORGANIC SUBSTANCES

Atomic absorption spectroscopy (AAS) 2 — 92

- The development of **atomic absorption spectroscopy** (AAS) has allowed chemists to detect and rapidly measure the concentration of metal ions in water systems (particularly those that may be polluted with heavy metals such as lead, mercury or cadmium) and in the tissues of animals and plants.
- Using **AAS to monitor the lead concentration in the air** or in soil that is suspected of being contaminated is important for the health of people because high lead concentrations in the environment are associated with nervous system disorders and brain damage. According to the NHMRC *Australian Drinking Water Guidelines*, drinking water must have a lead concentration of less than 0.01 mg/L (0.01 ppm).
- By **using AAS as an analytical tool**, chemists have discovered that trace elements have a variety of essential roles in our bodies. These include copper, which acts as a catalyst in the formation of haemoglobin and chromium, which is required for insulin action.

MODULE 8: APPLYING CHEMICAL IDEAS

See ➲ *Excel* Year 12 Chemistry p. 203

ANALYSIS OF INORGANIC SUBSTANCES

UV-visible spectrophotometry 2 — 90

- Knowledge of the **absorption peak** maxima (λ_{max}) allows chemists to quantitatively determine the concentration of coloured ions in solution using visible absorption spectrophotometry.
- The **hexaaquacobalt (II) ion** ($Co(H_2O)_6^{2+}$) is pink in aqueous solution and the ion shows a strong absorption in the blue-green end (511 nm) of the visible spectrum. This wavelength is used to determine the concentration of the ion in water.
- The **absorbance of five dilution standards of the $Co(H_2O)_6^{2+}$** ion are measured and a calibration graph is constructed. This graph can be used to determine the concentration of unknown solutions, following the determination of the absorbance of each solution.

MODULE 8: APPLYING CHEMICAL IDEAS

See ➲ *Excel* Year 12 Chemistry p. 201

ANALYSIS OF INORGANIC SUBSTANCES

The Beer–Lambert law — 88

- The colourimetry analysis of the concentration of a coloured species relies on the **Beer–Lambert law**, which states that the light absorbance (A) is directly proportional to the concentration (c) of the coloured species and the path length (l) of the solution the light passes through. This law is expressed mathematically:

  $$A = \varepsilon c l$$

  where ε is a constant called the molar absorptivity
  As long as l is constant, then the absorbance (A) is directly proportional to concentration (c).
- **The absorbance (A) is determined** by measuring the intensity of the light (I_o) passing through a reference cell, with the light intensity (I) passing through the sample cell. The following equation shows that absorbance is a logarithmic function:

  $$A = \log_{10}\left(\frac{I_o}{I}\right)$$
- The **greater the concentration of the coloured species**, the more light that is absorbed and the less light that is transmitted to the detector.

MODULE 8: APPLYING CHEMICAL IDEAS

See ➲ *Excel* Year 12 Chemistry p. 198

CHEMICAL SYNTHESIS AND DESIGN

The Haber process—ammonia manufacture — 98

- The **production of fertiliser** accounts for over 80% of the worldwide use of ammonia.
- **Nitrogen and hydrogen** are the feedstocks. The nitrogen is derived from the air and the hydrogen from natural gas and water. The Haber plant is usually located near natural gas supplies:

  $$N_2(g) + 3H_2(g) \leftrightarrows 2NH_3(g); \Delta H = -92 \text{ kJ/mol}$$
- **High pressures maximise ammonia** yield as the equilibrium shifts to the right. A compromise temperature increases reaction rate but does not reduce yield too much. A catalyst increases the reaction rate. Continuous removal of ammonia drives the reaction to the right.
- **Haber plants** are located near ports so that liquefied ammonia can be exported.

MODULE 8: APPLYING CHEMICAL IDEAS

See ➲ *Excel* Year 12 Chemistry p. 226

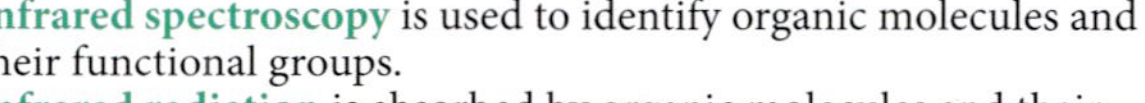

ANALYSIS OF ORGANIC SUBSTANCES

Infrared spectroscopy — 96

- **Infrared spectroscopy** is used to identify organic molecules and their functional groups.
- **Infrared radiation** is absorbed by organic molecules and their bonds stretch and bend at specific frequencies, as shown in the table below. The IR spectrogram is analysed to match the recorded frequencies to specific functional groups.

Compound type	Bond vibration	Frequency range (cm^{-1})	Compound type	Bond vibration	Frequency range (cm^{-1})
alkane	C–H stretch	2850–2960	alcohol	C–O stretch	1050–1150
alkane	C–H bend	1370–1480	carboxylic acid	O–H stretch	2500–3000
alkene	C–H stretch	3010–3090	carboxylic acid	C=O stretch	1700–1725
alkene	C=C stretch	1620–1680	ester	C=O stretch	1735–1750

MODULE 8: APPLYING CHEMICAL IDEAS

See ➲ *Excel* Year 12 Chemistry p. 217

ANALYSIS OF ORGANIC SUBSTANCES

Carbon-13 NMR and proton NMR — 94

- The **carbon-13 nucleus** has a property called *spin*. Spinning C-13 nuclei behave like tiny magnets as they create their own magnetic fields.
- The **spinning C-13 nucleus** can absorb a photon of radio frequency radiation and move from a low energy state to a high energy state. Different chemical environments affect the radio frequencies absorbed by the C-13 nuclei in organic compounds.
- The **C-13 NMR spectra resonance** peaks at different chemical shift (δ) values. The following chemical shift peaks are typical of various functional groups in organic compounds:
  C=C (90–150 ppm); C–OH (alcohols)(50–90 ppm);
  C=O (ketones)(205–220 ppm);
  C=O (esters and alkanoic acids)(160–185 ppm);
  C=O (aldehydes)(190–200 ppm)
- **Proton NMR** can also be used to identify the different chemical environments of hydrogen atoms in organic compounds.

MODULE 8: APPLYING CHEMICAL IDEAS

See ➲ *Excel* Year 12 Chemistry p. 211

CHEMICAL SYNTHESIS AND DESIGN

Electrolytic industries — 104

- The **aluminium industry and the chlor-alkali industry** use electrolysis to produce elements or compounds from raw materials. Aluminium is extracted from bauxite (hydrated aluminium oxide) ore using electrolysis. Impurities are removed from the ore and the aluminium oxide is dissolved in molten cryolite (Na_3AlF_6) at 1000 °C.
- The **mixture is electrolysed** and the molten aluminium formed at the cathode is regularly tapped off and run into moulds. The electrolytic refinery must be located near electrical power stations to supply the large amounts of electricity used in the refining process.
- **Sodium hydroxide and chlorine** are manufactured in the chlor-alkali industry. Salt is extracted from sea water and after purification it is electrolysed in electrolytic cells that use a polymer membrane to separate the two compartments. The sodium hydroxide produced is very pure. These industries must be located on the coast to extract the salt from the ocean.

MODULE 8: APPLYING CHEMICAL IDEAS

See ➲ *Excel* Year 12 Chemistry p. 233

CHEMICAL SYNTHESIS AND DESIGN

Renewable fuels—production of biodiesel — 102

- Biodiesel is a general term referring to many types of fatty esters that can be used as **renewable fuels** in diesel vehicles.
- The **feedstock is typically beef fat** (tallow) or recycled vegetable oil. The oil or fat is mixed with an excess of an alkanol such as methanol or ethanol and the mixture is heated and undergoes alkaline hydrolysis. The long-chain fatty acid esters are removed from the reaction mixture and form the final biodiesel product.
  Fatty acid ester example: $CH_3(CH_2)_{16}COOCH_3$
- There is a **social and environmental benefit** in using waste fats and oils instead of dumping them. Biodiesel fuel use reduces greenhouse gas emissions.

MODULE 8: APPLYING CHEMICAL IDEAS

See ➲ *Excel* Year 12 Chemistry p. 232

CHEMICAL SYNTHESIS AND DESIGN

Aspirin manufacture — 100

- **Aspirin** is a pharmaceutical containing acetylsalicylic acid. It is used for pain relief and as an anticoagulant to help prevent strokes and heart attacks. Acetylsalicylic acid is manufactured by a reaction between salicylic acid and acetic anhydride feedstocks.
- The **acetic anhydride** is dissolved in liquid toluene, salicylic acid is then added and the mixture is heated under reflux for about 20 hours at 90 °C. The heating increases the reaction rate. Excess acetic anhydride is used to drive the equilibrium to increase product yield.
- **Large crystals of acetylsalicylic acid** form during this time. Some acetylsalicylic acid remains dissolved and also has to be recovered. The unreacted salicylic acid and acetic anhydride are recovered from the solvent and recycled to minimise production costs. Pharmaceutical factories are located in urban areas to assist in the sales and distribution of the product.

MODULE 8: APPLYING CHEMICAL IDEAS

See ➲ *Excel* Year 12 Chemistry p. 229